Ecological Studies, Vol. 126

Analysis and Synthesis

Edited by

M.M. Caldwell, Logan, USA
G. Heldmaier, Marburg, Germany
O.L. Lange, Würzburg, Germany
H.A. Mooney, Stanford, USA
E.-D. Schulze, Bayreuth, Germany
U. Sommer, Kiel, Germany

Ecological Studies

Volumes published since 1992 are listed at the end of this book.

Springer

Berlin
Heidelberg
New York
Barcelona
Budapest
Hong Kong
London
Milan
Paris
Santa Clara
Singapore
Tokyo

W.J. Junk (Ed.)

The Central Amazon Floodplain

Ecology of a Pulsing System

With 137 Figures and 72 Tables

Springer

Dr. Wolfgang J. Junk
Max-Planck-Institut für Limnologie
Postfach 165
24302 Plön, Germany

QH
112
.C46
1997

ISSN 0070-8356
ISBN 3-540-59276-8 Springer-Verlag Berlin Heidelberg New York

CIP data applied for

Die Deutsche Bibliothek – CIP–Einheitsaufnahme
The **central Amazon floodplain**: with 72 tables/Wolfgang J.
Junk (ed.). – Berlin; Heidelberg; New York; Barcelona;
Budapest; Hong Kong; London; Milan; Paris; Santa Clara;
Singapore; Tokyo: Springer, 1997
   (Ecological studies; Vol. 126)
   ISBN 3-540-59276-8
NE: Junk, Wolfgang J. [Hrsg.]; GT

© Springer-Verlag Berlin Heidelberg 1997
Printed in Germany

The use of general descriptive names, registered names, trademarks, etc. in this publication does not imply, even in the absence of a specific statement, that such names are exempt from the relevant protective laws and regulations and therefore free for general use.

Cover design: Design & Production GmbH, Heidelberg

Typesetting: Best-set Typesetter Ltd., Hong Kong

SPIN 10484060   31/3137/SPS – 5 4 3 2 1 0 – Printed on acid-free paper

Foreword

For a long time wetlands have received little attention from ecological researchers worldwide. In part, the reason may be that neither the terrestrial nor the aquatic ecologists, the limnologists, felt themselves competent to study areas which combine within themselves terrestrial as well as aquatic conditions of life, and instead they preferred to concentrate, or to limit, their efforts on ever more detailed investigations of their traditional subjects. In part, too, this neglect may be due to the fact that the word "wetland" is a collective name which comprises many, very distinct landscape units, from swamps and peatbogs to river floodplains. Even certain arctic regions with permafrost soil must be taken as "wetland" since the summer sun melts the frozen surface of the ground, creating temporary pools of high bioproductivity in the form of the well-known mosquito nuisance.

The ecosystems of wetlands thus show in themselves terrestrial and aquatic characteristics, either spatially intermingled or temporally alternating. Floodplains, for example, are seasonally inundated for periods of up to many months and are during those periods aquatic biotopes, whereas in the more arid months they lay dry and represent terrestrial habitats. In the equatorial lowlands where the temperature remains practically constant over the course of the whole year, that alternation is caused by the changing water levels of the rivers and their discharges, and thus indirectly by the pluvial climates of the catchment basins. The arctic summer pools, however, follow the thermic climate of their environment. As we see, "wetland" is not an unequivocal concept.

It is only in recent decades that ecological research has started to turn its activities to the wetlands. The reason was a new, purely scientific interest, and equally the prospect of utilizing them for the benefits of the alarmingly increasing human population of the Earth – and/or the awareness of the threatening destruction of those peculiar ecosystems and their generally high biodiversity and bioproductivity.

Some voluminous publications on wetlands in general that have come out in recent years – important and valuable as they are – have the inevitable disadvantage that they have to present, in the same volume or series of books, all too many different objects of research comprised under the term "wetland" with their many different problems and aspects. The various scientific endeavours are not all directed to one common supreme goal or point of reference but are individually selected by the scientific or regional specialties of the authors. The whole vast field of the many different types of "wetlands" with their, again, different sets of problems and aspects, if they had already been explored, would yet fill immense rows of books!

It is now of great merit for Dr. Wolfgang J. Junk, Head of the Working Group of Tropical Ecology of the Max-Planck-Institute of Limnology at Plön, Germany, to edit a book that for the first time presents a complete survey of the existing knowledge on the ecology of one type of tropical wetland, a huge floodplain, that of the biggest river on Earth, the Amazon. Just the size of that ecosystem would justify the dedication of a voluminous monograph to one particular section of the biosphere. That is equally justified by its geographical position in the equatorial girdle since it reduces thermic seasons and their effects on life to a minimum and all the more stresses the dominating role of the seasonal "flood pulse" for the annual life rhythm of the organisms. Finally, the Amazonian floodplains are to a great extent still scarcely or not at all altered by modern man's activities, nowadays almost an exception even in the tropics.

Nobody other than Dr. Junk could be called upon to conceive the idea of this book. It is no question that he has the greatest, longest and most diversified experience in that unique ecosystem. As my doctoral student he worked for over 2 years on the "floating meadows" of Amazonian rivers, lagoons, and flooded terrains, studying their flora and fauna in an ecological context. These floating meadows are closely connected on one hand with the water – many species of their plants and animals depend on certain pH values and chemical compositions – and, on the other hand, with the terrestrial substratum of their biotopes, i.e., the floodplains, since the life cycles of their plants and animals are exposed to the seasonal alternations of aquatic and terrestrial conditions. Subsequently, the peculiar and, until then, scientifically scarcely investigated biotopes and ecosystems of the Amazonian floodplains became the main object of Dr. Junk's research. He studied them and several of the problems they set under various aspects and questionings, but with the final goal of understanding the connections between the parts of the system which make the whole function.

To meet the wide-ranging demands placed by the production of a book on such a complex matter with its many different components, Dr. Junk was able to call on a group of highly competent colleagues. The staff of his Working Group of Tropical Ecology and Brazilian scientists who worked together on problems related to the Amazonian floodplains.

It is a wide goal that Dr. Junk as editor and all the authors of the chapters have set themselves, and they have achieved it. The book is a decisive step forward in the endeavour to enhance the comprehension of such a huge and peculiar ecosystem as that of the Amazonian floodplains. The research of such a vast complex is like composing a great mosaic picture. Many stones, large and small ones, have to be set in the right places. The collocation of those stones which are to act as basic points of reference gives a concept of the whole picture. That is what has been achieved by this book; the filling in of the remaining lacunas with more details will be a task for the future. The tropical ecologists are grateful for this great achievement.

Plön, Germany, Winter 1996   Harald Sioli

Acknowledgments

This book is the result of a long and fruitful cooperation between the Max-Planck-Institute for Limnology, Plön, Germany, and the Instituto Nacional de Pesquisas da Amazônia (INPA), Manaus, Amazonas, Brazil. The cooperation started in 1962 with an agreement between the directors of both institutes, Prof. Dr. Harald Sioli and Prof. Dr. Djalma Batista. In the following years, both institutes stimulated scientific cooperation on Amazonian ecosystems and gave as much technical assistance as possible to German and Brazilian scientists to achieve the work. Since 1980, the studies have concentrated on a holistic approach to floodplain ecology. We acknowledge with deep gratitude the help of both institutes, the Max-Planck-Society and the Brazilian Research Council (CNPq), which provided financial assistance for the cooperation over several decades. During the last 5 years, the team received additional assistance from the SHIFT Program "Studies on Human Impact on Forests and Floodplains in the Tropics" financed by the German Ministry for Science and Technology (BMBF), the Brazilian Research Council (CNPq), and the Brasilian Environmental Agency (IBAMA).

Fieldwork in Amazonia was only possible because of the invaluable experience and friendly help of local technicians, fieldworkers, and fishermen. I cannot name all of them, but I am especially indebted to Mr. Celso Rabelo Costa and Mr. Uwe Thein, who have assisted us in Amazonia for many years.

For the typing of the manuscripts, I gratefully acknowledge the help of my secretaries, Mrs. Gerda Lemke and Mrs. Sabine Meier. Many of the figures were prepared by my technician, Mrs. Elke Busdorf. English corrections were made by Mrs. Nancy Weider. Prof. Dr. Rosemary Lowe-McConnell made final corrections of the English in all chapters and gave valuable comments. Prof. Dr. Otto Lange helped to prepare the book for the publishers.

Contents

Part III: Plant Life in the Floodplain

Part V: Conclusions

**23   Structure and Function of the Large Central Amazonian
         River Floodplains: Synthesis and Discussion**

Contributors

Adis, Joachim

Max-Planck-Institut für Limnologie, Arbeitsgruppe
Tropenökologie, Postfach 165, 24302 Plön, Germany

Darwich, Assad

Instituto Nacional de Pesquisas da Amazônia (INPA), c.p. 478,
69.011-970 Manaus/AM, Brazil

Franklin, Elisabeth

Instituto Nacional de Pesquisas da Amazônia (INPA), c.p. 478,
69.011-970 Manaus/AM, Brazil

Furch, Karin

Max-Planck-Institut für Limnologie, Arbeitsgruppe
Tropenökologie, Postfach 165, 24302 Plön, Germany

Gauer, Ulrich

Landessammlung für Naturkunde, Zoologische Abteilung,
Erbprinzenstr. 13, 76133 Karlsruhe, Germany

Höfer, Hubert

Staatliches Museum für Naturkunde, Postfach 6209,
76042 Karlsruhe, Germany

Irion, Georg

Senckenberg-Institut, Schleusenstr. 39a,
26382 Wilhelmshaven, Germany

Junk, Wolfgang J.

Max-Planck-Institut für Limnologie, Arbeitsgruppe
Tropenökologie, Postfach 165, 24302 Plön, Germany

Kern, Jürgen

Institut für Agrartechnik Bornim, Abt. Bioverfahrenstechnik,
Max-Eyth-Allee 100, 14994 Potsdam, Germany

Martius, Christopher

Staatliches Museum für Naturkunde, Postfach 6209,
76042 Karlsruhe, Germany

Mello, José A.S.N. de

Instituto Nacional de Pesquisas da Amazônia (INPA), c.p. 478,
69.011-970 Manaus/AM, Brazil

Messner, Benjamin

Zoologisches Institut, Ernst-Moritz-Arndt Universität,
Johann-Sebastian-Bach-Str. 11/12, 17489 Greifswald, Germany

Petermann, Peter

World Wildlife Fund – Aueninstitut, Josefstr. 1, 76437 Rastatt,
Germany

Piedade, Maria T.F.

Instituto Nacional de Pesquisas da Amazônia (INPA), c.p. 478,
69.011-970 Manaus/AM, Brazil

Putz, Rainer

Hauptstraße 76, 79104 Freiburg, Germany

Robertson, Barbara

Instituto Nacional de Pesquisas da Amazônia (INPA), c.p. 478,
69.011-970 Manaus/AM, Brazil

Saint-Paul, Ulrich

Zentrum für Marine Tropenökologie, Fahrenheitstr. 1,
28359 Bremen, Germany

Silva, Vera M.F. da

  Instituto Nacional de Pesquisas da Amazônia (INPA), c.p. 478,
  69.011-970 Manaus/AM, Brazil

Soares, Maria G.M.

  Instituto Nacional de Pesquisas da Amazônia (INPA), c.p. 478,
  69.011-970 Manaus/AM, Brazil

Wassmann, Reiner

  IRRI, Div. Soil and Water Sciences, P.O. Box 933, 1099 Manila,
  Philippines

Weber, Gerhard E.

  UFZ-Umweltforschungszentrum, Permoserstr. 15,
  04318 Leipzig, Germany

Woas, Stefan

  Staatliches Museum für Naturkunde, Postfach 6209,
  76042 Karlsruhe, Germany

Worbes, Martin

  Forstbotanisches Institut, Außenstelle: Büsgenweg 2,
  37077 Göttingen, Germany

Part I
Introductory Remarks

1 General Aspects of Floodplain Ecology with Special Reference to Amazonian Floodplains

Wolfgang J. Junk

1.1 Introduction

Floodplains are wetlands which oscillate between terrestrial and aquatic phases. This makes them alternately suitable for aquatic and terrestrial organisms, but makes utilization by humans difficult. For this reason, the extensive floodplains of Europe and North America were largely eliminated or strongly modified early on by human activities. For example, flood control measures by Tulla brought about extensive changes in the large floodplain of the upper Rhine valley in the middle of the last century (Kunz 1975) long before the study of limnology was established. Therefore, floodplains were neglected by limnologists and terrestrial ecologists for a long time (Bayley 1980; Junk 1980).

With the growing awareness of the ecological and economic significance of wetlands in general and floodplains in particular, the number of studies undertaken increased considerably. Tropical floodplains became of interest because they were considered to be useful land reserves for agriculture, animal husbandry and forestry as well as areas for settling rapidly increasing populations. The construction of large impoundments for the production of hydroelectric energy modified the flood regime and affected the floodplains upstream and downstream of the dams. During the past few years growing objections have been raised to the destruction of floodplains because of their multiple functions, e.g., for water storage during floods, for fisheries or as habitats of a rich and characteristic flora and fauna. Thus, conflicts of interest are inevitable (Junk and Welcomme 1990).

Most studies relating to floodplains are oriented toward either limnological or terrestrial aspects. Limnologists concentrate on permanent water bodies in the floodplain, treating them as classical lakes, periodically disturbed by the flood. Terrestrial ecologists concentrate on the dry area, considering flooding as a periodic disturbance of a terrestrial system. Thus, the spacio-temporal heterogeneity that is central to the functioning of

Ecological Studies, Vol. 126
Junk (ed) The Central Amazon Floodplain
© Springer-Verlag Berlin Heidelberg 1997

these systems is placed in the wrong perspective. The studies do not answer the main questions about the functioning of the system as a whole and the roles of the organisms within the system. Attempts to link the terrestrial with the aquatic phase have been undertaken in Europe by Dister (1980); Amoros et al. (1986) and Bravard et al. (1986); in the USA. by Gosselink et al. (1981); Larson et al. (1981); Patrick et al. (1981); Wharton et al. (1981); Ward (1989) and Ward and Standford (1995); and for tropical systems by Sioli (1965, 1984); Welcomme (1979, 1985); Irmler (1981); Adis (1984); Junk (1980, 1983a,b, 1984b, 1985, 1986); Davies and Walker (1986); Ellenbroek (1987); Junk et al. (1989) and Junk and Welcomme (1990).

In Amazonia, at the beginning of European colonization, rivers were the only available routes for travelling through the dense rain forest. Christobal de Acuña, who accompanied the Portuguese expedition led by General Pedro Teixeira in 1637–1638, reports sightings of large populations of Indians on the shores of the Amazon River and the presence of a great variety of fishes, turtles, and other aquatic animals. Since then all naturalists who have visited the region have made some contribution to the knowledge of the large Amazonian rivers and their floodplains with descriptions, paintings, and collections, e.g., Spix and von Martius who visited Amazonia from 1817 to 1820; Bates (1848–1859); Wallace (1848–1852), who lost his collections when his ship caught fire and was destroyed, Spruce (1849–1864), Coudreau (at the turn of the nineteenth century) and many others (Spix and von Martius 1823–1831; Bates 1864; Wallace 1889; Spruce 1908; Coudreau 1897a,b,c, 1899, 1900, 1901, 1903a,b,c, 1906). At the end of the last century, scientists drew attention to the overexploitation of some resources provided by the large rivers and their floodplains. Verissimo (1895) criticized the inadequate regulation of fisheries, which he considered irresponsible, and predicted the eradication of turtles, manatees, and the osteoglossid fish pirarucú *Arapaima gigas* within a century. In 1895 and 1896, Goeldi (1904) in two memoranda addressed to the Governor of the State of Pará protested against the inconsiderate extermination of herons in Lower Amazonia, all for the sake of their feathers.

In 1870, scientific work was strengthened with the foundation of the Museu Paraense (actually called Museu Paraense Emilio Goeldi). Scientists at the museum contributed substantially to the knowledge of archaeology and natural history of the lower Amazon basin in general and the large river floodplains and the estuary in particular. In 1954, the Instituto Nacional de Pesquisas da Amazônia (INPA) was founded in Manaus. Reconstructed in 1973, it is now a premier institution for scientific research in

Central Amazonia. Most of the studies reported in this book were made in cooperation with INPA.

Modern limnological studies in Amazonia started with Harald Sioli after World War II. Inspired by the holistic view of Bluntschli (1921), Sioli treated rivers and their floodplains as parts of the Amazonian landscape. After returning to Germany as a Director of the Max-Planck-Institute for Limnology in Plön in 1957, he established a long-term cooperation between his institute and INPA. In 1980, both institutions started a multi-disciplinary research project to investigate comprehensively the ecology of the floodplains in the middle Amazon River. This project gave equal emphasis to organisms and processes during the terrestrial and aquatic phases and concentrated on their interactions. Quantitative aspects of bio-mass and energy flux as well as the morphological, anatomical, physiologi-cal, and ethological adaptations of organisms to the change between terrestrial and aquatic phase have been investigated.

At the same time, many other Brasilian and foreign scientists were conducting research on the Amazon River and Negro River and their floodplains, locally called "várzea" and "igapó" respectively. The Univer-sity of Seattle, with Brazilian counterparts, concentrated on the river dis-charge, transport of sediments and dissolved solids, and biogeochemical processes in the Amazon River and its main affluents between Benjamin Constant at the Peruvian border and Santarem on the lower Amazon. The Universities of California and Maryland worked with Brazilian counter-parts on the limnology of Lago Calado, a floodplain lake near Manaus. Brazilian counterparts with French scientists from ORSTOM worked on fish ecology and fisheries.

Because of the paucity of information and the great complexity of the system's structure and dynamics, researchers were forced to address the studies in a very general, often descriptive way. Plant and animal groups were studied on a functional level because taxonomic problems slowed down the progress of the work. Many species are still new to science.

Maintenance problems with sophisticated equipment in Amazonia, and other hazards that are characteristic of scientific work in remote areas throughout the world, often hampered the research and often allowed only restrictive use of field methods, rather than the application of sophisticated methods and technologies. Therefore the information is not uniform for the different aspects of Amazonian floodplain ecology. Many questions are still open for further studies. Nevertheless, the available data permit in-sights into the structure, function, and multiple interactions within a very complex system.

1.2 Definition and General Characterization

North American hydrologists define the "active floodplain" of a river as "the area that is flooded by the greatest flood of the century" (Bhowmik and Stall 1979). However, this definition is not satisfactory because there is no compelling reason why a period of 100 years should be chosen, not to mention the fact that records going back so far are not available for many rivers. Additional ecological parameters are necessary for distinguishing floodplains from areas exposed to random flooding. Junk et al. (1989) proposed the following definition for floodplains:

"Floodplains are areas that are periodically inundated by the lateral overflow of rivers or lakes and/or by direct precipitation or groundwater; the resulting physico-chemical environment causes the biota to respond by morphological, anatomical, physiological, phenological, and/or ethological adaptations and produces characteristic community structures."

This ecological definition takes into consideration the fact that the flooding has specific effects on the organisms, which react by developing specific adaptations. It further implies that the effect of the flooding on the organisms is independent of the factors causing the flooding. Therefore, basic similarities among floodplains along rivers and beside lakes and impoundments can be expected.

Floodplains can be classified according to amplitude, frequency, predictability, and the source of flooding. For instance, floodplains of large rivers are characterized by a predictable monomodal flood pulse and a large flood amplitude (e.g., the floodplains of the Amazon River and its major tributaries, the Orinoco River, the Parana–Paraguay River). Floodplains in large depressions or in insufficiently drained areas are periodically inundated during the rainy season and are characterized by a predictable, monomodal flood pulse and a small flood amplitude (e.g., the savannas of Branco River, Pantanal of Mato Grosso, Bananal, Llanos bajos of the Orinoco River, savannas in Bolivia between the Beni River, Mamoré River and Guaporé River, some woodlands on the middle and upper Negro River catchment area). Floodplains of small creeks and rivers are characterized by an unpredictable polymodal flood pulse. Small- and medium-sized depressions in savannas and forests are also flooded by rain with an unpredictable polymodal flood pulse. Mangroves and coastal wetlands are characterized by a predictable polymodal flood pulse resulting from tidal influence.

Floodplains can also be characterized by the amounts of dissolved and suspended substances introduced by the floods (for instance, nutrient-rich floodplains of white-water rivers versus nutrient-poor floodplains of black-water rivers or floodplains inundated by rainwater). The vegetation cover can characterize a floodplain (e.g., forested floodplains versus floodable savannas). The connection with permanent water bodies strongly influences the colonization by aquatic plant and animal species (floodplains connected with large lakes or river systems versus floodplains isolated from major permanent aquatic systems e.g., many floodable savannas).

These categories are artificial and transitions and gradients exist between them. A large river passes from a polymodal, unpredictable flood pulse in the headwaters to a monomodal, predictable flood pulse in its lower course. In the Pantanal of Mato Grosso or on the Ilha do Bananal, floodplain areas near the main rivers correspond more to river floodplains, whereas areas far away from them correspond more to floodable savannas. Some of the remote areas are isolated from permanent waterbodies for long periods, thus lacking an exchange of aquatic organisms. The edges of the nutrient-rich Amazon floodplain are in part nutrient-poor because of a strong influence of nutrient-poor affluents from the uplands.

The powerful dynamics resulting from the flood pulse distinguish floodplains from most other transitional areas between land and water, which have relatively constant water levels e.g., peat bogs, swamps, or the littoral zones of many lakes (Fig. 1.1).

In recent years transitional areas have attracted considerable attention in ecology (Wiens et al. 1985; di Castri et al. 1988; Naiman and Décamps 1990; Holland et al. 1991; Ward and Wiens 1995). The concept of ecotones, first used by Clements (1905) to describe tension zones between communities, has now become scale-independent and is also applied to large transition areas at a landscape scale. According to the definition of Hansen and di Castri (1992), large river floodplains, which have the status of specific ecosystems (Junk 1980; Odum 1981), are considered ecotones.

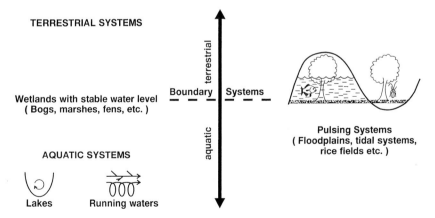

Fig. 1.1. The position of floodplains in relation to other aquatic and terrestrial ecosystems

1.3 The Flood Pulse Concept

The most comprehensive approach to deal with floodplains is the flood pulse concept (Junk et al. 1989). This concept is based on hydrological considerations of the river, its catchment area and its floodplain. From a hydrological point of view the river and its floodplain must be considered as an indivisible unit because of their common water and sediment budget. The same holds true for many biological aspects, such as the exchange of organisms, biomass, and energy. Junk et al. (1989) designated such complexes as "river–floodplain systems" encompassing both the rivers and their floodplains. The floodplain includes permanent lotic and lentic habitats as well as areas that are periodically exposed to advancing and receding floods. The areas oscillating between a terrestrial and an aquatic status are designated as the Aquatic/Terrestrial Transition Zone (ATTZ; Junk et al. 1989).

The degree of connection between the river and its floodplain depends on the water level of the river. At low water, the floodplain is disconnected from the river. The biocenoses on land and in the remaining water bodies begin to develop independently from the river. This development suffers a setback to the original stage when the river floods the area again. In low-lying areas this setback occurs every year. In high-lying areas or in areas separated from the river by levees, it may occur only once a decade during extremely high floods.

In their metabolic budgets, floodplains assume an intermediate position between open, transporting systems, and closed, accumulative systems. The periodic accumulation of substances followed by their transport out of the system may occur over short periods associated with the rhythm of the flood pulse, e.g., in the case of organic matter produced in the floodplain, or at intervals of hundreds or thousands of years as in the case of the displacement of sediment. During times of low water levels, backwaters and oxbow lakes are lacustrine systems and accumulative in nature. As the water level in the river rises, they assume the function of supplemental reservoirs; however, during high-water periods they may become water transport channels. During the terrestrial phase, organic litter accumulates in many habitats and is washed away during the aquatic phase to be either moved to another part of the floodplain or carried away in the river and later into the ocean. Aquatic and terrestrial plant and animal populations periodically drift away or are reduced, while others appear in increased numbers in the short term.

The spectrum of organisms inhabiting the floodplains ranges from completely aquatic to completely terrestrial species. Many species show various degrees of morphological, anatomical, physiological, phenological, and ethological adaptation to the terrestrial or the aquatic phase, which makes survival possible during whichever phase is less suitable to the species. Permanent water bodies, e.g., river channels and permanent floodplain lakes, serve as important refuges for aquatic animals during the low-water period. The nonfloodable uplands influence the floodplain via the inputs by the drainage system, as temporary refuge for terrestrial organisms, and as a species pool for facultative or potential colonizers of the floodplain.

Plants and animals use the available nutrients during their active growth phase and transfer some of them into the less active phase, thus fuelling an internal nutrient and energy cycle within the floodplain. For instance, plants that grow during the terrestrial phase take up nutrients from the sediments, store them in their tissue, and release them into the water when decomposing during the aquatic phase. Aquatic organisms can directly use the organic material produced during the terrestrial phase, e.g., some fishes, which feed on fruit of the floodplain forest, detritus and terrestrial

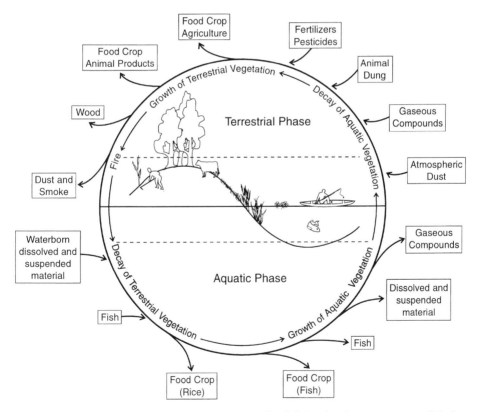

Fig. 1.2. Diagram of nutrient and energy cycles on floodplains. (Welcomme 1979, modified)

invertebrates. Bacteria, aquatic algae, and macrophytes take up the nutrients released from the decomposing terrestrial organic material. Organisms living during the terrestrial phase make use of the stranded aquatic material and of the nutrients released by them during decomposition. This exchange of energy and nutrients between the two phases by different groups of organisms is the principal reason for the high productivity of most floodplain systems (Fig. 1.2).

The river channel itself supplies the floodplain with water and dissolved and suspended solids from the catchment area. For many plant and animal species, it serves mainly as a "highway" for transport or migration. Organic material derived from the catchment area is of low importance for the food webs because the floodplain provides large amounts of high quality organic material. This distinguishes large river–floodplain systems from streams and small rivers. According to the River Continuum Concept (Vannote et al. 1980), food webs of small rivers are based to a large extent on allochthonous material, which is processed by different consumer communities during the transport downriver. Consumers in downstream reaches rely on the inefficiency of upstream consumers to process the organic material.

The short presentation of the flood pulse concept given above serves to offer a general basis for the interpretation of the central Amazon floodplain data given in the following chapters. The data will amplify and summarize the available databases. They are discussed in Chapter 23 to verify the hypotheses derived from the flood pulse concept, to formulate new hypotheses, and to stimulate further studies.

1.4 Floodplains and Limnological Terminology

Several limnological terms cannot be adequately applied to floodplains. This holds true mainly for terms related to the description of the morphology of lakes. In the early studies of limnology, research concentrated on lakes that have a rather stable water level. To describe such "classical" lakes, parameters were used such as surface area, maximum depth, mean depth, length of shoreline, extension of the littoral and profundal zone, etc. All these terms are static and do not apply to the dynamic situation of water bodies in a floodplain. Permanent and temporary lakes increase in size and become connected to each other with rising water level. When the level of the Amazon River near Manaus reaches 29 m above sea level, the river flows over its floodplain and the lakes become part of the river channel, transporting water and sediments.

The littoral zone is a very important border which separates terrestrial from aquatic habitats. The subdivision into epi-, supra-, eu-, and infralittoral (depending upon water depth) demonstrates this importance. In floodplains, each location can be arranged on a gradient extending from permanently aquatic to permanently terrestrial conditions. During the course of a flood cycle, a low-lying site can be subject to conditions that are typical for a moist terrestrial habitat, a shallow water body, a lake with an aphotic and anoxic zone, and a river channel. During falling water it is subject to a reverse transition. In contrast, a location on the top of a levee may undergo a transition from a terrestrial habitat stressed by desiccation to one with moist soil, and finally to a shallow water body. The littoral zone changes position along the flood gradient throughout the entire floodplain according to the water level, Junk et al. (1989) referred to it as a "moving littoral".

In most cases the flood gradient corresponds to the elevation gradient indicated in meters above sea level (a.s.l.). However, in some cases it is independent of the elevation gradient. In high-lying depressions and small isolated lakes at low water, aquatic conditions may prevail at 24 m a.s.l., because of poor drainage, whereas most of the floodplain is dry down to the river level at 18 m a.s.l.

The periods of time during which an area inside the Aquatic/Terrestrial Transition Zone passes through the terrestrial and the aquatic stages are called the "terrestrial" and "aquatic phases". We do not use the term "terrestrial and aquatic period", because a period suggests a defined length of time. The dry and wet period can be used as a synonym for the dry and the rainy seasons, which are determined by rainfall and describe the climate in a large area around Manaus. In the Amazon floodplain, the terrestrial phase varies from a few days to several years within a distance of a few tens of meters depending upon the position of the habitat on the flood gradient. Even this position is not fixed because high annual sedimentation rates can strongly modify the length of the terrestrial phase in a rather short period.

The dynamic change between land and water in floodplains makes it difficult to differentiate between aquatic and terrestrial organisms because many of them show adaptations to both environments. This is not only a problem of semantics, it is also related to functional aspects. For instance, the differentiation between allochthonous and autochthonous material is very important for the description of nutrient and energy fluxes in lakes and rivers. How do we deal with organic material produced during the low water period on the dry sediments of floodplain lakes? It can be allochthonous organic material because it is of terrestrial origin, produced by terrestrial plants, but what about the nutrients released during decomposition into the lake water? In a budget these nutrients can be considered to be autochthonous, because they have been taken up by the plants from the lake sediments.

The analysis is even more complicated because many herbaceous plants occupy an intermediate position between terrestrial and aquatic plants, e.g., wild rice and many other highly productive aquatic and semiaquatic grasses. They begin to grow on the dry lake bottom at low water taking up nutrients from the sediments. When the water level rises, they continue to grow in water up to 10 m deep taking up huge amounts of dissolved nutrients from the water by adventitious roots on the nodes. When the water recedes, they are stranded on the drying lake bottom, decompose, and release the nutrients to the dry sediments where the nutrients are taken up by plants that can grow there during the terrestrial phase.

This bias becomes clear as well in the description of the oxygen budget of floodplain lakes. As shown in Section 4.4.3 most floodplain lakes are very deficient in oxygen. Melack and Fisher (1983) showed for Lago Calado that diffusion from the air and autochthonous production by phytoplankton are not sufficient to supply the lake with oxygen. They relate the high oxygen demand to the input of large amounts of organic material produced by free-floating emergent aquatic macrophytes at rising and high water levels. A large portion of the organic material is decomposed in water, consuming oxygen. Moorhead and Reddy (1988) and Jedicke et al. (1989) found that these macrophytes release oxygen by exudation from the roots into the water, but a considerable portion is released into the air. Do these emergent aquatic macrophytes enter in the oxygen–carbon balance of the lake as autochthonous or allochthonous sources and sinks?

It is not the aim of this chapter to introduce new terms because they would not resolve these problems. The aim is to show the need to change limnological approaches when dealing with floodplain ecosystems. Boundaries between land and water in floodplains are fluid, as are the boundaries between terrestrial and aquatic organisms and processes.

1.5 Distribution and Size of Neotropical Floodplains

On a worldwide scale, the total area covered by floodplains has been strongly underestimated because many of them have not been recognized as such. This holds true as well for the Neotropics. Until now, knowledge about the distribution and the extent of neotropical floodplains is insufficient and has concentrated on the most conspicuous ones (Fig. 1.3, Table 1.1).

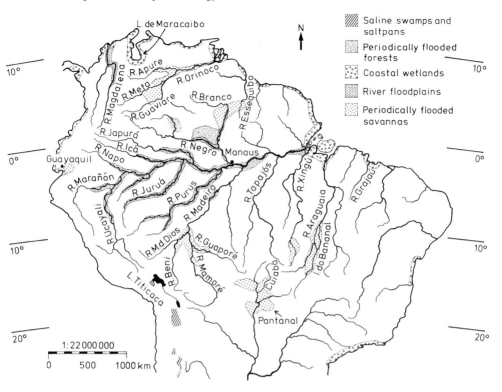

Fig. 1.3. Distribution and characterization of major neotropical wetlands. (Junk 1993)

The floodplains of the large neotropical rivers, e.g., the Amazon River and its affluents have been quite well surveyed (Table 1.2). However, many small creeks and small rivers are also accompanied by floodplains. Creek density near Manaus is higher than one creek per kilometer. Assuming an average width of 100 m for every floodplain, 10% of the area is covered by floodplains of small creeks. Falesi et al. (1971) estimated that about 40% of the soils north of Manaus are hydromorphic.

Due to the flat topography of large parts of the tropical South American lowlands, the poorly drained soils, the pronounced seasonality of precipitation and the high rainfall, large areas are periodically flooded during the rainy season. Precipitation of 1100 to 2000 mm concentrated over a few months of rainy season in the savanna belts in the north and south of the Amazonian rain forest is sufficient to periodically flood large areas. Some of them are well known, e.g., the Llanos bajos of the Orinoco basin, the Pantanal of Mato Grosso, the Bananal of the Araguaia River, or the savannas of the northeastern Bolivian lowlands between the rivers Madre de Dios, Beni, Mamoré and Guaporé (Llanos dos Mojos). In all other savanna

Table 1.1. Estimate of the extent of major natural floodplains in tropical South America. (Klinge et al. 1990; Junk 1993)

Floodplain area	Extent km 10^3	Prevailing vegetation
A. Polymodal unpredictable hydrological regime		
Along low-order streams within Amazonian lowlands	ca.1000	Forest
Andean footzone	>100	Forest
Border area Colombia/Venezuela/Brazil	50	Forest (Caatinga, Bana) savanna
B. Monomodal hydrological regime		
Along the Amazonas River and its big tributaries in Brazil	300	Forest, grassland
Beni River, Mamoré River, Guaporé River	150	Savanna, forest
Pantanal, Brazil	140	Savanna, forest
Central Brazil	ca.100	Savanna, scattered in the Cerrado
Venezuelan Llanos	80	Savanna, forest
Araguaia River (Bananal and adjacent lowlands)	65	Savanna, forest
Middle Negro River	50	Forest and shrubs, major portion of palms
Roraima lowlands and Rupununi lowlands, Guayana	33	Savanna
C. Bimodal hydrological regime		
Magdalena River floodplain	20	Forest, grassland
D. Coastal region, tidal regime		
Atlantic-Caribbean coast	120	Mangrove, tidal forest, and grasslands
Amazonas River estuary	50	
Orinoco River delta	22	Mangrove, tidal forest, and grasslands
Magdalena River delta	5	
Pacific coast	1	Mangrove, tidal forest, and grasslands

areas there are many small depressions which are periodically flooded during the rainy season but are not recorded on maps. According to Janssen (1986), the savannas of Humaitá (about 700 km south of Manaus) cover an area of about 630 km². About 40% of the area is periodically flooded during the rainy season, shows specific plant community structures, and fulfills the function of floodplains.

The survey of coastal wetlands concentrates mostly on mangrove systems. There are, however, large freshwater palm swamps behind the mangrove belt covered by *Mauritia flexuosa* or *Mauritia minor*. These swamps are not subject to the tidal pulse, but are subject to seasonal water-level fluctuations according to the dry and rainy seasons. Junk (1993) estimates that approximately 20% of the South American tropical lowlands are at

Table 1.2. Floodplains of the large Amazonian rivers in Brazilian territory, excepting the delta area of the Rio Amazonas and the Bananal of the Rio Araguaia. The areas include river channels and floodplain lakes. River channels occupy up to 10% of the total area of most white-water river floodplains, but up to 90% in some clear-water river floodplains (e.g., Rio Tapajós). The areas indicated in brackets include floodplains along minor affluents. (Source; geomorphological maps of RADAM Brazil)

River	km²	km²	From	To
Rio Amazonas	87 600	(107 700)	Frontier of Brazil–Peru–Colombia	Mouth of Rio Xingú
Rio Negro	11 600	(29 000)	Frontier of Brazil–Venezuela	Confluence with Rio Amazonas
Rio Branco	3 300	(4 200)	Boa Vista	Confluence with Rio Negro
Rio Icá	2 500	(3 300)	Frontier of Brazil–Colombia	Confluence with Rio Amazonas
Rio Jutaí	5 200	(8 200)	Near source	Confluence with Rio Amazonas
Rio Juruá	19 700	(27 900)	Frontier of Brazil–Peru	Confluence with Rio Amazonas
Rio Japurá	4 000	(7 100)	Frontier of Brazil–Colombia	Confluence with Rio Amazonas
Rio Tefé	1 400	(1 800)	Near source	Confluence with Rio Amazonas
Rio Javari	3 700	(9 100)	Frontier of Brazil–Peru	Confluence with Rio Amazonas
Rio Iriri	1 600	(2 600)	Near source	Confluence with Rio Xingú
Rio Madeira	9 800	(12 800)	Confluence Rio Mamore–Rio Beni	Confluence with Rio Amazonas
Rio Uatuma	3 300	(4 600)	Near source	Confluence with Rio Amazonas
Rio Tapajós	4 500	(13 600)	Confluence Rio Juruena–Rio Teles Peres	Confluence with Rio Amazonas
Rio Xingú	8 500	(17 700)	Confluence Rio Culuene–Rio Sete de Setembro	Confluence with Rio Amazonas
Rio Araguaia	4 800	(5 100)	Downstream of Bananal Island	Confluence with Rio Tocantins
Rio Tocantíns	8 800	(9 200)	Near source	Confluence with Rio Amazonas
Rio Purús	21 100	(40 600)	Frontier of Brazil–Peru	Confluence with Rio Amazonas
Rio Coarí	1 700	(2 800)	Near source	Confluence with Rio Amazonas
Total:	203 100	(307 300)		

least periodically wetlands and most of them belong to the floodplain category.

1.6 Human Impact

Since early history, humans have been attracted to floodplains because of the availability of water, wildlife, fish, and fertile soils. Interactions with floodplains, however, varied strongly. In temperate regions, e.g., in Europe and North America, flooding was not very predictable and was often not dependent on the local precipitation. Some of the most catastrophic floods of the Rhine River happened in the winter, when unpredictable short periods of warm weather led to sudden, rapid snow melting. Often floods overlapped with crop growth. Utilization of floodplains was a risky task. Therefore, flooding was not considered as a specific stage of the system but as a catastrophe that hindered conventional utilization. As soon as man developed the technology he made great advances in flood control. Many floodplains had already been eliminated by the end of the last century. Only a few are left in a natural state.

In the tropics, floodplains of large rivers were often centers of cultural development. Man was able to take advantage of the periodical flooding and drying, e.g., on the Euphrates and Tigris, Nile, Mekong or Indus and Ganges Rivers. The utilization of the floodplains for agricultural and animal farming in the tropics was favored by the predictability of the flood pulse that was associated with the dry and rainy seasons. Water availability and nutrient input by the floods combined well with the growth period of the crops. Some important crop plants and domestic animals, e.g., rice and water buffalo, were well adapted to floodplain conditions. Therefore adequate management of the floodplains became part of the cultural development of many tropical regions. In ancient Egypt taxes were calculated according to flood levels: taxes were waived when the peak flood level was below a certain level, because of the negative impact on crop yields (Hammerton 1972).

In the Neotropics, floodplains were also centers of human colonization (Junk 1995). Denevan (1976) estimates an average pre-Columbian population density of 14.6 persons per square kilometer in the várzea, in comparison with 1.2 persons in the nutrient-poor uplands. The Indians colonized elevated places on the edges of the floodplain and built major earthworks in areas with small water level fluctuations, for instance at the Ilha de Marajó at the mouth of the Amazon River. The communities were characterized by high levels of social organization. They practised fishing, hunting, and farming. Intensive agriculture concentrated on the levees because of the short flood period. The populations of wild rice (*Oryza glumaepatula, O. grandiglumis*), that cover large areas point to their importance as a food crop. Animals were not farmed except for domesticated Muscovy ducks (*Cairina moschata*) (Denevan 1966; Meggers 1984, 1987, 1992; Roosevelt 1991).

Today, the várzea contributes to fiber and wood production and supplies the local market with vegetables, fruit, milk, cheese, and meat (Petrick 1978). Cattle farming is becoming increasingly important (Sternberg 1960). In the State of Pará buffalo farming is common (Ohly 1987). Agricultural production is low because most crops and cultivation methods are little adapted to floodplain conditions (Junk 1982, 1995; Ohly and Junk, in press).

Fish is the most important product of Amazonian river floodplains and the main source of animal protein for the local population. Bayley and Petrere (1989) indicate an inland fishery potential of about 1 million tons year^{-1}. Conflicts of interests arise between the different users. Animal farming results in large-scale deforestation of the várzea to increase pastures, but reduces food for fruit-feeding fish. There exist serious conflicts between commercial fisherman and the local population which uses the fish for subsistence fishing. The development of specific multipurpose management strategies is required in order to optimize utilization and to avoid serious damage of the floodplain systems.

1.7 Objectives and Structure of This Book

The studies of Amazonian floodplains, among the largest tropical river floodplains in the world, have contributed substantially to a better understanding of the structure and function of these systems. Considerable data have now been accumulated which provide an important scientific background for sound management and the planning of further studies. A major problem, however, is that the existing information is dispersed in many reports, master's and doctoral theses, and journals which are not readily accessible. There is a need for a comprehensive overview and analysis of the data for the formulation of hypotheses which can stimulate further studies. The dynamic character of floodplain ecosystems requires an interdisciplinary discussion of the results, which is still lacking.

The following chapters provide a sound data base for a better understanding of the ecological conditions of the middle Amazon floodplain. Their purpose is to improve the theoretical framework for further studies of floodplain ecology and for the development of methods for sustainable multipurpose management. They also reveal the large gaps that still exist and that require studies for a better understanding of the system.

The book is subdivided into five parts. Part I provides information about large river floodplains in general and Amazonian floodplains in particular. It also reviews the efforts of the researchers and institutions who have studied Amazonian floodplains until now.

Part II (five chapters) considers the physical and chemical environment in the central Amazon floodplains. Chapter 2 presents data on climate, hydrology, mineralogy, and the geomorphological structure of the area investigated. This links the present state of the Amazon floodplain with its geological and paleoclimatic history. Chapter 3 presents information about the nutrient status of the soils. It compares the availability of some major nutrients of the nutrient-poor Negro River floodplain and the nutrient-rich Amazon River floodplain with the requirements of their respective flood-plain forests. Chapter 4 gives data on the chemical composition of Amazon River water, Negro River water, water of forest streams, rainwater, and groundwater influencing the limnology of the floodplain lakes. It summarizes the limnological conditions in these lakes and discusses the impact of the oscillation between terrestrial and aquatic phases on their water chemistry. In Chapter 5 a mathematical model is presented that quantifies the impact of the flood pulse on the concentration of some selected bioelements (Ca, Mg, K), evaluating also the uptake and release of bioelements by higher plants in the aquatic terrestrial transition zone. Chapters 6 and 7 describe the impact of the flood pulse on the nitrogen cycle and methane production in the floodplain.

In Part III, five chapters consider the impact of the flood pulse on plant life in the floodplain. Chapter 8 is an introduction to general aspects of plant distribution on the flood gradient; it concentrates on terrestrial and aquatic herbaceous plants, presenting data on species numbers, plant distributions, adaptations to the flood pulse, primary production, and photosynthesis. Chapter 9 collates data on chemical composition and food value of the herbaceous vegetation and the leaf litter of the floodplain forests. Decomposition processes on land and in water are exemplified and the release of nutrients from the decomposing material is quantified. Similar information is given about the floodplain forests in Chapters 11 and 12. Chapter 10 summarizes the information on aquatic algae (phytoplankton and periphyton).

The purpose of Part IV is to present data on animal life in the floodplain. Chapter 13 summarizes information about aquatic invertebrates (zooplankton, benthos and epi- and perizoon) including data on species numbers, biomass, life cycles and adaptations to the flood pulse. Terrestrial invertebrates are considered in Chapters 14 to 19. These represent the most species-rich group of animals worldwide and therefore they require special attention in floodplains too. Chapter 14 gives a general view of survival strategies, group and activity patterns, comparing the invertebrate faunas of the nutrient-rich Amazon River floodplain with the nutrient-poor Negro River floodplain. The subsequent chapters deal with specific groups of invertebrates showing the impact of the flood pulse on the animals, analyzing the reactions of the animals to the periodic flooding, and whenever possible verifying their role in the ecosystem. Respective chapters deal with tiger beetles and millipedes (Chap. 15), Oribatids (Chap. 16), collembola (Chap. 17), termites (Chap. 18), and spiders (Chap. 19). Restriction to these groups of invertebrates is due to the lack of studies about other groups important to the Amazon floodplain, such as ants and beetles.

Chapters 20, 21, and 22 deal with vertebrates. Because of their large numbers and biomass, fish, mammals, reptiles and birds play, or have played, an important role in the central Amazonian floodplain, occupying key positions in nearly all levels of the food webs. The Amazonian fish fauna is well-studied and known for its species richness. Chapter 20 concentrates on the impact of the flood pulse on the fish, on their food and feeding habits, growth, reproduction, migration and adaptations to hypoxia. These data allow a discussion of the advantages of different survival strategies in floodplain ecosystems. Mammals and reptiles were formerly very abundant in the Amazon floodplain, but in recent centuries their number has been greatly reduced by man. They provide good examples of the potential and limits of K-strategies in the floodplain (Chap. 21). In spite of the large number and species richness, the avifauna of the central Amazon floodplain has been little studied. Chapter 22 presents data on life cycles and adaptations of birds to the flood pulse and compares the avifauna of the floodplain with that of the nonflooded upland.

Part V presents the conclusions. Chapter 23 discusses the results of the previous chapters in the framework of the flood pulse concept. Emphasis is given to the importance of the flood pulse on habitat and species diversity. The impact of global paleoclimatic changes, which are indicated by the sedimentological, geological and biogeographical studies, is also considered. The impact of the flood pulse on primary production and decomposition of organic material is analyzed, and the interactions between the terrestrial and the aquatic phase on the transfer of bioelements and on different trophic levels are discussed.

Acknowledgments. I am very grateful to Prof. Harald Sioli, Max-Planck-Institut für Limnologie, Plön, Dr. James V. Ward, Eidgenössische

Anstalt für Wasserversorgung, Abwasserreinigung and Gewässerschutz, Dübendorf/Zürich, Dr. Brij Gopal, Jawaharlal Nehru University, New Delhi, and Dr. Rosemary Lowe-McConnell, Sussex, for critical comments on the manuscript and the correction of the English.

Part II
The Physical and Chemical Environment

2 The Large Central Amazonian River Floodplains Near Manaus: Geological, Climatological, Hydrological and Geomorphological Aspects

Georg Irion, Wolfgang J. Junk, and José A.S.N. de Mello

2.1 Formation of Várzea and Igapó

The Amazon River and its large tributaries are accompanied along their middle and lower courses by large fringing floodplains which cover an area of about 300 000 km² (Sect. 1.5). Their local names are: "várzea" and "igapó". These terms have been adopted in scientific literature, but until now they have not been defined. On the basis of hydrochemical investigations, Sioli (1950, 1951, 1956, 1965a, 1968) used the term várzea for floodplains along white-water rivers, which are rich in nutrients and suspended matter, and igapó for those along black-water rivers, which are poor in both. This nomenclature is supported by the zoological and botanical investigations of Irmler (1977) and Prance (1979), and it will be adopted here as well. The term "floodplain" will be used when referring to the general category. The periodically flooded várzea and igapó are contrasted with the upland that is not flooded, referred to as the "terra firme". The transition between flooded and nonflooded areas can be very abrupt, being formed by steep bluffs as high as 100 m.

The formation of várzea and igapó is closely related to fluctuations in sea level during Pleistocene times and is therefore a direct result of global climatic changes. Because of the low elevation of the Amazon Basin, where more than a million square kilometers lie less than 100 m above present sea level, the fluctuations in the level of the sea alternately promoted either an incision of the middle and lower Amazon and its tributaries during periods of low sea level, or a blockage of their outflow when the sea level was high. During cold periods, when sea levels were low, the rivers eroded deep valleys in the soft Cretaceous/Tertiary freshwater sediments of central Amazonia and in the rivers' own deposits from previous periods. With the rise in the sea level during the warm interglacial periods, the valleys were again filled with fluvial sediments. These effects influenced the river system at least 2500 km upstream from the Amazon mouth and in the areas north and south of the main river as far as the rims of the Guyana and Brazilian Shields.

Ecological Studies, Vol. 126
Junk (ed) The Central Amazon Floodplain
© Springer-Verlag Berlin Heidelberg 1997

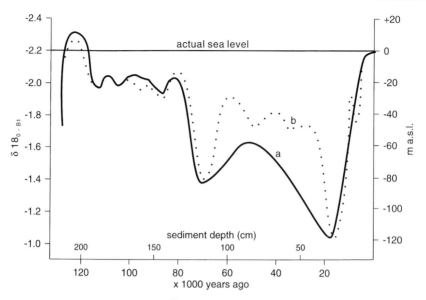

Fig. 2.1. *a* The correlation of the δ ^{18}O content of planktonic foraminifera in a sediment core from the Pacific and the course of the recent Quarternary sea level fluctuations. The amplitudes of the sea level fluctuations were calculated from the terraces on Barbados and New Guinea (Shackleton and Opdyke 1973). *b* Curve of Johnsson (1983) modified from data obtained on the Pacific West Coast of the United States

The paleo-sea-level curve, which is of essential importance for understanding the formation of the Amazonian Pleistocene várzea, is known in detail only for the very last period of the Pleistocene, the time from 10 000 to 100 000 years B.P. (before present), and for the Holocene (0 to 10 000 years B.P.) (Fig. 2.1). The sea-level curve of the rest of the Pleistocene going back to 2.4 million years B.P. is only incompletely known. However, it can be assumed that at least during the last 200 000 to 400 000 years there have been three warm periods with sea level stages either similar to or 10 to 20 m higher than those of today. During cold periods the sea level may have dropped more than 100 m below its present stage.

The distribution of the Pleistocene várzea, may be recognized from side-looking airborne radar maps published in 1971/1972 (Projeto Radambrasil 1972). These maps show that the paleo-várzea occupies an area of more than 1 million km^2 in the western Amazonian lowlands (Irion 1976a, 1982, 1984b; Klammer 1984). In some areas sequences of Pleistocene várzeas of different ages may be observed. In the area south and southwest of Manaus and west of Manacapurú, Pleistocene floodplains border the recent floodplain. Only a few meters above the recent várzea the so-called terra alta may be found in some areas near the Amazon. This may have formed 5000 B.P. when the sea level was 5 m higher than today or 100 000 B.P. during the last Pleistocene warm period.

During the maximum of the Würm cold period, about 20 000 years B.P., the sea dropped to about 120 m below its present level. It began to rise again 18 000 years B.P. and in the period from 12 500 years to 8000 years B.P. the sea level rose at an average rate of 2 cm year^{-1} (Fig. 2.2). Off the Brasilian coast the present m.s.l. (mean sea level) mark had already been reached 7000 years B.P., arriving at 5 m a.s.l. 5000 years B.P. then dropping smoothly to the 0 mark of today (Suguio et al. 1988). The sediment load of the Amazon and its tributaries during the period of rapidly rising water level was insufficient to keep the drowned valleys filled with sediment. Within these regions, large lakes were formed which have gradually been filled by materials from the upper courses of the rivers during the past several millennia (Irion 1976a, 1984b). In drowned lower sections of Amazon tributaries the sediments of the last three warm Pleistocene periods can be differentiated (Fig. 2.3; Müller et al. 1995) by 3.5 KHz high-resolution seismic equipment. The investigations of Damuth and Fairbridge (1970) on sediment cores from the protodelta of the Amazon showed that the deposition of sediment from the Amazon Basin was interrupted 10 000 to 11 000 years ago. Deltaic deposition was again resumed to a considerable extent

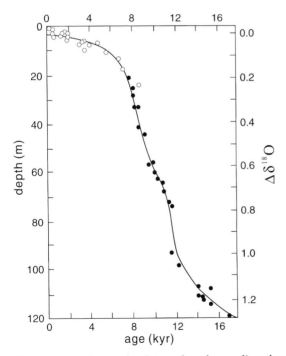

Fig. 2.2. Barbados sea-level curve based on radiocarbon-dated Caribbean reef-crest coral *Acropora palmata* (*filled circles*) compared with *A. palmata* age–depth data (*open circles*) for four other Caribbean island locations. (Fairbanks 1989)

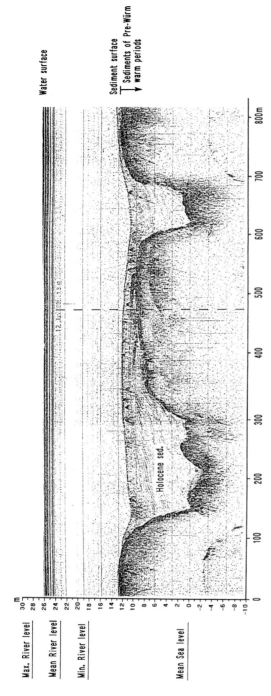

Fig. 2.3. 3.5 KHz profile from the mouth of Preto da Eva River showing two major acoustic units. The lower represents sediments of older sea level high stands, most probably 100 000 years old or even older. The upper unit has been deposited during the Holocene (the last 10 000 years) in a river system eroded during the last glaciation, when sea level has been low

about 5000 years ago after the lakes that had been formed in the Amazon valley had become filled with sediment.

The sedimentation processes in the lower courses of the black- and clear-water rivers characterized by a very low concentration of suspended material, such as the Negro River, Tapajós River, and Xingú River, have still not been completed today, as indicated by the large river beds near their confluences with the Amazon. The channel of the Negro River near Manaus can be as much as 30 km wide, and in other narrower places as much as 100 m deep. Its size and depth greatly exceed the dimensions required by its present water discharge. Klammer (1984) also determined that the sedimentation in the floodplain of the lower Amazon River is still not completed. The present várzea areas near Manaus have maximum ages from 6000 to 5000 years. Sternberg (1960b) determined the ages of fragments of Indian pottery on Careiro Island at the mouth of the Negro River and found that they were 1000 to 2000 years old. The dating of the sediment near Manaus by Irion and Junk (unpubl.) revealed a maximum age of somewhat more than 5000 years.

Most of the sediments of white-water rivers come from the region of the Andes and their foothills. Therefore, the várzea can be considered as a geochemical extension of the Andes and their foothill zone (Fig. 2.4;

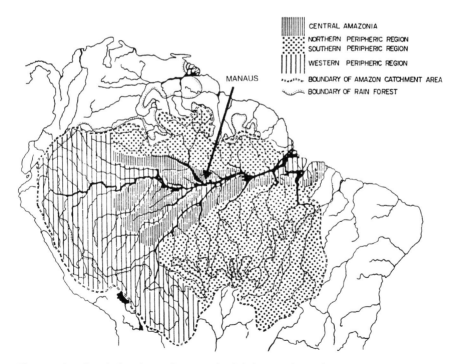

Fig. 2.4. Geochemical regions of Amazonia. (Fittkau et al. 1975)

Fittkau et al. 1975). Yet it has to be taken into consideration that the fine-grained Andean sediments are extensively altered, especially with respect to their clay mineral compositions, when deposited in the sub-Andean lowlands because of the intensive tropical weathering (Irion 1976a; Johnsson and Meade 1990). During this process, the concentration of smectite is greatly increased. Following this the sediments are remobilized and they form the largest portion of the suspended load of the Amazon River.

The sediments deposited in the igapó of the Negro River derive from the Guyana Shield and the Cretaceous sediments of central Amazonia and are poor in nutrients. A specific case of sedimentation is present downstream of the confluence of the Negro River and Branco River. The mixture of very acid water from Negro River with less acid water from Branco River leads to a rise in pH and to a flocculation of suspended and colloidal matter (Leenheer and Santos 1980) forming the vast archipelago of Anavilhanas in the Negro River upstream of Manaus. These sediments consist more than 90% quartz and kaolinite. They derive from tropical weathering in the catchment area that has occured during the last 10 or more million years (Irion 1984a).

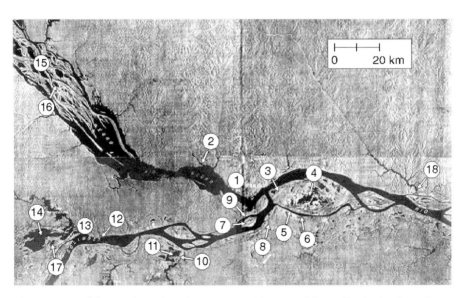

Fig. 2.5. Map of the area investigated. *1* Manaus; *2* Tarumã Mirim; *3* Ilha do Careiro; *4* Lago do Rei; *5* Paraná do Careiro; *6* Lago Jacaretinga; *7* Ilha de Marchantaria; *8* Paraná de Curarí; *9* Costa do Baixio and Lago Janauari; *10* Lago do Castanho; *11* Lago Janauacá, *12* Lago Calado and Lago Parú; *13* Manacapurú; *14* Lago Cabalhana; *15* Archipelago Anavilhanas; *16* Lago do Prato; *17* Lago Inácio; *18* Preto da Eva River

2.2 Characterization of the Study Area

Most of the investigations in the várzea were conducted in a section of the Amazon River located from 100 km upstream to about 50 km downstream of its confluence with the Negro River. Studies in the igapó concentrated on the mouth bay of Tarumã Mirim, a small affluent to the Negro River, about 10 km upstream of Manaus (Fig. 2.5). The investigations were limited to this region because of logistic problems and because of the necessity to correlate the observations with the flood cycle. Long-term records of the water level are maintained by the Manaus Harbor Authority. The data for the Negro River are recorded about 20 km upstream from its confluence with the Solimões River. (In Brazil, the Amazon River upstream of the confluence with the Negro River is called Solimões River.) Because of the

a

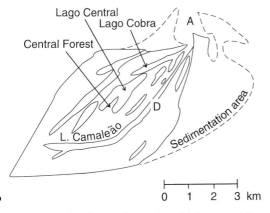

b

Fig. 2.6. a Ilha de Marchantaria at high water (view downstream). Floodplain forest and aquatic macrophytes indicate land but the whole island is inundated. b Indication of the principal study sites

small downstream gradient of 1 to $2\,cm\,km^{-1}$, the water level recorded at Manaus can be considered as representative of both rivers about $100\,km$ up- and downstream. By comparing the data in Lago Castanho, about $60\,km$ upstream from the confluence of the Solimões River and Negro River, with those from Manaus, Schmidt (1973a) demonstrated that the Manaus water gage provides reliable information on the flood cycle.

Since 1980, the INPA/Max-Planck team has concentrated its studies on Ilha de Marchantaria, the first island in the Amazon River upstream of the confluence with the Negro River (Fig. 2.6). The island was selected because of the short distance from Manaus. Furthermore, on islands the influence of the terra firme is eliminated or strongly reduced. This facilitated the interpretation of some of the results about water chemistry and soils, but it affected biological aspects. Many links exist between várzea and terra firme that are much stronger in the transition zone than on islands, e.g., in respect of species diversity. These questions remain open for further studies.

2.3 Climate and Hydrology

The climate of central Amazonia is hot and humid and is characterized by a weak thermal periodicity during the course of the year. The mean annual temperature is 26.6 °C. The warmest months are August, September, October, and November, during which the average temperatures range from 27.2 to 27.6 °C. The coolest months are January, February, March, and April, when the average temperatures are 25.9 to 26.1 °C. The diurnal fluctuations are greater than the annual fluctuations in temperature and can exceed 10 °C. Minimum temperatures below 20 °C occur once each year when cold, southern polar air masses move northward and influence the weather in central Amazonia for 1 to 3 days in May. These cold spells are locally called friagens. The relative humidity remains high for the entire year, averaging 75.6% in September and as much as 86.7% in April (Salati and Marques 1984; Ribeiro and Adis 1984). During the dry months, evaporation can exceed precipitation (Fig. 2.7).

Precipitation is clearly periodic. A rainy season from December to April alternates with a dry season lasting from June to October. A shorter second rainy season is usually neglected as it lasts only several days to a few weeks and is variable in date (October to December). Though meteorologically insignificant, the short rainy season has some ecological impact, e.g., on growth of herbaceous plants (Chap. 8) and the breeding behavior of some bird species (Chap. 22).

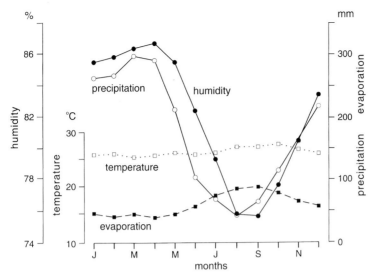

Fig. 2.7. Mean monthly precipitation, evaporation, relative humidity and temperature at Manaus. (Ribeiro and Adis 1984)

The total rainfall averages 2100 mm year^{-1}. There is evidence for pronounced local differences in the distribution of the rainfall in the region investigated (Ribeiro and Adis 1984; Molion and Dallarosa 1992). According to Ribeiro and Adis (1984), there was 45% less rainfall on Ilha de Marchantaria in 1981 and 1982 than there was in the forest reserve Reserva Ducke, which is located on the terra firme only some 40 km away. Rainfall amounts at both localities differ significantly from those at Manaus. The observed differences in the amounts of precipitation are explained by differences in the local air circulation. These, in turn, are due to differences in the albedo and roughness of the city, river, and forest areas. Geomorphic peculiarities, such as for instance its position on a valley floor and the low position in comparison with the surrounding terra firme may also influence the climate in the várzea. To what extent these findings are characteristic for the várzea in general is not yet clear because the data sets from Ilha de Marchantaria are too short. However, 45% less in rainfall on the várzea than on the surrounding terra firme certainly has a strong impact on plant growth and animal life in the várzea.

The seasonal distribution of the rainfall in the Amazon basin produces great fluctuations in the water levels of the rivers and streams. The maximum high water occurs in Manaus after the maximum rainfall period because the flood wave from the headwaters of the Andes arrives with a delay of 4 to 6 weeks. There is a similar 4 to 6 weeks delay before the river begins to rise after the beginning of the rainy season (Fig. 2.8). The mean

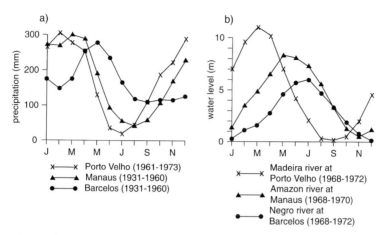

Fig. 2.8a,b. Monthly rainfall (**a**) and water level fluctuations (**b**) in the Madeira River at Porto Velho, the Amazon River at Manaus, and the Negro River at Barcelos, corresponding to southern, central, and northern parts of the Amazon basin. (Junk 1984b)

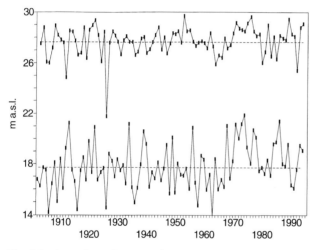

Fig. 2.9. Annual maximum and minimum water levels in the mouth of Negro River at Manaus. (Data from the Manaus Harbor Authority)

flood amplitude in the Amazon at Manaus is 9.95 m, calculated from data gathered from 1902 to 1994. The average maximum high water level is 27.68 ± 1.16 m (n = 92), and the average minimum level is 17.73 ± 1.83 m (n = 93). The absolute maximum water level was 29.70 m, recorded in 1953, whereas in 1926, the maximum was only 21.77 m. The lowest absolute water level of the river was 13.64 m, recorded in 1963, whereas in 1974, the lowest level reached was 21.84 m (Fig. 2.9).

The discharges of the Amazon and Negro Rivers near Manaus are determined by the precipitation in their large catchment areas. With the regular

fluctuations between dry and rainy seasons, river levels display monomodal flood curves. From an ecological point of view, it is important to
remember that the water-level changes in the Amazon and its main tributaries are rather predictable. The probability that the high-water peak will
be reached in the second half of June is 55%, and the probability that the
low-water level will be reached between October 15 and November 15 is
58% (Fig. 2.10). The level of low water varies more than that of high water,
even though the absolute range of water level remains rather similar at
about 8 m (Fig. 2.11). The probability that the high-water level will be
between 26 and 29 m is 88%, whereas the probability of a low-water level
between 16 and 19 m is 64%. The available data from Manaus Harbour
Authorities also show periods of extreme water levels lasting several years.
From 1963 through 1965, the high-water level remained below 27 m,
whereas it rose above the 28 m level every year from 1970 through to 1979.
From 1935 to 1937 and from 1965 to 1967 the low-water level was less than
17 m, whereas from 1971 to 1974 it did not fall below 20 m. The local
weather in the vicinity of Manaus has no influence on the river level. Heavy
local rainfall, however, can bring about rapid, short-term increases in the
levels of lakes in the floodplain. In such cases, water from the lakes flows
into the river, carrying floating vegetation, wood, and plankton with it.

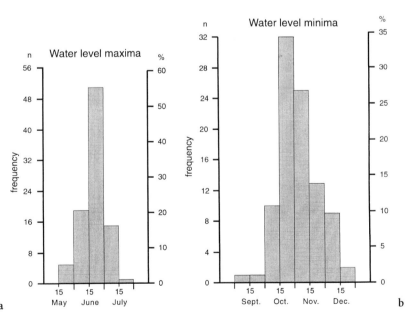

Fig. 2.10a,b. Frequency distribution of the water level maxima (a) and minima (b) in the
mouth of Negro River at Manaus from 1902 to 1994 according to the time of year. (Data
from the Manaus Harbor Authority)

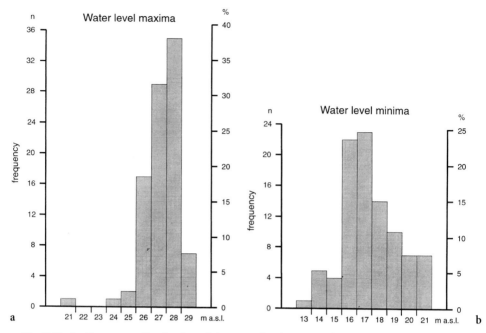

Fig. 2.11a,b. Frequency distribution of the water level maxima (**a**) and minima (**b**) in the mouth of Negro River at Manaus from 1902 to 1994 according to the absolute water level. (Data from the Manaus Harbor Authority)

The height of the water level is correlated with the discharge of the river. Meade et al. (1979b) and Richey et al. (1986) reported a flow of $220\,000\,\text{m}^3\,\text{s}^{-1}$ during the mean high water and $120\,000\,\text{m}^3\,\text{s}^{-1}$ during the mean low water at Obidos in the lower Amazon. According to Milliman and Meade (1983) and Milliman (1990), the water discharge of the Amazon River accounts for 19% of all riverine water delivered to the oceans. Upstream of the junction with Negro River and Madeira River, the two largest tributaries of the Amazon, the discharge of the Amazon River amounts to only half of that at Obidos.

Speculations about the relationship between the destruction of the Amazonian rain forest and a period of high floods at the beginning of the 1970s led to a discussion of possible consequences of anthropogenic activities on the discharge of the Amazon (Gentry and Lopez Parodi 1980; Nordin and Meade 1982; Sternberg 1987). Apparently, there are not yet enough data available for a reliable statement. The temporal differences between the high-water periods of the northern and southern tributaries as well as the large size of the watershed mask effects of the anthropogenic changes along any one of the tributaries on the water level in the main river. In addition, the El Niño phenomenon, a climatic cycle in South America may influence the quantity of rainfall and thereby also the discharge of the Amazon River

(Richey et al. 1989). The water level of the main river is therefore less suitable for judging the effects of human activities than that in the individual tributaries. Locally, land-use practices are reflected in the rivers by slight hydrochemical and biological changes such as, for instance, by a slight increase in the amount of suspended material and dissolved bioelements and an increase in the occurrence of aquatic macrophytes (Junk and Howard-Williams 1984; Santos et al. 1984). Mining activities and an increase in deforestation will in future accelerate anthropogenic impacts upon the rivers and their floodplains (Hettler and Irion 1994).

2.4 Grain Size and Mineralogical Composition

The suspended sediment concentration of the Amazon near Manaus is about $100\,mgl^{-1}$ (Gibbs 1967a; Irion 1976a; Meade 1985) (Sect. 4.3.1). The suspended matter is relatively fine grained and contains only minor amounts of fine sands. Silt (grain size of 2 to 63 μm) and clay (grain size <2 μm) are the predominating grain classes (Meade et al. 1979a; Irion et al. 1983; Meade 1985; Irion and Zöllmer 1990). The sandy bed material consists of grain sizes between 100 and 1000 μm (Mertes and Meade 1985). The fine sand and coarse silt fraction consists mainly of quartz with minor amounts of feldspar and clay minerals. Calcite, which is often a component of river sediments, is present only in the affluents of the Andes and the sub-Andean lowlands, e.g., the Ucayali and Marañón Rivers (Irion 1976a), but is dissolved on its way down the Amazon River. The clay and fine silt fractions (<6.3 μm) contain predominantly smectite and illite followed by chlorite and kaolinite. In the middle silt-sized grain class (6.3–20 μm) quartz is predominant.

These fine-grained suspended solids (or sediments when deposited) are relatively rich in elements essential for plant growth (Table 2.1). Smectite has a high ion exchange capacity, while chlorite and, to a lesser extent, illite weather moderately under tropical conditions and release magnesium, potassium, and a variety of trace elements that are essential for plant growth. In contrast to the relatively high magnesium and potassium content, the amount of sodium is small (Irion 1984a).

Because of the intense weathering in a tropical climate, in areas with low or no sedimentation chemical alteration of the fine-grained material can be observed. Soil profiles of levees of Holocene and Pleistocene age in the triangle between Negro River and Solimões River at their confluence near Manaus show an increasing loss of some elements in the uppermost decimetres of the profiles with increasing age (Table 2.1). This decrease in

Table 2.1. Chemical composition (mg kg^{-1}) of the fine fraction (<2 μm) of various soils and sediments (surface samples) from Amazonian Lowland. (Irion 1976b)

	Na	K	Ca	Mg	Fe	Mn	Zn	Cr	Co
Barreira formation	160	225	350	100	27 600	33	27	122	1.5
Guyana shield	600	700	700	280	66 000	84	50	24	3
Cretaceous-Tertiary sediments	1630	15 100	1300	5 000	55 800	330	126	77	5
Pleistocene várzea	1650	15 200	940	5 600	51 000	98	115	67	8
Recent várzea	3200	17 800	9800	11 700	55 025	970	135	62	16

element concentrations is probably related to the alteration of clay minerals. The amount of alteration in clay mineral compositions in Holocene sediments is less than the detection limit, except in the clay mineral chlorite (Fig. 2.12). In várzea sediments of Pleistocene age, however, alteration can be shown for most of the clay minerals (Irion 1984a).

Table 2.2 shows the contents of potassium, zinc, iron, manganese and organic carbon (C$_{org.}$) in four sediment profiles from várzea sections of different age formed from suspension and bottom load of the Amazon River. Potassium, an important constituent of the clay mineral illite, does not show significant variations in the Late Pleistocene and Holocene profiles. Only in the oldest sediment, in which illite is partly transformed into other clay minerals, is the content of potassium reduced. The trace metal zinc is more strongly affected by relatively high alterations in surface sediments. Its reduction can already be observed in sediments which are only some 100 years old (profile 17 in Table 2.2). Manganese is very easily released from the sediments. At the surface of profile 17 only about 12% of the original content has been found. In the Pleistocene surface sediments the manganese contents are reduced to about 4% and less. Iron follows the same trend as manganese but the dissolution process is less well established. The intense dissolution of these metals can be observed during low-water periods on the river banks, where water rich in iron and manganese drains from the sediment along water-impermeable horizons. The C$_{org.}$ content in the <2 μm fraction decreases as a consequence of the tropical climate relatively soon after deposition. In the surface sediments of profile 17 only half of the original C$_{org.}$ content is present, and in profile 20 and 12 it has decreased to about 25%.

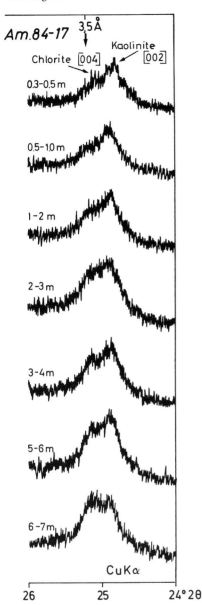

Fig. 2.12. The relative contents of the clay minerals chlorite and kaolinite (x-ray diffraction patterns of <2 µm fraction) in a sediment profile taken from a levee of the Amazon River near Manaus. The levee may have an age of at least several hundred years. As a result of tropical weathering the chlorite content is decreasing in the upward direction

2.5 Sediment Transport and Deposition

The total sediment load of the Amazon River at Obidos is estimated to be between 1.1 and $1.3 \times 10^9\,t\,year^{-1}$ (Meade et al. 1985; Richey et al. 1986; Meade 1994). Most of it is suspended material. Sandy bed load is transported in gigantic sand waves, 6 to 8 m high (as much as 12 m high at

Table 2.2. Content of potassium, zinc, iron, manganese (all from HNO_3 digestions) and $C_{org.}$ in the <2 μm fraction of profiles taken from várzea levees in the triangle between Negro River and Solimões River near Manaus. Profile 10 is taken from recently settled sediments, the sediments from profile 17 may have an age of some 100 years, profile 20 is of Late Pleistocene age, most probably formed during the last Pleistocene warm period 100 000 years ago, whereas profile 12 is significantly older. The content of the elements decreases with increasing age of the sediment

Profile No.	10	17	20	12
Depth (m)	K in mg kg^{-1} in the fraction <2 μm			
0–0.5	2500	2260	2100	1350
0.5–1	2500	1850	1880	1350
1–2	2900	2630	2510	1500
2–3	2900	2330	2180	1630
	Zn in mg kg^{-1} in the fraction <2 μm			
0–0.5	135	84	76	39
0.5–1	135	106	91	39
1–2	135	124	124	34
2–3	129	160	115	31
	Fe in % of the fraction <2 μm			
0–0.5	5.5	3.3	2.8	3.6
0.5–1	5.5	6.1	–	3.9
1–2	–	7.1	4.8	4.1
2–3	5.7	5.9	–	–
	Mn in mg kg^{-1} in the fraction <2 μm			
0–0.5	1400	176	56	19
0.5–1	1400	545	138	19
1–2	1400	1030	640	23
2–3	950	1120	818	30
	$C_{org.}$ in % of the fraction <2 μm			
0–0.5	1.9	0.9	0.5	0.6
0.5–1	1.9	0.7	0.8	0.6
1–2	1.9	0.7	0.3	0.4
2–3	2.0	1.1	1.3	0.3

Obidos; Mertes and Meade 1985), along the bottom of the main channel of the river (Sioli 1965b). During rising water, the river deposits sediments in its bed that are mobilized when the water level falls (Meade et al. 1985).

Detailed investigations on Ilha de Marchantaria show that sediments about 10–12 m below the high-water mark consist predominantly of sand mixed with small amounts of the finer fractions representing the bed load of the river (Fig. 2.13). As the elevation increases and sand bars are formed, the content of fine sediments increases. The upper layers of sediments on the bars consist of sand mixed with large amounts of finer material fractions. The amount of finer material is considerably greater on the parts of the bars facing away from the river channel. The finest sediments with the largest clay fractions were found in the depressions between bars (swales) and in the central lakes on the island (Irion et al. 1983).

In alluvial areas, sediment layers over 1 m thick can be deposited during a single year. Such spectacular events can actually be observed at Ilha de Marchantaria. However, in general sediment deposition in the middle Amazon is less conspicuous. Mertes (1994) calculated from field measurements and from Landsat data the transport and fate of water and sediment in the 200 km reach of the Amazon River upstream of its confluence with the Negro River. During high water, between 3 and 18 t of sediments are

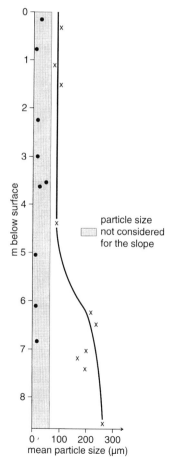

Fig. 2.13. Mean grain size of sandy sediments in the levee on the northwestern border of Ilha de Marchantaria near the upstream end of the island. (Irion et al. 1983)

transported per meter of river bank per day onto the várzea, of which in general far more than 50% is deposited. Sediment discharge diminishes strongly with increasing distance from the main river channel. It reaches about $90\,t\,day^{-1}\,m^{-1}$ in the channel and decreases successively to $10\,t$ $day^{-1}\,m^{-1}$ on the bank, to $5\,t\,day^{-1}\,m^{-1}$ at 500 m distance from the bank and to $1\,t\,day^{-1}\,m^{-1}$ within a distance of 1 km. These data show the great variability of sedimentation rates inside the várzea.

In most of the várzea lakes, sediments are stratified and deposited in the form of varvites (Fig. 2.14). The varvites consist of paired layers of clay to middle silt-sized minerals alternating with coarse silts and fine sands. The paired layers are 2–5 mm thick and are deposited annually. When the river rises, water enters through channels from the downstream end of the islands into the lakes transporting fine-grained suspended load into the lake basin. These particles make up about 75% of a layer. At high water, when most of the floodplain is inundated, current velocites are high and relatively coarse sediments are transported into the lake basins. When the water level drops and the water drains slowly through the downstream channels, the fine-grained suspended matter, still present in the lake water, is deposited. In the central lake of Ilha de Marchantaria, a 6 m thick layer of

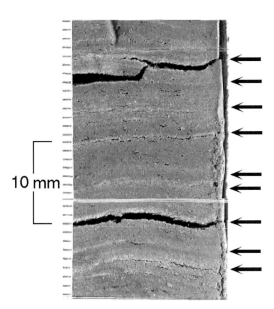

Fig. 2.14. Varvites from the central lake of Ilha de Marchantaria. The changes in brightness are due to the different grain sizes of the layers. The brighter horizons indicated by arrows are coarse silts and fine sands deposited during the high waters between May and July, whereas the darker layers are fine silts and clays settled in the period after high water and as the water slowly rises during the first 4 to 5 months of the year

lake sediments contains about 1800 varvites with an average thickness of 3.5 mm, but for a precise dating additional analyses are required. Similar varvites were found in várzea lakes at the mouth of Lago Janauacá and between Jurutí and Obidos and as well on Ilha das Onças (Irion, unpublished). In lakes farther away from the main river, e.g., Lago do Rei on Ilha do Careiro, very fine-grained sediments are deposited and no varvites are developed. It can be assumed that those lakes have lower sedimentation rates. In shallow lakes the development of varvites is disturbed by wind-induced turbulences and by bioturbation during low-water periods.

2.6 Geomorphological Aspects

The várzea is a small-scale mosaic of different structural elements of fluvial origin (Fig. 2.15). There are thousands of lakes of different shapes and fetches, separated by old levees, periodically interconnected with each other and the river channel. Melack (1984) indicates for the várzea, including the Peruvian part, about 8050 lakes with a fetch greater than 100 m. Most of these have fetches between 250–1000 m, 10% are smaller than 250 m across and 10% have fetches of 7–60 km. About 5010 have a round-oval shape, 1530 are levee lakes, 830 are dendritic, 140 are crescent, 270 are oxbow lakes and 270 are of a composite nature.

Large lakes are permanently connected with the main rivers through long, deep channels called "furos". These cause the water levels in the lakes to follow that in the river, except during the periods of extremely low water. During years of extreme low water, lakes measuring more than a hundred square kilometers are drained until they contain only a few square kilometers of water with a depth of just 1 m. Small lakes may become connected with the river only at high floods and maintain rather constant water levels. There are no oxbow lakes in the Amazon floodplain, because the Amazon River shows only little meandering in its middle and lower reaches.

At the edge of the várzea, imbedded in the terra firme, are dendritic branching lakes with steep banks called Ria lakes (Fig. 2.15, lagos de terra firme). These are the drowned valleys of the tributaries the mouths of which have been closed off, except for small connecting channels, by the alluvial deposits from the Amazon (Gourou 1950). These channels are always large enough to allow the lake level to fluctuate parallel to the level of the Amazon River.

Large channels (paranás) often run hundreds of kilometers along the terra firme, transporting the water that flows from the terra firme through numerous streams and rivers into the Amazon River and onto its floodplain.

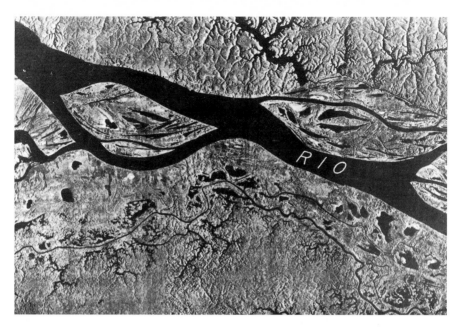

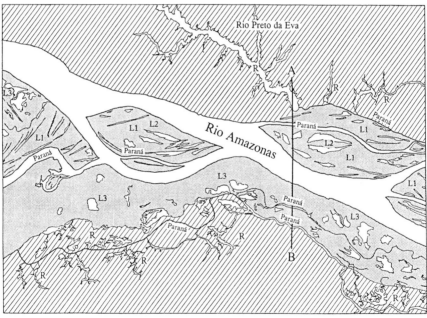

| | Várzea
Late Holocene | | Only some examples are indicated
R = Ria Lakes |

Várzea Lakes:

| | Terra Firme
Cretaceous and Pleistocene | L1 = formed between two levées
L2 = formed between two groups of levées
L3 = formed in lower areas of the várzea
where no levées are present |

10km

Fig. 2.15a

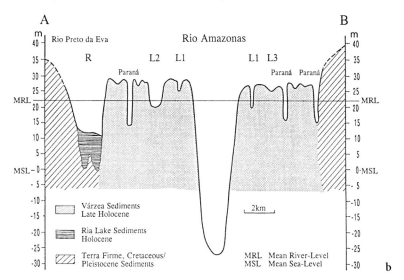

Fig. 2.15a,b. Morphology of the Amazonian floodplain east of Manaus (the center of the map is at 3°11.1'S and 59°23.5'W). **a** The *upper map* is taken from the side-looking airborne radar map of Amazonia (Projeto Radambrasil 1972), the *lower map* is the sketch of the same map which shows some principal geological and fluvial aspects. The area is separated into terra firme of Cretaceous (*upper part*) and Pleistocene (*lower part*) age. In the terra firme more than 15 Ria-lakes are interbedded of which the largest is Preto da Eva River. In the floodplain (várzea) of the Amazon River it is possible to distinguish different types of lake. There are several long channels (Paranás) of which one (in the lower third of the map) runs through the whole distance of the map from west to east. **b** Cross section (*A–B*) through the várzea and the neighboring terra firme

Just as there are a variety of lakes there are many terrestrial structures which differ in form, stability, grain size of the soil, position on the flood gradient, and vegetation cover.

Sand flats appear on the river bed during low water. They are formed by large sand dunes or by undifferentiated sand layers. During high water, the sand flats are submerged and their shape may change quickly due to deposition of new sediments or erosion by strong currents. Stable mud flats appear in the lakes and swales at low water. Sedimentation rates usually range from 1–5 mm year^{-1}; however, in protected areas near the river channel, deposition rates on mud flats may reach 50–100 cm year^{-1}. Bars are elevated, former river banks, formed by sediment deposition along the river channel. They often run parallel to one another for several kilometers within distances of some 100 m. The slopes of their sides may be either steep or gentle. Well-developed bars are submerged only during periods of very high water. Swales are depressions between the bars. At low water they often contain small lakes or temporary swamps. During high water they can serve as channels for supplemental water flow. In the bed of the Amazon and along its shores toward the terra firme there are isolated islands of laterite. Little is known about their sizes, distribution, or origin. They may have great importance for the position of the river channel because they form long-term impediments to the erosional forces of the river.

Near the main channel the várzea is subject to continuous structural rearrangements by erosion and sedimentation. Stretches of the shore several hundred meters long and as wide as 50 m can break off and fall into the water as the level in the river either rises (Bates 1864) or falls. These zones, called "terras caidas", were feared by early travellers whose boats were forced to hold to the channels near the shores and faced the danger of being swamped by falling earth or capsizing. In spite of these dynamics, the várzea along the middle Amazon as a whole is relatively stable. Old maps show that the river near Manaus has changed only slightly during the past 100 years. The Lago do Rei on Careiro Island was already an important fishing area 400 years ago, and it still is today. Large amounts of sediment are presently deposited in the region of Ilha de Marchantaria. The right bank of the river opposite this island and upstream from the mouth of the Paraná de Curarí is strongly eroded. In the foothills of the Andes, the channels of the Amazon tributaries move much faster, migrating distances of 30 to 50 m year^{-1}, and can relocate their entire alluvium within a few decades (Salo et al. 1986).

From the enormous water mass of the Amazon River results a river bed of an extraordinary shape. At Manacapurú, at a distance of 1600 km from the mouth, the maximal water depth is 40 m. At Obidós at a distance of 1000 km from the mouth, where the width of Amazon River is only 2.5 km, the maximal water depth is about 60 m (Sioli 1984). Mertes (1985) calculated values from 20 to 26 m for the average depth of the Amazon main stem between Manacapurú and Obidós. At Manacapurú the water level is only about 24 m a.s.l. Therefore more than half of the water body of the Amazon River downstream of Manacapurú is below m.a.s.l.

2.7 Discussion and Conclusions

The active floodplains of the lower courses of Amazonian rivers are only a few thousand years old and are the most recent structural elements in the sedimentary history of the Amazonian lowlands. However, in Amazonia floodplain systems as such are much older and have shaped large areas of the landscape. In contrast to most of the other floodplains of the world the formation of the central Amazonian floodplains is not only dependent on transport and deposition of particulate material from the headwaters into the river valleys; it is also related to sea level changes that caused changes in the slope of the river levels, modifying their sedimentation and erosion behavior. During cold Pleistocene periods, when sea level was low, the rivers and creeks of central Amazonia incised deeply into the sediment deposits. During that time the extension of the floodplains was relatively small. But during warm periods, when sea level rose again, large areas of the lowland valleys were drowned forming lakes that became filled with river sediments, depending on the sediment load of the respective rivers.

It is known that there have been at least six warm periods with high sea level stages during the last 2.4 million years. However, the variation of the sea level is known in more detail only for the last 100 000 years and therefore only generalized conclusions about the history of sedimentation in the Amazon basin can be drawn. It can be assumed that sea levels during Pleistocene (and Pliocene) warm periods have been in general some metres or even tens of meters higher than today. During the times of these higher sea levels large parts of the western Amazonian lowland were covered with fluvial sediments, forming today's paleo-várzea which occupies an area of more than 1 million km^2. Because of the higher sea level maxima, and hence the higher water levels of the Amazonian drainage system, the sediments of the paleo-floodplains are situated at levels several meters above the recent ones. They are inactive and not inundated anymore by the floods of central Amazonian rivers.

In the lower reaches of rivers poor in suspended load, the filling process was slower and extended lakes (Ria lakes) have formed in response to the last sea level rise. Black-water and clear-water rivers have not yet filled up their valleys, which form wide and deep bays in their lower courses which do not correspond to their discharge. 3.5 KHz seismic profiles on these lakes show sediments of at least three different Pleistocene sea level heights (Müller et al. 1995). These results are in a good agreement with the supposed effects of the changing water levels during Quaternary time in the Amazon drainage system. During its rise following the Würm glaciation mean sea level did not reach today's height until 7000 B.P. The várzea, therefore, could only have formed on its present level during the last 7000 years. Before this time the surface of the várzea must have been lower.

The sedimentation in the várzea is controlled by the formation of levees which generally run parallel for several kilometers at distances of some 100 m. Several levees may enclose depressions occupied by lakes. Because of the annual fluctuation of the water levels of the rivers, sedimentation in the lakes is discontinuous. During times of rising water levels clays and fine silts are deposited, and during high-water periods coarse silt and fine sand are deposited. From these periodic sediment depositions (varvites) an minimum age of the lakes can be calculated. The results show that many lakes are several thousand years old and that the levees which form the largest part of the várzea near Manaus are comparatively stable. Major changes are concentrated in relatively small areas near the main channel.

The majority of the sediments of the Amazonian várzea derive ultimately from the Andes. Initially most of the Andean sediments are deposited in the Sub-Andean lowlands where they undergo an extensive alteration because of tropical climate conditions. There is a strong increase in the content of the clay mineral smectite and a decrease in the water-soluble mineral calcite. But there is no major additional alteration of the material when it is later deposited in the central Amazonian várzea. From a geochemical and mineralogical point of view the várzea is a specific element in central Amazonia. Whereas the soils on the Cretaceous, Paleozoic, and Precambrian formations of the Amazonian lowlands are extremely poor in nutrients and in alkali and alkaline-earth elements, the content of these elements is comparatively high in the várzea.

Global sea level changes were accompanied by global climatic changes, which have affected the Amazon basin as well. Supposedly, during cold periods the climate was cooler and dryer than today. This may have affected the extent of the Amazonian rain forest. Some authors even postulate a fragmentation of the forested area inside the basin. According to the refuge theory (Haffer 1982) the fragmentation allowed an exchange of savanna species between the Cerrado belts north and south of the basin and the acceleration of the speciation process in the remaining forest refuges because of the isolation of plant and animal populations as summarized by Prance (1982). Such climatic changes should have modified the total discharge of the Amazon River and its affluents and their sedimentation and erosion pattern. The investigations of Amazonian alterites (weathering horizons/soils) and of Ria lake sediments near Manaus (Irion 1989; Müller et al. 1995) do not support the existence of extreme changes in vegetation density in central Amazonia during the Late Tertiary and Quaternary. In any case it seems unlikely that there was a fundamental change in the discharge pattern of the large Amazonian rivers because of their large catchment areas in the lowlands north and south of the equator. The geological history of the Amazonian lowlands shows that large river floodplains must have existed for at least the last 2.4 million years but probably for a much longer period, probably 10 million years or more.

Acknowledgments. We are very grateful to Prof. Dr. German Müller, University of Heidelberg, Dr. Robert Meade, United States Dep. of the Interior, Geological Survey, Denver, Colorado, and Dr. Rosemary Lowe-McConnell, Sussex, for critical comments on the manuscript and the correction of the English.

3 Chemistry of Várzea and Igapó Soils and Nutrient Inventory of Their Floodplain Forests

KARIN FURCH

3.1 Introduction

There is general agreement that the alluvial soils of the várzea, which are derived from the settled suspended load of the Amazon River, are rich in nutrients sustaining a high natural productivity (Sioli 1954a, 1969, 1975; Irion 1978; Cochrane and Sanchez 1982; Nascimento and Homma 1984; Sombroek 1984; Fearnside 1985; Meggers 1985; Lima 1986; Furch and Klinge 1989; Junk et al. 1989; Martinelli et al. 1989). Sediments originate in the Andes and the pre-Andean zone, they contain clay minerals, i.e., montmorillonite, illite and chlorite, with relatively high cation exchange capacities (Gibbs 1967; Irion 1976b, 1984a; Sect. 2.4). In contrast, the soils of the black-water floodplains, the igapó, are considered to be poor in nutrients and of low production potential (Sioli 1954a, 1969, 1975; Irion 1978; Fearnside 1985; Singer and Aguiar 1986; Furch and Klinge 1989). These soils partly originate from erosional processes of strongly weathered and lixiviated tertiary sediments containing mainly kaolinite which has a low cation exchange capacity (Irion 1976b, 1984a; Irion and Adis 1979).

Detailed information about the chemistry of várzea and igapó soils of the middle Amazon region is scarce (Stark and Holley 1975; Worbes 1986; Ohly 1987; Furch and Klinge 1989; Meyer 1991; Martinelli et al. 1993). A few years ago we started a soil chemical programme (Furch, in prep.) to specify the differences between the floodplain soil types with respect to the most important chemical soil parameters (Fig. 3.1). From a biogeochemical perspective the main objectives of this study are: (1) to evaluate the nutrient regime of the floodplain soils; (2) to study the impact of the flood pulse on the soil nutrient levels; and (3) to explore the relationships between the nutrients of the soil and the natural forest vegetation. The initial results of this study are presented in this paper.

Ecological Studies, Vol. 126
Junk (ed) The Central Amazon Floodplain
© Springer-Verlag Berlin Heidelberg 1997

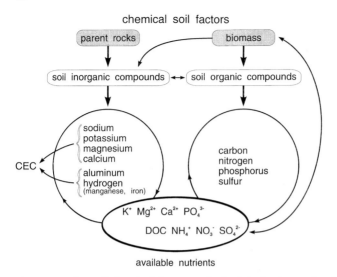

chemical soil factors

available nutrients

Fig. 3.1. Major soil chemical factors and schedule for the study of soil chemistry

Table 3.1. Characterization of floodplain forest soils. (Meyer 1991)

Igapó (Entisol, Quartzi-Psamment)					Várzea (Inceptisol, Typic Tropaquept)				
Horizon[a]	Depth (mm)	Thickness (mm)	d (g cm^{-3})	Root density[b]	Horizon[a]	Depth (mm)	Thickness (mm)	d (g/cm^3)	Root density[b]
L	20–0	20	0.06	+	L	10–0	10	0.05	–
A$_{hh}$	0–30	30	0.72	+++	A$_{hh}$	0–40	40	0.81	+++
(A$_h$) G$_r$	30–180	150	1.09	+	(A$_h$)G$_{or}$	40–90	50	1.02	+
G$_{(o)r}$	180–480	300	1.29	+/–	G$_{ro}$	90–230	140	1.18	+
G$_o$B$_v$	>480	>520	1.26	–	IIB$_v$G$_o$	230–670	440	1.25	+/–
					G$_{ro}$	>670	>330	1.20	–

[a] Müller (1982).
[b] +++ very high; ++ high; + medium; +/– low; – no roots.

3.2 Methods

Soil sampling at the várzea (Ilha de Marchantaria) and at the igapó
(Tarumã Mirím) was carried out during the low-water period (October to
December). Samples were taken from semiterrestrial forest soils (5–8
months year^{-1} flooded), semiaquatic shoreline sediments (8–12 months
year^{-1} flooded) and aquatic sediments (permanently flooded). Additionally, suspended sediments of the Rio Solimões were sampled. A detailed
description of the floodplain forest soils, given by Meyer (1991), including
the taxonomic classification (US Soil Taxonomy 1990) is summarized in
Table 3.1. For the individual profiles, spatial variations of the data given in

Table 3.1 are observed in the field for the upper horizons depending on the position of the altitude of the sampling locality on the flood gradient, as reported by Meyer (1991). Profiles of semiaquatic and aquatic sediments were not separated clearly into respective horizons, so, due to reasons of comparability, soil and sediment profiles were sampled as follows:

1. Humic soil layers, i.e., the A_{hh} horizons of the forest soils and the uppermost thin layer (1–10 mm thick) of the semiaquatic sediments (a mixture of homogeneous felt-like plant residues and fine sediments).
2. Mineral soils (A and G horizons) divided into layers of 10 (20) cm thickness. For most profiles sampling depth was restricted to 30 (40) cm. Values for bulk density (g dry weight cm^{-3}) of mineral forest soils were somewhat different from those given by Meyer (1991): 0.94 (0–10 cm) and 1.08 (10–30 cm) in the várzea and 1.03 (0–10 cm), 1.59 (10–20 cm) and 1.39 (30–40 cm) in the igapó.
3. The uppermost 20 cm of the aquatic sediments were sampled with a sediment corer and were not divided into separate layers. The number of samples varied between 2 and 10 for each soil layer (Tables 3.2–3.4).

Soil samples were air-dried at temperatures below 40 °C and ground to pass a 2 mm sieve; coarse organic fractions such as roots, leaves and twigs were removed.

Chemical analysis. Wet combustion of oven-dried soils (105 °C) with an HF/HNO$_3$ mixture in teflon pressure bombs (microwave oven) was done to determine the total amounts of Na, K, Mg, Ca (atomic absorption spectrometry, Perkin Elmer, type 1100 B) and P (molybdate blue method, Auto-Analyzer, type AA 2). Total C, N, and S were determined chromatographically after heat combustion at 1030–1800 °C using a CNS-analyzer (Carlo Erba instruments, type NA 1500). Extraction of air-dried soil samples (diameter $\leq$2 mm) with 1 M ammonium acetate solution (pH 7) was performed to determine the exchangeable basic cations and extraction with 1 M KCl solution was used to determine exchangeable ammonium and exchangeable acidity (titration using NaOH with and without masking of Al ions with NaF). Extraction with a H_2SO_4/HCl solution was performed to determine available PO_4. Aqueous soil solutions were prepared by mixing 15 g air-dried soil with 400 ml deionized water for 2 h. After filtration through paper filter (S&S, type 597 1/2, pore size 7–12 µm, Ref. No. 311847), the following dissolved components were determined: Na, K, Mg, Ca by atomic absorption spectrometry (1100 B); PO_4 by the molybdate blue method (AA 2); NH_4 by the indophenol blue method (AA 2); SO_4 turbimetrically as $BaSO_4$ (AA 2); DOC by infrared detection (carbon analyzer, Beckman, type 915 A); NO_3 by the sulfanile amid method; and Cl by

Table 3.2. Mean content of total sodium, potassium, magnesium, calcium, carbon, nitrogen, sulfur, and phosphorus in várzea soils on Ilha de Marchantaria exposed to different annual flood periods, in the suspended sediments of the Rio Solimões and in the forest soils of the igapó Tarumã Mirim

	Sampling depth (cm)	Number of samples (n)	Na (g kg⁻¹)	K (g kg⁻¹)	Mg (g kg⁻¹)	Ca (g kg⁻¹)	C (g kg⁻¹)	N (g kg⁻¹)	S (g kg⁻¹)	P (g kg⁻¹)
Várzea										
Semiterrestrial										
Humic	2–0	8	6.2 (0.8)	18.2 (1.2)	8.4 (0.5)	7.8 (0.3)	91.2 (25.0)	7.6 (2.5)	1.29 (0.46)	0.81 (0.09)
Mineral	0–10	8	8.3 (0.6)	21.6 (0.5)	9.9 (0.2)	6.7 (0.5)	10.4 (2.2)	1.4 (0.3)	0.22 (0.05)	0.61 (0.09)
Mineral	10–30	7	8.4 (0.6)	21.3 (1.0)	9.8 (0.2)	7.0 (0.9)	7.0 (1.4)	1.1 (0.2)	0.13 (0.03)	0.64 (0.10)
Semiaquatic										
"Humic"[a]	0.3–0	8	7.1 (1.1)	20.7 (0.9)	9.3 (0.2)	7.6 (0.7)	39.4 (15.9)	4.0 (1.2)	1.10 (0.55)	1.12 (0.43)
Mineral	0–10	10	7.5 (1.1)	20.2 (1.2)	9.5 (0.3)	7.7 (0.9)	28.9 (15.5)	3.1 (1.2)	0.91 (0.60)	0.89 (0.16)
Mineral	10–30	10	9.1 (1.0)	20.5 (1.0)	9.7 (0.6)	8.6 (0.7)	10.7 (5.8)	1.3 (0.4)	0.22 (0.11)	0.76 (0.05)
Aquatic										
Mineral	0–20	3	7.1 (0.4)	19.0 (1.1)	9.6 (0.5)	10.7 (3.1)	21.8 (5.8)	2.4 (0.4)	0.93 (0.53)	1.20 (0.48)
Suspended sediments	–	2	8.1 (1.0)	21.8 (0.2)	10.3 (<0.1)	9.5 (0.4)	12.3 (0.1)	3.3 (0.1)	0.35 (0.05)	0.81 (0.02)
Igapó										
Semiterrestrial										
Humic	2–0	10	0.11 (0.03)	1.12 (0.24)	0.68 (0.11)	0.39 (0.05)	75.0 (19.5)	5.1 (1.2)	0.42 (0.08)	0.21 (0.03)
Mineral	0–10	10	0.10 (0.02)	1.03 (0.16)	0.69 (0.16)	0.33 (0.02)	39.5 (3.6)	3.1 (0.3)	0.27 (0.04)	0.15 (0.02)
Mineral	10–20	2	0.08 (0.01)	0.75 (0.03)	0.53 (<0.01)	0.26 (0.05)	10.1 (0.4)	1.1 (0.1)	0.13 (0.01)	0.08 (0.01)
Mineral	30–40	2	0.09 (0.01)	0.65 (0.05)	0.61 (<0.01)	0.27 (0.02)	6.4 (0.1)	0.8 (<0.1)	0.16 (0.03)	0.06 (0.01)

Standard deviations are given in parentheses.

[a]The uppermost sediment layer, rich in organic material.

Table 3.3. Mean content of exchangeable base cations (Na, K, Mg, and Ca), cation exchange capacity (CEC), base saturation (BS) and mean pH value in várzea soils on Ilha de Marchantaria exposed to different annual flood periods and in the forest soils of the igapó Tarumã Mirím

	Sampling depth (cm)	Number of samples (n)	Na (cmol kg⁻¹)	K (cmol kg⁻¹)	Mg (cmol kg⁻¹)	Ca (cmol kg⁻¹)	CEC (cmol kg⁻¹)	BS (%)	pH (KCl) (range)
Várzea									
Semiterrestrial									
Humic	2–0	8	0.19 (0.07)	0.51 (0.27)	3.40 (0.51)	17.1 (2.1)	23.3 (2.4)	91 (3)	3.8 (3.8–4.3)
Mineral	0–10	8	0.18 (0.07)	0.34 (0.13)	2.84 (0.64)	12.5 (3.6)	19.9 (3.5)	79 (11)	3.6 (3.5–4.0)
Mineral	10–30	7	0.21 (0.09)	0.29 (0.10)	3.63 (0.50)	12.9 (3.1)	20.0 (2.9)	85 (11)	3.6 (3.5–4.6)
Semiaquatic									
"Humic"ᵃ	0.3–0	8	0.36 (0.26)	0.50 (0.09)	3.80 (0.41)	16.5 (2.3)	23.9 (3.1)	89 (3)	4.0 (3.7–4.6)
Mineral	0–10	10	0.31 (0.15)	0.45 (0.10)	3.77 (0.49)	16.2 (2.6)	23.1 (3.3)	90 (4)	3.9 (3.7–4.6)
Mineral	10–30	10	0.24 (0.09)	0.30 (0.06)	3.45 (0.90)	14.9 (2.6)	20.3 (3.0)	93 (5)	5.4 (4.8–6.1)
Aquatic									
Mineral	0–20	3	0.28 (0.12)	0.43 (0.03)	4.13 (0.42)	24.7 (7.8)	31.0 (8.2)	95 (3)	5.1 (4.7–7.0)
Igapó									
Semiterrestrial									
Humic	2–0	10	0.10 (0.05)	0.27 (0.06)	0.31 (0.09)	0.31 (0.15)	5.7 (1.2)	17 (4)	3.5 (3.5–3.6)
Mineral	0–10	10	0.08 (0.05)	0.18 (0.07)	0.14 (0.04)	0.08 (0.07)	3.8 (0.4)	12 (4)	3.7 (3.6–3.7)
Mineral	10–20	2	0.02 (0.01)	0.04 (<0.01)	0.06 (<0.01)	0.07 (0.02)	2.1 (<0.1)	8 (1)	3.8 (3.7–3.9)
Mineral	30–40	2	<0.01 (<0.01)	0.02 (0.01)	0.04 (<0.01)	0.08 (0.01)	2.4 (0.1)	6 (<1)	3.9 (3.9–3.9)

Standard deviations are given in parentheses.
ᵃ The uppermost sediment layer, rich in organic material.

Table 3.4. Mean content of water-soluble sodium, potassium, magnesium, calcium, organic carbon (DOC), ammonium, nitrate, sulfate, and phosphate in várzea soils on Ilha de Marchantaria exposed to different annual flood periods and in the forest soils of the igapó Tarumã Mirím

	Sampling depth (cm)	Number of samples (n)	Na (mg kg⁻¹)	K (mg kg⁻¹)	Mg (mg kg⁻¹)	Ca (mg kg⁻¹)	DOC (mg kg⁻¹)	NH$_4$-N (mg kg⁻¹)	NO$_3$-N (mg kg⁻¹)	SO$_4$-S (mg kg⁻¹)	PO$_4$-P (mg kg⁻¹)
Várzea											
Semiterrestrial											
Humic	2–0	8	46.0 (19.3)	51.6 (11.3)	29.3 (16.4)	99.2 (25.7)	691 (645)	35.3 (29.2)	1.02 (1.13)	124 (50)	1.82 (2.37)
Mineral	0–10	8	27.6 (11.5)	17.2 (7.0)	5.5 (2.2)	20.8 (4.2)	78 (50)	3.4 (3.3)	0.87 (0.84)	17.2 (8.0)	0.03 (0.06)
Mineral	10–30	7	23.7 (8.5)	9.0 (5.1)	3.4 (2.6)	11.8 (8.2)	46 (28)	3.6 (7.3)	0.60 (0.48)	8.5 (6.4)	0.21 (0.37)
Semiaquatic											
"Humic"[a]	0.3–0	8	59.8 (25.5)	52.8 (15.6)	44.8 (30.5)	196 (138)	287 (178)	32.5 (31.1)	3.08 (4.86)	257 (161)	0.16 (0.35)
Mineral	0–10	10	53.6 (30.8)	37.1 (14.9)	33.4 (33.9)	135 (109)	243 (324)	27.7 (41.9)	1.52 (1.92)	206 (188)	0.11 (0.23)
Mineral	10–30	10	39.1 (10.8)	15.2 (5.5)	6.8 (2.6)	24.0 (8.0)	76 (57)	9.4 (9.5)	0.15 (0.12)	22.6 (17.6)	0.32 (0.56)
Aquatic											
Mineral	0–20	3	64.6 (27.8)	38.7 (4.3)	37.8 (21.2)	295 (212)	222 (41)	26.9 (4.7)	0.19 (0.24)	248 (88)	0.58 (0.16)
Igapó											
Semiterrestrial											
Humic	2–0	10	11.5 (2.3)	46.1 (12.9)	6.1 (1.9)	3.4 (1.4)	612 (176)	52.0 (21.1)	0.78 (0.31)	14.6 (4.5)	4.49 (2.07)
Mineral	0–10	10	9.2 (2.3)	29.7 (5.9)	4.4 (1.5)	2.6 (1.8)	379 (180)	37.7 (15.1)	0.92 (1.32)	8.8 (8.7)	1.75 (1.04)
Mineral	10–20	2	2.3 (<1)	8.2 (0.8)	2.8 (1.1)	2.6 (2.6)	148 (115)	5.5 (3.0)	3.43 (1.04)	4.3 (3.5)	0.12 (0.16)
Mineral	30–40	2	2.3 (<1)	3.5 (0.4)	1.7 (0.7)	3.4 (2.6)	28 (28)	2.4 (1.7)	3.22 (0.01)	3.5 (0.2)	<0.01 (<0.01)

Standard deviations in parentheses.

[a] The uppermost sediment layer, rich in organic material.

the mercury rhodanide method. For further details see Furch (1976) and Furch and Junk (1985, 1992, 1993).

3.3 Total Element Content

3.3.1 Metals (K, Na, Mg, and Ca)

Potassium, sodium, magnesium and calcium belong to the most important basic cations in soils. Except for sodium they are essential bioelements that accumulate in large quantities in the vegetation. The total contents of the respective elements allow us to estimate the amount potentially available in the soil and are important for the characterization and the evaluation of the soil. Direct conclusions about the short-term availability of the elements for plant growth from the total amounts are not possible. The origin of these elements is mostly geogenous. Their amounts and distribution depend upon the composition of the parent material, soil genesis, and the level of weathering and lixiviation.

According to the available information about the total contents of alkali and alkaline-earth metals in soils (Baumeister and Ernst 1978; Schachtschabel et al. 1982), várzea soils can be classified as rich to very rich. The contents of Na, K, and Ca are in the upper range of soils, the content of Mg is even higher. In the soils on the levees of Ilha de Marchantaria, flooded for 5–8 months year^{-1} and covered by floodplain forest (semiterrestrial soils), the lowest values of total Na, K, and Mg are found in the uppermost humic horizons (Table 3.2). Values of Ca are similar to those found in the mineral horizons. There are only small differences in the total contents of alkali and alkaline-earth metals between the uppermost mineral soil layers. A reduction due to weathering and uptake by the vegetation is not detectable. This is remarkable because potassium is taken up and stored by the vegetation in large quantities (Klinge et al. 1983).

The total amounts of Na, K, Mg, and Ca in sediments flooded for 8–12 months year^{-1} (semiaquatic sediments) differ only slightly from those of the forest soils. Aquatic sediments of Lago Camaleão, covered permanently with water, contain lower concentrations of total Na and K but a considerably higher level of Ca than the mineral horizons of the periodically dried sediments (Table 3.2). Lake sediments show a gradient in the content of alkali and alkaline-earth metals depending upon the distance from the lake entrance. Contents of Na, K, and Mg decrease in the sediments with increasing distance from the lake entrance, content of Ca increases. This points to a differentiation in the distribution of metals depending upon the level of contact with the river (Furch, in prep.).

The suspended material in the main stem of the Solimões, which can be considered as parent material for the formation of the topsoil layer of the várzea, does not show significant differences with respect to the total contents of Na, K, Mg, and Ca in comparison with the várzea soils (Table 3.2). In the soils of the igapó, collected at Tarumã Mirím, total amounts of Na, K, Mg, and Ca are more than one order of magnitude lower than described for várzea soils. The low calcium content is remarkable (Table 3.2).

3.3.2 Nonmetals (C, N, S, and P)

Soil carbon, nitrogen, sulfur, and phosphorus are often significantly related to biotic processes. Carbon and nitrogen are derived mostly from the fixation of CO_2 and N_2 by plants. Sulfur and phosphorus are mostly of geogenous origin, except for the inputs from the atmosphere (see Sect. 4.3.4). An overwhelming amount occurs in the soil organic material. Consequently, carbon, nitrogen, sulfur, and to some extent phosphorus are concentrated in the humic surface soil layers and show a strong depletion with soil depth. This phenomenon is more pronounced in the várzea soils than in the igapó soils (Table 3.2).

The carbon content in the humic surface layers is high in the várzea forest soil as well as in the igapó forest soil, corresponding with the classification of "strongly humous" (humus content between 8 and 15%) and "rich in humus" (humus content between 15 and 30%). Also the nitrogen content in the humic soil layers is high in the várzea and in the igapó (Table 3.2). Significant differences in the total carbon and nitrogen content between várzea and igapó forest soils were not observed. The top mineral soil layers of the igapó are even somewhat higher in C and N than those of the várzea (Table 3.2).

Remarkable amounts of total sulfur occur only in the humic surface layers of both várzea and igapó forest soils. Várzea soils are significantly richer in sulfur than igapó soils (Table 3.2); however, sulfur content in the igapó soils is still within the range generally reported for soils of the humid climate (Baumeister and Ernst 1978; Schachtschabel et al. 1982).

There are significant differences in total phosphorus contents of várzea and igapó soils (Table 3.2) especially with respect to the mineral soil layers. While the humic surface layers of the várzea soils are about four times higher in P than those of the igapó their mineral layers are on average between six to ten times richer in P than those of the igapó. The highest concentrations of P are observed in the top layers of the semiaquatic and aquatic sediments of Lago Camaleão (Table 3.2). They are even richer in P than the fresh sediment load of the Rio Solimões.

3.4 Cation Exchange Capacity

The cation exchange capacity (CEC) of soils is a measure of the soil's capacity to retain and supply major cations and is strongly influenced by the product of specific surface and surface charge density of the soil colloids (Uehara and Gillman 1981). CEC is therefore directly related to the quantity and quality of clay minerals, organic material, and sesquioxides, as well as pH. The actual amount of sorbed cations at the charged exchanger surface depends on the ionic composition of the ambient solution and the electrostatic forces of the cations, i.e., their valence, hydration and size (Bolt and Bruggenwert 1976). The following gradient indicates the adsorption strength of soil particles with respect to some of the most frequent cations: $Na^+ < K^+ < Mg^{2+} < Ca^{2+} < Al^{3+}$ (Schachtschabel et al. 1982). Exchangeable cations are supposed to be plant available (Schachtschabel et al. 1982).

The short-term flooded semiterrestrial várzea forest soils of Ilha de Marchantaria have a high cation exchange capacity. The sum of exchangeable cations (CEC, Table 3.3) varies on average between 20 and 23 cmol kg^{-1} soil. The base saturation is very high with average values 79–91% of the CEC. Among the basic cations (Na, K, Ca, and Mg) the alkaline-earth cations calcium and magnesium occupy the largest portions (76–88% of the CEC). The content of exchangeable potassium is therefore relatively low: between 0.3 and 0.5 cmol kg^{-1} soil (Table 3.3). Due to the high base saturation, the exchangeable acidity (mainly the exchangeable hydrogen and aluminum ions) is low, with an average between 2.1 and 4.0 cmol kg^{-1} soil. Since the overwhelming majority of roots occurs in the uppermost soil layers (Table 3.1) the mineral soil layers from 0–30 cm depth are supposed to be the most important ones for the nutrient supply. Compared to the humic surface soil layer and mineral soil layers that are deeper than 30 cm the mineral soil layers from 0–30 cm depth are impoverished in exchangeable calcium and magnesium and to a lesser extent in potassium and are enriched in exchangeable aluminum (Furch, in prep.). As mentioned above a respective decrease in the total contents of Ca and Mg was not observed.

In spite of the generally high base saturation and the generally low exchangeable acidity the forest soils have to be characterized as strongly acidic with pH(KCl) values of 3.5–4.6. The long-term inundated semi-aquatic sediments of the Lago Camaleão, which are dry for 1 to 4 months each year, generally show a CEC similar to that of the short-term inundated forest soils. A depletion of exchangeable Ca and Mg in the mineral layers was not observed and, due to the long-term submergence (reducing conditions), pH(KCl) values are higher on average (3.9–5.4) than those of the forest soils. The permanently inundated aquatic sediments of Lago Camaleão show a considerably higher content of exchangeable calcium and higher pH(KCl) values (up to 7) than the topsoils of the periodically dry habitats of Ilha de Marchantaria (Table 3.3).

The periodically flooded semiterrestrial igapó forest soils of Tarumã Mirím have a low CEC with average values of 2.1–5.7 cmol kg^{-1} soil and a very low base saturation with average values of 6–17% of the CEC. Among the basic cations the exchangeable alkaline-earth cations, especially calcium, occur in extremely low concentrations. The content of exchangeable potassium with average values between 0.02 and 0.27 cmol kg^{-1} soil is low, but not extremely low compared with that of the várzea soils (Table 3.3). The depletion of exchangeable basic cations below the humic surface soil layer with increasing depth of the mineral layer is very strong. The exchangeable acidity of these soils is very high (83–94% of the CEC). Soil reaction is more acidic than in the várzea soils (average pH(KCl) values 3.5–3.9).

3.5 Content of Soluble Substances in Aqueous Soil Extracts

Water-soluble substances are directly and easily available for plant growth and are easily transferred between soil horizons or lost from the soil by lixiviation. In soil science, water-soluble substances are normally determined in saline soils only. However, considering the impact of heavy rainfall and periodic flooding, this aspect of soil chemistry is also important for várzea and igapó soils. The amount of water-soluble substances depends strongly on the biological activity in the soil or the soil horizons, respectively, e.g., intensity of decomposition processes, N-dynamics, uptake and release rates of microorganisms and vegetation.

The content of soluble organic carbon (DOC) is high in the humic surface layers of the forest soils in the várzea as well as in the igapó. Lower contents are observed in the mineral soil layers; the igapó soils are richer in DOC than the várzea soils (Table 3.4).

The content of soluble inorganic nitrogen is high in the humic surface layers of the forest soils in the várzea as well as in the igapó. In both soil types ammonium-nitrogen occupies a major portion of the inorganic nitrogen (Table 3.4). Both the amount of total nitrogen and the amount of soluble ammonium-nitrogen decrease with soil depth. However, content of soluble inorganic nitrogen compounds shows a large variation within a soil layer (Table 3.4), indicating a high mobility and an easy transfer to deeper soil layers.

As expected from the total sulfur content, concentrations of soluble sulfate are much higher in the várzea soils than in the igapó soils, and are particularly high in the long-term inundated semiaquatic várzea sediments. The very high sulfate concentrations (Table 3.4) indicate a high mineralization rate of aquatic macrophytes, which are absent in the igapó.

In spite of the low total phosphorus content of igapó soils the soluble portion is remarkably high and even higher than those of the respective várzea soil layers (Table 3.4). It is assumed that these findings are not due to a low content of available phosphate in the várzea soils, but may be related to a higher anion sorption capacity of várzea soils than of igapó soils. This assumption was supported by the results of an acid extraction procedure. The acid extractable amounts of phosphate were significantly higher in the mineral várzea soils than in the respective igapó soils (Furch, in prep.).

Among the soluble major cations that occur in the different soil layers (Table 3.4), calcium and potassium are the ones that show the most contrast. The largest differences between várzea and igapó soils are represented by the content of soluble calcium with high concentrations in the várzea and extremely low values in the igapó. The differences between the two inundation areas concerning soluble potassium in the soils are insignificant (Table 3.4).

Figure 3.2 shows the distribution of base cations in the soils of blackwater and white-water floodplains and its relationship to the length of the annual flood period. Generally, as the length of the flood period increases there is an increase in the amount of cations and/or an increase in the Ca portion of the four cations. This trend is much stronger in the soils of the várzea than in the soils of the igapó, mainly with respect to the exchangeable and water-soluble cations. The results show that in spite of a nearly identical total amount of the four cations in all soil types of the várzea, the amount of exchangeable and soluble base cations is positively related to the length of the flood period. Thus it is suggested that the flood pulse stimulates the nutrient dynamics in the soils, i.e., their release and accumulation. A comparison between the composition of the easily mobilized base cations in the soils (Fig. 3.2c) with that of the dissolved base cations in flood (river) and lake water (Fig. 3.2d) suggests that flood conditions have an impact on the similarity between the sediment and water: the longer the flood period the stronger the similarity (Furch, in prep.).

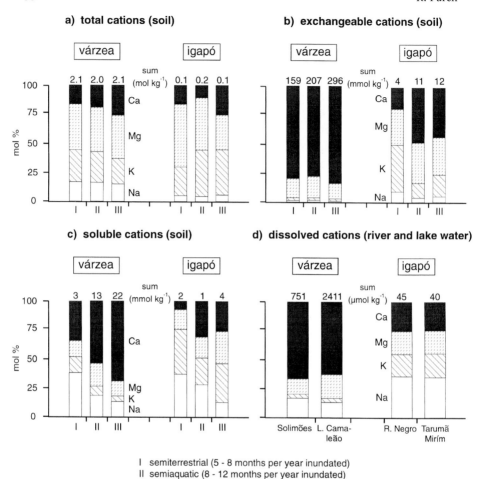

Fig. 3.2. Total content and content of exchangeable and water-soluble base cations (Na, K, Mg, and Ca) and their relative distribution in the soils (0–10 cm depth) exposed to different annual flood periods, and the content of base cations in river and lake water (0 m, average of a two year period, n = 24) in contrasting Amazonian floodplain systems

3.6 Bioelement Stock in the Systems and Bioelement Fluxes Between Soil and Forest Vegetation

Studies on the chemical composition of leaves, bark, and wood of numerous tree species from different Amazonian forest types indicate that there are great differences between the forest types (Klinge et al. 1983, 1984) and that the amounts of bioelements in the vegetation are reflected by the chemical conditions in soils and natural waters of the locality. The biomass of the várzea floodplain forest is rich in bioelements. It is chemically more similar to forests from nutrient rich habitats, e.g., of Central America (Golley and Richardson 1977) or of temperate regions, than to the forests of terra firme and igapó (Klinge 1976, 1985; Furch and Klinge 1978, 1989; Klinge and Furch 1991). The forests of terra firme and igapó show in their aboveground biomass extremely low contents mainly of Ca, Mg, K, and P, but are only slightly different from the várzea forest with respect to N (Sect. 9.3.1).

On the other hand, the floodplain forest of the igapó is less productive than the várzea forest. Average leaf litter production in the igapó is about $5\,t\,ha^{-1}\,yr^{-1}$ whereas in the várzea forest it is about twice as high (Adis et al. 1979; Furch and Klinge 1989; Worbes 1989, 1994b; Meyer 1991). There are no data available regarding primary production in the igapó. However, litter production suggests that total production in the igapó is only about half as much as total production in the várzea (data summarized in Sect. 11.4).

The following bioelement budget for both forest types is based on four assumptions. We assume the aboveground biomass (standing crop) in the várzea forest corresponds to $250\,t\,ha^{-1}$ wood, $5\,t\,ha^{-1}$ bark and $10\,t\,ha^{-1}$ leaves, and in the igapó forest to $250\,t\,ha^{-1}$ wood, $5\,t\,ha^{-1}$ bark and $5\,t\,ha^{-1}$ leaves. Although it is assumed that leaf biomass in the igapó forest is higher due to a longer life span of leaves, an estimate of 5 t corresponding to the annual leaf fall was chosen for reasons of comparability.

The annual biomass production corresponds in the várzea forest to $17.4\,t\,ha^{-1}$ (7.4 t wood and bark and 10 t leaves) and in the igapó forest to $8.7\,t\,ha^{-1}$ (3.7 t wood and bark and 5 t leaves). The soil nutrient stock was calculated per unit area from the humic surface layer and from the upper 30 cm mineral layer. Nutrient inputs from fresh leaf litter and from rainwater were not considered. The amounts of nutrients in the standing crop and those necessary for the annual production of aboveground biomass are calculated using mean concentrations given for leaves, bark and wood by Klinge et al. (1983, 1984).

The calculation shows that the amount of total nutrients in the soil and the amount stored in the vegetation are much greater in the várzea than in the igapó (Fig. 3.3). The várzea forest uses higher amounts of bioelements for the production of a standing crop unit than the igapó forest. On the other hand the igapó forest needs more time than the várzea forest for the production of a standing crop unit. In the várzea forest only a small proportion of the total soil nutrient reserves is stored in the biomass (Fig. 3.3). The ratio of nutrients in soil to nutrients in standing crop is very high for Mg (77) and K (39), high for Ca (20) and relatively low for P (10) and N (4). At least Mg, K, and Ca seem to be available in excess in the várzea soils.

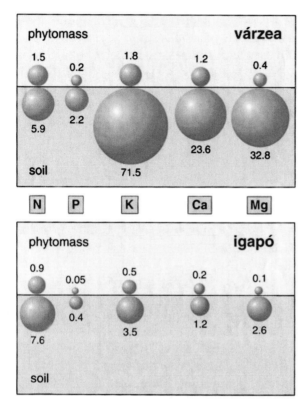

Fig. 3.3. Distribution of major nutrients in the soils and in the aboveground forest vegetation of two contrasting Amazonian floodplain systems, várzea and igapó (t ha^{-1}). Nutrient content of the soils is calculated for the top layers (30 cm depth), including the humic surface layer and excluding the litter layer

Also in the igapó forest larger amounts of nutrients are stored in the topsoil than in the vegetation. However, of the ratio of nutrients in the soil to nutrients in standing crop, only magnesium (30) is found in large excess. Ratios for all other nutrients are much smaller corresponding to seven for potassium, five for calcium, nine for phosphorus and eight for nitrogen. Except for N, with a ratio that is twice as high as in the várzea, ratios for the nutrients in the igapó are smaller than in the várzea indicating that the native forest vegetation is at a disadvantage due to the unfdavourable nutrient conditions of the soil. Differences in the soil/vegetation nutrient ratios between várzea and igapó are much smaller than differences in the soil nutrient content, suggesting that the two systems differ in their strategies of nutrient use.

The annual requirements for biomass production are much lower in the igapó forest than in the várzea forest because the annual biomass production is considerably lower (Fig. 3.4). Furthermore the nutrient content in the biomass of the igapó forest is much smaller. Compared with the amounts required by the várzea forest (100%) the igapó forest requires for the production of one unit biomass only 68% N, 32% P, 43% K, 15% Ca and 29% Mg. This indicates a greater nutrient use efficiency of the igapó forest vegetation (Furch and Klinge 1989). Considering the total annual biomass production, the igapó forest requires only 34% N, 16% P, 22% K, 8% Ca and 14% Mg compared with the amounts required by the várzea forest (100%). This indicates the remarkably low nutrient demand of the igapó forest vegetation.

The budget shows that the amounts of total nutrients in the soils of várzea and igapó are sufficient to maintain the respective floodplain forest. Since large amounts of total nutrient reserves are not quickly removable

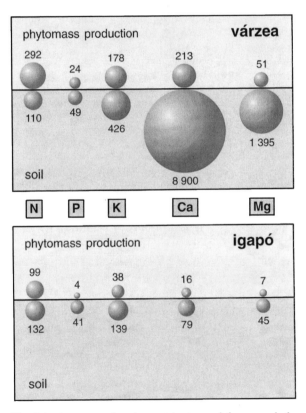

Fig. 3.4. Amounts of major nutrients used for annual aboveground forest biomass production (kg ha^{-1} year^{-1}) and content of exchangeable (K, Mg, Ca, and NH$_4$-N) and acid-extractable (PO$_4$-P) nutrients in two contrasting Amazonian floodplain systems, várzea and igapó (kg ha^{-1}). Nutrient content of the soils is calculated for the top layers (30 cm depth), including the humic surface layer and excluding the litter layer. For further explanations, see text

and therefore cannot be used for uptake by plants, available nutrients
should be considered for budgeting of nutrient requirements. In Fig. 3.4,
the amounts of available nutrients, i.e., the amounts of exchangeable K,
Mg, Ca and NH_4-N and acid-extractable PO_4-P, are shown. Available nitrogen is underrepresented by means of exchangeable ammonium, because
soluble nitrate, corresponding to 2% and 8% of the exchangeable amount
of ammonium nitrogen of várzea and igapó soils, respectively, was
not considered. The amount of exchangeable K underrepresents the
total amount of available K because a considerable portion of the non-
exchangable K stored in layer silicates is available for plant growth, too
(Mengel and Kirkby 1978; Schachtschabel et al. 1982).

   The available nutrients in the soils of the igapó meet the annual requirements of the igapó forest but would not satisfy the requirements of the
várzea forest (Fig. 3.4). In the soils of the várzea there are enough nutrients
available for the annual requirements of the várzea forest except for nitrogen. However, as mentioned above, the N pool is greater and is permanently replenished by decomposition of litter and organic soil material,
and probably by nitrogen fixation (Chap. 6).

3.7 Discussion and Conclusions

There are considerable differences in soil nutrient conditions in the two
floodplain systems. The conditions are favorable for plant growth in the
várzea with respect to all aspects of soil chemical characterization. There
are high nutrient reserves, reflected by a high content of total nutrients, a
high base saturation, indicating high buffer capacity and a high content of
available nutrients with respect to exchangeable, extractable and water-
soluble amounts. In contrast, soil nutrient conditions in the igapó are
unfavorable with low nutrient reserves except for nitrogen, low base saturation and low content of available base cations. With respect to the
amounts of available nitrogen and phosphorus differences between the two
soil types are negligible.

   The participation of the aboveground forest biomass in total nutrient
stock is small in both floodplain systems, corresponding to 3% in the
várzea and 10% in the igapó (based on the sum of the five nutrients).
Amounts and percentages for the individual nutrients stored in the forest
biomass are given in Table 3.5. Only a small portion of the total soil
nutrient pool is available for plant growth. Except for N and P, the amounts
of available nutrients are considerably higher in the várzea soils than in the
igapó soils (Table 3.5). The portion of the total nutrient pool which is

Table 3.5. Nutrient inventory of the floodplain forest system ($kg\,ha^{-1}$). **a** Calculated for total nutrient content in the soil. **b** Calculated for available nutrient content in the soil

a Calculated for **total** nutrient content in the soil

	N	P	K	Mg	Ca
Várzea					
Aboveground biomass[a]	1 479	224	1 837	427	1 183
(content of total nutrients)					
Soil (30 cm depth)	5 907	2 180	71 493	32 804	23 550
(content of **total** nutrients)					
Aboveground biomass (%)	20	9	3	1	5
Igapó					
Aboveground biomass[b]	919	49	493	86	240
(content of total nutrients)					
Soil (30 cm depth)	7 610	430	3 470	2 601	1 237
(content of **total** nutrients)					
Aboveground biomass (%)	11	10	12	3	16

b Calculated for **available** nutrient content in the soil

	N	P	K	Mg	Ca
Várzea					
Aboveground biomass[a]	1 479	224	1 837	427	1 183
(content of total nutrients)					
Soil (30 cm depth)	110	49	426	1 395	8 900
(content of **available** nutrients)					
Aboveground biomass (%)	93	82	81	23	12
Igapó					
Aboveground biomass[b]	919	49	493	86	240
(content of total nutrients)					
Soil (30 cm depth)	132	41	139	45	79
(content of **available** nutrients)					
Aboveground biomass (%)	87	54	78	66	75

[a] Corresponding to a standing crop of 265 t ha^{-1}.
[b] Corresponding to a standing crop of 260 t ha^{-1}.

available for plant growth varies between less than 1% for potassium and 38% for calcium (Table 3.5). It is different in the two soil types, except for nitrogen (2%). Considering the amounts of available nutrients as a soil nutrient pool, we find that in the igapó the majority of the five nutrients (N, P, K, Mg, Ca) are stored in the aboveground forest biomass, whereas in the várzea the majority of calcium and magnesium is stored in the soil (Table 3.5). The differences between the two systems regarding the distribution of nutrients between soil and vegetation are not as great as would be expected from the differences in nutrient content and productivity and therefore are not a good indicator for the nutrient status of várzea and igapó.

It should be noted that the biomass nutrient inventories are not complete, because data for the below-ground biomass and the litter layer are not available for both systems. As reported for rain forests of the upper Rio Negro region, and summarized by Medina and Cuevas (1989), root biomass is high, contributing 16–20% of the total living biomass in a mixed terra firme forest and even 33% in the frequently flooded tall Amazon *caatinga*. The participation of the below-ground biomass in the nutrient pool of the total living biomass is even higher: 31% in the mixed forest and 54% in the Amazon *caatinga*, when calculated for the sum of the five nutrients. Nutrient stock stored in the fine litter and dead wood of these forests is much lower, corresponding to 12% of the total nutrient stock stored in the living biomass of the mixed forest and 9% of the total nutrient stock stored in the living biomass of the Amazon *caatinga*.

In várzea and igapó leaf litter dynamics show considerable differences. Leaf litter fall is high in the várzea forest (about $10 t ha^{-1} year^{-1}$) and low in the igapó forest (about $5 t ha^{-1} year^{-1}$) (Adis et al. 1979; Furch and Klinge 1989; Meyer 1991). Nutrient content of várzea leaf litter is on average more than twice as high as that of igapó leaf litter (based on the sum of the five nutrients) corresponding to an annual nutrient input by leaf litter that is four to five times higher in the várzea than in the igapó (Furch and Klinge 1989). In the várzea the annual leaf litter fall represents a nutrient stock corresponding to about 12% of the nutrient stock in the above-ground biomass, in the igapó it corresponds to about 7% of the nutrient stock. The mass of the leaf litter layer is similar in várzea and igapó during the low-water period (Meyer 1991), indicating a higher decomposition rate in the várzea than in the igapó. This was demonstrated by decomposition experiments with leaf material of tree species from the várzea and the igapó (Furch et al. 1989). The decomposition of 95% of the litter layer needs about 1 year in the várzea and about 2 years in the igapó (Meyer 1991; Sect. 9.3). These findings show the great importance of the litter layer regarding both the nutrient stock and the nutrient turnover in the floodplains.

In the várzea and in the igapó the supply of available nutrients, i.e., exchangeable or extractable nutrients, is high enough to meet the demands of the individual forest vegetation. For the high nutrient requirements of the várzea forest, which are reflected by a high annual biomass production and by a high nutrient content in the biomass, a large nutrient pool in the soil is available. The igapó forest must compensate for the unfavorable nutrient status of its soils and certain strategies are required. Nutrient requirements are reduced by a low biomass production, which corresponds to one half of the production in the várzea, and nutrient content in the biomass is low, indicating a high nutrient-use efficiency (Furch and Klinge 1989). It is assumed that the igapó forest provides physiological and morphological adaptations to the nutrient poverty of the soils. Meyer (1991) suggests that the dense superficial root mat traps nutrients via mycorrhiza from decomposing plant material as soon as they are released. For Amazonian terra firme forests growing on soils that are comparably poor in nutrients, "direct nutrient cycling" had already been shown (Medina and Cuevas 1989). The importance of such a mechanism for nutrient retention is obvious since the majority of required nutrients are incorporated in leaf material which is annually returned to the ground as litter.

Rainwater and floodwater are additional sources of nutrients for the floodplains. The amounts of nutrients contributed annually by rainwater are small (see Sect. 4.3.4). In the igapó, the nutrients in the rainwater, if used quantitatively for forest growth, meet only a few percent of the annual requirements (Table 3.7). In the várzea, rainwater is a negligible nutrient source (Table 3.6). A substantial contribution to the nutrient stock of both systems is potentially given by the dissolved nutrient load of the floodwater. In the várzea the amounts calculated for a mean water depth of 4 m at high water would be high enough to meet the total Ca and Mg requirements and about 20% of the K requirements of the floodplain forest for its annual aboveground biomass production (Table 3.6). Even in the igapó the amounts of dissolved Ca, Mg, and K in the floodwater would meet a high

Table 3.6. Nutrient import and nutrient requirements in the várzea ($kg\,ha^{-1}\,year^{-1}$)

	N	P	K	Mg	Ca
Sediment deposition (total)	41–148	33–37	855–979	414–464	428–464
Sediment deposition (available)[a]	0.2–4.7	0.2–4.0	4.3–6.9	15–21	150–172
River water (dissolved)[b]	6.2[d]	1.0[e]	35	54	390
Rainwater (dissolved)[c]	0.3–8.0[d]	<0.1–0.2[e]	0.7–5.8	0.1–3.3	0.2–4.3
Requirements by the forest vegetation[f]	292	24	178	51	213

[a] Estimated amounts of exchangeable K, Mg, Ca, and NH_4-N and extractable PO_4-P in deposited sediments.
[b] Assuming a mean water depth of 4 m at high water level (Furch and Junk 1992).
[c] See Table 4.4 in Chapter 4.
[d] Sum of NH_4-N and NO_3-N.
[e] PO_4-P.
[f] Corresponding to an annual aboveground biomass production of $17.4\,t\,ha^{-1}$.

Table 3.7. Nutrient import and nutrient requirements in the igapó (kg ha^{-1} year^{-1})

	N	P	K	Mg	Ca
River water (dissolved)[a]	2.9[c]	0.2[d]	13.6	4.7	8.9
Rainwater (dissolved)[b]	0.3–8.0[c]	<0.1–0.2[d]	0.7–5.8	0.1–3.3	0.2–4.3
Requirements by the forest vegetation[e]	99	4	38	7	16

[a] Assuming a mean water depth of 4 m at high water level; calculated with data for N and P from Anonymous (1972b) and with data for K, Mg, and Ca from Furch (1984c), see Tables 4.1 and 4.2 in Chapter 4.
[b] See Table 4.4 in Chapter 4.
[c] Sum of NH_4-N and NO_3-N.
[d] PO_4-P.
[e] Corresponding to an annual aboveground biomass production of 8.7 t ha^{-1}.

percentage of the annual requirements of the floodplain forest (Table 3.7). In contrast, the supply of dissolved nitrogen and phosphorus in the floodwater is low in the igapó and in the várzea. There is no information about the potential use of the dissolved load in the floodwater by the floodplain forests and no information about the amount of dissolved nutrients that annually leave the systems during receding floodwater. In addition, the nutrient loss caused by the export of floating litter during this time may be considerable, especially in the igapó (Irmler 1979b).

In contrast to the dissolved load of the floodwater, the suspended load annually settled in the systems causes a permanent enhancement of the nutrient stock. Due to the very low content of suspended solids in the Rio Negro (about 6 mg/l, Gibbs 1967a) this aspect of nutrient import may be negligible for the igapó but important for the várzea, annually flooded by the Rio Solimões rich in suspended mineral solids (see Sect. 4.3.1). In Table 3.6 the amount of total nutrients is calculated assuming an annual settlement of a sediment layer with a thickness of 3 mm and a volume weight of 1.5 g cm^{-3}, corresponding to an annual sediment increment of 45 t ha^{-1}. This calculation shows that even the available portion of nutrients in the settled sediment load represents a remarkable contribution to the nutrient stock of the várzea. Nitrogen input is very small, corresponding to the annual nitrogen input by rainwater. Thus the conditions for nitrogen supply are more critical in the várzea than in the igapó. This points to the importance of nitrogen fixation in the várzea forest (see Sect. 6.7).

The large nutrient reserves in the várzea soils, the high content of available nutrients, the periodical nutrient input by floodwater, and the high productivity in the várzea allow the utilization of the várzea. Because of their slow growth and their importance for nutrient retention and nutrient conservation, igapó floodplain forests should be protected.

Acknowledgments. I am grateful to Prof. Dr. W. Zech, Bayreuth, for his critical comments on the manuscript and to S. Bartel, E. Bustorf, N. Dockal and U. Thein for technical assistance.

4 Physicochemical Conditions in the Floodplains

Karin Furch and Wolfgang J. Junk

4.1 Introduction

The different water colors of Amazonian rivers are documented in the river names e.g., Rio Negro (black river), Rio Branco (white river), Rio Claro (clear river), Rio Verde (green river), and indicate differences in water quality. Sioli (1950) related water color to specific conditions in the catchment areas and recognized three main water types based on water color, load of suspended solids, pH, and load of dissolved minerals, indicated by the specific conductance.

- White-water rivers have a muddy color derived from the high sediment concentration. They are relatively rich in dissolved minerals and have a near-neutral pH value. Sediments and dissolved minerals are derived from erosion in the Andes and the Andean foothills. Water and flood-plain soils are relatively fertile.
- Black-water rivers have dark transparent water due to high amounts of dissolved humic substances. These substances are formed mainly in podzolic soils which occur in small patches throughout the Amazon basin, but are concentrated in the catchment area of the Negro River (Klinge 1967). The water is acidic and the amounts of dissolved inorganic substances are low. Water and floodplain soils are of low fertility.
- Clear-water rivers drain areas where there is little erosion. Their water is transparent and may be greenish. The pH varies from acidic to neutral depending upon the geology of the catchment area. The amounts of dissolved minerals are small to intermediate. The fertility of the waters and floodplain soils is variable, but it is lower than that of white-water rivers.

Based on the geology of the Amazon basin and using mainly the pH value, specific conductance, and the amount of Ca, Mg, and PO_4 as parameters, Fittkau (1971) proposed an ecological classification. He differentiated the catchment area of the Amazon River into three geochemical provinces: the geochemically rich Andean and pre-Andean region, the geochemically

Ecological Studies, Vol. 126
Junk (ed) The Central Amazon Floodplain
© Springer-Verlag Berlin Heidelberg 1997

poor shields of Guiana and Central Brazil, and the very poor central Amazonian region. The floodplains of the Amazon River and its large white-water tributaries were labeled as part of the Andean and pre-Andean region. A detailed investigation into the relationship between the chemistry of major ions in rivers of the Amazon system and the geology, topography, and soils of their catchments was presented by Stallard and Edmond (1983). Comparing the specific conductance as a measure of the ionic content of Amazonian waters with that of other rivers of the world, Furch et al. (1982) found that the ionic content of white water from the Solimões River (the Amazon River upstream of its confluence with the Negro River is called Solimões River) near Manaus contains only one third of the world average. The ionic content of the Negro River and the forest streams from the central basin is almost as low as the ionic content of rainwater, but water from the Negro River contains large amounts of dissolved organic compounds. Some clear-water rivers, i.e., Curuá Una River have an ion content intermediate between that of the Amazon River and Negro River. The pH of Amazon water lies within the range of global average. The pH of black-water rivers varies between 4 and 5, whereas the pH of clear-water rivers and streams is variable but normally occupies an intermediate position (Fig. 4.1a,b).

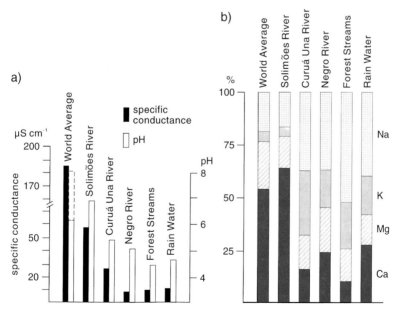

Fig. 4.1a,b. Specific conductance (20 °C), pH value (a), and the distribution of alkali and alkaline-earth metals (b) of Amazonian rivers and streams in comparison to rainwater near Manaus and to the world average water. (According to Junk et al. 1981; Furch 1984)

Sioli's and Fittkau's classifications are useful for a very general ecological classification of the Amazon basin and its water bodies. However, heterogeneous geology of the Guiana shield and the Central Brazilian shield is reflected in a much greater chemical heterogeneity of their streams and small rivers than was previously assumed (Furch and Junk 1980; Furch 1985, 1986).

In recent years many papers have been published by scientists of the CAMREX project at the University of Seattle about the chemical composition of the water of the Amazon River and its major tributaries with the aim of establishing a biogeochemical and hydrological model of the river system and its catchment area (Richey 1983; Meade et al. 1985; Ertel et al. 1986; Richey et al. 1980, 1986, 1989; and others). These studies will be considered here only as far as they directly contribute to the understanding of the processes going on in the central Amazon floodplain.

4.2 Light Regime

The light regime in water bodies is related to the content of dissolved and particulate organic matter and inorganic solids (Kirk 1983). Transparency in Amazonian white-water and black-water rivers is low. In the Amazon River, the Secchi depth is about 30–50 cm mainly because of a high load of suspended solids. In the Negro River, the Secchi depth is about 2 m because of coloured humic substances (Sects. 4.3.1, 4.3.2). When the Amazon River water enters the floodplain, the suspended load settles down and the transparency can rise to 2 m (Sect. 4.4). When the water level in the lakes drops to a depth of less than 4 m, the sediments can become stirred up by wind induced turbulence and by activities of animals, e.g., fishes. Transparency declines, and at the lowest water level, the minimum values can be about 10 cm.

The light climate recorded as PAR (photosynthetically active radiation) with cosine and scalar corrected light sensors, has been studied in the Ria Lake Tarumã Mirím, which is dominated by black water from the Negro River and in a bay in front of Lago Camaleão, which is dominated by white water from the Amazon River (Furch et al. 1985; Furch and Otto 1987). Light intensity near the equator is very high. At noon the quantum flux density above the water surface reached maximum values of up to $3000\,\mu mol\,m^{-2}\,s^{-1}$ (scalar) and $1800\,\mu mol\,m^{-2}\,s^{-1}$ (cosinus) or 180 000 lx (photometric), respectively (see also Sect. 8.5.4). Comparative studies on the Schöhsee in northern Germany showed maximum quantum flux densities of only up to $900\,\mu mol\,m^{-2}\,s^{-1}$ (scalar) and $750\,\mu mol\,m^{-2}\,s^{-1}$

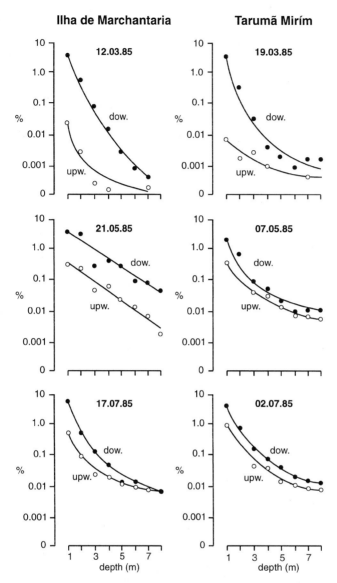

Fig. 4.2. Downwelling (*dow.*) and upwelling (*upw.*) irradiances at depths of 1–8 m in percent of the incident radiation at the surface in a white-water biotope (Ilha de Marchantaria) and a black-water biotope (Tarumã Mirím) during three different months at rising and high water levels. According to the changing water level and sediment load the light regimes in both lakes differ markedly. Mean Secchi depth = 1.5 m in black water and 0.3 m in white water. Mean attenuation coefficient K_D in black water = 1, in whitewater = 5. (According to Furch et al. 1985)

(cosinus) corresponding to 50 000 lx (photometric) above the water surface at noon.

In the water, radiation decreases with increasing depth. The extinction coefficient of the black-water lake Tarumã Mirím was about 1.2, that of the white-water bay in front of Lago Camaleão was about 5 (Fig. 4.2). In Tarumã Mirím, the high extinction coefficient was the result of the dark brown water color corresponding to 130–150 units on the HAZEN scale; the amount of suspended solids was only 1–5 mg l^{-1}. In the bay in front of Lago Camaleão the amount of suspended solids was 70–100 mg l^{-1} and the water color was only slightly brownish, corresponding to 30 units on the HAZEN scale. High extinction coefficients as a result of large amounts of colored humic substances and suspended inorganic solids indicate a narrow euphotic zone and a low primary production of algae and submerged aquatic macrophytes (Sect. 8.3.3, Sects. 10.2.2, 10.3.2 and 10.4).

In spite of the low Secchi depth of 1.5 m in Tarumã Mirím, a quantum flux density of up to 2 μmol m^{-2} (scalar) was measured on sunny days at noon at a water depth of 8 m corresponding to about 0.01% of the total radiation above the water surface. In the white water of the bay in front of Lago Camaleão with a Secchi depth of about 0.3 m, no photons could be detected at a depth of 3–5 m. The irradiance reflectance, i.e., the ratio between upward and downward radiation at any given depth, demonstrates that in black water the light climate shifts to diffuse light conditions as depth increases (Furch et al. 1985; Furch and Otto 1987).

4.3 Chemistry of Different Water Types

4.3.1 The Solimões/Amazon River Near Manaus

The Solimões/Amazon River is a classic representative of Amazonian white-water rivers (Sioli 1950). Its turbid loamy color originates from a high content of suspended mineral solids (about 100 mg/l near Manaus). However, large differences in the content and the particle-size distribution of total suspended solids have been recorded (see also Sects. 2.4, 2.5). They depend on season (Gibbs 1967a,b; Schmidt 1972a; Mertes et al. 1993; Furch unpubl.), water depth (Curtis et al. 1979; Meade et al. 1979a,b) and distance from the river bank (Meade et al. 1979a). The content of suspended solids in the surface water (0–1 m depth, varied from 31 to 247 mg l^{-1} in six individual samples within one day from a cross section of 2.5 km (Meade et al. 1979a). Up to 822 mg l^{-1} of suspended solids were found in deep water layers (0.5 m above the bottom). A strong downstream depletion in the

Table 4.1. Chemical composition of the surface water (mid-channel) of the Solimões River near Manaus, above the confluence with the Negro River

Source (study period)		I (1969/1970)		II (1974/1977)		III (1976/1978)		IVa (1980/1981)		IVb (1981/1982)	
		n		n		n		n		n	
Range			50.0–93.5		47.4–80.4				60.8–95.5		59.1–96.9
Mean	Cond.	12	66.0	34	64.1		–	12	74.9	12	74.1
SD	(μS cm^{-1})		12.9		8.7				11.0		11.6
Range			6.50–7.50		6.00–7.90		6.65–7.20		6.25–7.51		6.20–7.38
Mean	pH	12	6.91	34	6.73	5	6.85	12	6.78	12	6.79
SD			–		–		–		–		–
Range			0.03–0.32		0.01–1.00		0.06–0.22		0.03–0.56		0.04–0.63
Mean	H	12	0.12	34	0.19	5	0.14	12	0.16	12	0.16
SD	(μeq l^{-1})		0.09		0.19		0.07		0.15		0.17
Range					45–172		86–109		101–167		95–218
Mean	Na		–	34	101	5	96	12	131	12	131
SD	(μeq l^{-1})				37		11		17		37
Range					9–34		24–27		20–26		20–26
Mean	K		–	34	24	5	25	12	22	12	23
SD	(μeq l^{-1})				6		1		2		2
Range			95–238		44–142		97–122		94–134		88–156
Mean	Mg	12	138	34	91	5	106	12	111	12	111
SD	(μeq l^{-1})		41		20		10		11		21
Range			272–494		161–554		378–500		384–574		333–742
Mean	Ca	12	361	34	373	5	418	12	483	12	489
SD	(μeq l^{-1})		77		83		48		62		114
Range							0.4–0.6		0.3–11.1		1.3–9.9
Mean	NH_4		–		–	2	0.5	12	3.0	12	5.3
SD	(μeq l^{-1})						0.14		3.2		2.3
Range			300–700		420–710		475–590		500–710		390–910
Mean	HCO_3	12	489	29	556	5	514	12	590	12	588
SD	(μeq l^{-1})		139		64		45		71		146
Range			56–141		21–140		42–59		46–90		68–140
Mean	Cl	12	92	28	75	5	50	12	65	12	90
SD	(μeq l^{-1})		24		38		7		12		23
Range			18–228[a]				40–80		35–115		31–131
Mean	SO_4	12	74[a]		–	5	56	12	82	12	63
SD	(μeq l^{-1})		63[a]				16		20		34
Range			1.29–6.00				3.50–11.60		0.54–14.74		2.36–10.37
Mean	NO_3	12	3.40		–	5	5.66	12	7.69	12	6.84
SD	(μeq l^{-1})		1.66				3.36		3.37		2.57
Range			25.7–60.0								
Mean	N_{tot}	12	43.1		–		–		–		–
SD	(μmol l^{-1})		12.0								

I, surface waters sampled monthly from Aug. 1969 to July 1970 calculated from data given by Schmidt (1972b);
II, surface waters sampled monthly from June 1974 to May 1977 recalculated from data given by Furch (1984c);
III, individual surface waters sampled from June 1976 to April 1978 calculated from data given by Stallard and Edmond (1983);
IVa, IVb, surface waters sampled monthly from Nov. 1980 to Oct. 1981 and from Nov. 1981 to Oct. 1982 respectively (Furch, unpubl. data).

content of suspended solids was observed (Gibbs 1967a,b; Meade et al. 1979b; Martinelli et al. 1989). Mean annual values for surface waters sampled every month at one locality in mid-channel, about 10–20 km above the mouth of the Negro River, were 69 mg l^{-1} (Schmidt 1972a) and 82 and 88 mg l^{-1} (Furch unpubl. data).

The water of the Solimões River is relatively rich in nutrients and

Table 4.1. *Continued*

Source (study period)		I (1969/1970) n		II (1974/1977) n		III (1976/1978) n		IVa (1980/1981) n		IVb (1981/1982) n	
Range			<0.01–0.84				0.35–0.43		0.30–1.53		0.34–1.09
Mean	PO_4	12	0.37		–	3	0.38	12	0.82	12	0.67
SD	(μmol l^{-1})		0.23				0.04		0.39		0.28
Range			0.84–4.39		0.46–0.84						
Mean	P_{tot}	12	2.00	28	3.38		–		–		–
SD	(μmol l^{-1})		0.95		1.86						
Range									197–633		212–966
Mean	DOC		–		–		–	12	401	12	431
SD	(μmol l^{-1})								107		249
Range							91–315[b]		40–220		80–260
Mean	CO_2		–		–	5	199[b]	12	134	12	163
SD	(μmol l^{-1})						98[b]		54		51
Range			128–160		48–277		141–146		126–192		32–160
Mean	Si	12	140	28	148	5	144	12	156	12	90
SD	(μmol l^{-1})		10		40		2		20		47
Range					0.36–5.45		0.10–2.90				
Mean	Al		–	35	1.52	5	0.88		–		–
SD	(μmol l^{-1})				1.27		1.15				
Range			5.97–21.49		0.34–5.35		<0.01–7.30				
Mean	Fe	12	12.93	36	1.75	5	1.62		–		–
SD	(μmol l^{-1})		4.22		1.29		3.18				
Range					0.01–0.43						
Mean	Mn		–	36	0.11		–		–		–
SD	(μmol l^{-1})				0.10						
Range					0.08–0.26						
Mean	Ba		–	34	0.17		–		–		–
SD	(μmol l^{-1})				0.04						
Range					0.22–0.63						
Mean	Sr		–	34	0.44		–		–		–
SD	(μmol l^{-1})				0.10						
Range					0.02–0.06						
Mean	Cu		–	35	0.04		–		–		–
SD	(μmol l^{-1})				0.01						
Range					0.02–0.10						
Mean	Zn		–	36	0.05		–		–		–
SD	(μmol l^{-1})				0.02						

[a] Calculated from the carbonate portion given by Schmidt (1972b).
[b] Calculated from the ratio of dissolved CO_2 to atmospheric CO_2 given by Stallard and Edmond (1983). Cond., specific conductance at 25 °C; n, number of samples; –, not analyzed.

electrolytes (Table 4.1). With one exception (Stallard and Edmond 1983), the data represent means based on monthly sampling for at least one year. The chemical quality of the Solimões River water has the following characteristics:

- Mean specific conductance between 64 and 75 μS cm^{-1}.
- Mean pH values between 6.7 and 6.9; as a consequence protons are a minor cationic component (Furch et al. 1982).
- Relatively high content of major cations, corresponding to on average 35–48% (Na), 37–41% (K), 27–41% (Mg), 48–65% (Ca) and 42–53%

(sum of major cations) of the respective contents in the world average freshwater (Livingstone 1963).

- Relatively high amounts of calcium among the major cations reaching 56–65 eq% which is slightly higher than the world average for freshwater (53 eq% Ca).
- Relatively high alkalinity, corresponding on average to 51–62% of the respective amount in the world average freshwater. HCO_3 is the predominant major anion in the Solimões River water and with 74–83 eq% of the total major anions (HCO_3, SO_4, Cl) it is more dominant than in the world average freshwater (67 eq%).
- Balanced amounts of major cations and anions. In contrast to Amazonian black waters, clear waters and rainwater, low in electrolytes, concentrations of NH_4, NO_3, and organic ions can be ignored for the ionic balance.

The values of the major dissolved components in the Solimões River, do not differ significantly among the data sets (Table 4.1) except that the chloride content reported by Stallard and Edmond (1983) is significantly lower than that recorded by Schmidt (1972a) and Furch (unpubl.; column IVb in Table 4.1). For the minor dissolved nutrients and trace elements there is only one discrepancy among the data sets and that is the high mean value for iron, given by Schmidt (1972a). The agreement among the data sets is much closer than those given for the chemistry of Negro River and terra firme waters.

The close chemical relationship between the Solimões River and the world average freshwater is reflected more by the ionic composition than by the ionic content (Fig. 4.1). Classic Amazonian white waters are characterized as "carbonate waters", indicating freshwaters not only rich in carbonates but also rich in calcium which is the dominant major cation. During its course from the Andes to the ocean the content of dissolved solids in the Amazon River decreases significantly due to mixing with the large water masses from its tributaries, which are generally much poorer in dissolved solids (Gibbs 1967a,b, 1972; Irion 1976; Wissmar et al. 1981; Kempe 1982; Stallard and Edmond 1983; Furch 1987). However, ionic composition, i.e., the proportions of major cations and major anions, shows no significant change (Furch 1987). Gibbs (1967b) concluded "that the physical weathering dominant in the Andean mountainous environment controls both the overall composition of the suspended solids discharged by the Amazon and the amount of dissolved salts and suspended solids discharged". This is also valid for the floodplain lakes adjacent to the river (várzea lakes). Even if lake water becomes strongly diluted by local inputs of electrolyte-poor water from the terra firme its ionic composition identifies it as typical white water, i.e., typical carbonate water (Furch 1982, 1984c).

4.3.2 The Negro River Near Manaus

The Negro River, the largest tributary of the Amazon River, is a typical representative of the Amazonian black waters (Sioli 1950). Its transparent red-brown colour originates from a high content of dissolved humic substances and is reflected by analytical data for the color (mean: 159 mg Pt l^{-1}, Anonymous 1972b), for the KMnO$_4$ consumption (mean: 54 mg l^{-1}, Anonymous 1972b) and for the content of dissolved organic carbon (10.8 mg DOC l^{-1}, Ertel et al. 1986 and between 10 and 14 mg DOC l^{-1} in a shore lake permanently connected with the Negro River, Rai 1978). The humic acid (HA) content is about ten times higher and the content of fulvic acids (FA) is about three times higher than in the Solimões River, with values of about 4.5 mg HA l^{-1} and about 7 mg FA l^{-1} respectively in the Negro River (Ertel et al. 1986). Even higher values of humic matter (26.6 mg l^{-1}) were reported by Schmidt (1972a). Corresponding to its geological origin on the Precambrian shield of the northern region of the Amazon basin, the water of the Negro River is poor in nutrients and electrolytes (Table 4.2). With one exception (Stallard and Edmond 1983) data represent mean values based on monthly sampling for at least one year. There are differences between the data of Rai (1981) and Stallard and Edmond (1983) but the following general characterization of the Negro River water is confirmed by most studies:

- Low ionic content, expressed as specific conductance varying between 9 and 10 μS cm^{-1}.
- Low pH values with means between 4.8 and 5.1; as a consequence protons are major cations (Furch et al. 1982).
- Low content of major cations, corresponding to on average 15–18% (K), 7–11% (Na), 3–4% (Mg) and 2–3% (Ca) of the respective content in the world average freshwater (Livingstone 1963).
- Dominance of sodium among the major cations (28–64 eq%), while calcium, the classic dominant cation in freshwaters, occupies only 19–28% of the major cations.
- Low alkalinity; HCO$_3$ is frequently absent or below the detection limit. In studies presenting data of the three major anions (HCO$_3$, Cl, SO$_4$) it is shown that HCO$_3$ never occupies a major portion among the anions (Rai 1981; Stallard and Edmond 1983; Furch 1987).
- Large portion of silica among the dissolved solids.

Table 4.2. Chemical composition of the lower Negro River and adjacent black-water bodies near Manaus

Source (study period)		Rio Negro I (1966/1968)		Rio Negro II (1974/1977)		Rio Negro III (1974/1975)		Rio Negro IV (1976/1978)	
		n		n		n		n	
Range			7.6–11.6		7.3–15.8				
Mean	Cond.	27	9.3	25	9.9		–		–
SD	(µS/cm)		0.9		2.6				
Range			4.60–5.23		4.40–6.20				4.95–5.36
Mean	pH	27	4.84	27	4.95		–	4	5.08
SD			–		–				–
Range			5.9–25.1		0.6–39.8				4.4–11.2
Mean	H	27	14.6	27	11.3		–	4	8.4
SD	(µeq/l)		5.1		11.7				2.9
Range			18.9–59.0		5.0–27.0				17.3–20.8
Mean	Na	20	31.2	31	17.2	12	141.8	4	18.6
SD	(µeq/l)		10.6		5.7				1.9
Range			6.0–15.4		2.5–13.7				7.3–10.1
Mean	K	20	10.7	31	8.7	12	19.7	4	8.9
SD	(µeq/l)		2.4		2.9				1.4
Range			8.9–20.9		3.5–15.4				9.2–13.6
Mean	Mg	27	13.7	31	9.7	12	33.8	4	11.6
SD	(µeq/l)		3.2		3.1				1.9
Range			11.6–22.5		3.1–18.8				12.2–25.4
Mean	Ca	27	16.5	31	11.1	12	24.9	4	18.8
SD	(µeq/l)		2.9		3.7				5.7
Range			1.2–5.6						
Mean	NH_4	27	2.7		–		–	1	1.0
SD	(µeq/l)		1.0						
Range			70.0–100.0		<10.0–220.0				8.3–12.0
Mean	HCO_3	27	86.7	26	106.9	12	30.0	4	9.5
SD	(µeq/l)		10.0		73.7				1.7
Range			50.8–74.5		22.6–79.8				6.8–15.2
Mean	Cl	12	56.6	26	47.3	12	50.8	4	9.8
SD	(µeq/l)		4.5		18.0				3.8
Range									3.2–8.8
Mean	SO_4		–		–	12	155.5	4	5.5
SD	(µeq/l)								2.6
Range			1.1–3.8						1.7–5.3
Mean	NO_3	27	2.4		–		–	4	3.3
SD	(µeq/l)		0.6						1.5
Range			21.6–41.1						
Mean	N_{tot}	27	27.2		–		–		–
SD	(µmol/l)		4.2						
Range			0.10–0.28						0.02–0.11
Mean	PO_4	27	0.18		–		–	2	0.07
SD	(µmol/l)		0.04						0.06
Range			0.15–0.45		0.03–2.19				
Mean	P_{tot}	27	0.27	26	0.81		–	1	0.22
SD	(µmol/l)		0.07		0.54				
Range									
Mean	DOC		–		–		–		–
SD	(µmol/l)								
Range									164–703[a]
Mean	CO_2		–		–		–	4	463[a]
SD	(µmol/l)								231[a]
Range			71.2–97.5		23.0–102.1				65.3–72.0
Mean	Si	11	86.6	26	71.4		–	4	69.0
SD	(µmol/l)		7.5		15.9				3.1

Table 4.2. *Continued*

	bay (0 m) Va (1967/1968)		bay (bottom) Vb (1967/1968)		Lago Tupé VI (1974/1975)		Tarumã (0 m) VII (1975/1977)		Tarumã (bottom) VIII (1975/1976)	Source (study period)	
n		n		n		n		n			
	7.6–12.5		7.4–9.8				7.3–13.4		9.5–17.4		Range
12	9.4	12	8.7		–	22	9.8	17	12.1	Cond.	Mean
	1.2		0.7				1.8		2.1	(µS/cm)	SD
	4.80–5.40		4.80–5.40		4.40–5.80		4.50–6.20		4.00–5.30		Range
12	5.02	12	5.04	12		22	4.88	17	4.46	pH	Mean
	–		–				–		–		SD
	4.0–15.8		4.0–15.8		1.60–39.8		0.6–31.6		5.0–100.0		Range
12	9.6	12	9.1	12		22	13.1	17	34.6	H	Mean
	3.5		3.4				9.1		27.2	(µeq/l)	SD
					13.0–56.5		4.1–22.5		2.2–17.0		Range
	–		–	12	21.7	28	15.8	17	9.8	Na	Mean
					–		4.0		3.6	(µeq/l)	SD
					5.1–10.2		1.7–15.7		0.9–10.2		Range
	–		–	12	7.7	28	8.4	17	4.9	K	Mean
					–		2.8		2.7	(µeq/l)	SD
					4.9–12.3		2.2–13.2		1.1–8.5		Range
	–		–	12	8.2	28	8.1	17	4.4	Mg	Mean
					–		2.9		2.2	(µeq/l)	SD
					5.0–25.0		0.6–14.3		0.9–9.0		Range
	–		–	12	5.0	28	9.4	17	4.5	Ca	Mean
					–		4.1		2.4	(µeq/l)	SD
					0.4–5.9						Range
	–		–	12	2.3		–		–	NH$_4$	Mean
					–					(µeq/l)	SD
	20.0–130.0		10.0–100.0		0.0–120.0		10.0–180.0		<10.0–120.0		Range
12	50.0	12	51.4	12	80.0	23	77.0	17	28.8	HCO$_3$	Mean
	34.2		30.3		–		72.0		46.9	(µeq/l)	SD
					160.8–431.6		21.7–84.6		21.2–80.4		Range
	–		–	12	256.7	23	51.4	17	50.8	Cl	Mean
					–		18.5		19.3	(µeq/l)	SD
											Range
	–		–	12	212		–		–	SO$_4$	Mean
					–					(µeq/l)	SD
					0.8–2.7						Range
	–		–	12	1.5		–		–	NO$_3$	Mean
					–					(µeq/l)	SD
											Range
	–		–		–		–		–	N$_{tot}$	Mean
										(µmol/l)	SD
					0.02–0.79						Range
	–		–	12	0.30		–		–	PO$_4$	Mean
					–					(µmol/l)	SD
							0.03–2.19		0.03–2.54		Range
	–		–		–	23	0.70	17	1.01	P$_{tot}$	Mean
							0.68		0.73	(µmol/l)	SD
					717–1208						Range
	–		–	12	940		–		–	DOC	Mean
					–					(µmol/l)	SD
	64–261		46–271		200–334						Range
12	123	12	130	12	256		–		–	CO$_2$	Mean
	59		66		–					(µmol/l)	SD
					28.3–79.9		20.4–76.5		16.2–74.1		Range
	–		–	12	39.9	23	60.7	17	52.8	Si	Mean
					–		14.7		16.2	(µmol/l)	SD

Table 4.2. *Continued*

Source (study period)		Rio Negro I (1966/1968)		Rio Negro II (1974/1977)		Rio Negro III (1974/1975)		Rio Negro IV (1976/1978)	
			n		n		n		n
Range				0.37–0.89		1.46–5.56			2.40–4.00
Mean	Al	4	0.56	31	4.20	–		4	3.23
SD	(µmol/l)		0.25		0.99				0.67
Range				3.62–8.45		0.98–6.80			1.90–3.80
Mean	Fe	27	5.74	31	2.92	–		4	2.88
SD	(µmol/l)		1.41		1.15				0.88
Range				0.09–0.35		0.06–0.27			
Mean	Mn	15	0.17	31	0.17	–		–	
SD	(µmol/l)		0.07		0.05				
Range						0.02–0.09			
Mean	Ba	–		31	0.06	–		–	
SD	(µmol/l)				0.02				
Range						0.02–0.07			
Mean	Sr	–		31	0.04	–		–	
SD	(µmol/l)				0.01				
Range						0.01–0.04			
Mean	Cu	–		31	0.03	–		–	
SD	(µmol/l)				0.01				
Range						0.02–0.12			
Mean	Zn	–		31	0.06	–		–	
SD	(µmol/l)				0.02				

I, surface water of the mid-channel sampled monthly from Feb. 1966 to April 1968 calculated from data of Anonymous (1972b);

II, surface water of the mid-channel sampled monthly from Oct. 1974 to April 1977 recalculated after Furch (1984c);

III, surface water of the mid-channel sampled monthly from May 1974 to April 1975 (Rai 1981; monthly values were not given);

IV, individual surface waters of the mid-channel sampled from June 1976 to April 1978 calculated from data of Stallard and Edmond (1983);

Va, Vb, surface and bottom water of a bay of the Negro River sampled monthly from Nov. 1967 to Dec. 1968 calculated from data of Schmidt (1976);

VI, Lago Tupé, a Ria Lake, located about 30 km northwest of Manaus at the left-hand bank of the Negro River sampled monthly from May 1974 to April 1975 calculated for the whole water body from data given in the text, tables, and figures (Rai 1978 1981; standard deviation was not calculated);

VII, surface water in the mouth area of Tarumã Mirím, a Ria lake, located about 20 km northwest of Manaus at the left-hand bank of the Negro River sampled monthly from Jan. 1975 to April 1977 recalculated after Furch (1984c);

VIII, bottom waters in the mouth area of Tarumã Mirím sampled monthly from Jan. 1975 to July 1976 (Furch, unpubl. data).

[a] Calculated from the ratio of dissolved CO_2 to atmospheric CO_2 given by Stallard and Edmond (1983); Cond., specific conductance at 25 °C; n, number of samples: –, not analyzed.

Among other black-water bodies only the Lago Tupé, a Ria Lake with permanent connection with the river, is chemically very different from the river. Comparing column VI with column III in Table 4.2, lake water shows much lower concentrations of major cations and much higher concentrations of major anions than the river water (Rai 1981).

The Tarumã Mirím at its mouth area is a Ria Lake. There is little difference chemically between the lake and the river. Its surface water shows lower concentrations of ions (Table 4.2) due to the mixing with very electrolyte-poor water from the terra firme region (Furch 1976, 1984; Furch et al. 1982). Dilution is more pronounced for the bottom water (columns VII and VIII in Table 4.2) indicating that mixing with terra firme water is incomplete resulting in a chemical stratification. As a consequence, a seasonality of ion concentrations can be observed, lower for surface water and

Table 4.2. *Continued*

bay (0 m) Va (1967/1968)	bay (bottom) Vb (1967/1968)	Lago Tupé VI (1974/1975)		Tarumã (0 m) VII (1975/1976)		Tarumã (bottom) VIII (1975/1976)		Source (study period)	
n	n	n		n		n			
					2.25–8.34		2.51–6.26		Range
–	–		–	28	4.53	17	4.40	Al	Mean
					1.38		1.24	(μmol/l)	SD
			2.52–5.84		0.50–4.19		0.60–4.06		Range
–	–	12	4.15	28	2.21	17	2.14	Fe	Mean
			–		1.09		0.95	(μmol/l)	SD
					0.04–0.26		0.02–0.21		Range
–	–		–	28	0.14	17	0.10	Mn	Mean
					0.05		0.05	(μmol/l)	SD
					0.02–0.09		0.01–0.08		Range
–	–		–	28	0.05	17	0.04	Ba	Mean
					0.02		0.02	(μmol/l)	SD
					0.01–0.05		<0.01–0.04		Range
–	–		–	28	0.03	17	0.02	Sr	Mean
					0.01		0.01	(μmol/l)	SD
					0.02–0.06		0.02–0.06		Range
–	–		–	28	0.02	17	0.04	Cu	Mean
					0.01		0.01	(μmol/l)	SD
					0.02–0.12		0.05–0.13		Range
–	–		–	28	0.07	17	0.09	Zn	Mean
					0.03		0.02	(μmol/l)	SD

higher for bottom water of Tarumã Mirím. Ion concentrations are higher and more similar to those of the Negro River during the high-water period and lower (i.e., similar to those of the terra firme streams), during the low-water period (Furch, unpubl. data). Dilution by terra firme water is not indicated by mean values of chloride, total phosphorus, silicon, trace metals, and specific conductance (Table 4.2). Mean values for specific conductance tend to be higher than those of the Negro River, probably due to the higher H-ion concentration in the terra firme waters. Although data sets for the water of a bay of the Negro River are restricted to four parameters (Schmidt 1976), the water can be identified as river water; this is also true for the bottom water of the bay (Table 4.2).

Significant differences among the data sets are observed for the contents of major anions. The mean concentrations, given for all black-water bodies (Table 4.2), vary between approximately $10–110\,\mu eq\,l^{-1}$ for HCO_3, $10–260\,\mu eq\,l^{-1}$ for Cl, and $5–210\,\mu eq\,l^{-1}$ for SO_4. Thus, a general statement about the dominant major anion in black-water bodies cannot be given. However, Rai's very high values for Cl and SO_4 contradict all studies and may point to methodological problems (Table 4.2). Stallard and Edmond (1983) and Furch (1987) found chloride to be dominant. With regard to the mean concentrations of major anions and cations in the Negro River, a strong anion excess (up to ten times) was observed in all studies except for that of Stallard and Edmond (1983) who reported a significant cation excess.

A strong anion excess was also reported by Johnson (1967) in Malaysian waters. Although he discussed these findings as "an important factor in determining the acid nature of these waters", we tend to assume that the anion excess observed in the Amazonian black waters is probably not real for many samples and may be caused by the use of chemical methods with often unsatisfactory detection limits for the analysis of very dilute waters.

4.3.3 Terra Firme Affluents

The physical and chemical properties of the Central Amazonian forest streams are well documented (Sioli 1950; Fittkau 1964, 1967, 1971; Schmidt 1972b; Furch 1976, 1984; Furch et al. 1982; Stallard and Edmond 1983; Lesack 1988, 1993). They derive from deeply weathered tertiary sediments and are characterized as follows: transparent, ranging in color from colorless to greenish and brown (clear and black-water streams according to the classification of Sioli (1950)), very low concentrations of inorganic solutes in some cases below the detection limits of the applied analytical methods (e.g., for HCO_3, SO_4, Mg, Ca, K, P), low nutrient content and low pH values.

Data given in Table 4.3 represent streams sampled within a radius of up to 150 km from Manaus. The general characterization of terra firme affluents given above is confirmed by all studies. Significant differences in the content of the anions Cl and SO_4 are probably related to an overestimation of these anions by methodological problems. The chloride content is often higher than the sulfate content and alkalinity often was not detectable. These findings differ from the results of Stallard and Edmond (1983) who found alkalinity levels higher than $40\,\mu eq\,l^{-1}$ in rivers draining upper Tertiary and Quaternary sediments comparable to those near Manaus. According to the studies of Schmidt (1972b) and Furch (unpubl. data) NO_3 occurs in the streams north of Manaus in small quantities; however, larger quantities of NO_3 were observed in the affluents of Lago Calado (Lesack 1988, Table 4.3).

The cation content of terra firme streams was used by Furch (1976, 1984) and Furch et al. (1982) for a hydrochemical classification of terra firme streams. The content of alkali ions is considerably greater than the content of alkali-earth ions. Because of the low pH value, protons are often more dominant than the alkali ions. With increasing dominance of the protons among the major ions, specific conductance becomes less relevant as an indicator for the total content of the other inorganic ions in the water (Furch et al. 1982).

Table 4.3. Chemical composition of central Amazonian rainforest streams in the terra firme near Manaus

Source (study period)		I	(1968)	II	(1974/1975)	III	(1980/1981)	IV	(1984/1985)
		n		n		n		n	
Range			4.7–12.1		6.0–22.3		6.0–34.1		
Mean	Cond.	12	8.1	27	11.5	19	19.2		–
SD	(μS cm^{-1})		2.3		3.9		10.2		
Range			4.40–5.30		4.10–4.75		3.98–4.90		4.4–4.7
Mean	pH	12	4.60	27	4.40	19	4.31	27	4.61
SD			–		–		–		–
Range			5.0–39.8		20.0–79.4		12.6–104.7		20–40
Mean	H	12	25.3	27	39.9	19	49.1	27	24.5
SD	(μeq l^{-1})		9.8		17.9		32.3		1.1*
Range			8.3–45.7		5.4–31.0		10.6–24.7		5–9
Mean	Na	12	17.2	27	10.9	19	15.4	27	6.3
SD	(μeq l^{-1})		14.1		5.9		3.6		0.4*
Range			3.9–35.3		0.5–13.3		0.7–7.5		0.4–0.7
Mean	K	12	10.1	27	4.9	19	3.1	27	0.6
SD	(μeq l^{-1})		12.9		3.3		1.9		<0.1*
Range			<0.8–26.5		1.3–14.3		1.6–6.5		1–2
Mean	Mg	12	4.2	27	4.2	19	3.7	27	1.4
SD	(μeq l^{-1})		0.8		2.8		1.2		0.1*
Range			<1.0–17.8		<0.3–14.6		2.1–7.1		1–2.5
Mean	Ca	12	2.8	27	3.3	19	3.9	27	1.5
SD	(μeq l^{-1})		5.6		3.7		1.5		0.3*
Range							0.6–4.0		0.4–1.2
Mean	NH$_4$		–		–	19	1.8	27	0.7
SD	(μeq l^{-1})						1.2		0.1*
Range									
Mean	HCO$_3$		–		–		–		–
SD	(μeq l^{-1})								
Range			28.2–98.7		53.6–69.1		10.2–26.2		8–17
Mean	Cl	12	52.2	2	61.4	19	15.9	27	13.0
SD	(μeq l^{-1})		17.8		11.0		3.7		0.8*
Range							7.1–14.2		3–7
Mean	SO$_4$		–		–	19	8.9	27	3.9
SD	(μeq l^{-1})						1.9		0.2*
Range			0.14–0.5				0.07–1.21		6–15
Mean	NO$_3$	12	0.27		–	19	0.28	27	11.5
SD	(μeq l^{-1})		0.13				0.28		0.97*
Range			8.6–46.4						10–25
Mean	N$_{tot}$	12	21.3		–		–	27	17.2
SD	(μmol l^{-1})		11.6						0.5*

I, sampled along the road BR-174 (Manaus–Caracaraí) in September and October 1968 (Schmidt 1972);
II, sampled along the road BR-174 (Manaus–Caracaraí) in November 1974 and March 1975 (Furch 1984c) and along the road BR-319 (Manaus–Porto Velho) in March 1975 (Furch unpubl. data);
III, sampled along the road BR-174 (Manaus–Caracaraí) from November 1980 to July 1981 (Furch, unpubl. data);
IV, sampled in the surroundings of Lago Calado from February 1984 to February 1985 (Lesack 1993).
Cond., specific conductance at 25 °C; –, not analysed; s.d., standard deviation; * denotes standard error (being about five times lower than the respective standard deviation).

Table 4.3. *Continued*

Source (study period)		I (1968)		II (1974/1975)		III (1980/1981)		IV (1984/1985)
		n		n		n		n
Range							0.03–0.82		0.01–0.06
Mean	PO_4		–		–	19	0.22	27	0.03
SD	(μmol l^{-1})						0.18		<0.01*
Range			0.06–0.39		<0.03–0.77				0.06–0.2
Mean	P_{tot}	12	0.18	27	0.31		–	27	0.13
SD	(μmol l^{-1})		0.11		0.26				0.01*
Range					333–958		398–2321		
Mean	DOC		–	27	524	19	1200		–
SD	(μmol l^{-1})				200		719		
Range							110–356		
Mean	CO_2		–		–	19	196		–
SD	(μmol l^{-1})						66		
Range			72.6–140.6		37.0–96.1		45.6–101.5		
Mean	Si	12	91.5	27	72.6	19	68.4		–
SD	(μmol l^{-1})		18.5		18.2		18.2		
Range					0.58–7.26				
Mean	Al		–	27	3.21		–		–
SD	(μmol l^{-1})				1.63				
Range			0.70–5.68		0.48–3.83				
Mean	Fe	12	2.08	27	1.72		–		–
SD	(μmol l^{-1})		1.57		0.84				
Range					0.02–0.18				
Mean	Mn		–	27	0.06		–		–
SD	(μmol l^{-1})				0.03				
Range					0.02–0.19				
Mean	Ba		–	27	0.06		–		–
SD	(μmol l^{-1})				0.04				
Range					<0.01–0.03				
Mean	Sr		–	27	0.01		–		–
SD	(μmol l^{-1})				0.01				
Range					<0.01–0.24				
Mean	Cu		–	27	0.09		–		–
SD	(μmol l^{-1})				0.07				
Range					0.01–0.34				
Mean	Zn		–	27	0.04		–		–
SD	(μmol l^{-1})				0.07				

The considerable variations in the content of ions of central Amazonian streams are not related to specific geological conditions because Tertiary sediments covering central Amazonia are not sufficiently differentiated in geological maps (Stallard and Edmond 1983). However, small differences in the input from the catchment area, e.g., sediments and organic matter, can result in relatively large chemical changes of the extremely electrolyte-poor water. Man-made modifications in a catchment of Amazonian streams can also cause significant changes in solute concentrations. Williams (1993) reported that deforestation in the catchment of Mota brook, an affluent to Lago Calado, led to an increase in nitrate and phosphate concentrations. Campos (1994) found slightly higher levels of major solutes and nutrients in the streams north of Manaus than had been reported in studies of other authors 10 to 20 years ago, probably because of heavy deforestation in the catchment during the last 10 years.

Lesack (1988) points to the unbalanced relationship between the inorganic anions and cations. Deficits of anions vary between 14 and 55% of the sum of the cations. A similar situation has been shown by Lesack and Melack (1991) for rainwater. The authors explain that these imbalances are caused by the influence of organic anions in the ionic balance of the rainwater. A similar explanation is given by Stallard and Edmond (1983) for electrolyte-poor Amazonian streams. The dissolved fraction, which does not occur or only partially occurs in ionic form, is dominated by organic carbon, carbon dioxide, and silicate. They occur in much larger quantities than all the inorganic ions (Table 4.3).

4.3.4 Amazonian Rainwater

The chemical composition of rainwater and the deposition of solutes in the Amazon ecosystems is of wide interest. Since most Amazonian lowland soils are very poor in nutrients it is generally assumed that rainwater provides an important contribution to the nutrient supply of the forest vegetation (i.e., Jordan et al. 1980; Jordan 1982). In addition, it is widely accepted that in Amazonia the atmospheric contribution to dissolved materials in running waters can be substantial (Stallard and Edmond 1981). In Amazonia, rainwater and its nutrient load are believed to play a more important role in the cycle of chemical elements in ecosystems than in ecosystems of temperate regions.

Detailed information about rainwater chemistry in the Amazon region is still rare, possibly due to the high effort needed for a representative rainwater study (Galloway and Likens 1976; Lesack and Melack 1991). Among the studies available, only few provide data sets based on a representative

Table 4.4. Concentrations of dissolved and particulate substances in rainwater of the Amazon basin; comparison of data previously reported

Source (study period)		I (1965/1968) n	I (1965/1968)	II (1974/1975) n	II (1974/1975)	III (?) n	III (?)	IV (1976/1977) n	IV (1976/1977)	V (1976/1977) n	V (1976/1977)	VI (1980/1981) n	VI (1980/1981)	VII (1980/1981) n	VII (1980/1981)	VIII (1983–1985) n	VIII (1983–1985)
Cond. ($\mu S\,cm^{-1}$)	Mean	48	8.6		–	30	7.0	22	–	25	7.9	25	8.1		–		–
	SD		5.5								4.6		8.7				
pH	Mean	48	4.56		–	30	4.70	22	5.12	25	5.30	90	4.30	14	4.77 (4.81)	123	4.89
	SD																
H ($\mu eq\,l^{-1}$)	Mean	48	27.3		–	30	20.0	22	7.5	25	5.0	90	50.1	14	17.0 (15.5)	123	12.9
	SD		22.8						3.1		3.7		37.5		10.0		
Na ($\mu eq\,l^{-1}$)	Mean		–	25	5.2		–	22	9.5	13	21.7	30	17.0	14	2.7 (1.8)	123	2.5
	SD				4.2				5.0		47.4		15.7		2.0		
K ($\mu eq\,l^{-1}$)	Mean		–	25	2.6		–	22	1.0	13	2.6	30	5.9	14	1.1 (0.8)	123	0.7
	SD				2.7				0.6		1.5		17.1		1.7		
Mg ($\mu eq\,l^{-1}$)	Mean	53	10.8	25	1.7	30	0.4	22	1.8		–			14	0.8 (0.5)	123	1.5
	SD		10.4		1.4				1.1						0.9		
Ca ($\mu eq\,l^{-1}$)	Mean	53	8.6	25	3.6		–	22	1.4		–			14	0.5 (0.3)	123	2.1
	SD		8.4		3.9				1.1						0.7		
NH$_4$ ($\mu eq\,l^{-1}$)	Mean	53	12.9		–	30	4.9	9	0.5	23	16.1		–	14	2.3 (2.3)	123	6.6
	SD		9.2						0.6		16.1				3.4		
HCO$_3$ ($\mu eq\,l^{-1}$)	Mean	46	74.7		–		–		–		–		–		–		–
	SD		16.3														
Cl ($\mu eq\,l^{-1}$)	Mean	47	55.4		–	30	8.5	22	10.3	25	28.8	30	18.9	14	4.3 (2.5)	123	4.7
	SD		10.9						5.1		17.8		39.8		3.4		
SO$_4$ ($\mu eq\,l^{-1}$)	Mean		–		–		–	22	9.0		–	20	61.0	14	3.3 (2.9)	123	4.5
	SD								4.6				43.9		2.5		
NO$_3$ ($\mu eq\,l^{-1}$)	Mean	53	6.7		–	30	0.3	19	1.7		–		–	14	3.5 (2.6)	123	3.5
	SD		5.9						1.8						3.6		

Parameter (μmol l⁻¹)	Study	n	Mean	SD
N_{tot}	I	39	31.6	13.6
PO_4	I	53	0.05	0.05
PO_4	V	19	0.16	0.16
PO_4	VI	60	0.26	0.36
PO_4	VII	14	0.60 (0.60)	1.29
PO_4	VIII	123	0.06	–
P_{tot}	I	53	0.37	0.27
P_{tot}	—	30	0.48	–
Si	—	30	<0.001	–
Si	—	22	<0.3	–
Al	II	23	0.37	0.30
Fe	I	53	0.40	0.28
Fe	II	23	0.47	0.56
Mn	II	23	0.03	0.01
Ba	II	23	0.03	0.02
Sr	II	23	0.01	0.01
Cu	II	23	0.05	0.03
Zn	II	23	0.07	0.05

I, Anonymous (1972a) Manaus, April 1965–March 1968;
II, Furch (1984c) Manaus and adjacent lake areas, Oct. 1974–May 1975;
III, Brinkmann (1983) Manaus and adjacent rainforest areas, study period not known;
IV, Stallard and Edmond (1981) different localities in the central Amazon basin, June 1976 and May 1977;
V, Franken and Leopoldo (1984) Barro Branco watershed, forest reserve "Adolfo Ducke" (20 km from Manaus), Sept. 1976–Sept. 1977;
VI, Franken and Leopoldo (1984) Bacia Modelo watershed, forest reserve (60 km from Manaus), Feb. 1980–Feb. 1981;
VII, Galloway et al. (1982) San Carlos de Rio Negro, Venezuela, Sept. 1980–March 1981 (volume-weighted mean concentrations in parentheses);
VIII, Lesack (1988) Lago Calado (about 80 km from Manaus), Sept. 1983–Oct. 1985 (data given are volume weighted mean concentrations).
Cond, specific conductance at 25°C; n, number of samples; –, not analyzed.

sample collection for at least one annual cycle (Anonymous 1972a; Lesack 1988; Lesack and Melack 1991; Williams et al. in press); the other studies are related to shorter periods and/or refer to a low number of samples (Stallard and Edmond 1981; Galloway et al. 1982; Brinkmann 1983; Franken and Leopoldo 1984; Furch 1984; Forti and Neal 1992). Only Galloway et al. (1982), Lesack (1988), Lesack and Melack (1991), and Williams et al. (in press) considered the volume-weighted means for all common major solutes and inorganic nutrients.

Mean Concentrations and Chemical Type

The data sets for the chemistry of rainwater sampled in the central region of the Amazon basin are listed in Table 4.4. In general, rainwater is acidic and poor in electrolytes, corresponding to mean pH values of 4.3–5.3 and specific conductance (25 °C) of 7.0–8.6 μS cm^{-1}. Variations in solute concentrations were high as indicated by standard deviations frequently exceeding 100% of the mean (Table 4.4). There are considerable differences between the mean values reported in different studies. The ratios of highest mean value to lowest mean value observed for rainwater concentration varied between approximately 8 for potassium and approximately 32 for ammonium (Table 4.4). Considering the unusually high content of solutes and nutrients in the rainwater of San Carlos, Venezuela, studied in 1975 and 1976 (Jordan et al. 1980), differences between data sets would even be higher.

As reported for acidic and ion-poor natural waters in Amazonia, hydrogen and sodium ions are dominant cations in rainwater (Furch et al. 1982). Ammonium is also a major cation, and its content in rainwater is usually higher than that of potassium, calcium, and magnesium. For major anions a dominance was observed for chloride, obviously derived from marine aerosols (Andreae et al. 1990). For the data sets given in Table 4.4, a clear dominance of one anion species is not given due to the lack of investigations and to the heterogeneity of results. In more recent rainwater studies organic anions, mainly those of formic and acetic acid, were discussed as major components in rainwater of Amazonia as well as in rainwater of Australia and USA, presenting up to 65% of the total anion load (Galloway et al. 1982; Keene and Galloway 1984; Likens et al. 1987; Andreae et al. 1988, 1990; Lesack and Melack 1991).

Numerous rainwater studies report that solute concentrations during the dry season can be considerably higher than solute concentrations in the wet season due to the enrichment of ions and dust particles in the atmosphere (Andreae et al. 1990). For Amazonian rainwater (Table 4.5) this phenomenon is reflected by calcium and magnesium only (Anonymous

Table 4.5. Mean concentrations of dissolved and particulate substances in rainwater of the Amazon basin during the dry and the wet season. For further explanation see Table 4.4

Source (study period)	Season	I (1965/1968)				VIII (1983/1985)			
		Dry		Wet		Dry		Wet	
		n		n		n		n	
Mean SD	Cond. ($\mu S\,cm^{-1}$)	10	8.7 1.9	38	8.6 6.2		–		–
Mean SD	pH	10	4.49	38	4.59	34	4.54	89	5.03
Mean SD	H ($\mu eq\,l^{-1}$)	10	32.2 25.3	38	26.0 23.0	34	28.9	89	9.3
Mean SD	Na ($\mu eq\,l^{-1}$)		–		–	34	5.1	89	1.9
Mean SD	K ($\mu eq\,l^{-1}$)		–		–	34	1.4	89	0.6
Mean SD	Mg ($\mu eq\,l^{-1}$)	10	14.9 20.3	43	9.8 6.5	34	2.1	89	1.3
Mean SD	Ca ($\mu eq\,l^{-1}$)	10	15.6 13.0	43	7.0 7.0	34	2.9	89	1.9
Mean SD	NH_4 ($\mu eq\,l^{-1}$)	10	13.1 4.6	43	12.9 10.2	34	13.3	89	5.2
Mean SD	HCO_3 ($\mu eq\,l^{-1}$)	10	73.0 24.5	36	75.1 14.0		–		–
Mean SD	Cl ($\mu eq\,l^{-1}$)	9	56.4 9.1	38	55.1 11.5	34	8.2	89	4.0
Mean SD	SO_4 ($\mu eq\,l^{-1}$)		–		–	34	8.9	89	3.5
Mean SD	NO_3 ($\mu eq\,l^{-1}$)	10	3.9 0.5	43	7.4 6.4	34	9.2	89	2.2
Mean SD	N_{tot} ($\mu mol\,l^{-1}$)	10	36.3 15.9	29	30.0 13.3		–		–
Mean SD	PO_4 ($\mu mol\,l^{-1}$)	10	0.05 0.01	43	0.05 0.07	34	0.16	89	0.04
Mean SD	P_{tot} ($\mu mol\,l^{-1}$)	10	0.50 0.26	43	0.34 0.28		–		–
Mean SD	Fe ($\mu mol\,l^{-1}$)	10	0.56 0.36	43	0.37 0.26		–		–

1972a) or by all dissolved ion species, including the nutrients (Lesack 1988, Lesack and Melack 1991). Mean concentrations given by Lesack (1988) and Lesack and Melack (1991) also indicate that wet season rainwater is not a diluted dry season rainwater, i.e., proportions of dissolved components differ between the dry and wet seasons (Table 4.5). Phosphate, nitrate, and hydrogen ions are more enriched in the dry season than the other solutes. Nevertheless, during the wet season more solutes were deposited than during the dry season (Lesack and Melack 1991).

Ionic Balance, pH, and Acidity

Among the rainwater studies presenting data sets for all major inorganic solutes (columns IV, VII and VIII in Table 4.4) only one shows that anions and cations are well-balanced (Stallard and Edmond 1981). The other data sets reveal a considerable excess of cations: on average between 52–62% of the cations are not balanced by inorganic anions. However, Galloway et al. (1982) and Lesack and Melack (1991) demonstrated that this anion deficit is compensated by an equivalent amount of organic anions, mainly those of formic and acetic acid.

If the pH value of natural rainwater is lower than the pH derived from atmospheric equilibrium with CO_2 (pH 5.65), rainwater can be termed acid rain. In general, sources of rainwater acidity are from a surplus of inorganic anions of strong mineral acids (sulfuric, nitric, hydrochloric and perhaps phosphoric acid). Chloride ions are mainly of marine origin. Nitrate and sulfate ions come from oxidized N and S compounds in the atmosphere, which are derived from terrestrial inputs. Since industrial emissions are only of minor or local importance in the Amazon basin, three categories of terrestrial input into the atmosphere can be identified: biological emissions, soil dust, and combustion products from burning vegetation (Stallard and Edmond 1981). Substantial emissions of reduced N and S gases into the atmosphere by rain-forest soils and canopy have been reported for Amazonia (Andreae and Andreae 1988; Harriss et al. 1988, 1990). Haines et al. (1983) suggested that the rain forest is a source of volatile P compounds. It is likely that the low pH values of Amazonian rainwater are based in part on weak organic acids derived from biogenic processes (Galloway et al. 1982; Church et al. 1984; Andreae et al. 1990; Lesack and Melack 1991).

Input of Rainwater Solutes

From a biogeochemical point of view, the interest in Amazon rainwater chemistry is based on the nutrient yields annually supplied by rainwater to

Table 4.6. Annual wet deposition of major solutes and nutrients in the Amazon basin calculated for a mean annual rainfall of 2500 mm, and annual nutrient requirements of Amazon floodplain forests calculated for an aboveground biomass production of 17.4 t year^{-1} (várzea forest) and 8.7 t year^{-1} (igapó forest)

(kg ha^{-1} year^{-1})	Wet deposition		Requirement	
	A	B	Várzea	Igapó
H	0.13–1.25	0.20	–	–
Na	1.04–12.5	1.44	2.9	1.6
K	0.69–5.76	0.69	178	38.3
Mg	0.13–3.28	0.46	50.9	7.3
Ca	0.15–4.31	1.05	213	15.9
Cl	2.22–50.0	4.17	–	–
SO$_4$-S	1.16–24.5	1.80	–	–
NH$_4$-N	0.18–5.65	2.31	–	–
N$_{tot}$	–	–	292	98.6
NO$_3$-N	0.10–2.35	1.23	–	–
PO$_4$-P	0.04–0.21	0.05	–	–
P$_{tot}$	–	–	23.7	3.8

A, range, calculated with mean values given in Table 4.4; B, calculated with values for volume weighted mean concentrations given by Lesack (1988).

the forest ecosystems (Ungemach 1971; Jordan et al. 1980, 1982; Franken and Leopoldo 1984; Andreae et al. 1990). In Table 4.6 the annual amounts of wet-deposited solutes are calculated with mean rainwater concentrations given in Table 4.4. Since annual rainfall in the Amazon basin is high (between approximately 2000 and 8000 mm, Salati and Marques 1984), the yield of annually deposited solutes and nutrients is considerable despite their low concentrations in the rainwater. The conspicuous differences observed for the nutrient content in rainwater are reflected in correspondingly wide ranges of the calculated amounts for wet deposition (column A in Table 4.6). Therefore, the most representative data set for rainwater (Lesack 1988) is used for a separate calculation (column B in Table 4.6).

Rainwater contributes a small part of the annual nutrient requirements even in the igapó forest which has a low nutrient demand (Sect. 3.6). In addition, a large portion of nutrients observed in rainwater may be derived from the natural vegetation itself. This attempt to roughly estimate the nutrient supply only provides insight into the order of magnitude involved. Detailed studies are required for a budget of nutrient cycles between vegetation, soil, and atmosphere.

4.3.5 Groundwater

The chemistry of groundwaters is similar to the chemistry of surface waters in that it is predominantly influenced by the geology and biogeochemistry of the catchment area. The physical and chemical conditions of the soils or parent rocks, hydrological and geomorphological conditions, quality and quantity of the sources feeding the groundwater, and the type of vegetation cover are the most important factors determining the quality of groundwater in natural ecosystems. In the Amazon floodplain, factors controlling groundwater quality are complex due to small-scale changes in grain size of the sediments, the age of the sediments, and the periodic change between oxic and anoxic conditions in the soils. This results in a high spatial and temporal variability of the chemical composition of the groundwater.

There are few data available for groundwater of the terra firme (Nortcliff and Thornes 1978; Brinkmann 1983, 1989; Furch et al. 1989; Forti and Neal 1992; Williams 1993). The data show that groundwater from Central Amazonian terra firme forests are very poor in electrolytes and typically low in alkaline-earth ions. However, the ionic content is higher than in rainwater

Table 4.7. Chemical compositions of Amazonian groundwaters

		Terra firme[a] n = 24	L. Calado[b] n = 13			L. Camaleão[c] n = 17		
		Mean	Range	Mean	SD	Range	Mean	SD
Cond.	($\mu S\,cm^{-1}$)	26	–	–	–	203–1100	646	254
pH		4.50	5.47–6.78	6.03		4.17–7.55	5.37	
H	($\mu eq\,l^{-1}$)	31.6	0.2–3.4	0.9	0.8	<0.1–67.6	4.2	16.3
Na	($\mu eq\,l^{-1}$)	46.1[d]	28.8–74.1	53.6	15.0	313–1130	763	212
K	($\mu eq\,l^{-1}$)	8.7[d]	6.1–33.2	15.4	9.2	35.8–358	102	73.9
Mg	($\mu eq\,l^{-1}$)	19.7	20.6–85.2	55.8	18.3	412–4967	1512	965
Ca	($\mu eq\,l^{-1}$)	10.0[d]	100–452	273	103	883–9830	4420	2316
NH$_4$	($\mu eq\,l^{-1}$)	3.6	2.2–160	40.6	40.3	1.4–199	59.2	56.6
HCO$_3$	($\mu eq\,l^{-1}$)	<10.0[d]	113–524	340	131	50.0–11960	5616	3366
Cl	($\mu eq\,l^{-1}$)	45.1	32.8–206	69	50.6	31.9–454	165	117
SO$_4$	($\mu eq\,l^{-1}$)	4.6[d]	4.0–124	27.9	32.5	128–8890	1126	2079
NO$_3$	($\mu eq\,l^{-1}$)	26.4	0.3–2.8	1.5	0.8	<0.1–56.0	7.3	8.3
PO$_4$	($\mu mol\,l^{-1}$)	0.97[d]	0.13–10.40	2.97	3.42	0.10–6.85	1.35	1.52
CO$_2$	($\mu mol\,l^{-1}$)	410	–	–	–	230–5430	2300	2080
Si	($\mu mol\,l^{-1}$)	13.2	–	–	–	295–5625	584	248

n, number of samples; Cond., specific conductance at 25 °C; –, not determined.
[a] From a latosolic podsol soil (Brinkmann 1983).
[b] From the surroundings of Lago Calado (Lesack 1988).
[c] From the shoreline of Lago Camaleão (Furch, unpubl. data).
[d] From well water at Manaus (Furch and Junk 1992).

(Brinkmann 1989) and considerably lower than in through-fall water (Forti and Neal 1992).

A detailed study of groundwater influx into Lago Calado shows that the catchment area receives inputs of varying quantities from terra firme and várzea. Chemical composition is characterized by a high variability in ionic content and ionic composition (Lesack 1988; Table 4.7). As "mixed" groundwaters they usually show an ionic content lower than that of the Amazon River water, but a higher ionic content than groundwater from the terra firma except for nitrate. The ionic composition of major cations and anions is more similar to that of the Amazon River than to that of terra firme streams and rainwater. In contrast to the water in Lago Calado (Fig.

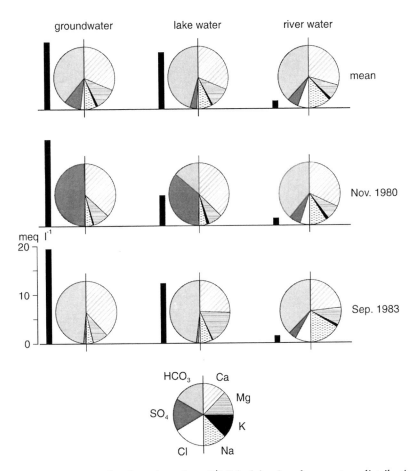

Fig. 4.3. Content of major solutes (meq l^{-1}) (*black bars*) and percentage distribution (eq%) of major anions (*left-hand half of the circle*) and major cations (*right-hand half of the circle*) in groundwaters near the shore line of Lago Camaleão, in lake water and in river water (Solimões River)

4.4), no clear seasonal change in solute content of the groundwaters was observed.

The total content of dissolved minerals in groundwater on Marchantaria Island is much greater than that in the Amazon River (Furch, unpubl.; Fig. 4.3, Table 4.7). An enrichment can occur due to evapotranspiration, weathering from the mineral soil, increased ion exchange during reducing conditions, decomposition of organic material (Furch et al. 1988, 1989; Furch and Junk 1992) and dissolution of soluble salts in the soil (Chap. 3). Obviously all factors are involved in the enrichment process in varying strength because the groundwater varies considerably in chemical composition (Table 4.7, Fig. 4.3).

The release of exchangeable cations can be quick and is probably more important than the release by weathering. Major changes in NO_3 and H content point to strongly varying redox conditions. According to Ponnamperuma (1972), the change from oxidizing to reducing conditions is the most important process, stimulating the increase of conductance and content of major cations in the soil solutions of paddy fields.

Evaporation of oxidized soil solutions and solutes released during decomposition of dry macrophytes and leaf litter enrich the upper soil layer with acidic salts containing up to 80% sulfates. In aqueous soil extracts up to 4.5 g soluble salts per kilogramm soil were recorded on the upper soil layer of Lago Camaleão (Furch, unpubl.). During rainfall they are leached and may contribute considerably to the dissolved load of the groundwater.

High NH_4 concentrations indicate the role of organic material in the release of nutrients (Chap. 6 and Sect. 9.2.3). Lesack (1988) reported that seepage water contributed significant amounts of NH_4 (approximately 40%) and PO_4 (approximately 20%) to the total annual input into Lago Calado. The contribution of seepage water to the annual input of NO_3 and the major solutes Na, K, Mg, Ca, H, HCO_3, SO_4, and Cl was negligible in Lago Calado. However, the input of some ions, e.g., Na, Ca, and Mg by groundwater into Lago Camaleão may be essential in explaining the element dynamics of the lake during falling- and low-water periods (Chap. 5).

4.4 Physicochemical Attributes of Different Várzea Lakes

4.4.1 The Influence of Different Waters on the Chemistry of Várzea Lakes

A major factor controlling the chemical composition of the water in floodplain lakes is the input of water masses from different sources and of different chemical compositions. The importance of these sources can be evaluated by chemical compounds that show a conservative mixing pattern. Because of the large differences in chemical composition between Amazon River water, water from terra firme affluents, and rainwater, specific conductance and the contents of Na, K, Ca, Mg, and HCO_3 have been used to describe mixing patterns.

Schmidt (1973a) described seasonal changes in chemical composition of the water of Lago Castanho, a floodplain lake bordering the terra firme. Specific conductance reached a maximum of about $60 \mu S \, cm^{-1}$ at the end of February as the water level rose and Amazon water dominated the lake. Lowest values (less than $20 \mu S \, cm^{-1}$) were found at low-water periods. A similar pattern was found for Ca and bicarbonate. It was explained by continuous dilution of the Amazon water by rainwater and electrolyte-poor surface runoff from the adjacent terra firme. In Lago Calado, a Ria Lake more influenced by direct water inputs of terra firme streams than Lago Castanho, seasonal fluctuations of major solutes were even more pronounced (Furch 1982, Fig. 4.4).

Forsberg et al. (1988) used alkalinity as a tracer and showed that the influence of terra firme water could be determined by the ratio between lake catchment area and lake basin area. Ria Lakes with a ratio of less than 20 showed at the end of the low-water season that they were a mixture of river water and local water with a variable nutrient composition. When the ratio was >20, terra firme water dominated and nutrient composition was similar to that of the terra firme affluents. Using the relative Ca proportion of major cations as a tracer, Furch (1982) demonstrated that Lago Calado water strongly influenced by terra firme affluents was chemically more similar to Ca-dominated Solimões River water than to alkali-dominated terra firme water, even at conditions of maximum dilution. In Fig. 4.5 the ratio of Ca in lake water to Ca in river water (Solimões River) is used to describe the impact of local waters other than river water on seasonal variation in different floodplain lakes. Lakes strongly influenced by local waters, such as Lago Calado and Lago Castanho, show lower Ca contents than the river during long periods of the year. Calcium-ion content reaches a peak at the beginning of the floods, when there is a strong influx from the river. Dilution by terra firme water and rainwater reduces the Ca content. In small lakes without a direct connection with the terra firme, e.g., Lago Jacaretinga, Ca levels are lower than the respective river-water level only during short periods of the year, due to dilution by rainwater during the low-water period. Lakes totally isolated from the terra firme and with a very large Aquatic Terrestrial Transition Zone (ATTZ), e.g., the Lago Camaleão, have Ca levels that never fall significantly below the Ca level of the river but are up to 25 times higher than the level in river water during the low-water period.

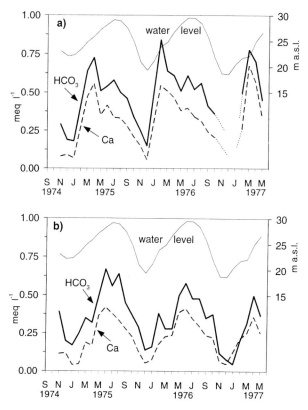

Fig. 4.4a,b. Water-level fluctuations and seasonal variations of the Ca and HCO$_3$ content in the water of Lago Calado. **a** Near the channel connecting the lake with the river. **b** Near the inflow of forest stream water. (Furch 1982)

For most solutes, influx of terra firme water results in dilution. An exception can be nitrogen compounds. Throughout a 1-year study period (1983/1984) the inputs of N from precipitation and terra firme affluents to Lago Calado were about three times greater than those from the Amazon River (Lesack 1988). Most studies of the chemical composition of water from terra firme streams show low levels of NH$_4$ and NO$_3$. High values are reported only by Lesack (1988, 1993). Total N content in streams is high; however, it is lower than in the water of the Amazon River and Negro River (Tables 4.1, 4.2, 4.3). In rainwater, contents of NO$_3$ and NH$_4$ and probably also N$_{tot}$ are relatively high (Table 4.4).

The possibilities and also the limits of the applicability of hydrochemical tracers are shown by Fig. 4.5. In all várzea lakes, even in those strongly influenced by local waters, Ca peaks were observed which exceeded the Ca

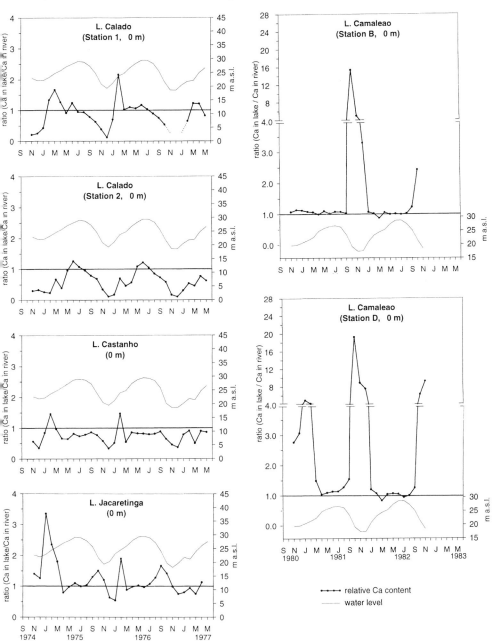

Fig. 4.5. Seasonal variation of the relative Ca content in the surface waters of different várzea lakes expressed as the ratio of Ca content of lake water to Ca content of Solimões River water. *Station 1* of Lago Calado, near the river inflow; *station 2*, near forest stream inflow; *station B* of Lago Camaleão, near the river inflow; *station D*, middle of the lake. (Furch unpubl. data)

content of the Amazon River. We postulate that lake's internal processes, e.g., nutrient uptake and release by the vegetation, nutrient exchange between sediments and water, and groundwater dynamics, can cause considerable seasonal changes in the chemical composition of the lake water including the content of tracers (Sect. 4.4.5; Chaps. 5 and 9).

4.4.2 Thermal and Chemical Stratification and Mixing Patterns

During the daytime most várzea lakes have a thermal stratification of 1–3 °C. The position of the thermocline varies between 2–4 m in depth. Cooling during the nighttime and windy weather destroy the stratification. Even so, total mixing of small lakes occurs only sporadically at high water level. The frequency and depth of mixing is greater in large lakes than in small protected ones because of the greater impact from wind induced turbu-

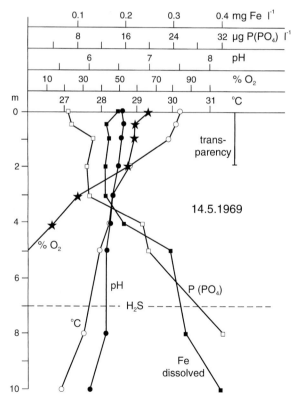

Fig. 4.6. Transparency and the vertical variation of temperature, pH, and the concentration of dissolved oxygen, phosphate and iron in Lago Castanho on 14.5. 1969 (samples taken at 13:00 h). (Schmidt 1973a)

lences (Melack 1984; McIntyre and Melack 1988). Local currents resulting from the inflow of river water or affluents from the terra firme can modify the stratification pattern (Alves 1993).

In medium sized várzea lakes (1–10 km² lake area) a pronounced chemical stratification is established, when the water is deeper than about 4–6 m. Lago Castanho showed a rather stable chemical stratification from March to September when water depth was greater than 5 m. During this period, deeper water layers became anoxic, H₂S appeared, and concentrations of

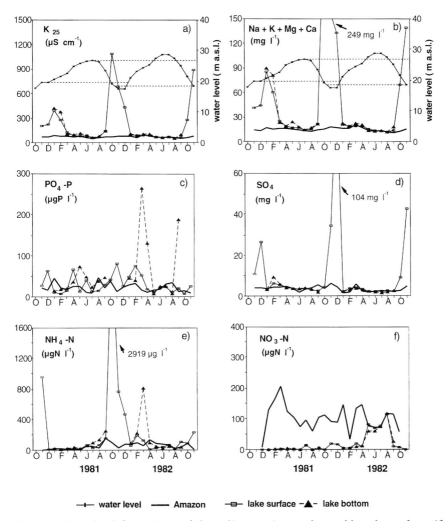

Fig. 4.7. Water-level fluctuations of the Solimões River and monthly values of specific conductance at 25 °C (K_{25}), major cations (*sum of Na, K, Ca, and Mg*), and the content of dissolved nutrients (NH_4, SO_4, PO_4, and NO_3) in the main channel and in Lago Camaleão (surface and bottom water) from November 1980 to November 1982. (Furch and Junk 1993)

free carbon dioxide, dissolved phosphate, and iron increased (Fig. 4.6, Schmidt 1973a). A temperature-related distinct epi-, meta- and hypolimnion were usually found during daytime only. Nighttime and often also daytime clinograde temperature curves were not found. Holomixis occurs only during periods of unusually cool and windy weather, e.g., when cold fronts from the south break into the Amazon basin, locally called "friagem".

When the water level falls below a depth of 4 m, frequent holomixis and homothermy resulted in the resuspension of the upper sediment layer and a decrease of transparency from about 2 m to 10 cm. During that time, an increase in total iron from $0.25\,\mathrm{mg\,l^{-1}}$ to $2.89\,\mathrm{mg\,l^{-1}}$ and total phosphorus from $5\,\mu\mathrm{g\,l^{-1}}$ to $154\,\mu\mathrm{g\,l^{-1}}$ in the surface water layer was observed (Schmidt 1973a). In Lago Calado daily mixing from top to bottom occurred when water depth was lower than 3–4 m. When water depth exceeded 5 m, a thermocline developed at a depth of about 3 m and an anoxic water layer was established (Melack and Fisher 1983).

In Lago Camaleão no stable thermocline developed (Rodrigues 1994; Kern 1995). An increase in solutes near the bottom was observed during short periods only, probably because the influx of river water inhibited the establishment of a stable chemical stratification. During very high water levels differences between lake and river chemistry began to disappear because of increasing replacement of lake water by river water when the river passed over the levees (Fig. 4.7).

4.4.3 The Oxygen Concentration in Várzea Lakes

Due to currents and turbulence, the water in the Amazon itself is well supplied with oxygen. The concentrations in the surface water vary from 4.0 to $5.5\,\mathrm{mg\,l^{-1}}$, corresponding to 53–73% saturation. The biochemical oxygen demand for one day (BOD_1) is low and reaches 0.1 to $0.3\,\mathrm{mg\,l^{-1}}$. It increases to $0.5\,\mathrm{mg\,l^{-1}}$ when the river level recedes (Fig. 4.8).

For all studied várzea lakes pronounced oxygen deficiency has been reported. In Lago Calado long-term anoxia occurred at 3–5 m water depth (Melack and Fisher 1983; McIntyre and Melack 1984). During calm weather, a pronounced stratification resulted in a diel change of oxygen concentration at 2 m water depth from 0 to $4\,\mathrm{mg\,l^{-1}}$. Oxygen input by phytoplankton was modest because of low levels of primary production. Air–water exchanges, vertical mixing, and perhaps advection were considered as important factors controlling the oxygen concentration in the lake. Air–water exchanges were low except during rainstorms, but resulted in a net input of about $0.1\,\mathrm{g\,O_2\,m^{-2}\,h^{-1}}$ to the lake because of undersaturation of surface water with oxygen (Melack and Fisher 1983).

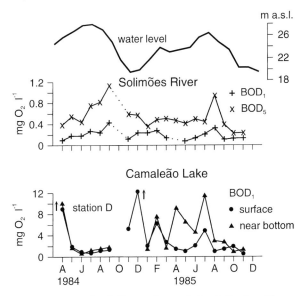

Fig. 4.8. Water-level fluctuations of the Solimões River and seasonal variations of the biochemical oxygen demand after one and five days in river water and after one day in water of Lago Camaleão (surface and bottom). *Arrows* indicate total consumption of the available oxygen during exposure and underestimation of the demand. (Junk 1990)

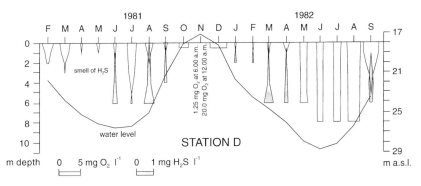

Fig. 4.9. Water-level fluctuations and seasonal variation of the oxygen and hydrogen sulfide content in Lago Camaleão (*station D*). Until May 1981 the presence of H_2S was noted by smell only. (Junk et al. 1983)

Extremely low oxygen concentrations were recorded by Junk et al. (1983) from Lago Camaleão, where anoxia and the presence of H_2S were observed periodically at a depth of 1–2 m (Fig. 4.9). The biochemical oxygen demand for 1 day (BOD_1) was high throughout the year and reached a maximum value of more than $12\,mg\,O_2\,l^{-1}$ at rising water level. During the very high flood of 1984, BOD_1 decreased at high water to $1–2\,mg\,O_2\,l^{-1}$, because the river water with low BOD_1 values surpassed the levees and replaced the lake water (Fig. 4.8). Generally, the BOD_1 values of water samples taken near the surface were lower than those of samples taken near the bottom, indicating a greater concentration of labile organic material near the lake bottom.

The oxygen content of lake water inside the large floating aquatic macrophyte communities depends on the structure of the floating layer (Fig. 4.10). In the upper few centimeters of dense plant masses high oxygen concentrations are observed during the day because of the photosynthetic

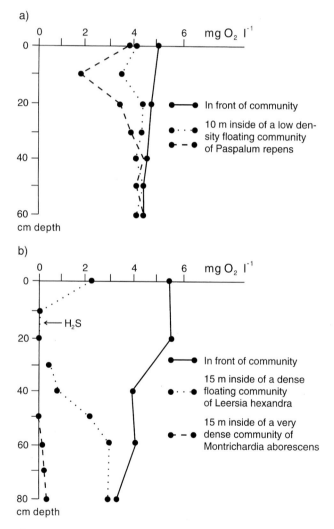

Fig. 4.10. Oxygen content inside floating vegetation mats of different density in Lago Manacapurú (**a**) and Lago Parú (**b**). (Junk 1983a)

activity of periphyton. Beneath this layer, there is a zone of distinct oxygen depletion and hydrogen sulfide formation because dense root masses and stems of the macrophytes hinder the diffusion of oxygen from the surface. Oxygen production by algae is hindered because of reduced light penetration. Exudation of oxygen through the root system has been shown for *Eichhornia crassipes* and *Pistia stratiotes* (Moorhead and Reddy 1988; Jedicke et al. 1989) but the amounts are not large enough to meet the high oxygen demand of the decomposing detritus in the floating layer. Below the floating layer the oxygen content can rise again due to horizontal water exchange with areas not covered by plants.

The strong depletion of oxygen in várzea lakes can be explained by the large amounts of organic material produced by terrestrial and aquatic herbaceous plants and the floodplain forest. Considering a mean diffusion rate of $0.1\,g\,O_2\,m^{-2}\,h^{-1}$ (Melack and Fisher 1983), a mean oxygen concentration of $5\,mg\,l^{-1}$ in the inflowing river water, a water depth of 6 m and a flood period of 240 days, availability of oxygen corresponds to $30\,g\,m^{-2}$ from the Amazon and $576\,g\,m^{-2}$ from diffusion. Considering a net primary production of phytoplankton of $1.6\,g\,m^{-2}\,day^{-1}$ (Schmidt 1973b), of which 90% is completely decomposed, contribution of phytoplankton to the oxygen content of the lake reaches $41\,g\,m^{-2}$.

The total amount of $647\,g\,O_2\,m^{-2}$ introduced into the lake during the flood period by river water, phytoplankton production, and diffusion from the air is sufficient to oxidize 605 g organic material. The biomass produced by annual herbaceous plants during the terrestrial phase in the ATTZ varies between 320 and $720\,g\,m^{-2}$ dry weight, of which more than 90% is decomposed during the flood period (Howard-Williams and Junk 1976). Perennial plants (*Paspalum fasciculatum*) produce $1560\text{–}5760\,g\,m^{-2}$ dry material (Junk and Piedade 1993c), of which up to 50% is decomposed during the flood period. Leaf-litter production of the várzea forest reaches $780\text{–}1360\,g\,m^{-2}\,year^{-1}$ dry weight (Sect. 11.4) and production of woody litter $600\,g\,m^{-2}\,year^{-1}$ (Martius 1989), of which a considerable part is decomposed during the flood period (Chaps. 9, 11, 12). Hence, the oxygen demand for decomposition of the organic material produced during the terrestrial phase in a várzea lake with a large ATTZ is theoretically large enough to consume in the aquatic phase the total amount of available oxygen (Sect. 1.4). It also shows that anaerobic decomposition processes play a major role in várzea lakes (Chap. 7 and Sect. 9.2.3).

We postulate that oxygen conditions are improved in floodplain lakes with a small ATTZ in comparison with the lake basin area (LBA), with a low biomass production in the ATTZ during the terrestrial phase and with small amounts of emergent aquatic macrophytes growing during the aquatic phase. In lakes of the Negro River floodplain, oxygen depletion is less pronounced than in comparable várzea lakes (Fig. 20.4 in Chap. 20) because the low nutrient status of water and sediments results in a low production of organic material by herbaceous plants and the floodplain forest. According to Geisler (1969) oxygen concentrations in the Negro River varied between 43% saturation at high water and 83% at low water. Highest BOD_2 values reached $0.9\,mg\,l^{-1}$, highest BOD_{20} values $1.9\,mg\,l^{-1}$.

4.4.4 Processes Within Lakes

Hydrochemical conditions in floodplain lakes are influenced by internal processes, e.g., the uptake and release of nutrients by primary producers and consumers and the exchange between water and sediments. Parameters related to photosynthesis by phytoplankton, e.g., pH, oxygen, and free carbon dioxide, show greater diurnal fluctuations than seasonal fluctuations in the surface water layer (Schmidt 1973a,b). Comparative nutrient enrichment experiments performed in an igapó lake and a várzea lake indicated phosphorus limitation in black water and nitrogen limitation in white water (Forsberg 1984). Experiments during different river stages in Lago Calado (summarized in Fisher et al. 1991) point to phosphorus limitation during rising- and high-water periods and nitrogen limitation during falling- and low-water levels for phytoplankton growth (Setaro 1983; Setaro and Melack 1984; Pinheiro 1985). Approximately 60–90% of the uptake of phosphate in surface water of Lago Calado occurred in the 0.2– 3.0 μm fraction, which indicates the importance of very small organisms in the phosphorus cycling in Lago Calado. Only 20% of the regenerated N and 5% of the regenerated P were required by phytoplankton. Since there is no increase of available N and P it is assumed that the majority of these nutrients are consumed by heterotrophic organisms, indicating a high supply of organic material derived from other primary producers. Changes in the nitrogen cycle of the epilimnion of Lago Calado were related to changes in mixing pattern (Morrissey and Fisher 1988). Compared to small epilimnetic heterotrophs, e.g., rotifers, protozoans, fungi, and bacteria, regeneration by the macrozooplankton of Lago Calado was small, because macrozooplankton density was low (Lenz et al. 1986).

Studies on fluxes of oxygen, ammonium, nitrate, phosphate, iron, and silicate across the sediment–water interface show that at low water in the presence of oxygen, fluxes were not as pronounced as they are under anoxic conditions at high water (Smith-Morrill 1987). Studies of Smith and Fisher (1985), Figueiredo (1986), and Smith-Morrill (1987) suggest sequestering of phosphate by the sediments; studies of Schmidt (1973a) and Setaro and Melack (1984) indicate release of phosphates during resuspension of the sediments at low water. Nitrogen fluxes are discussed in Chapter 6, and methane production in Chapter 7.

4.4.5 Land–Water Interactions and the Bioelement Cycles

The large biomass produced in the ATTZ by herbaceous plants and the floodplain forest raises the question of its impact on the nutrient pool of várzea lakes. Howard-Williams and Junk (1977) pointed out that aquatic macrophytes store per unit area much more nitrogen, phosphorus, and potassium than is introduced by the dissolved load of the river water as long as there is no throughflow. Calculations of Piedade et al. (1992) and Furch and Junk (1992) using data of Lago Camaleão showed similar results. In Lago Camaleão during rising water level aquatic macrophytes cover about 50% of the lake surface. Even so, nutrient content in the water was not depleted in comparison with the river water except for nitrate. Additional nutrient sources are required to explain the observed plant growth.

One important nutrient source for the lake is the sediment that is trans- ported by the floods into the lake basin. Experiments of Engle and Sarnelle (1990) indicate that particulate matter of the Amazon contains 16–38% of the algal-available P in river water. However, considering the negative impact of increased sediment load on phytoplankton growth by a reduc- tion of the eupthotic zone, the importance of algae for uptake and recycling of the sediment-bound P-fraction is supposed to be small. A major path- way for sediment-bound nutrients into the lake water is via higher plants growing in the ATTZ. The vegetation which grows during the terrestrial phase in the ATTZ takes up nutrients from the sediment. During decompo- sition during the aquatic phase, herbaceous plants, leaf litter and woody litter of the floodplain forest release large portions of its nutrient stock into the water (Chap. 9). Total amounts of N, P, and K stored per unit area in the plants and also the amounts potentially released from the decomposing plant material into the water during a 4-month period are several times larger than those introduced per unit area in dissolved form by the river (Furch and Junk 1992; Junk and Weber in press; Table 4.8a,b). In contrast, river water contributes more Na, Ca, and Mg (Table 4.8a,b). The impact of the ATTZ on the nutrient budget of a lake depends on the ATTZ/LBA ratio, the amount of organic material produced per unit area during the terres- trial phase in the ATTZ, and the nutrient content of the produced material. The impact is largest when the ATTZ/LBA ratio is one, that means when the lake dries completely at low water.

Studies of Furch et al. (1983), Furch (1984a,b) and Furch and Junk (1993) on Lago Camaleão, a lake with a ATTZ/LBA ratio of nearly one, show that

Table 4.8a. Chemical composition of various bioelement sources in the várzea. The amounts of dissolved bioelements released to the water from 1 kg dry plant material correspond to a decomposition period of 4 months

Bioelements	Na	K	Ca	Mg	N	P
Amazon water[a] (mg kg^{-1})	2.99	0.87	9.72	1.35	0.16*	0.03**
Susp. sediments[b] (g kg^{-1})	8–10	19–22	9–10	9–10	1–3	0.7–0.8
Fresh leaf litter[a] (g kg^{-1})	0.18	4.33	18.58	3.02	13.50	1.02
Released (g kg^{-1})	0.12	3.41	10.65	1.93	3.73*	0.62**
Fresh wood[c] (g kg^{-1})	0.11	6.59	3.25	1.45	4.75	0.82
Released (g kg^{-1})	0.06	4.00	0.73	0.68	0.85*	0.30**
Fresh herbs[d] (g kg^{-1})	0.07	14.76	3.47	1.61	12.60	1.69
Released (g kg^{-1})	0.06	12.48	3.25	1.23	4.04*	1.10**

*, dissolved inorganic nitrogen (sum of NH_4-N, NO_3-N, NO_2-N); **, PO_4-P.
[a] Furch et al. (1989).
[b] Furch (Chap. 3).
[c] Furch and Klinge (1989 and unpubl. data).
[d] Furch and Junk (1992).

Table 4.8b. Amounts of bioelements stored per square meter (g m^{-2}) in a hypothetical várzea lake

Bioelements	Na	K	Ca	Mg	N	P
Amazon water (g 4 m^{-3})	11.96	3.48	38.88	5.40	0.62*	0.104**
Sediment layer (g m^{-2})	36–45	86–98	43–46	41–46	4–15	3–4
Fresh leaf litter (g m^{-2})	0.10	2.29	9.85	1.60	7.16	0.54
Released (g m^{-2})	0.06	1.81	5.64	1.02	1.98*	0.33**
Fresh wood (g m^{-2})	0.07	3.95	1.95	0.87	2.85	0.49
Released (g m^{-2})	0.04	2.40	0.44	0.41	0.51*	0.18**
Annual herbs (g m^{-2})	0.04	8.86	2.08	0.97	7.56	1.01
Released (g m^{-2})	0.04	7.49	1.95	0.74	2.42*	0.61**
Perenn. herbs (g m^{-2})	0.21	22.24	5.20	2.42	7.56	2.54
Released (g m^{-2})	0.18	18.72	4.88	1.85	6.06*	1.52**

The following assumptions were used for calculation: 4 m water depth, 4.5 kg m^{-2} sediment deposition, corresponding to an annual layer increment of 3 mm. The amounts of dissolved bioelements released to the water per square meter from dry plant material correspond to a decomposition period of 4 months.
Leaf litter, 530 g m^{-2} corresponding to 50% of the annual leaf fall; coarse woody litter, 600 g m^{-2}; annual terrestrial herbaceous plants, 600 g m^{-2}; perennial terrestrial herbaceous plants, 1500 g m^{-2}.
For further explanations, see Table 4.8a.

at low water values of specific conductance and all ions except PO_4 and NO_3 were ten to twenty times higher than in the Amazon River (Fig. 4.7). Maximum concentrations at low water are explained in part by the release of solutes from decomposing plant material, swept into the remaining shallow pool by surface runoff, and by the inflow of seepage water. The decrease in concentrations during rising water periods is explained by dilution with river water and by nutrient uptake of aquatic macrophytes. The nutrient model presented in Chapter 5 shows that the release of nutrients from the plants might be sufficiently large to explain the peak in potassium at low water in Lago Camaleão, but not the peaks observed in Na, Ca, and Mg. Other sources (e.g., inflow of groundwater or seepage water) are required to explain the high concentrations of these elements at low water in the lake.

4.5 Discussion and Conclusions

Physicochemical conditions in várzea lakes are strongly influenced by the inflow of water with different concentrations of dissolved and suspended solids from the parent river, affluents from the terra firme, precipitation, and groundwater. Due to the large water level fluctuations of the parent river, the flood pulse (Sect. 1.3) controls the impact of the other water sources.

Low concentrations in dissolved and suspended solids are found in rainwater, terra firme affluents, groundwater from the terra firme, and the Negro River. High concentrations occur in Amazon River water and the highest concentrations are found in groundwater of the várzea, which is not affected by terra firme affluents. Amazon water and várzea groundwater are typical carbonate waters containing calcium and hydrogen carbonate as dominant ions. These ions behave conservatively and were used as hydrochemical tracers to describe the mixing patterns in várzea lakes influenced by local water inputs other than river water (Furch 1982; Forsberg et al. 1988). The variable alkalinity levels in mixed várzea lakes are related to the ratio between the size of the catchment area and the size of the lakes (Forsberg et al. 1988). However, internal processes in lakes can modify the concentrations of major ions indicating that tracers do not behave as conservatively as expected. Stream water and rain water can be comparatively rich in nitrogen and they can therefore be important sources of nitrogen for floodplain lakes. This input can strongly influence the food webs because nitrogen is considered to be a limiting factor for phytoplankton production in várzea lakes (Forsberg 1984; Setaro and Melack 1984). The impact of várzea groundwater and seepage water on lake chemistry is important but has not been quantified for all types of floodplain lakes.

Size and fetch of the lakes strongly influence hydrochemistry because they influence wind-induced turbulence. Strong turbulence results in an increasing oxygen input by diffusion, transport of oxygen into deeper water layers, and transport of nutrients into the euphotic layer by deeper mixing of the water body (Melack 1984). Strong waves also reduce the growth of aquatic macrophytes and favour phytoplankton development.

A major factor determining the physicochemical conditions in floodplain lakes is the ATTZ/LBA ratio. Lakes with a large ATTZ/LBA ratio are subject to large changes in environmental conditions and show a pronounced annual cycle according to the flood pulse. With exposure of the lake bottom at low water, nutrients bound to the sediment become available for plants growing during the terrestrial phase and are subsequently released into the water when the plants decompose during the aquatic phase (Howard-Williams and Junk 1977; Furch and Junk 1992; Piedade et al. 1992). This additional input of large amounts of nutrients into the várzea lakes during the aquatic phase enhances productivity (Junk et al. 1989; Furch and Junk 1993). The amount of organic material can be sufficient to make the lakes strongly hypoxic.

We postulate that lakes with a small ATTZ/LBA ratio behave similarly to classical lakes. During isolation from the parent river, plant nutrients in the euphotic layer are quickly depleted. Nutrients are accumulated in the hypolimnion and buried in the sediment. Major changes in the physicochemical conditions occur only when the river invades the lakes at high water.

Nutrient export from the floodplain during falling water is counteracted by nutrient sequestering aquatic macrophytes. The decomposition of these aquatic macrophytes during the terrestrial phase helps to fill up the nutrient reservoir in the sediment. In habitats dominated by highly productive aquatic and semiaquatic species e.g., *Echinochloa polystachya*, sequestering and storage of dissolved nutrients by the plants during rising and high water can be greater than nutrient loss from the system during falling water. We postulate that nutrient import and export of várzea lakes varies considerably with annual differences in the flood amplitude. Because of the nutrient transfer between the terrestrial and the aquatic phase the várzea system functions on a high nutrient level. Major changes in the flood pulse will modify the nutrient status of the lakes because they change the ATTZ/LBA ratio. In the igapó, the nutrient status in water and soil is low. Sequestering of nutrients from the water is reduced because of the absence of aquatic macrophytes.

Acknowledgments. We are grateful to Dr. John Melack, University of California, for critical comments on the manuscript and correction of the English and to S. Bartel, E. Bustorf, N. Dockal and U. Thein for technical assistance.

5 Modelling Nutrient Fluxes in Floodplain Lakes

GERHARD E. WEBER

5.1 Introduction

Only a few years ago Mitsch and Fennessy (1991) stated that modelling nutrient exchanges in wetlands was "now" developing. However, their statement applies to temperate wetlands rather than tropical floodplain lakes, for which still very few modelling studies exist. Their hydrological and hydrochemical complexity, often ill-defined boundaries, as well as complex and little understood interface processes (e.g. sediment–water exchange of chemicals, free water solutes and flooded terrestrial and aquatic/semiaquatic vegetation) distinguish them from lake ecosystems that have been subject to modelling efforts for several decades. Due to these distinctions, the question remains open whether mathematical approaches derived from "established" lake models are appropriate for wetlands as well (Mitsch and Fennessy 1991). For these "established" models, Reckhow (1994) concluded that "limited observational data and limited scientific knowledge are often incompatible with the highly detailed model structures" of water quality simulation models, and thus shed further doubt on their potential usefulness for floodplain ecosystems. Provided that data and knowledge limitations are overcome, multidimensional hydrodynamic models (Mertes 1994), and approaches including Geographic Information Systems (Costanza and Maxwell 1991) might become appropriate for future modelling of nutrient fluxes in floodplain lakes. Given the present state of knowledge, we consider highly aggregated point models, rather than complex spatial simulation models, as useful tools to identify causalities and improve our understanding of nutrient fluxes in floodplain lakes. However, the validity of describing floodplain lakes with spatially zero-dimensional models needs to be established for every case study, and may limit their general application.

5.2 Conceptual Considerations

In the Central Amazon the annual flood pulse as the principal driving force for its floodplain ecosystems (Junk et al. 1989) shows a mean amplitude of

Ecological Studies, Vol. 126
Junk (ed) The Central Amazon Floodplain
© Springer-Verlag Berlin Heidelberg 1997

10 m (Chap. 2). Thus, the water budget and dissolved and suspended loads of its constituents set the stage for the floodplain's nutrient dynamics. Although this stage is modified by other factors to a varying degree, an understanding of nutrient fluxes in Amazon floodplains primarily requires a sound knowledge of their hydrology. Whereas this key role of hydrology in wetland ecosystem research is widely acknowledged, LaBaugh (1986) concluded that it remains one of the least understood components of wetland ecosystems. Recently, Lesack and Melack (1995) have concluded from a detailed hydrological survey that the general paradigm attributing a dominant role to flooding from the river channel does not hold for Lake Calado, a floodplain lake considered representative for Ria lakes, an important class of lakes in the Central Amazon, where local runoff was the dominant source of water during later stages of rising water. Hence, the notion of a flood pulse largely dominated by river water requires evaluation for each case study, and mixing patterns differing from site to site add further variability to floodplain nutrient dynamics (Chap. 4).

Our understanding of nutrient fluxes in a floodplain ecosystem is conceptualized in Fig. 5.1, which shows relevant factors and processes grouped around the lake water body. Depending on local hydrology, solutes and suspensoids are advected between floodplain and river. Other fluxes directly linked to hydrology depend on precipitation, runoff, and groundwater movements. Further abiotic fluxes originate from exchange processes between the water column and sediment. Whereas these abiotic fluxes are common to river floodplains in general, the biotic fluxes depicted in Fig. 5.1 are based on the assumption that higher vegetation largely dominates, and phytoplanktonic, or periphytic nutrient fluxes, are comparatively negligible except for nitrogen fluxes. Figure 5.1 also highlights the most characteristic property of a floodplain ecosystem, the transition between emerged and submerged, or terrestrial and aquatic phases. This transition is dependent on the geographical location in the floodplain and the hydrograph, and triggers a functional switch of its major biota. For example, for terrestrial macrophytes inundation terminates growth and initiates biomass decomposition as an additional source of nutrients to the water column. With rising water, rooted and free floating forms of aquatic macrophytes grow abundantly, the former deriving an increasing fraction of their total nutrient uptake from the water column, thus both forming a nutrient sink for the water body. Subsequently, re-emergence is followed by decomposition of aquatic macrophytes, and growth of terrestrial vegetation (Sects. 1.3 and 4.4.5, Chaps. 8 and 9).

For all of these factors or processes, the extent to which they contribute to nutrient dynamics depends on the local conditions of the study site and therefore shows large variation. Considering bicarbonate as a conservative

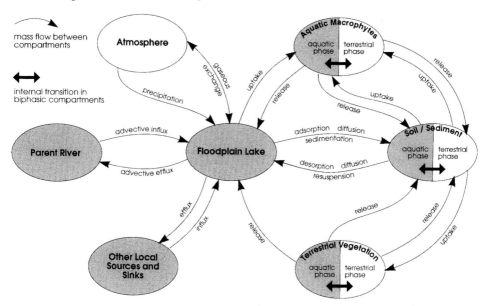

Fig. 5.1. Conceptual model of nutrient fluxes in a floodplain ecosystem in the Central Amazon

tracer, Forsberg et al. (1988) attempted a qualitative prediction of floodplain lake water quality at low water based on the ratio of catchment area to water surface area. A lake with a ratio of more than 20 would be expected to contain primarily local water, whereas lakes with a ratio of less than 20 would hold a mixture of local and river water. However, Lesack and Melack (1995) showed that with a ratio of 7, Lake Calado was dominated by local waters at low stage, and asked for a re-evaluation of the findings of Forsberg et al. (1988). This shows that care should be taken with generalizations on hydrochemistry and nutrient fluxes in floodplain lakes. Due to the large variation between sites, results from, or a model of, a particular study site hold limited potential for inductive conclusions on general principles of floodplain lake functioning. The challenge of identifying possibly a few key factors and their causal relations to ecosystem properties calls for comparative studies of floodplain lakes. However, research as yet remains mostly focused on studies of particular single sites.

For a lake on the Orinoco River floodplain, Hamilton and Lewis (1987) used a mass balance model to study causes of hydrochemical seasonality. Based on the water budget, including exchanges with the parent river and an adjacent lake, precipitation onto and evaporation from the lake surface, hydrochemical quality of the water fluxes, and the assumption of conservative mixture, they calculated the expected annual evolution of concentrations of solutes and suspended matter in the lake water. Hamilton and

Lewis (1987) used their model to show effects of mass transport and water balance, and thereafter identified factors like sediment-water interactions, and biotic processes which could account for the observed deviations from conservative expectation. For example, they showed that increasing potassium concentrations during falling water could be caused entirely by decomposition of aquatic macrophytes.

In the most elaborate hydrological study so far reported for an Amazon floodplain lake, Lesack (1988) used a mass balance model to evaluate retention characteristics of Lago Calado, a dendritic floodplain lake near Manacapurú on the Solimões River. All components of the lake's water budget were measured except exchange with the parent river, which was calculated as its residual (Lesack and Melack 1995). Predictions for nutrient stocks in the lake water were in good accordance with the observations until the lake stopped receiving water from the Solimões River. However, subsequently, observed solute masses exceeded the predictions, and Lesack (1988) concluded that either an internal source of solutes within the lake or one or more inputs were underestimated. As one of the potential sources of solutes he considered leaching from vegetation in the flooded forest and from macrophyte stands. In both of these studies (Hamilton and Lewis 1987; Lesack 1988), modelled nutrient fluxes were exclusively based on the water balance, and biotic nutrient fluxes were not included in the models.

5.3 Modelling Cation Seasonality in Lago Camaleão

The following modelling approach was applied by Weber et al. (1996) to annual cation seasonality in Lake Camaleão, a channel lake located on Ilha de Marchantaria 15 km upriver of the confluence of the Negro and Solimões Rivers (Sect. 2.2). Their purpose was to reveal causalities underlying hydrochemical seasonality and particularly to study the role of the macrophyte vegetation. If we consider a solute subject to advective influxes in_{adv} and effluxes out_{adv} only, the rate of change in storage M is given by the mass balance equation

$$M' = in_{adv} - out_{adv},\qquad\qquad(1)$$

where M' denotes the temporal derivative dM/dt. If in- and efflux rates are calculated from the water balance and the solute concentrations of its constituents, and initial conditions are given, the mass balance equation can be integrated numerically to calculate the expected evolution of mass, or concentration of a conservative solute over a period of interest. If we additionally consider a biogenic component f_{bio} denominating rates of biogenic in-/effluxes, the mass balance equation becomes:

$$M' = in_{adv} - out_{adv} + f_{bio}. \tag{2}$$

We assume that biogenic solute fluxes originate predominantly from macrophyte vegetation, and/or the inundation forest (Sect. 4.4.5). This implies that we are not targeting nitrogen fluxes, the modelling of which would have to include biotic nitrogen fixation, as well as denitrification (Chap. 6).

We distinguish the two functional groups aquatic macrophytes and terrestrial vegetation (Fig. 5.1). According to local conditions either group can be further differentiated. For terrestrial vegetation in Lago Camaleão, Weber et al. (1996) discriminated between the subgroups terrestrial annuals, semiaquatic perennials and the inundation forest.

Although for notational brevity an index to distinguish between the functional groups is omitted, it should be kept in mind that the following biological state, rate, and auxiliary variables are group specific. Each biological component is described by the biphasic state variables A, the area covered, and N, the nutrient stock in the biomass. We assume time-invariant distinct elevational ranges for the biological components and project their spatial distribution on the stage surface relation S of the floodplain lake. This is an oversimplification for any real floodplain lake. However, the approach is open to refinement, and where more detailed data on spatial distribution of the individual groups are available, their respective hypsographic relations can be substituted. Except for the floodplain forest, the assumption of a time-invariant spatial distribution is possibly more critical (Sect. 8.2) but was nevertheless considered a valid first approximation for single hydroperiods in Lago Camaleão. Let l denote the lower, and u the upper elevational boundary of an area A covered by a particular component. Emerged, and submerged areas, A_e, and A_s, for any of the biological components at any time are approximated by:

$$A_e = \begin{cases} S(u) - S(l) & \forall \quad G < l \\ S(u) - S(G) & \forall \quad l \leq G \leq u \\ 0 & \forall \quad G > u \end{cases} \tag{3}$$

$$A_s = S(u) - S(l) - A_e,$$

where G denotes stage height given by the hydrograph. The areal transition between emerged and submerged state is approximated by:

$$A_e' = \begin{cases} -S' & \forall \quad l \leq G \leq u \\ 0 & \forall \quad (G < l) \vee (G > u) \end{cases} \tag{4}$$

$$A_s' = -A_e'.$$

With the respective phase subscripts, mean biogenic areal nutrient densities D are given by:

$$D = \begin{cases} N/A & \forall \quad A > 0 \\ 0 & \forall \quad A = 0. \end{cases} \tag{5}$$

Both phases of the nutrient stock N in any component are subject to a growth/decomposition process h, and the phase transition p. The phase transition in any component is given by:

$$p_e = \begin{cases} A'_e \cdot D_e & \forall \quad G' \geq 0 \\ A'_e \cdot D_s & \forall \quad G' < 0 \end{cases} \tag{6}$$

$$p_s = -p_e.$$

For the terrestrial group, growth is not considered here since it is limited to the terrestrial phase with no direct impact on lake water quality. With a time horizon of a single hydroperiod, we treat the emerged nutrient stock in the terrestrial group as a standing stock input in the beginning of the hydroperiod. For the same reason we consider decomposition and the related nutrient release during the aquatic phase only. The growth/decomposition process h is then given by:

$$h_e = 0$$
$$h_s = -r_s \cdot N_s, \tag{7}$$

where r_s is the exponential release rate per unit time. The assumption of exponential release dynamics is again a simplification that may be replaced by more realistic models once sufficient empirical data are available. The nutrient flux from terrestrial vegetation into the lake water is given by:

$$f_{bio,terrestrial} = -h_s. \tag{8}$$

As opposed to the terrestrial group, aquatic macrophytes characterized by *Echinochloa polystachya* as the dominating species in Lake Camaleão (Sects. 8.2 and 8.5) are assigned to growth as well as to decomposition given by:

$$h_e = \begin{cases} g \cdot A_e \cdot C & \forall \quad G' \geq 0 \\ 0 & \forall \quad G' < 0 \end{cases}$$

$$h_s = \begin{cases} g \cdot A_s \cdot C & \forall \quad (G' \geq 0) \vee ((G' < 0) \wedge (z \geq z_{min})) \\ -r_s \cdot N_s & \forall \quad (G' < 0) \wedge (z < z_{min}), \end{cases} \tag{9}$$

where g denotes a constant growth rate of biomass per unit area and time, C gives the nutrient content per unit biomass, z gives mean water depth, and z_{min} a water depth threshold below which aquatic decomposition, and thus nutrient release, take place. Corresponding to the terrestrial group, decomposition during the emerged phase is not considered. The biogenic nutrient flux between lake water and aquatic macrophytes is then given by:

$$f_{bio,\,aquatic} = \begin{cases} -h_s \cdot w(z) & \forall \quad h_s \geq 0 \\ -h_s & \forall \quad h_s < 0, \end{cases} \tag{10}$$

where $w(z)$ gives the water-derived fraction of nutrient uptake as a function of water depth, as described in Weber et al. (1996).

With the respective phase subscripts changes in the biological nutrient stocks are given by:

$$N' = h + p, \tag{11}$$

with growth/decomposition process h, and phase transition p as described in Eqs. (6), (7), and (9). The initial conditions for the biogenic nutrient stocks are: $N_e = A_e \cdot i \cdot C$, and $N_s = 0$, where i denotes initial areal biomass density and C gives nutrient contents per unit biomass. Equations (11) and (2) can be integrated numerically to calculate the expected evolution of a solute over a period of interest.

In their study of Lake Camaleão, Weber et al. (1996) simulated cation concentrations for the annual hydroperiod based on Eq. (1) as a neutral approach (sensu Caswell 1976) with the water budget consisting solely of exchanges with the parent river and precipitation onto and evaporation from the lake surface. We refer to Eq. (1) as the "conservative" model, and the respective simulation gives the "conservative expectation". With Eqs. (2) to (11) the contribution of macrophytes and leaf litter of the inundation forest to the observed cation seasonality was evaluated. This approach is referred to as the "non-conservative" model. Figure 5.2 depicts results for calcium and potassium for one of the two monitored hydroperiods. Except for approximately two months during the falling-water period, when observed concentrations increased by an order of magnitude, predictions of the conservative model generally agreed with the observed seasonal patterns for calcium (Fig. 5.2B) as well as for magnesium, and sodium, which were closely correlated with calcium. Biogenic calcium fluxes did not account for the observed increase during falling water, and predictions of the non-conservative model differed only marginally from those of the conservative model (Fig. 5.2B). For potassium, the observed seasonality differed from conservative expectation during most of the hydroperiod, and the deviations were generally well accounted for by the biogenic fluxes of the non-conservative model (Fig. 5.2C). Compared to the lake inventory at high water, the submerged biogenic stock of calcium was smaller by an

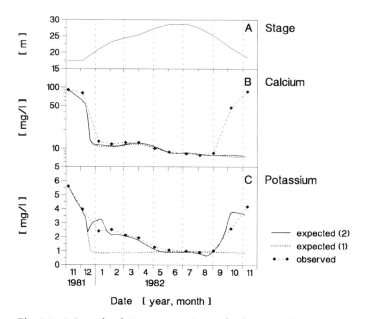

Fig. 5.2. A Stage level; **B** concentrations of calcium; and **C** potassium in Lago Camaleão during the 1982 hydroperiod. *Expected (1)*, simulated with conservative model – no biogenic fluxes; *expected (2)*, simulated with non-conservative model – including biogenic fluxes: (Weber et al. 1996; through-flow submodel: "low"; precipitation: "data set 1")

order of magnitude whereas the biogenic potassium stock exceeded the lake inventory by a factor of four. Weber et al. (1996) concluded that the considered biogenic cation fluxes are insignificant for the seasonal evolution of sodium, calcium, and magnesium, whereas they could comprise a major share of internal potassium loading in Lago Camaleão. Although the key role attributed to biogenic potassium fluxes during the rising-water period was unequivocal, their contribution to potassium loading during the falling-water period remained questionable. The latter was due firstly to the unidentified non-potassium source, which might also contribute to potassium loading, and secondly to limited knowledge of decomposition dynamics of aquatic macrophyte stands.

5.4 Discussion

Weber et al. (1996) showed that the observed increase in concentrations of non-potassium cations during the falling-water period, could not possibly be caused by a biotic factor. Alternatively, they suggested that groundwater seepage during the rapidly falling stage level might account for the observed cation loading. The authors had neglected groundwater fluxes due to a lack of data. However, local groundwater was found to show very high cation concentrations (Sect. 4.3.5) and the rapid change in stage level during the falling-water period might cause relatively high seepage rates from the surrounding groundwater system (Lesack 1995; Weber 1996). These speculations take us back to the key role of hydrology, thus highlighting the need to complement both data base and model by groundwater fluxes. Nevertheless, the presented approach to modelling of major cation fluxes proved a useful analytical tool to identify causalities and enabled a first quantitative evaluation of the role of higher vegetation for hydrochemical seasonality and the underlying nutrient fluxes in the studied floodplain lake.

Weber et al. (1996) did not include nitrogen in their modelling study since it shows a much more complex and less regular seasonal evolution (Chaps. 4 and 6). Due to their regular and simple seasonality, major cations qualified as tracers for detection of basic causal processes. While these causalities apply to nitrogen as well, they are complemented by additional factors, for example N_2-fixation and denitrification as important nitrogen-pathways in the floodplain (Chap. 6). Hence, for major cations, the presented approach could serve as a framework that is easily tailored to local conditions at different sites and which could thus facilitate comparative studies. However, Reckow's (1994) criticism concerning limited data and scientific knowledge for lake water quality models also holds for "minimal" models of floodplain lakes. Although a modelling approach is readily available, considerable effort, with emphasis on floodplain hydrology, will have to be made to establish the data base for comparative studies.

6 Nitrogen Turnover in the Várzea

Jürgen Kern and Assad Darwich

6.1 Introduction

Nitrogen is one of the most important macronutrients for all organisms. It reaches the floodplain in the dissolved and particulate load of the Amazon River during floods and from the atmosphere as dry and wet deposition and via nitrogen fixation. Anthropogenic sources are negligible in most parts of the várzea. From a few studies on nitrogen fluxes in Amazonian environments there is some evidence that nitrogen can be a limiting factor for primary production in the várzea, at least in the aquatic phase (Forsberg 1984; Setaro and Melack 1984; Furch and Junk 1993). For more insight into the pathways of nitrogen input and output, nitrogen fluxes were measured in various ecotopes at Lago Camaleão. This lake belongs to the class of várzea lakes that are not hydrochemically affected by the terra firme, in contrast to Lago Calado, which is a dendritic várzea lake influenced by the runoff from the terra firme. Both lakes have been examined intensively during the last two decades for the nitrogen exchange between river and lake, the input by precipitation and nitrogen fixation, and the output by denitrification (Furch 1982, 1984b; Melack and Fisher 1988; Junk and Piedade 1993c; Lesack 1993; Lesack and Melack 1995; Kern et al. 1996). The study at Lago Camaleão was focused on the gaseous nitrogen exchange between atmosphere, hydrosphere, and pedosphere, and provided new results for nitrogen fixation and denitrification in the Amazon floodplain.

6.2 Methods

For each 2- to 3-week interval the nitrogen flux at Lago Camaleão is defined as:

$$M = \int_{t_i}^{t_f} \left(P \cdot C_P + F - D \right) \cdot A + \left(Q_V \cdot C_W \right) dt \tag{1}$$

Ecological Studies, Vol. 126
Junk (ed) The Central Amazon Floodplain
© Springer-Verlag Berlin Heidelberg 1997

M   = input and output of nitrogen (t N)
t_i, t_f = initial time and finishing time of interval
P   = local measurements of daily precipitation rate $(m\,day^{-1})$
C_p = nitrogen concentration $(NH_4 + NO_3)$ in precipitation $(mg\,N\,l^{-1})$
F   = rate of nitrogen fixation $(g\,N\,m^{-2}\,day^{-1})$
D   = rate of denitrification $(g\,N\,m^{-2}\,day^{-1})$
A   = total area of the lake basin = $6.5 \times 10^6\,m^{-2}$
Q_V = in- and efflux rate of river and lake water; flux based on stage change
      of Rio Negro in Manaus $(10^6\,m^3\,day^{-1})$
C_W = nitrogen concentration in surface water of Lago Camaleão near the
      river $(mg\,N\,l^{-1})$

The variables F and D represent the average rate of N_2 fixation and denitrification within the 650-ha lake basin.

   Each variable is derived from different ecotopes at Lago Camaleão with
varying sizes (A_1-A_7) where rates were measured or assumed to be zero.
The gaseous nitrogen fluxes are calculated using:

$$F = \frac{1}{A}\sum_i F_i A_i \tag{2}$$

and

$$D = \frac{1}{A}\sum_i D_i A_i. \tag{3}$$

The areas of the ecotopes A_1-A_7 were estimated by means of aircraft pho-
tographs linked with water stage measurements (Petermann, pers. comm.).
In the aquatic habitat of Lago Camaleão gaseous nitrogen turnover was
related to an area varying between 50 and 650 ha. In the terrestrial habitat
of the lake basin, the gaseous nitrogen turnover was related to exposed
areas that did not change in size during the low-water period (Fig. 6.1).

F_1 = rate of N_2 fixation in plankton related to an area $A_1 = 0.5-6.5 \times 10^6\,m^2$.
F_2 = rate of N_2 fixation in periphyton related to an area $A_2 = 0-1.5 \times 10^6\,m^2$.
F_3 = rate of N_2 fixation in flooded sediment related to an area $A_3 = 0.5-6.5$
      $\times 10^6\,m^2$.
F_4 = rate of N_2 fixation in exposed sediment without perennial macrophytes
      related to an area $A_4 = 0-0.9 \times 10^6\,m^2$.
F_5 = rate of N_2 fixation in exposed sediment with perennial macrophytes at
      a lower elevational boundary related to an area $A_5 = 0-1.5 \times 10^6\,m^2$.
F_6 = rate of N_2 fixation in exposed sediment with perennial macrophytes at
      an upper elevational boundary related to an area $A_6 = 0-1.5 \times 10^6\,m^2$.
F_7 = rate of N_2 fixation in exposed sediment with floodplain forest related to
      an area $A_7 = 0-2.1 \times 10^6\,m^2$.

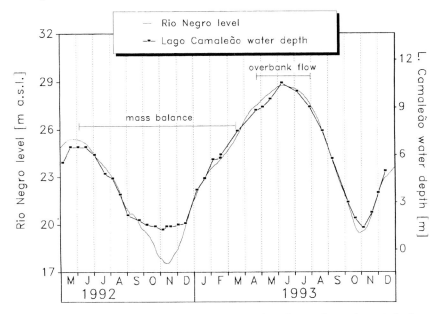

Fig. 6.1. Water stages of Rio Negro and Lago Camaleão during the study period. The mass balance ran from 2 June 1992 until 14 March 1993, when the river stage was 25.35 m a.s.l.

The rate of denitrification (D_1-D_7) was calculated using the same formulas as were used for calculating N_2 fixation.

There are still deficiencies in the nitrogen flux data of exposed sediment with macrophyte vegetation at an upper elevational boundary and exposed sediment with floodplain forest. Therefore, gaseous nitrogen fluxes in the ecotopes A_6 and A_7 are given by F_6, F_7, D_6 and $D_7 = 0\,g\,N\,m^{-2}\,day^{-1}$. The nitrogen balance ran from June 2, 1992 until March 14, 1993, when Lago Camaleão still had a distinct lake basin. At the beginning and the end of the nitrogen balance the river level was 25.35 m a.s.l. (Fig. 6.1). Surfaces and lake volumes of Lago Camaleão were extrapolated according to Weber (Chap. 5). The nitrogen balance at Lago Camaleão was not restricted to the water body but to the whole lake basin, which covers an area of 650 ha; that is the lake surface area at a high water stage (>28.4 m a.s.l.). Information about the study area is given in Chapter 2.

Processes occurring within the lake were calculated from empirical data and from measurements in 1992/1993. They do not mean a gain or a loss for the entire lake basin and have therefore not been taken into consideration for the mass balance. For a comparison, internal processes, i.e., sedimentation and diffusion, were fed into a figure that included all nitrogen fluxes measured up to 14 March 1993 (Fig. 6.4). Seepage has not been measured at Lago Camaleão as yet. However, studies on the seepage exchange at Lago

Calado have shown that the hydraulic regime and deeply weathered soils of the lake basin provides good conditions for a water exchange, ranging from 13.95 to $0.051 m^{-2} h^{-1}$, between the lake and adjacent groundwater system (Lesack 1995).

6.3 Exchange of Nitrogen Compounds Between the River and the Floodplain

Biogeochemical processes within the floodplain result in different hydrochemical properties between the floodplain lakes and the Solimões River (Chap. 4). Concentrations of NH_4^+, NO_2^-, and dissolved organic nitrogen (DON) were two to three times higher and NO_3^- concentrations were 4 times lower in Lago Camaleão than in the Solimões River on average over a one year period (Table 6.1). That means that the lake acts as a sink for NO_3^- (Furch and Junk 1993) and as a source for other dissolved nitrogen compounds due to the water flow. Considering the load, NO_3^- is the main component of dissolved nitrogen imported from the river during rising water (Sect. 4.4). During receding water, the várzea lakes export more NH_4^+ than NO_3^- to the river (Table 6.2, Fig. 6.1). The exchange of DON between the river and the várzea appears to be balanced. Lago Camaleão was a sink for nitrogen during the period of study from 2 June 1992 to 14 March 1993. The input of nitrogen was 4.0 t and the output 3.0 t (Table 6.2).

Particulate nitrogen (PN) is exchanged between Solimões River and Lago Camaleão (Fig. 6.2). Total suspended solids (TSS) in the river contain about 0.3% nitrogen (Sect. 4.3.1). The concentrations of TSS in the river

Table 6.1. Average concentrations of dissolved nitrogen compounds in rainwater, in Solimões River water and water of Lago Camaleão from June 1992 to May 1993 with their corresponding standard errors (n = 24)

	Rain	Solimões River Surface	Lago Camaleão	
			Surface	Bottom
NH_4^+ μmol/l	6.6[a]	2.2 ± 0.7	4.4 ± 1.7	18.5 ± 7.3
NO_3^- μmol/l	3.5[a]	9.2 ± 1.1	2.4 ± 0.6	2.4 ± 0.5
NO_2^- μmol/l	0.2[b]	0.1 ± 0.0	0.2 ± 0.1	0.4 ± 0.1
DON μmol/l	6.9[a]	4.7 ± 1.1	12.5 ± 2.0	10.7 ± 2.4

[a] After Lesack and Melack (1991).
[b] After Anonymous (1972a).

Table 6.2. Exchange of nitrogen compounds between the water of Solimões River and Lago Camaleão. The balance was conducted from 2 June 1992 to 14 March 1993 when the river level ranged between 17.56–25.36 m a.s.l. (Kern 1995)

	Input		Output		Σ
	(t)	(%)	(t)	(%)	(t)
NH_4^+	0.5	7.3	0.9	23.1	−0.4
NO_3^-	2.0	29.4	0.7	17.9	1.3
DON	1.5	22.1	1.4	35.9	0.1
Σ	4.0	58.8	3.0	76.9	1.0
PN	2.8[a]	41.2	0.9[b]	23.1	1.9
Total N Σ	**6.8**		**3.9**		**2.9**

[a] Calculated according to Engle and Melack (1993) because of high suspended load: µM PN = 0.044 (TSS) + 7.37.
[b] Derived from persulfate digestion according to Grasshoff (1976).

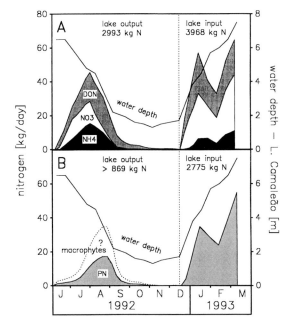

Fig. 6.2. **A** Flux of dissolved nitrogen within the water body between Solimões River and Lago Camaleão; **B** flux of particulate nitrogen within the water body between Solimões River and Lago Camaleão

water vary with the river stage and the water depth (Schmidt 1972c; Meade et al. 1979b; Devol et al. 1995). During low water and rising water, when the percentage of water of Andean origin exceeds the percentages of water of tributaries and local origin, the Solimões River has the highest TSS concentrations (Devol et al. 1995). In the surface water of the Solimões River near Marchantaria Island it ranged between 35 and 220 mg l^{-1} and had an average value of 85 mg l^{-1} over a 2-year period (Chap. 4).

During the rising water at the beginning of 1993, when the river water had a TSS concentration of about 130 mg l^{-1}, the concentration of PN was calculated according to Engle and Melack (1993) using the following equation:

$$\mu M\ PN = 0.044\ (TSS) + 7.37. \tag{4}$$

The Lago Camaleão received an input of 2.8 t PN during the period of our study. With increasing distance from the river channel the TSS concentration drops very quickly within the floodplain (Mertes 1994). With a sedimentation rate of about 3 mm year^{-1} (Sect. 2.5), 2.6 t nitrogen was stored in the lake sediment of Lago Camaleão during the period of study (Kern 1995).

During low water, sediment particles may be resuspended and exported with the receding water (Schmidt 1972c; Meade et al. 1985). In Lago Camaleão nitrogen loss promoted by resuspension of sediment was not very likely because the lake was protected from the wind and hydrologically isolated during low water. The measured output of 0.9 t PN, therefore, primarily derived not from sedimentary solids but from seston.

During and after high water stages there occurs an output of coarse organic detritus and macrophyte stands. During the study period in 1992/1993, the high water reached the very low stage of 25.42 m a.s.l. There was no overbank flow and the export of plants into the Solimões River could be neglected.

A considerable nitrogen output has to be assumed by fish migration from the lake to the river and by fishing. The mean fish biomass in the Lago Camaleão is 160 g fresh weight m^{-2} (Bayley 1983) and corresponds to a nitrogen stock of about 3 t at middle water stage. This indicates a high production and thus a high probability of nitrogen output by fishes from the lake.

6.4 Nitrogen Input by Dry and Wet Deposition

Aerosols and fine particles are transferred from the atmosphere to the pedosphere and hydrosphere by wet deposition and to a lesser extent by dry deposition (Lesack and Melack 1991).

Nitrogen in precipitation may be derived from different sources, i.e.,

Table 6.3. Comparison of the concentrations of ammonium and nitrate in the precipitation and their fluxes in the várzea and the city of Manaus

	Dry season			Wet season			N-flux
	NH_4	NO_3	N Input	NH_4	NO_3	N Input	$(NH_4 + NO_3) - N$
	(μmol/l)		(kg N/ha)	(μmol/l)		(kg N/ha)	(kg N/ha a)
Várzea							
Lesack and Melack (1991)	13.3	9.2	1.6[a]	5.2	2.2	1.2[a]	2.8[a]
Lago Calado							
(n = 123)							
Kern (1995)	17.8	11.9	2.0	1.2	1.2	0.6	2.6
Lago Camaleão							
(n = 6)							
Manaus							
Anonymous (1972a)	10.4	3.7	0.4	12.1	7.9	5.3	5.7
(n = 53)							
Andreae et al. (1990)				4.7	1.3	2.4	>2.4
Hotel Tropical							
(n = 13)							

[a] Calculated by precipitation data from 1992/1993.

lightning and biogenic and anthropogenic emissions. The concentrations of nitrogen in rainwater may be higher than those in Amazonian terra firme affluents (Junk and Furch 1985). Bakwin et al. (1990a, b) reported biogenic emission of NO from oxisol soils of the terra firme, probably due to denitrification, with high emission rates in the dry season and low emission rates in the wet season. Reactive nitrogen oxides are assumed to be transported from urban areas like Manaus or from sources outside the Amazon basin into the terra firme forest (Andreae et al. 1990; Jacob and Wofsy 1990).

Quantity and quality of the Central Amazon precipitation is subject to spatial and temporal variability (Anonymous 1972a; Ribeiro and Adis 1984; Andreae et al. 1990; Lesack and Melack 1991) (Sect. 4.3.4). Precipitation in the urban area of Manaus led to greater nitrogen input than did rainfall in the várzea (Table 6.3). In the várzea during the dry season both at Lago Calado and Lago Camaleão concentrations of NH_4^+ and NO_3^- found in the precipitation were much higher than during the wet season. They ranged up to $30 \mu mol N l^{-1}$. The total wet deposition accounted for 1.3 t nitrogen from June 1992 to March 1993 in the 650 ha large lake basin of Lago Camaleão. During the dry season (June–November) 11 t was deposited compared with 0.2 t deposited during the wet season (December–March).

The amount of organic nitrogen in the rain, which can account for a nitrogen input of about $2 kg ha^{-1} year^{-1}$ according to Lesack and Melack

(1991), was not listed and not fed into the balance. Consequently, the real nitrogen flux will be higher than 2–3 kg ha^{-1} year^{-1}.

6.5 Nitrogen Flux at the Sediment–Water Interface

In both the lake water and the aquatic sediment of Lago Camaleão, plant material undergoes continuous decomposition, which leads to a significant enrichment of NH_4^+ in the sediment interstitial water during the high water phase (Fig. 6.3). An accumulation of NH_4^+ in the sediment interstitial water is favored by a high biomass production during high water (Junk and Piedade 1993c) in combination with anaerobiosis, which prevents nitrification. Between the water column and sediment an NH_4^+ gradient is evident. It decreases in Lago Camaleão and in Lago Calado during low water (Smith-Morrill 1987). In some cases, as reported from the subtropical Lago Infernão, the NH_4^+ supply in the sediments may be essential for macrophytes (Howard-Williams et al. 1989).

Several physical and biological processes, namely resuspension, diffusion and bioturbation, lead to a release of NH_4^+ from the sediment. Diffusive flux ranged between 20–50 µmol N m^{-2} h^{-1} in Lago Camaleão and Lago Calado (Smith-Morrill 1987; Kern 1995). In Lago Camaleão twice as much diffusive flux of NH_4^+ was found in lake sediments that were 3 km further from the river than in the sediments that were nearer to the river (0.7 km). 4.5 t nitrogen moved by diffusion from the sediment to the overlying water of Lago Camaleão during the period of study. However, advective flux of

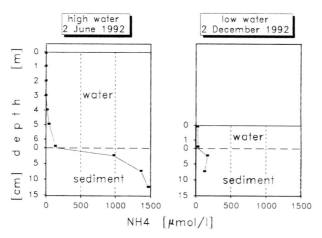

Fig. 6.3. Profile of ammonium concentration in the water and the sediment of Lago Camaleão

NH_4^+, may exceed the diffusive flux by several times, as found in Lago Calado sediments (Smith-Morrill 1987).

6.6 Nitrogen Flux Within Higher Vegetation

Higher plants drive a powerful mechanism circulating nitrogen between the sediment and the water. During the terrestrial phase the trees from the floodplain forest and herbaceous plants rooted in the sediment take up nitrogen from the sediment, transform it into organic material and release it during the aquatic phase into the water by decomposition. Herbaceous plants growing during the aquatic phase take up some of these nutrients from the water and recycle them during the terrestrial phase to the sediment, where they decay on the drying lake bottom. When the river level rises very high, detritus and large parts of living plant material can drift into the river channel (Chap. 8).

The nitrogen content in plant material varies considerably. The mean values are $12.6 \, g \, kg^{-1}$ DW (dry weight) for herbaceous plants, $13.5 \, g \, kg^{-1}$ DW for fresh leaf litter and $4.8 \, g \, kg^{-1}$ DW for wood (Furch and Junk 1992; Chap. 9).

The decomposition of *Echinochloa polystachya* led to a release of 5.4 t of nitrogen and the decomposition of leaf litter from the flooded forest led to a release of about 20.1 t of nitrogen into Lago Camaleão during the study period of 9.5 months. Similar amounts were taken up by the plants that grew during the aquatic phase pointing to a high internal nitrogen cycle. These values are rough estimates because both the biomass of macrophytes in the lake basin and the export rate into the river vary considerably between years according to the river level.

However, the nitrogen demand of macrophytes and woody plants growing in Lago Camaleão is not satisfied by the inflow of river water. Therefore, we assume that either N_2 fixation contributes a substantial amount to the nitrogen budget of the várzea, or lake internal nitrogen fluxes are much higher than the nitrogen flux between the lake and the river channel.

6.7 Nitrogen Input by N_2 Fixation

The nitrogen reservoir of the atmosphere is available to only a few procaryotic bacteria and cyanobacteria. All forms of N_2 fixation occur in the várzea, ranging from free living N_2 fixers to loose associations (periphyton, rhizocoenoses) and to obligatory symbioses (*Azolla*, legumes).

Table 6.4. Maximal rates of N_2 fixation in tropical and subtropical biocoenoses

	Substrate	N_2 fixation	Authors
Plankton		(μmol N_2-N mg Chl-a^{-1} h^{-1})	
Lago Calado (1980/1981)		0	Melack and Fisher (1988)
Lago Calado (1989/1990)		2.7	Doyle and Fisher (1994)
Lago Camaleão (1992/1993)		0	Kern (1995)
Periphyton attached to roots		(nmol C_2H_4 g DW^{-1} h^{-1})	
Lago Calado	*Pistia* (light)	90.5	Melack and Fisher (1988)
Lago Calado	*Pistia* (dark)	24.2	Melack and Fisher (1988)
Lago Calado	*Echinochloa*	32.0	Melack and Fisher (1988)
Lago Calado	*Eichhornia* (light)	2.4	Melack and Fisher (1988)
Lago Calado	*Eichhornia* (dark)	0.1	Melack and Fisher (1988)
Lago Calado	submersed leaves	10000	Doyle (1991)
Lago Infernão	*Eichhornia*	492.0	Howard-Williams et al. (1989)
Lago Camaleão	*Echinochloa* (light)	1 296.0	Kern (1995)
Lago Camaleão	*Echinochloa* (dark)	1.3	Kern (1995)
Symbioses with higher plants			
Lago Central	*Aeschynomene* (nodules)	353.8	Kern (1995)
Lago Camaleão	*Azolla* (plants)	4.4	Kern (1995)
Lago Camaleão	*Echinochloa* (rhizosphere)	0	Kern (1995)

N_2 fixation was measured by applying the C_2H_2 reduction method (Hardy et al. 1973; Dudel et al. 1990). All calculations of N_2 fixation at Lago Camaleão were made on a molar basis of $3:1$ for $C_2H_2:N_2$.

Aquatic Phase. Among all biocoenoses in the várzea studied for N_2 fixation, periphyton exhibited the highest rates (Table 6.4). Periphyton in Lago Camaleão consists primarily of filamentous algae and aquatic invertebrates which are attached to floating roots of macrophytes and to adventitious roots of semiaquatic grasses such as *Echinochloa polystachya*.

In contrast to the plankton of Lago Camaleão and Lago Calado, where N_2 fixation only played a minor role (Table 6.4), there are optimal conditions in the periphyton for N_2 fixers.

Floating roots represent a suitable substratum for the settlement and growth of algae. The drift of the water current is reduced and the nutrient supply is enhanced by the filtering of the suspended solids from the water (Engle and Melack 1990, 1993). The development of periphytic algae increases after the lake water has reached its highest level. At that time photosynthetic conditions improve due to increased insolation following the sedimentation of suspended solids and the lack of vertical growth of *Echinochloa polystachya*, which is fixed with its basal root in the sediment (Sect. 10.3).

In all studies listed in Table 6.4, there was a light dependence for N_2 fixation. Incubating periphyton at night led to a reduction in N_2 fixation, which means that nitrogen is primarily fixed by autotrophic N_2 fixing cyanobacteria. In flooded sediments no N_2 fixation could be observed (Table 6.5).

Terrestrial Phase. With receding water, large areas with perennial vegetation and without vegetation are exposed in the várzea. At Lago Camaleão in sediments without macrophytes, N_2 fixation started in the early dry phase within 1 or 2 days after exposure. In the course of a progressive establishment of N_2 fixers, the rate of fixation increased up to $87\,\mu mol\,N\,m^{-2}\,h^{-1}$. The lack of C_2H_2 reduction in dark incubations showed that N_2 was fixed exclusively by photoautotrophic organisms. In the exposed sediment without macrophyte vegetation, N_2 fixation had an average value of $10.1\,\mu mol\,N$ $m^{-2}\,h^{-1}$ over 3 months, whereas in exposed sediment with perennial macrophytes N_2 was fixed at an average rate of only $2.7\,\mu mol\,N\,m^{-2}\,h^{-1}$ during a period of 2 months. Progressive water loss results in cracks where nitrogen can be fixed even at low levels, i.e., below $30\,\mu mol\,m^{-2}\,s^{-1}$ (Kern 1995).

Balance. On the basis of a conversion factor of $3:1$ for $C_2H_4:N_2$ the input of nitrogen by periphytic N_2 fixation was $1.2\,t$ during the aquatic phase of Lago Camaleão (Fig. 6.4).

Table 6.5. Chemical and physical properties of flooded and exposed sediments from Lago Camaleão at different depths (cm) and distances from the water line (m), respectively, between October 1992 and January 1993. Data are means with their corresponding standard deviations (n = 3)

	Flooded sediment		Exposed sediment with macrophyte vegetation		Exposed sediment without macrophyte vegetation	
	0–5 cm	5–10 cm	1 m	5 m	1 m	20 m
	Sediment depth		Distance from water line		Distance from water line	
N_2 fixation (µmol/m² day)	0.00 ± 0.00		57.20 ± 99.07	20.80 ± 36.03	603.20 ± 771.50	238.88 ± 153.24
Denitrification (µmol/m² day)	107.20 ± 161.38		473.60 ± 517.93	1947.60 ± 415.78	107.60 ± 126.73	1042.16 ± 977.64
N_2O evolution (µmol/m² day)	0.00 ± 0.00		447.20 ± 358.50	960.00 ± 3.39	114.80 ± 120.30	363.76 ± 473.09
C tot (% DW)	2.16 ± 0.19	1.88 ± 0.29	5.22 ± 0.53	3.20	1.24 ± 0.05	1.21 ± 0.05
N tot (% DW)	0.22 ± 0.03	0.18 ± 0.03	0.48 ± 0.03	0.33	0.13 ± 0.01	0.13 ± 0.00
C/N	9.82 ± 0.36	10.79 ± 0.50	10.87 ± 0.42	9.70	9.34 ± 0.67	9.41 ± 0.55
Water content (% FW)	67.90 ± 1.04	55.83 ± 2.89	62.73 ± 10.56	56.90 ± 9.90	46.45 ± 0.92	32.83 ± 1.45
Temperature (°C)	28.00 ± 0.00	28.00 ± 0.00	29.63 ± 4.77	30.15 ± 6.29	33.30 ± 1.56	25.83 ± 2.04
Pore water analyses						
Conductivity (µS/cm)	402.00 ± 4.24	351.50 ± 16.26	622.67 ± 287.88	2120.00		768.00 ± 73.54
NH_4^+ (µmol/l)	178.10 ± 30.35	205.19 ± 100.33	98.45 ± 71.12	345.18 ± 205.31	253.21 ± 151.52	360.46 ± 240.67
NO_3^- (µmol/l)	0.08 ± 0.14	0.32 ± 0.28	0.50 ± 0.47	0.50	6.17 ± 5.11	3.93 ± 2.32

DW, dry weight; FW, fresh weight.

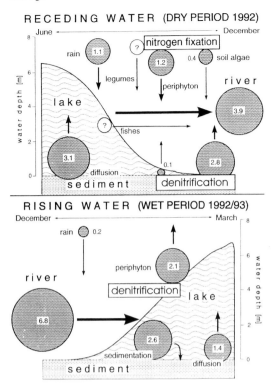

Fig. 6.4. Input and output of nitrogen (tons per 650 ha) at Lago Camaleão during the period of study from 2 June 1992 until 14 March 1993

In contrast to the low rate of N_2 fixation in the *Azolla* symbiosis, the amount of nitrogen fixed by the nodules of *Aeschynomene rudis* was comparatively high. This semiaquatic legume was found in the Lago Central on Marchantaria Island but not in the Lago Camaleão where the nitrogen balance was conducted. Here other perennial legumes might have been more important in fixing nitrogen. Worbes (1986b) reported that most species from the flooded forest of Lago Camaleão are legumes, contributing 17% to the total number of individuals. Few data on the fixation potential of semiterrestrial legumes in the várzea exist (Sylvester-Bradley et al. 1980; Magalhães 1986). At Lago Camaleão N_2 fixation was indicated by low $\delta^{15}N$ values in the leaves of perennial legumes (Kern 1995).

In culture experiments, Magalhães and Döbereiner (1984) found associations between the roots of *Echinochloa polystachya*, *Paspalum repens*, and other C_4 grasses and the N_2 fixing bacterium *Azospirillum amazonense*. In situ incubations on Marchantaria Island could not substantiate N_2 fixation in rhizocoenoses of C_3 and C_4 plants. We assume that the diffusion barrier in the waterlogged sediments was much too great to measure N_2 fixation by means of the C_2H_2 reduction method (Kern 1995).

Obvious nitrogen enrichment in exposed sediment was enabled by N_2 fixation by soil algae. On a surface area of 240 ha, the nitrogen input was 0.4 t during the period of study.

6.8 Output of Gaseous Nitrogen Compounds (N_2, N_2O, NO_x) by Denitrification

Denitrification, the reduction of NO_3^- to N_2O and N_2 is the only appreciable process, besides nitrification and volatilization, that releases nitrogen into the atmosphere. Indications of denitrification are sparse in tropical regions. Garcia-Méndez et al. (1991) and Matson et al. (1990) reported a natural N_2O release (N_2O evolution) from tropical soils. Other investigations of tropical and subtropical lake sediments succeeded in measuring denitrification potential after the amendment with NO_3^- (Viner 1982; Howard-Williams et al. 1989; Whitaker and Matvienko 1992). However, none of these studies were conducted continuously over a period of time that included seasonality and the change between terrestrial and aquatic environments. The present results from Lago Camaleão give a first insight into the extent of nitrogen loss via denitrification along the aquatic–terrestrial transition zone on the Amazon floodplain.

Aquatic Phase. Denitrification can occur in aquatic lake sediments of the várzea if there is a sufficient NO_3^- supply. In Lago Camaleão at the end of low water in 1992 denitrification increased up to a rate of 12.2 μmol $N m^{-2} h^{-1}$, corresponding to a maximum NO_3^- concentration of 4.9 μmol l^{-1} in the overlying water. High NO_3^- concentrations may be explained by nitrification that was enabled by an increase in O_2 input into the diminishing water column. In the sediment interstitial water, NO_3^- concentrations never surpassed 0.5 μmol l^{-1} indicating rapid consumption of NO_3^-. High denitrification potentials measured after the amendment with 100 μmol l^{-1} NO_3^- in the sediments of Lago Camaleão and Lago Calado substantiated a continuing NO_3^- limitation (Kern et al. 1996). With the inflow of river water into várzea lakes, NO_3^- concentrations increase significantly and support denitrification. Before NO_3^- is transported into the aquatic sediments it can contact denitrifiers in other habitats, for example the adventious roots of the semiaquatic macrophyte *Echinochloa polystachya* in Lago Camaleão (Kern 1995). During receding water, a rich N_2 fixing periphyton led to a positive nitrogen balance. During rising water an average of 0.1 μmol N h^{-1}

was released by denitrification in the adventious roots of each macrophyte node. This corresponded to a daily nitrogen loss of about 30 kg in Lago Camaleão.

Terrestrial Phase. Various factors like soil moisture, pH, NO_3^- supply and the availability of organic matter control denitrification. Due to elevation and water level sediments are exposed during August. Sediments from levees of Lago Camaleão covered with floodplain forest did not seem to have an influence on denitrification (Kern, unpubl.). From the lower elevational boundary with a muddy sediment, the amount of released N_2 and N_2O increased as time after exposure increased. However, denitrification does not undergo a diurnal rhythm, as shown for N_2 fixation. In exposed sediments of Lago Camaleão, denitrification occurred in the daytime as well as at night. Nocturnal rates of denitrification were often higher than daytime rates. This was probably due to the high oxygen sensitivity of the N_2O reductase which may be inhibited under photosynthetic activity in the daytime. The average denitrification rate in exposed sediments without macrophytes was $25.9 \, \mu mol \, N \, m^{-2} h^{-1}$ at Lago Camaleão.

Progressively higher rates of denitrification, up to $100 \, \mu mol \, N \, m^{-2} h^{-1}$, were measured in exposed sediment within a stand of *Echinochloa polystachya* that provided large amounts of rotting organic material. Because the simultaneous N_2 fixation did not play an important role in this ecotope, the exposed sediment with macrophytes represented an obvious source for gaseous forms of nitrogen within the várzea.

Balance. Nitrogen loss by nitrification and denitrification plays a minor role as long as NO_3^- concentrations in the water body are low. In the aquatic sediment of Lago Camaleão denitrification resulted in a nitrogen output of 0.1 t during the period of study (Fig. 6.4). With receding water and with the transition of anaerobic zones to aerobic zones a considerable release of gaseous nitrogen compounds occurs. During the period of study, in 90 ha of exposed Lago Camaleão sediment without vegetation 0.7 t nitrogen was released by denitrification, about double the amount that was fixed by cyanobacteria in this ecotope. In contrast, the nitrogen release was much higher in exposed sediments with macrophyte vegetation. In this ecotope, covering about 150 ha at Lago Camaleão, a large supply of decomposible organic material offered excellent conditions for denitrification. Approximately 2.1 t of nitrogen was released from this ecotope into the atmosphere. Consequently in the entire unforested exposed sediment at Lago Camaleão nitrogen loss accounted for 2.8 t during the period of study (Fig. 6.4). This nitrogen flux is quite high compared with those from the flooded forest at Lago Camaleão and the terra firme forest (Jordan et al. 1983).

In the periphyton 2.1 t of nitrogen was released at Lago Camaleão. This corresponded to 90% of the NO_3^- input by river water. Similar intensive denitrification in the epiphyton of macroalgae were found in an English estuary (Law et al. 1993).

6.9 Mass Balance of Nitrogen at Lago Camaleão

During a balance period of 9.5 months there was a total input of 9.7 t and a total output of 8.9 t nitrogen in the lake basin of Lago Camaleão (Fig. 6.4). A positive balance was primarily the result of a higher nitrogen influx into the Lago Camaleão compared with the nitrogen efflux into the Solimões River. In three months Lago Camaleão received 6.8 t nitrogen from the Solimões River, whereas during 6.5 months the lake exported only 3.9 t to the river. This points to a retention of nitrogen in Lago Camaleão, as shown by Hamilton and Lewis (1987) in the Venezuelan floodplain Lago Tineo. In contrast to Lago Calado with its main nitrogen input by runoff from terra firme (Melack and Fisher 1990), at Lago Camaleão the Solimões River water was the main source of nitrogen. The riverine nitrogen contributed 70% to the total nitrogen input.

Gaseous nitrogen turnover plays an important role at Lago Camaleão. At low water, and as long as the sediments on Marchantaria Island are not covered by water, there was a predominance of gaseous nitrogen output in the flux between atmosphere and pedosphere.

During the aquatic phase periphyton plays a very important role in the exchange of gaseous nitrogen compounds between atmosphere and hydrosphere. We found that 1.2 t nitrogen was imported by N_2 fixation and 2.1 t nitrogen was exported by denitrification. The amount of total fixed N_2 was 1.6 t N contributing 16% to the total nitrogen input at Lago Camaleão,

Table 6.6. Nitrogen flux in the lake basin of Lago Camaleão on an area of 650 ha. The balance was conducted from 2 June 1992 to 14 March 1993. (Kern 1995)

	Input		Output		Σ
	(t)	(%)	(t)	(%)	(t)
Water flow (from Table 6.2)	6.8	70.1	3.9	43.8	2.9
Rain	1.3	13.4			1.3
N_2 fixation	1.6	16.5			1.6
Denitrification			5.0	56.2	−5.0
Total nitrogen flux	**9.7**	**100.0**	**8.9**	**100.0**	**0.8**

which is twice the percentage of nitrogen input at Lago Calado (Doyle and Fisher 1994). The total amount of nitrogen released was 5.0 t N, contributing 56% to the total nitrogen output at Lago Camaleão (Table 6.6).

Figure 6.4 presents a preliminary model for balancing different pathways of nitrogen input and output at Lago Camaleão. Because not all ecotopes and nitrogen fluxes could be included, this model still is provisional. Only when data from evapotranspiration, seepage, N_2 fixation by perennial legumes in the flooded forest, and animal migration (included fishing) have been included will it be possible to specify individual pathways of nitrogen flux and evaluate their importance in the budget of nitrogen within the várzea.

The unusually low flood of 1992 allowed for a study of the mass flux of nitrogen for 9.5 months without the occurrence of flooding throughout the whole island. Therefore, the balance reflects the mass flux of nitrogen with an input of 9.7 t and an output of 8.9 t at Lago Camaleão only through the specific hydrological conditions of the study period in 1992/1993.

The Amazon flood pulse affects the nitrogen budget at Lago Camaleão in two ways. Our current understanding is that Lago Camaleão obtains most of its nitrogen from the Solimões River.

Considerable amounts are released by microbial processes directly into the atmosphere. The extent of the gaseous nitrogen flux is related to the transition between the aquatic and terrestrial phase and is thus a result of the flood pulse of the Solimões River.

Acknowledgments. We are grateful to Prof. Gerd Dudel, Institut für Allgemeine Ökologie und Umweltschutz, Tharandt and to Dr. John Melack, University of California, Santa Barbara for critical comments and linguistic improvements to the manuscript.

7 Methane Emissions from the Amazon Floodplain

Reiner Wassmann and Christopher Martius

7.1 Introduction

Methane (CH_4) is an important greenhouse gas that also affects the chemistry and oxidation capacity of the atmosphere (Cicerone and Oremland 1988). The current burden of methane in the atmosphere is about 4700 Tg (1 Tg = 1 million tons) (Wahlen et al. 1989), and the global annual emission is ca. 505 Tg CH_4 year^{-1} (Crutzen 1991). Ca. 80% of the total methane emission is of modern biogenic origin, whereas only 20% is due to fossil carbon sources (Wahlen et al. 1989). The increase in the tropospheric methane concentration from 0.7 ppm in preindustrial times to the present value of 1.7 ppm can be attributed mainly to expanding agricultural activities, such as rice cultivation and animal husbandry (Bouwman 1989). Natural wetlands are the largest source of atmospheric methane; the estimated global source strength is ca. 110 Tg CH_4 year^{-1} of which ca. 60% is attributed to tropical wetlands (Bartlett and Harriss 1993). However, the strengths of individual sources of atmospheric methane can only be estimated with broad ranges of uncertainty.

Methane is generated in the last step of anaerobic fermentation of organic matter. The inundation of wetlands cuts off the oxygen supply to the soil resulting in anaerobic conditions and, thus, favoring methane production.

Methane can be released from submerged soils to the atmosphere by diffusion, ebullition, and through the aerenchyma of plants (Wassmann et al. 1992).

7.2 The Habitats of the Várzea Ecosystem

The water level of the Amazon and its tributaries encompass a distinct seasonal cycle; the water level in June/July is ca. 10 m higher than the level

Ecological Studies, Vol. 126
Junk (ed) The Central Amazon Floodplain
© Springer-Verlag Berlin Heidelberg 1997

in October/November (Sect. 2.3; Wassmann et al. 1992). These enormous fluctuations result in an extensive seasonally flooded area called "várzea". The várzea comprises an array of habitats which can be classified according to the inundation periods (Sect. 2.6). A typical section of the várzea consists of three distinct areas or habitats, which are generally present within one várzea lake:

1. *Open Water.* The permanent aquatic regions are dominated by phytoplankton and free of macrophytes. Due to the high sediment load of the Amazon water the phytoplankton community is restricted to a thin upper layer of the water column (1–4 m) (Sect. 10.2.2).

2. *Floating Vegetation.* Regions with a short nonflooded period during the annual low water phase are covered by herbaceous plant communities showing high photosynthetic activities throughout the year. The life cycles of these fast growing plants start during the nonflooded period; the upper parts of the floating plants remain emerged during flooding and the plants may become uprooted (Sect. 8.3).

3. *Flooded Forest.* Trees are restricted to areas with a relatively short flooded phase (less than 8 months) (Sect. 8.2).

7.3 Processes Involved in the Methane Budget of the Várzea

The methane budget of the várzea is an important component of the carbon cycle of this ecosystem and can be described as the interactive product of three processes: (1) microbial methane production by strictly anaerobic bacteria; (2) methane oxidation by methanotrophic bacteria; and (3) methane transfer by vertical flux from the methanogenic zone to the atmosphere.

The prerequisites for the onset of methanogenesis are anoxic conditions with a redox potential of less than $-300 \, mV$ and the availability of organic substrate (Conrad 1989) (Sect. 4.4.3). These conditions are generally found in habitats with high organic inputs and absence of oxygen, e.g., the soils of the Amazon floodplains. The incoming water covers substantial amounts of organic material, i.e., soil organic matter, debris, etc., creating favorable conditions for methane production during the flooded phase.

Methanogenic bacteria can utilize CO_2/H_2, acetate, and methanol, whereas the locally produced methane can be consumed by aerobic bacteria utilizing C_1 compounds as sole energy and carbon source (Conrad 1989). Anaerobic methane oxidation which occurs in marine environments can most likely be neglected in the várzea due to the limited availability of sulfate. Methane can be emitted from flooded soils by three different pathways: (1) diffusion through the water column; (2) ebullition; and (3) via plant-mediated transport.

The current knowledge on the mechanisms involved in methane emission in the várzea is fragmentary:

Production. The isotopic signature of methane entrapped in the várzea soil indicates acetate fermentation as the predominant mechanism of methane formation (Wassmann et al. 1992). The soil layer from 5–20 cm is the main methanogenic zone that can be derived from the methane concentration profile (Crill et al. 1988). The methane production in the water column is negligible even though the anoxic hypolimnion contains high methane concentrations due to methane transfer from the soil (Crill et al. 1988).

Oxidation. During high water levels, methane oxidation occurs in the water column (Crill et al. 1988) and inside the plant stems of the floating grasses *Paspalum* and *Echinochloa* (Chanton et al. 1989), while methane oxidation in the várzea soil is insignificant (Wassmann et al. 1992).

The internal oxidation of CH_4 during flooding is less than 15% of the locally produced methane due to limited oxygen availability in the water column (Crill et al. 1988). In the floating vegetation the magnitude of methane oxidation varies with the life cycle of the grass plants; the oxidation is low as long as the plants are rooted and is relatively high in uprooted plant stands (Chanton and Smith 1993). The intensity of methane oxidation during the aerobic periods of the várzea soils is unknown.

Transport. Ebullition is the predominant pathway for vertical methane transfer in the várzea. The emergence of gas bubbles contributed 73–85% to the overall emission of methane in the várzea (Crill et al. 1988; Devol et al. 1988; Wassmann et al. 1992) indicating only minor flux rates of methane by diffusion through the water column and plant aerenchyma. Large amounts of gaseous methane build up in the soil due to high production rates, low solubility of methane in water, and hydrostatic pressure.

7.4 Spatial and Seasonal Variation
in Methane Emission Rates

The available results on methane emission from each habitat show pronounced variations for each habitat (Table 7.1). The average emission rates obtained by different measuring campaigns and at various lakes of the várzea ranged from 12 to $88 \, mg \, CH_4 \, m^{-2} \, day^{-1}$ for open water, from 35 to $590 \, mg \, CH_4 \, m^{-2} \, day^{-1}$ for floating vegetation, and from 7 to $230 \, mg \, CH_4 \, m^{-2}$ day^{-1} for flooded forest. The comparisons between the different várzea habitats is not consistent: emissions from flooded forest are the highest in two investigations (Bartlett et al. 1988; Wassmann and Thein 1996) and are the lowest in two other studies (Devol et al. 1988, 1990). The emission rates in floating vegetation are generally high and represent maximum values in four studies. In general, emission from open water is lowest (Table 7.1).

The reasons for the inconsistencies in the emission data from the várzea are not clear. All investigations deployed closed chamber techniques (see Schütz et al. 1989). Wassmann and Thein (1996) conducted between 900 and 1100 individual emission measurements in each várzea habitat by using a semiautomatic sampling device. The other investigations conducted between 11 and 116 measurements per habitat. Diurnal, seasonal as well as interannual variations represent one constraint on the reliability of discontinuous measurements with a relatively low number of measurements. The spatial and temporal variations of emerging gas bubbles, which represent the main component of methane emitted from the várzea, imply a large standard error for flux measurements using chambers that cover $1 \, m^2$ or less (Wassmann et al. 1992). Another possible explanation for the large range in emission rates of one habitat may be due to spatial variations in the landscape. The complex hydrological conditions in the várzea create diverse site-specific flow patterns. The lake on the Ilha de Marchantaria studied by Wassmann and Thein (1996) is subjected to a temporary

Table 7.1. Methane emission rates ($mg \, CH_4 \, m^{-2} \, day^{-1}$) obtained by different studies

Reference	Observation period	Open water	Floating vegetation	Flooded forest
Bartlett et al. (1988)	July–Sept.	27	192	230
Devol et al. (1988)	July	12	590	110
Devol et al. (1990)	Nov–Dec.	40	131	7
Devol et al. (1990)	Annual	88	390	74
Bartlett et al. (1990)	Nov–Dec.	74	201	126
Wassmann and Thein (1996)	Annual	34	35	76

throughflow of river water during high water level. The input of oxygen from river water into these lakes will impede methane production and stimulate methane oxidation and, thus, can be expected to reduce methane emission in comparison to lakes without a throughflow of river water.

The shift from flooded to nonflooded conditions implies various biological and physical consequences for the methane pools and fluxes in the várzea. The high ebullition rates indicate the presence of large amounts of methane entrapped in the soil during flooding. The fate of this gaseous methane pool is not clear, the methane can either be oxidized or released to the atmosphere when the water recedes. Aerobic soils can act as a sink for methane, but the methane flux by dry várzea soils has not been investigated so far. The uptake rates determined in other tropical wetlands, e.g., $1.9\,mg\,CH_4\,m^{-2}\,day^{-1}$ in a forest in Central Africa (Tathy et al. 1992), are much lower than emission rates observed during the aquatic phase in the várzea. Considering the long inundation periods of the várzea, the methane uptake by upland forest soils ("terra firme") will only have a minor effect on the annual net flux of methane from the entire Amazon region.

7.5 Methane Emission from Termites

The quantity of methane emitted from termites is under debate. Initial results obtained with a laboratory study of termites suggests a global source strength in the range of $150\,Tg\,CH_4\,year^{-1}$ (Zimmerman et al. 1982). Measurements with intact termite nests reveal a considerably lower methane emission from termites which ranges ca. $26\,Tg\,CH_4\,year^{-1}$ (see Martius et al. 1993). However, all estimates are based on a relatively small number of field observations conducted with one or only a few species. The first emission measurements with termites from Amazonia were conducted with a wood-eating species of the genus *Nasutitermes* (Martius et al. 1993). The average emission per nest is $2\,mg\,CH_4\,day^{-1}$; the emission related to termite biomass was recorded with an average value of $3\,\mu g\,CH_4\,g\,(dry\ wt.)\,h^{-1}$. Preliminary measurements with humus-eating termites indicate slightly higher emission rates than wood-eating termites (Martius et al., unpubl. data). Using an estimated termite biomass of $1\,g\,(dry\ wt.)\,m^{-2}$ in várzea forest (Martius 1994b), the emission from termites is ca. $3\,\mu g\,CH_4\,m^{-2}\,h^{-1}$ which is 3–5 orders of magnitudes lower than the emission rates recorded for flooded soil (Table 7.1). Apparently, methane emissions from termites are insignificant in terms of the methane budget of the várzea, but can be regarded as a relevant component in the regional context by including upland forests (Martius et al. 1993).

7.6 Significance of Methane Emission from the Várzea in the Regional and Global Context

The formation of methane, which completes the anaerobic degradation of organic matter, is an essential component of the carbon cycle in aquatic ecosystems. The relatively high methane emission rates observed in the várzea give evidence that a considerable portion of the organic material is decomposed anaerobically and is converted to methane. The minimum and maximum values of methane emission in each habitat as compiled in Table 7.1 correspond to annual methane fluxes of 4.4–$30.1\,g\,CH_4\,m^{-1}$ for open water (with 12 months of inundation), 9.4–$158\,g\,CH_4\,m^{-2}$ for floating vegetation (with 9 months of inundation), and 1.2–$41.9\,g\,CH_4\,m^{-2}$ for flooded forest (with 6 months of inundation). The methane flux rates are converted to carbon fluxes and related to biomass production in Table 7.2. Carbon fixation in the floating vegetation is in the range of $2000\,g\,C\,m^{-2}$ (Junk and Piedade 1993; Sect. 8.5). Carbon fixation in open water and flooded forest (Table 7.1) are derived from the values for biomass production, i.e., $600\,g\,(dry\,wt.)\,m^{-2}$ in open water (Sect. 10.2.2; Junk 1985a) and

Table 7.2. Annual carbon and methane budget of the Brasilian várzea (minimum and maximum emission rates adopted from Table 7.1)

	Open water		Floating vegatation		Flooded forest	
	Min.	Max.	Min.	Max.	Min.	Max.
CH_4 emission ($g\,CH_4\,m^{-2}\,year^{-1}$)	4.4	30.1	9.4	158	1.2	41.9
$C_{(CH4)}$ emission ($g\,C_{(CH4)}\,m^{-2}\,year^{-1}$)	3.3	22.6	7.1	119	0.9	31.5
Biomass production ($g\,C\,m^{-2}\,year^{-1}$)	300[b]		2000[c]		1650[d]	
$C_{(CH4)}$ emission rel. to biomass prod. (1%)	1.1	7.5	0.36	6.0	0.1	1.9
Area (km^2)	App. 40 000		App. 40 000[a]		App. 100 000[a]	
Total CH_4 emission ($Tg\,CH_4\,year^{-1}$)	0.18	1.2	0.38	6.32	0.12	4.19

[a] Junk (1985a), see text.
[b] Junk and Piedade (1993c).
[b] Junk and Piedade (1993c).
[c] Worbes (this Vol.), see text.
[d] Junk (pers. comm.).

$3300\,g\,(dry\,wt.)\,m^{-2}$ in flooded forest (Sect. 11.4), by assuming a C content of 50%. The carbon flux via methane emission corresponds to 1.1–7.5% of the C fixation in open water and is similar in floating vegetation. Methane emissions in terms of the C budget is least significant in the flooded forest (0.1–1.9%).

The total methane emission from the várzea can be estimated by extrapolating the source strength of each habitat (Table 7.2). However, one of the main constraints on a reliable estimate of the methane source strength from the várzea is the insufficient knowledge on the vegetation cover. The total area of the várzea is known with a relatively high degree of accuracy, but the extent of the three várzea habitats can only be tentatively estimated. By ignoring the open water area of the river courses, the open water of várzea lakes is estimated to be ca. $40\,000\,km^2$ (Table 7.2). An identical area is estimated to be covered by floating vegetation; the area of flooded forest is estimated to be ca. $100\,000\,km^2$ (Table 7.2). The annual CH_4 emission from the three habitats of the várzea are in the range of $0.18–1.2\,Tg\,CH_4$ for the open water, $0.38–6.32\,Tg\,CH_4$ for the floating vegetation, and $0.12–4.19\,Tg\,CH_4$ for the flooded forest (Table 7.2). Based on these results, the estimated methane source strength of the várzea is $1–9\,Tg\,CH_4\,year^{-1}$. This flux corresponds to ca. 1–8% of the global source strength of wetlands and to ca. 2–14% of the wetlands located in the tropics.

The Brasilian várzea covers ca. 2% of the Amazon region and represents an annual emission of $1–9\,Tg\,CH_4$; the largest methane source of this region. The methane emission from termites in the entire Amazon (várzea and terra firme) accounts for ca. $1\,Tg\,CH_4\,year^{-1}$ (Martius et al. 1993). Due to the large area of uplands, a methane uptake by upland soils could substantially reduce the net flux of methane from the entire Amazon region. Goreau and de Mello (1988) determined mean uptake rates of ca. $0.8\,mg\,CH_4\,M^{-2}\,day^{-1}$ in clay soils and $1.7\,mg\,CH_4\,m^{-2}\,day^{-1}$ in sandy soils of Central Amazonia. However, the roles of várzea and terra firme as sources and sinks in the methane budget of the Amazon region are far from being understood and require more comprehensive investigations of net fluxes and the processes involved.

Part III
Plant Life in the Floodplain

8 Plant Life in the Floodplain with Special Reference to Herbaceous Plants

Wolfgang J. Junk and Maria Teresa F. Piedade

8.1 Introduction

In the Amazon floodplain, four major plant communities can be distinguished: algae (phytoplankton and periphyton), aquatic herbaceous plants, terrestrial herbaceous plants, and the floodplain forest. According to their specific requirements, these plant communities occupy different habitats. The distribution of the vegetation is influenced by several factors. The most important ones are:

- The duration of the aquatic and terrestrial phases.
- The physical stability of the habitat influenced by sedimentation and erosion processes, current and wave action.
- Successional processes, related to the life span of the plants and the age of the habitat.
- Human impact.

The interactions of these factors and the dynamics of the flood pulse (Sect. 1.3) make the situation very complex.

The life spans of the different plant communities vary from days in algal communities and months in herbaceous communities to decades or even centuries in forest communities. Algal and herbaceous plant communities react quickly to changes in environmental conditions and represent the actual status of the ecological conditions in a specific habitat. Responses of the floodplain forest are much slower. Because of the longevity of many tree species, the floodplain forest can be considered as a biological long-term memory of major ecological events in the specific habitat.

Herbaceous plants play a specific role in the floodplains because they have representatives in the aquatic and the terrestrial phases and are therefore especially well suited to couple together land and water. Because of their relatively short life cycles and high reproduction rates, herbaceous plants very quickly colonize the whole range of habitats in Amazonian floodplains when they are allowed to do so. Restrictions arise because of competition with the floodplain forest for light, shortage of nutrients (e.g.,

Ecological Studies, Vol. 126
Junk (ed) The Central Amazon Floodplain
© Springer-Verlag Berlin Heidelberg 1997

in black-water habitats), or strong water movements. Herbaceous plant communities are important habitats for aquatic and terrestrial animals and play an important role in the nutrient cycles and food webs because of their high biomass and primary production. Details about specific sampling sites mentioned in this chapter are given in Section 2.2. For hydrochemical characterization of water types see Chapter 4.

8.2 General Distribution of the Vegetation

8.2.1 Distribution According to the Duration of the Aquatic and Terrestrial Phases

During the terrestrial phase, várzea areas lower than 19.5 m above sea level (a.s.l.), corresponding to more than 300 days of average annual flooding, are colonized by annual terrestrial plants. From 19.5 m upward perennial grasses and herbs can be found. From 20.5 m upward, corresponding to less than 270 days of average annual flooding, the most flood tolerant shrubs begin to grow and from 22 m upward, corresponding to less than 230 days of average annual flooding, a well-developed floodplain forest occurs (Fig. 8.1a). There is some evidence that trees in the igapó can occupy even lower lying areas, probably because of higher oxygen concentrations in the water near the bottom of black-water lakes in comparison to white-water lakes (Sect. 4.4.3).

During the aquatic phase, the growth of algae and aquatic macrophytes increases with increasing availability of habitat and is therefore associated with the water level. Periphyton growth concentrates on substrates, e.g., leaves and branches of the floodplain forest and the free-floating root masses of aquatic macrophytes. Therefore it reaches maximum levels at high water.

The distribution of herbaceous plants and trees is the result of the impact of present and past flood events. Annual species are primarily influenced by the hydrograph of the respective year, perennial herbaceous plants by the most recent flood cycles, and trees by flood events of many decades or even centuries (Junk 1989).

Periods of prolonged low water levels lasting for several years are of special importance for the successful establishment of tree seedlings at the lower range of their occurrence on the flood gradient. This has been already demonstrated for the swamp cypress in North America (Demaree 1932). Several years of uninterrupted submergence, on the other hand, can eliminate plants already established on deep-lying habitats as shown by the

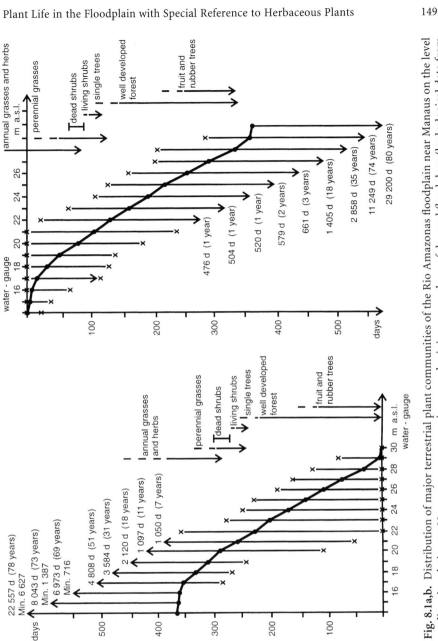

Fig. 8.1a,b. Distribution of major terrestrial plant communities of the Rio Amazonas floodplain near Manaus on the level gradient in relation to 80 years average, maximum and minimum numbers of dry and flood days (hydrological data from Manaus harbor). Total numbers of years without dry and flood periods are indicated in parentheses. Values are calculated according to the hydrological year. (Junk 1989)

prolonged flood period from 1970 through 1974. During that period, the water did not fall below the 20 m level, and all individuals of *Eugenia inundata*, *Myrcia* sp., *Symmeria paniculata*, *Cocoloba ovata*, and *Ruprechtia tenuifolia* growing on areas from 19.5 m to 20.5 m died in the surroundings of Manaus (Fig. 8.1b). To date, seedlings growing during the terrestrial phase in these areas were unable to become established permanently because the following flood periods were too long. Flooding of higher areas did not cause losses of adult plants.

8.2.2 Distribution According to Sedimentation and Erosion

Sedimentation rates within the floodplain vary strongly. On the banks, on sand and mud flats near the river, the deposition of sediments ranges from several centimeters to more than 1 m per year. The small grain size of the sediments of the Amazon River reduces oxygen supply by aeration or water exchange (Sects. 2.4, 2.5). Increased sediment deposition buries the tree roots; these suffer from the increasingly anoxic conditions and finally die. Some tree species, e.g., *Salix humboldtiana* and *Alchornea castaneifolia* can develop new roots in the topsoil layer. These species belong to the pioneer vegetation of the river banks. Species which do not tolerate high sedimentation rates, e.g., *Eugenia inundata* and *Symmeria paniculata*, do not occur, although they are able to survive prolonged periods of submergence. In rivers carrying little seston and a coarse bed load, such as the Rio Tocantins, sand flats with a grain size greater than 2 mm are colonized by closely related *Eugenia* species, which become temporarily buried under 1–2 m thick coarse sediments without being damaged.

Annual and perennial herbaceous plants are better adapted to high sedimentation rates than trees. Regrowth results from seeds transported in great number by the river or from parts of uprooted stems. Perennial grasses, e.g., *Paspalum fasciculatum*, *Echinochloa polystachya*, and *Paspalum repens*, form new shoots on the nodes in contact with the topsoil layer. They form extensive monospecific stands, stabilize the sediments, and accelerate the sediment deposition. Growth of semiaquatic and aquatic annual and perennial species is regulated during high water by the current and waves. Stands of *Paspalum repens* and *Echinochloa polystachya* growing in exposed locations are frequently carried away by the current.

During the terrestrial phase the growth of herbaceous plants depends on the water retention capacity of the soils and the amount and distribution of local rainfall. Total amount and species composition of annual terrestrial plants varies strongly from year to year depending on the rainfall pattern.

Erosional processes often cause the loss of undermined banks as high as 15 m ("terras caidas") and large-scale destruction of well-developed floodplain forest communities. In the foothills of the Andes, the headwaters of the Amazon shift with a mean rate of more than 50 m year^{-1} (Salo et al. 1986). The influence of erosion is less pronounced near Manaus (Sect. 2.6).

8.2.3 Successional Processes

In a dynamic system like the Amazon floodplain, successional processes strongly influence the plant communities. In many várzea lakes at low water a zonation of terrestrial and semiaquatic vegetation along the flood gradient can be observed. Annual plants occupy the lowest level, perennial plants begin at about 19 m a.s.l., and from 20–21 m a.s.l. different tree species of the floodplain forest become established. This zonation corresponds to an allogenic succession because sediment deposition of a few millimeters per year results in a slow rise of the habitats, reducing the flood stress. After some decades or centuries the terrestrial phase is sufficiently long to allow the substitution of one plant community by the next one. Final stage is the climax forest on the levees. An acceleration of the allogenic succession is observed on sediment bars with high deposition rates.

Inside the floodplain forest on different positions along the flood gradient different stages of an autogenic succession can be observed. Strong disturbances by the river, thunderstorms or by man, result in a set back of the forest communities favoring a secondary succession. Therefore communities may differ strongly in species composition (Sects. 11.2.4, 11.5).

During the terrestrial phase seedlings of species growing in late seral stages begin to grow all over the floodplain. However, they become eliminated during the next aquatic phase. This maintains the floodplain vegetation over large areas in an early seral stage. This holds true as well for aquatic macrophyte communities. Free-floating aquatic macrophytes become colonized by palustrine species. However, the flood pulse leads to a set back of the autogenic succession during the annual terrestrial phase, maintaining the aquatic community in an early successional stage.

Isolated floodplain lakes often undergo only small changes in water level. Bays of lakes extending into the terra firme often change very little in surface area because of the steep banks. Even when the water level is low, terra firme streams supply sufficient water to maintain permanent aquatic conditions. In these lakes floating islands, locally called "matupá" develop. They are advanced seral stages of an autogenic succession within the community of floating plants (Junk 1983).

At the beginning of the succession, species that only have a limited ability to float, e.g., *Ludwigia affinis, L. octovalvis, L. decurrens, Andropogon bicornis, Scirpus cubensis, Thelypteris* spp., *Aeschynomene* spp., *Phaseolus* spp., begin to settle on the beds of floating plants. The secondary settlers make the floating layer denser, fix it with their roots and stolons, and displace the pioneers to the edge of the floating vegetation islands. The water exchange within the floating island becomes reduced. Under strongly hypoxic or anoxic conditions, the decomposition of organic material is retarded, and the floating layer increases. At this stage, the islands often become so heavy that they sink to the bottom of the lake. The evolution of gas in the dead material can cause pieces of the islands to rise to the surface again, where they are quickly reoccupied by secondary settlers. The floating communities that develop in this way appear as heterogeneous mosaics of fragmented populations of different ages and levels of succession.

When the tree-like *Montrichardia arborescens* (family Araceae) invades the island, the succession continues because the rhizomes and stems serve as floats. The accumulation of organic material produces floating islands several meters thick, which may protrude as much as 30 cm

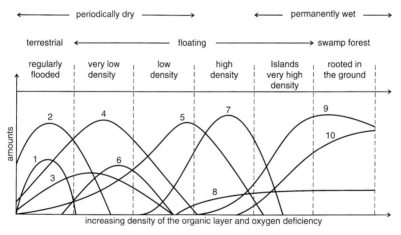

Fig. 8.2. The participation of some characteristic aquatic and semiaquatic macrophytes in the formation of rooted and free-floating plant communities of different densities and successional stages. Low density floating mats show reduction of O_2 in the middle of the layer but no H_2S. Dense floating mats show a strong reduction of O_2 and H_2S development in the middle of the layer. Very dense floating mats show complete reduction of O_2 and development of H_2S throughout nearly the whole layer. *1*, Annual terrestrial plants; *2*, perennial terrestrial plants (*P. fasciculatum*); *3–5*, annual and perennial semiaquatic plants (*Oryza* spp., *Echinochloa polystachya*, *Paspalum repens*); *6*, annual aquatic plants (*Pistia stratiotes, Salvinia* spp.); *7*, perennial aquatic plants (*Leersia hexandra, Scirpus cubensis*); *8*, vines; *9*, *Montrichardia arborescens*; *10*, trees. (Modified after Junk 1983)

above the water surface. Later, tree species, e.g., *Pseudobombax munguba*, *Senna reticulata*, *Cecropia latiloba*, and *Ficus* spp., can colonize the island (Fig. 8.2).

8.2.4 Anthropogenic Influences on the Vegetation Pattern

Parts of the Amazon floodplain have already been cultivated for several thousand years because of their fertility (Sect. 1.6). Therefore human impact has to be considered when discussing plant distribution and frequency. The influence of Indians on the vegetation is hard to evaluate because of insufficient archaeological evidence about the land use. We suppose that it was slight because wood consumption of the Indians was modest. They did not clear the forest to create pastures because animal husbandry was not practised on a large scale (Junk 1995).

With the European colonization the land use changed. Today anthropogenic influences on the distribution of the vegetation can be grouped into three categories that are closely interrelated: exploitation of the floodplain forest by selective logging, pasture cultivation for cattle and buffalo ranching, and agriculture. All activities reduce the floodplain forest in favor of the growth of terrestrial and aquatic herbaceous plants and algal communities. Selective wood cutting is the most moderate form of exploiting the floodplain forest. Trees are cut during the terrestrial phase and the logs are removed with boats during the aquatic phase. In the small clearings a temporally and spatially limited growth of herbaceous plants is favored. Large-scale forest clearings to produce pastures and fields are serious encroachments on the forest.

It can be assumed that fire formerly played a minor role for the vegetation of the middle Amazon floodplains because the forest is an efficient barrier. With the progressive destruction of the forest and the subsequent establishment of secondary aggregations of herbaceous plants, the susceptibility of the vegetation to fires during the dry season increased considerably. Fire is employed to clear the pastures and the land for crop plantations. It becomes a stress factor of increasing importance to which the native flora and fauna of the várzea are not adapted.

The areas used for agriculture are still relatively small and are concentrated on the levees near urban centers. They provide habitats for many herbaceous plants with little flood tolerance that would otherwise be eliminated by the floodplain forest. Furthermore, many weeds have been introduced with crop seeds and have successfully established on the levees of the várzea because flood stress is rather small (Sect. 8.3.2).

8.3 Site Conditions, Community Structure and Adaptations of the Herbaceous Vegetation of the Várzea

The alternation between well-defined terrestrial and aquatic phases makes it useful to classify the species as either aquatic or terrestrial. However, such a classification is problematic, because many species show adaptations, permitting their occurrence in terrestrial and aquatic habitats (Fig. 8.3). There exists a gradient from aquatic to terrestrial species. Figure

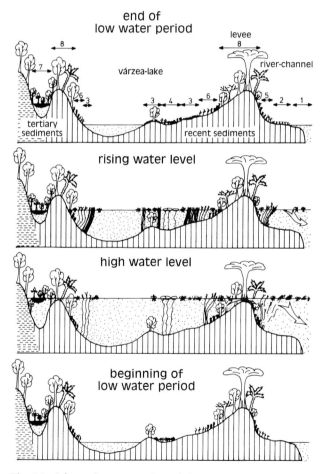

Fig. 8.3. Schematic presentation of the principal natural habitats in the várzea and their natural plant communities. *1*, River channel; *2*, sediment bar; *3*, low-lying stable mud flats; *4*, swale; *5*, stable river shore; *6*, lake shore; *7*, lake with little fluctuating water level; *8*, levee. (Junk 1986)

a

b

Fig. 8.4. a Center of a várzea lake (Lago do Reis) about 18 m a.s.l. at the beginning of the terrestrial phase. Seeds of annual terrestrial plants will begin to germinate with the beginning of the rainy season. **b** Várzea lake area (Lago do Reis) about 21 m a.s.l. at the beginning of the terrestrial phase. Shrubs (*Symmeria paniculata*) are partly covered with drying aquatic macrophytes (*Paspalum repens*). **c** Várzea lake area (Lago do Reis) about 21 m a.s.l. at the end of the terrestrial phase. The lake bottom is covered with terrestrial and semi-aquatic annual plants and shrubs (*Symmeria paniculata*). **d** Stand of the perennial grass (*Echinochloa polystachya*) during intermediate water level in Lago Camaleão. Lake bottom about 21 m a.s.l., water depth about 5 m. In the *background*, the várzea forest. **e** Permanently wet depression in a várzea lake (Lago Jacaretinga) with a species-rich aquatic and semi-aquatic herbaceous vegetation (*Eichhornia crassipes, Victoria amazonica*, etc.) at the beginning of rising water level. In the *background*, the várzea forest. **f** Várzea lake (Lago Camaleão) during high water. Free-floating aquatic macrophytes (*Eichhornia crassipes, Paspalum repens*) *in the foreground*, rooted perennial grasses (*Echinochloa polystachya*) *in the middle*, and the floodplain forest in the *background*

c

d   Fig. 8.4c,d

e

f

Fig. 8.4e,f

8.4a–f shows the development of the vegetation in the várzea during the hydrological cycle.

8.3.1 The Most Important Habitats in the Várzea and Their Herbaceous Plants

Habitats with Alternating Terrestrial and Aquatic Phases. The species composition of the herbaceous vegetation in the várzea depends mainly on habitat stability, light, length of the hydroperiod, and anthropogenic influences. Further important factors influencing the structure during the terrestrial phase are the amount and distribution of rainfall. The great variability in the environmental factors results in a large habitat diversity and quick shifts in the community structure annually as well as in successive years. Some of the communities are in a state of accelerated allogenic succession, while others are regarded as quite heterogeneous conglomerates of anthropogenic transition stages that are undergoing a rapid development toward the higher seral stages of a floodplain forest.

In spite of the dynamics inherent in the system, various groups of plants occupying large areas can be identified and associated with certain habitats. Which phytosociological rank each of these groups are assigned to, whether and to what extent they can be subdivided into smaller groupings, and which positions the individual species assume in such groups will not be given closer attention. In Table 8.1 the most important habitats are characterized according to several environmental parameters, and some of the most abundant species are listed.

The flood pulse determines the main growth period of the plants (Fig. 8.5). Depending on the position of the habitat on the flood gradient and the life cycle, it can be short or long. The annual *Sagittaria sprucei* (family Alismataceae) can only be found at low water level during a three month period in moist depressions. The perennial grass *Echinochloa polystachya* is common all year-round in large quantities with a slight decrease at the beginning of the low water period when it rejuvenates by vegetative propagation, but flowering occurs only at high water level.

In addition to the habitats subjected to the change between a terrestrial and an aquatic phase, there are habitats in the floodplain with permanent aquatic, terrestrial and palustrine conditions, e.g., permanent floodplain lakes and channels, the canopy of the floodplain forest and floating vegetation islands (Sect. 8.2.3). They represent important refuges for aquatic and terrestrial organisms during the unfavorable period. The floodplain forest also harbours in the canopy many epiphytes and a specific invertebrate canopy fauna (Chaps. 18, 19).

Table 8.1. The most important habitats of herbaceous plants in the várzea

Habitat	Environmental factors influencing species composition		Species
	Terrestrial phase	Aquatic phase	
Main channel	No terrestrial phase	Permanently aquatic; permanently strong current, waves, turbulences, high load of suspensoids	Free-floating plants, *Echinochloa polystachya*, *Paspalum repens*. Depending upon input from other habitats, species-poor
Low-lying alluvial deposits in the main channel	Short terr. phase, strong erosion and sediment deposition, sandy sediment	Partly strong current, waves, turbulences, high load of suspensoids	Annual terrestrial plants, *Echinochloa polystachya*, *Paspalum repens*, Intermediate in number of species
Low-lying lake beds	Short terr. phase, little erosion and sediment deposition, fine-grained sediment	Partly strong waves, normally low load of suspensoids	Annual terrestrial plants, *Paspalum repens*, *Oryza* spp., *Hymenachne amplexicaulis*, *Luziola spruceana*, species-poor
Stable river banks	Short to medium terr. phase, little erosion and deposition, steep slopes and unstable sediments due to small landslides	Strong current, waves and turbulences, high load of suspensoids	*Paspalum fasciculatum*, *Echinochloa polystachya*, *Paspalum repens*; species-poor
Unstable river banks	Short to medium terr. phase, strong land-slides	Strong current, waves and turbulences, high load of suspensoids	Short-living terrestrial and semiaquatic plants; species-poor
Inundation forest	Medium to long terr. phase, stable sediments, low light intensity	Little water movement, low load of suspensoids, low light intensity	Musaceae, Marantaceae, *Scleria microcarpa*, species-poor
Areas disturbed by man	Medium to long terr. phase, stable sediments, high light intensity	Little water movement, low load of suspensoids, high light intensity	Very variable, many vines, many ruderal species; species-rich
Sheltered bays	Short terr. phase, stable sediments, high light intensity	Little water movement, low load of suspensoids, high light intensity	All free-floating species aquatic and semiaquatic grasses, species-rich
Floating islands	Permanently waterlogged	Organic soil, stable, strongly hypoxic or anoxic	*Montrichardia arborescens*, *Scirpus cubensis*, *Thelypteris* spp. many vines, some trees species; intermediate in number of species
Perennial moist depressions	Waterlogged soil, fine sediments and organic material	Medium to short aquatic phase, low water movement, low load of suspensoids	*Montrichardia arborescens*, intermediate in number of species

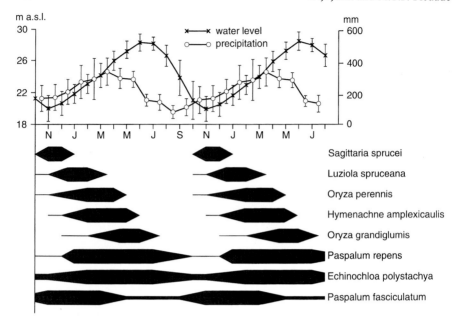

Fig. 8.5. Main growth period of some aquatic and semiaquatic herbaceous plants in relation to the water level. The *width of the bars* indicates the relative frequency of the species. (Junk 1986)

8.3.2 Species Richness and Distribution

In the várzea near Manaus 388 herbaceous plant species were identified (Junk and Piedade 1993a,b). They belong to 64 families and 182 genera (Table 8.2). The best represented family is Gramineae with 60 species. In second place is Cyperaceae with 37 species, followed by Leguminosae with 30 and Compositae with 27 species. A total of 42 families were represented by one to three species. It can be expected that the total number of species is about 10 to 20% greater than the value reported here. The occurrence of unreported species can be expected on the sites at the highest elevations, which are subject to flooding only at intervals of several years, and in areas used for agriculture.

According to their modes of existence, 330 species, or 85% of the total, are terrestrial, 34 species are aquatic (9% of the total) and only 20 species (5%) are considered to be swamp dwellers. Aquatic species with terrestrial phases number 17 and account for 4% of the total, whereas *Cynodon dactylon* and *Paspalum fasciculatum* have their main periods of growth during the terrestrial phase but survive for many months with their stems under water (Table 8.3).

Table 8.2. The most speciose families of herbaceous plants of the Amazon floodplain near Manaus

Family	n	Family	n	Family	n
Gramineae	60	Rubiaceae	13	Piperaceae	8
Cyperaceae	37	Malvaceae	13	Cucurbitaceae	8
Leguminosae	30	Solanaceae	12	Labiatae	8
Compositae	27	Onagraceae	11	Scrophulariaceae	8
Euphorbiaceae	20	Marantaceae	10	Lemnaceae	6
Convolvulaceae	14	Amaranthaceae	9	Polypodiaceae	6

n, Number of species.

Table 8.3. Classification of herbs according to frequency of occurrence, habitat, and mode of existence

Frequency		Mode of existence		Principle habitats	
type[a]	n	type[b]	n	type[c]	n
1	110	a	34	A	42
2	189	b	17	B	44
3	72	p	20	C	26
4	12	t	330	D	92
5	5	v	58	E	273
				F	25

n, Number of species.
Many species show a wide range of mode of existence and habitat preferences. Therefore totals of these categories are greater than total species numbers.
[a] 1, Rare; 2, widely distributed; 3, common; 4, forms monospecific stands; 5, dominant over large areas.
[b] a, Aquatic; b, aquatic with a terrestrial phase; p, palustric; t, terrestrial; v, vine.
[c] A, Várzea lakes; B, floating islands of organic material; C, low-lying lake beds; D, areas of intensive sedimentation; E, disturbed areas at high elevations; F, inundation forest.

According to site preferences, among the terrestrial species 273 (71% of the total) prefer disturbed locations at higher elevations. A total of 92 species (24% of the total) are characteristic of areas of fresh sediment deposition, but isolated individuals can also settle on sites at higher elevations. The low-lying beds of lakes are preferred by 26 species (7%), and the inundation forest supports 25 species (7%). Bays of the várzea lakes with standing water are colonized by 42 species (11% of the total) during the aquatic phase, while 44 species (11%) inhabit the floating islands.

An analysis of the species abundance shows that the great majority of them occur in relatively small numbers. The frequency values of 1, 2, or 3 were assigned to 111 or 29%; 189 or 49%, and 70 species or 18% of the total, respectively. Only 12 species, or 3% of the total, were very abundant and 5 were so dominant that they were frequently able to form large monospecific stands. Of the 17 very abundant dominant species, 4 are aquatic macrophytes; *Pistia stratiotes*, *Scirpus cubensis*, *Eichhornia crassipes*, and *Salvinia auriculata*. Six species are aquatic with a terrestrial phase; *Echinochloa polystachya*, *Hymenachne amplexicaulis*, *Leersia hexandra*, *Oryza perennis*, *Paspalum repens*, and *Montrichardia arborescens*. *M. arborescens* is also classified as a swamp-dwelling plant. Five species are terrestrial; *Alternanthera pilosa*, *A. brasiliana*, *Paspalum conjugatum*, *Ludwigia densiflora*, and *Sorghum arundinaceum*.

8.3.3 Adaptations to the Flood Pulse

Reproduction and Propagation. The simplest way to overcome the impact of periodical flooding and drying is the production of flood and drought resistant seeds or spores. It allows many ruderal species to exploit the conditions on the floodplains during the terrestrial phase. Long periods of dormancy permit the accumulation of seeds in the soil, making a rapid recolonization of an area possible even after several years of adverse conditions. On land, the seeds of *Portulaca oleracea* can germinate after a period as long as 40 years (Koch 1969).

Most plants growing during the terrestrial phase produce large numbers of diaspores. A single plant of *Paspalum conjugatum* can produce 1500 seeds, while *Eleusine indica* produces 40000 to 50000, and in exceptional cases, as many as 135000 seeds. *Portulaca oleracea* forms 10000 seeds (Holm et al. 1977) and according to Wehsarg (1954) it can even produce as many as 193000. A large number of seeds also guarantees the reproduction and dispersion of most annual and perennial aquatic grasses, such as *Luziola spruceana*, *Leersia hexandra*, *Paspalum repens*, *Hymenachne amplexicaulis*, *Echinochloa polystachya*, and *Pasplaum fasciculatum*.

Germination of the diaspores of the aquatic species occurs mostly during the terrestrial phase. The growth of the young plants can progress considerably before the time of the flood, as in the case of the aquatic grasses *Oryza* spp., *Echinochloa polystachya*, *Hymenachne amplexicaulis*, and *Luziola spruceana*, as well as the pontederiacean, *Pontederia rotundifolia*, and the fern, *Ceratopteris pteridoides*. The seed bank in the sediments of floodplain lakes is very large. On exposed sediments in Lago Camaleão as many as 10000 seedlings m^{-2} were counted. Seeds are also transported in large numbers by the river. On a sediment bank arising for the first time in the middle of the Rio Amazonas a seedling density of 10 to

$100\,\text{m}^{-2}$ was observed. There is still relatively little known about the seed dispersal of herbaceous plants in the várzea. It can be assumed that dispersal by water plays an important role. The small size of the seeds of most of the herbaceous plants in the region permits them to be carried by the current without special organs for flotation. Furthermore, drifting floating plants carry the seeds as well. For some species with very small seeds wind dispersal was observed, e.g., *Ludwigia densiflora*.

For trees from the inundation forest, a dispersal of the seeds by fishes has been described (Gottsberger 1978; Goulding 1980; Goulding et al. 1988). Seeds of many herbaceous plants in the várzea are also a preferred food of fishes. The tambaqui, *Colossoma macropomum*, feed on the grains of *Oryza* spp. The sardinha, *Triportheus* spp., and the matrincha and jatuarana, *Brycon* spp., consume the seeds of various species of the family Convolvulaceae. The stomachs of the armored catfishes, *Loricaria cataphracta* and *Loricariichthys* spp., contain large quantities of grass seeds (Py-Daniel 1984). To what degree the fishes contribute to seed dispersal must remain an open question because in many cases the seeds consumed had been fragmented. The dispersal through other kinds of animals, such as birds, also remains to be determined.

Seed dispersal of *Ludwigia densiflora* can be considered an adaptation to the flood pulse. Stems of adult plants with the closed seed-bearing capsules persist under water. The capsules open at the beginning of the next dry phase, release the seeds on the moist sediment, and guarantee a high number of seedlings. While the other *Ludwigia* species also occur in moist habitats on the terra firme, *Ludwigia densiflora* seems to be confined to the várzea.

Most aquatic and semiaquatic species propagate mainly by vegetative reproduction. The increased importance of vegetative as compared to sexual reproduction among water plants has already been mentioned by Arber (1920). Later it was discussed by various authors, as summarized by Sculthorpe (1985). An important argument was that conditions for the germination of seeds and spores in the aquatic system are fulfilled so seldom that many species could not support their populations exclusively by sexual reproduction. Floating species, such as *Salvinia* spp., *Azolla* cf. *microphylla*, *Ricciocarpus natans*, *Ceratophyllum demersum*, and *Najas* spp., reproduce by fragmentation, whereas *Eichhornia crassipes*, *Pistia stratiotes*, and *Limnobium stoloniferum* produce runners. *Ceratopteris pteridoides* produces frond buds. The perennial grasses regenerate their stands mainly by sprouting new shoots at the nodes, as in the cases of *Paspalum fasciculatum*, *Echinochloa polystachya*, and *Paspalum repens*. The great vegetative reproduction potential of many aquatic and semi-aquatic plants inhabiting tropical floodplains may be regarded as an adaptation to the increased loss rates resulting from periodic desiccation of the habitats. Several species from neotropical floodplains, e.g., the water hyacinth, *Eichhornia crassipes*, and the water fern, *Salvinia auriculata* have been introduced to other parts of the tropics. Under stable hydrological conditions, they have created serious problems due to the high rates of vegetative propagation. Some of the newly built reservoirs in Amazonia also suffer a mass development of aquatic macrophytes because nutrient input from the flooded reservoir basin is large and the flood pulse as a controlling factor is strongly reduced (Junk 1982; Junk and Mello 1987).

Adaptations of Plants to Flooding and Desiccation. Several strategies are used to adapt to the change between the terrestrial and aquatic phase. Many annual herbs in the várzea have short life spans. After a few weeks, even species that grow over 2 m high, such as *Ludwigia densiflora*, produce seeds and die (compare Sect. 8.5.2). This allows the successful colonization of sites that are dry for short periods of time only. A high tolerance to flooding is shown by the perennial grasses *Paspalum fasciculatum* and *Cynodon dactylon*. The stems can survive at least two years under water and assume the function of rhizomes, storing energy and nutrients. As soon as they emerge again they develop new shoots.

Aquatic species exhibit a wide spectrum of responses to drought. The rhizomes of *Victoria amazonica*, *Nymphaea amazonum* and *Montrichardia arborescens* survive in moist sediments. Many emergent species show astounding morphological and physiological plasticity, which often makes it difficult to recognize their land forms as conspecific with the better known water forms. Free-floating species, such as *Eichhornia crassipes*, *Pontederia rotundifolia*, *Ludwigia natans*, *Neptunia oleracea*, and *Paspalum repens*, show reduction in size in the areas of their leaf surfaces, in water content, and, in the cases of *Ludwigia* species and *Neptunia oleracea*, in the amounts of aerenchyma in the roots and stems.

The Floating Way of Life. Submerged plants rarely occur in várzea and igapó, because light conditions are unsuitable. In the relatively turbid water, which is often brownish from humic substances, the 1% limit of light penetration is reached at a depth of only 1 to 5 m (Sect. 4.2). The water level in the Amazon rises at an average rate of 5 cm day^{-1}. However, for periods of several weeks, rates of 10 to 15 cm day^{-1} have been recorded. This results in a rapid deterioration of the illumination near the bottom of newly flooded areas. The compensation point for photosynthesis will be reached after only a few weeks. The great majority of the macrophytes in

Table 8.4. Modes of existence of the aquatic macrophytes on the floodplains of the middle Amazon

Substrate contact	Position of the photosynthetic organs	Example
	Submerged	None present
Rooted in soil	Floating leaves	*Victoria amazonica, Nymphaea amazonum*
	Emergent leaves	*Echinochloa polystachya, Oryza* spp. *Hymenachne amplexicaulis Sagittaria sprucei*
Transition form: roots in the soil or free-floating	Plant with emergent leaves	*Leersia hexandra, Paspalum repens, Montrichardia arborescens, Pontederia rotundifolia, Luziola spruceana*
	Submerged below the surface of the water	*Utricularia foliosa, Ceratophyllum demersum Naias* sp., *Wolffiella* spp.
	Resting on the water surface	*Lemna* spp., *Spirodela, Salvinia* spp., *Azolla* cf. *microphylla, Ricciocarpus natans*
Free-floating	Transition form: leaves on the water's surface or extending above	*Phyllanthus fluitans, Limnobium stoloniferum, Ludwigia natans, Ceratopteris pteridoides*
	Leaves extending above the surface of the water	*Eichhornia crassipes, Pistia stratiotes, Scirpus cubensis, Alternanthera hassleriana*

the várzea are adapted to a mode of existence that allows them to maintain their organs of photosynthesis just below or above the water surface (Table 8.4).

Free-floating submerged plants are represented by a few species that form small populations as they are less competitive for light in comparison with emerged species. Plants rooted in the sediment and producing floating leaves are represented by two species. *Victoria amazonica* is abundant and *Nymphaea amazonum* is rare because of competition from the robust semiaquatic and aquatic grasses.

In spite of the large water-level fluctuations there are several species rooted in the sediment with emergent leaves. Some of them, such as *Echinochloa polystachya*, *Oryza perennis*, and *Hymenachne amplexicaulis*, must be considered very successful at colonizing the várzea because they often form large populations. Various species can be regarded as transition forms between floating plants and those rooted in the sediment. *Luziola spruceana, Oryza perennis, Leersia hexandra, Paspalum repens* and *Montrichardia arborescens* can remain rooted in the sediment in shallow water, but in deep water they pass to a free-floating mode of life, which *Paspalum repens, Leersia hexandra* and *Montrichardia arborescens* can maintain for many years.

The group of free-floating species is diverse. *Eichhornia crassipes, Scirpus cubensis, Pistia stratiotes* and *Salvinia auriculata* are very abundant. Several species, e.g., *Limnobium stoloniferum* and *Ludwigia natans*, form both floating and emergent leaves. In the case of *Ceratopteris pteridoides*, the young sterile fronds are usually formed to float on the surface whereas the fertile fronds with the sporangia emerge above the water surface. The leaves of *Phyllanthus fluitans* represent an interesting transitional form, which is arched and swollen, and come to rest only with their margins on the surface.

Adaptations to Hypoxia. Flooding has its initial effect on plants through the root system. The waterlogged soil becomes hypoxic or anoxic within a few hours (Ponnamperuma 1984). Thus the primary stress due to flooding is an oxygen stress. Plants that grow under such conditions have morphological, anatomical, or physiological adaptations (Crawford 1983; Kozlowski 1984).

Morphological adaptations include the formation of aerenchyma on the roots and the basal stem area, e.g., in all *Ludwigia* species, in *Neptunia oleracea, Sesbania exasperata,* and *Sphenoclea zeylanica.* In most species, the aerenchyma is formed or reinforced at the beginning of the flood. In several species, e.g., *Caperonia castaneifolia* and *Polygonum tomentosum,* the internodes become greatly elongated and thickened after flooding. Uprooted plants can exist floating free, flowering and fruiting for several weeks.

The formation of pneumatophores is observed in *Phaseolus* cf. *longifolius.* The stem that creeps over the floating vegetation forms lateral roots on submerged sections. In addition to the roots penetrating the water there are numerous roots, reaching five cm in length, extending into the atmosphere. They have thin walls and contain a well-developed lacunar system around the central cylinder (Junk and B. Furch, unpubl.). The orthotropic roots form pneumatic zones, as known from the neotropical hydrophilic palms *Mauritia flexuosa* and *Euterpe oleracea* (Granville 1974).

When exposed to flooding, many plants develop adventitious roots that float in the water and assume some or all functions of the roots in the soil. These roots are formed at the nodes, as in the aquatic grasses, or along the entire submerged stems, as in *Phaseolus* cf. *longifolius*, and many representatives of the Convolvulaceae and Cucurbitaceae. Frequently, the roots in the soil and lower sections of the stems die after submergence. Other species, such as *Echinochloa polystachya*, *Oryza* spp., and *Hymenachne amplexicaulis*, maintain their contact with the sediment until the end of their growth periods, even though they also possess well-developed masses of roots at the nodes. The functions of the primary root system, however, may be strongly reduced or fully eliminated after a few weeks of submergence.

The morphological and anatomical adaptations must be evaluated individually according to their significance for life in the várzea. The formation of aerenchyma by many rooted species within a few days of the soil becoming waterlogged enables the plants to tolerate shallow flooding, e.g., in depressions filled with rainwater during the terrestrial phase or on floating islands. The same applies to the formation of pneumatophores, which function only in shallow water. Pressurized ventilation is considered a special adaptation of wetland plants to anoxic conditions in the sediment and seems to be very important in the Amazon floodplain too. It permits the transport of oxygen from the surface to the roots through the lacunar system in the leaves and petioles by a temperature gradient-dependent one-way diffusion (Knudsen diffusion) of gas molecules into young leaves and a gas efflux from old leaves. In *Victoria amazonica* flux rates can reach about $5000\,\mathrm{ml\,h^{-1}}$ (Grosse et al. 1991).

The importance of the flood amplitude as a limiting factor for the distribution of plants rooted in the sediment is illustrated by *Eichhornia azurea* and *Thalia geniculata*. These species are abundant along the lower Amazon where the flood amplitude is only 4 to 6 m. In the vicinity of Manaus, where the flood amplitude is twice as high, these are very rare.

8.4 Herbaceous Plants of the Igapó

Studies of herbaceous plants have concentrated on the várzea because herbaceous vegetation in the Rio Negro floodplain is scarce. Most aquatic and semiaquatic species are absent. Experiments in wire-mesh cages showed high growth rates during the first two months for *Pistia stratiotes*, *Salvinia auriculata* and free-floating *Paspalum repens*. Later, the plants decreased in size and died (Sect. 8.5.3). *Eichhornia crassipes* did not grow at all. Low nutrient levels are considered to be the limiting factor for most free-floating species. Berg (1961) considers low electrolyte content in conjunction with low pH values toxic for *Eichhornia crassipes*. Exceptions are

Oryza perennis and *Utricularia foliosa,* which are common on the Rio Negro floodplain.

Herbaceous plant growth during the terrestrial phase on sandy soils is very scarce. When clay prevails, herbaceous vegetation is denser but much smaller and less diversified than in the várzea. The productive C_4 grasses *Paspalum repens, Paspalum fasciculatum* and *Echinochloa polystachya* do not occur. Transplantation experiments failed. Here as well, low nutrient status and, on sandy soils, drought stress are considered the limiting factors.

Since the early 1970s we have observed some aquatic macrophyte growth in the archipelago of Anavilhanas downstream of the confluence with the Rio Branco. This may be the consequence of an increased nutrient and sediment input because of increased erosion in the catchment area due to man-made disturbance.

8.5 Biomass and Primary Production

Plants rooted in the ground grow vertically increasing the biomass per unit area. Free-floating plants tend to grow horizontally, not only accumulating biomass per unit area but expanding the area occupied. Because of different methodological approaches, free-floating and rooted herbaceous plants are discussed separately.

8.5.1 The Relative Growth Rate (RGR) of Floating Species

Free-floating plant communities grow by vegetative propagation. When sufficient area is available they spread horizontally. For short periods of time plant growth is therefore best described by the relative growth rate (RGR), which is the increase in gram biomass per gram plant material per day. This unit can be transferred into a more comprehensive unit, the doubling time T, which indicates the time which the plants need to double the biomass or the occupied area (Mitchell 1974).

$$RGR = \frac{\ln x_t - \ln x_0}{t} \qquad T = \frac{\ln 2}{RGR}$$

Measurements in the várzea were made by harvesting the plants every month from floating wire-mesh cages exposed between the floating plant communities. Available figures for growth rates vary strongly in different years between lakes and even between different parts of the same lake. Studies of free-floating species *Eichhornia crassipes, Salvinia auriculata*

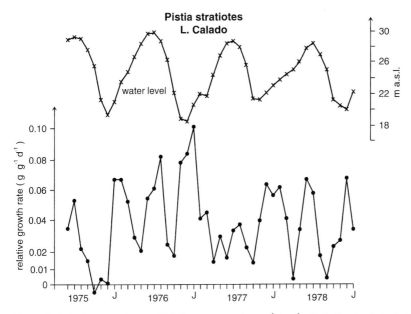

Fig. 8.6. Relative growth rate (RGR), measured in g g^{-1} day^{-1}, of *Pistia stratiotes* in the inner part of Lago Calado, in relation to changes in water level

and *Pistia stratiotes*, show maximum values of RGR of 0.08–0.1 g g^{-1} day^{-1} (Junk and Howard-Williams 1984; Fig. 8.6). At this rate, a doubling of the biomass occurs in 7.2–6.9 days. The results of cage experiments suffer methodological artifacts, e.g., by phytophagous animals trapped in the cages, protection of the plants against waves or reduced water and nutrient exchange with the surroundings. Local differences and differences from year to year are evident. Even so, two growth maxima can be recognized. The first occurs when the water level begins to rise. The second is a weaker one and occurs at high water level. Low growth rates were always found at the time of falling water level. Biomass losses of *Pistia stratiotes* were even larger than biomass accumulation, resulting in a negative RGR in Lago Castanho (Junk and Hoard-Williams 1984).

A comparison of the net primary production of free-floating populations with that of the rooted vegetation is difficult because free-floating plants increase biomass mostly by horizontal spreading, whereas rooted plants do it by vertical growth. *Salvinia auriculata* completely covers the water surface with a biomass corresponding to a dry weight of about 100–150 g m^{-2}. During a 4-week period with a doubling time of 7 days, the biomass per square meter would not change significantly, however, the area covered by the plant would increase 16-fold. Starting with 100 g biomass, this corresponds to a theoretical monthly primary production of

$1500\,\mathrm{g\,m^{-2}}$ and a total biomass of $1600\,\mathrm{g\,m^{-2}}$, which, in reality, cannot be reached because the maximum RGR can no longer be attained when space becomes limited.

The exponential growth potential of free-floating macrophytes is demonstrated when calculating the biomass increase during the rising- and high-water periods. Considering an average RGR of 0.069, which corresponds to a doubling time of 10 days, $1\,\mathrm{m^2}$ of *S. auriculata*, corresponding to $100\,\mathrm{g}$ dry weight biomass, would grow from first of December to the 30th of June to about $2\,\mathrm{km^2}$, corresponding to a biomass of $200\,\mathrm{t}$ dry weight. This calculation is not realistic because it does not consider major population losses, but it shows that a few plants remaining after the low-water period can reestablish large populations by vegetative propagation in a very short time when environmental conditions become favorable.

8.5.2 Biomass and Net Primary Production of Rooted Species

Data on the biomass and net primary production of rooted herbaceous plants in the várzea show a wide range of variation dependent upon the species or the community and the duration of the growth period. A limited growth period is characteristic for most populations in the aquatic terrestrial transition zone of the várzea. Under natural conditions, it lasts from 2

Table 8.5. Biomass (living and dead material) of annual herbaceous plant communities in sedimentation areas in the várzea near Manaus shortly before the beginning of flooding

Day	Locality	Dominating species	Biomass $\mathrm{t\,ha^{-1}}$ dry weight
07.3.74	Costa do Baixio	Cyperaceae	6.4 (3.4)
	Costa do Baixio	Cyperaceae, Gramineae	3.2 (2.2)
	Costa do Baixio	*Aeschynomene sensitiva*	8.7 (5.3)
15.1.82	Marchantaria Sedimentation area	Cyperaceae	5.7 (3.5)
	Marchantaria Sedimentation area	Gramineae	4.8 (2.4)
	Marchantaria Sedimentation area	*Shpenoclea zeylanica*	7.2 (3.3)
	Marchantaria Sedimentation area	Cyperaceae, Onagraceae	6.1 (2.6)
	Marchantaria Sedimentation area	Cyperaceae, Gramineae	5.5 (2.7)
	Marchantaria Sedimentation area	Cyperaceae	3.4 (1.7)

Harvested area, $1\,\mathrm{m^2}$; n = 10; ( ) = Standard deviation.

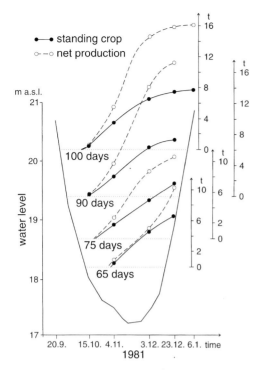

Fig. 8.7. Biomass (t dry wt. ha^{-1}) and net primary production of *Ludwigia densiflora* on Ilha de Marchantaria in relation to the duration of the terrestrial phase. Net production was calculated by adding the calculated weight of lost plantlets between two successive harvests. Fine rootlets are not considered. (Junk 1986)

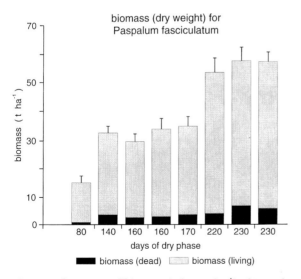

Fig. 8.8. Aboveground biomass (t dry wt. ha^{-1}) of *Paspalum fasciculatum* at different positions on the flood gradient shortly before inundation. Only the biomass produced in the actual growth period was considered. (n = 10, *standard deviation bars* in the columns. (Junk and Piedade 1993c)

to 3 months for many annual terrestrial species, to as long as 6 to 10 months for the perennial terrestrial grass *Paspalum fasciculatum* and for many floating aquatic species.

Annual species reach maximum biomass values of up to about $9\,t\,ha^{-1}$ in a 3-month growth period (Table 8.5). Net primary production can be about twice as high as shown for *Ludwigia densiflora* (Fig. 8.7). The highest biomass values were attained by the large perennial C4 grasses *Paspalum fasciculatum*, *P. repens* and *Echinochloa polystachya*. Values for *Paspalum fasciculatum* vary strongly depending upon the duration of the growth period (Fig. 8.8). The highest value of $57\,t\,h^{-1}$ was obtained in habitats on the levees with about 230 days of dry phase. However, under natural con-

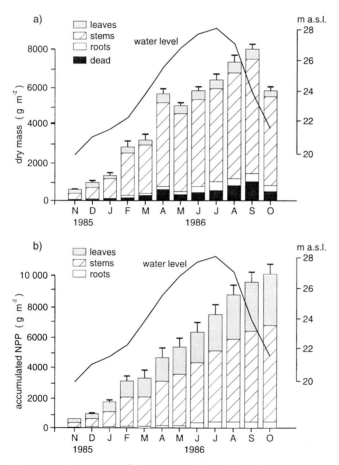

Fig. 8.9. a Biomass ($g\,m^{-2}$ dry wt.) of *Echinochloa polystachya* present in each month (mean + SD). Leaf mass includes both laminae and sheaths. **b** Accumulated net primary production ($g\,m^{-2}$ dry wt.) of *Echinochloa polystachya* since the appearance of new plants in November (mean + SD). (Piedade et al. 1991)

ditions such habitats are normally occupied by the floodplain forest. *Echinochloa polystachya*, which grows during both the terrestrial and the aquatic phase, attains a dry biomass of 80 t ha^{-1} year^{-1} (Fig. 8.9a).

In both species, the net primary production is about 25% greater than the maximal biomass values (Fig. 8.9b). Thus, the net productivity of *Echinochloa polystachya* reaches values of about 100 t ha^{-1} and it is among the most productive plant species in the world (Piedade et al. 1991, 1992). In natural plant populations, net production of *Cyperus papyrus* ranges from 48 to 143 t ha^{-1} year^{-1} dry weight (Thompson et al. 1979). Even the most productive crops seldom reach such high values. Westlake (1963) reported biomass values of 73 to 85 t ha^{-1} year^{-1} and a net production of 108 t ha^{-1} year^{-1} for *Saccharum officinarum* plantations on Hawaii and Java. Biomass values of other aquatic species are given in Table 8.6. The figures reported for *Paspalum repens* are considerably higher than those recorded by Junk (1970). This is due to the fact that *P. repens* in part behaves like a free-floating plant, expanding laterally with a low density at the outside of the stands.

Net primary production of all species is considerably higher than biomass; however, losses during the aquatic phase are difficult to measure because of very quick decomposition (Chap. 9). Junk and Piedade (1993c) assume a monthly loss rate of 10–25% of the available biomass but consider

Table 8.6. Biomass (living and dead material) of well-developed stands of *Paspalum repens*, *Oryza perennis*, *Hymenachne amplexicaulis* and *Luziola spruceana*

Period	Species	Lake	Biomass (t ha^{-1} dry wt.)
17.04.75	P. repens	Lago Janauari	22.1 (3.7)
16.06.75	P. repens	Lago Calado	18.3 (3.4)
17.06.75	P. repens	Lago Calado	12.7 (2.2)
02.06.75	P. repens	Lago Castanho	17.6 (3.6)
03.06.75	P. repens	Lago Castanho	19.2 (4.4)
08.06.75	P. repens	Costa do Baixio	17.4 (3.1)
02.07.83	P. repens	Marchantaria	21.4 (4.3)
03.07.83	P. repens	Marchantaria	14.2 (2.2)
18.03.83	P. repens	Marchantaria	13.7 (1.8)
28.02.75	O. perennis	Lago Janauari	19.1 (4.0)
15.04.75	O. perennis	Lago Calado	12.6 (2.7)
05.03.83	O. perennis	Lago Camaleão	17.2 (3.4)
10.03.83	O. perennis	Lago Central	15.3 (3.4)
15.04.75	H. amplexicaulis	Lago Janauari	22.7 (4.1)
28.04.81	H. amplexicaulis	Lago Camaleão	17.2 (3.4)
14.12.79	L. spruceana	Lago Camaleão	5.5 (1.2)
28.12.81	L. spruceana	Lago Camaleão	6.4 (1.3)

Harvested area, 1 m^2; n = 10; ( ), standard deviation.

25% more realistic. Under optimal conditions, for *Paspalum repens*, with a growth period of 8–9 months, net primary production can be twice as high as biomass, reaching up to 50 t ha^{-1}. However, stands are frequently disturbed by wind and current and are therefore difficult to study for such a long period. The high annual net primary production is also the result of the successive exploitation of the same area by several plant communities during the hydrological cycle. In sheltered habitats a community of terrestrial plants is replaced by one of aquatic plants rooted in the sediment, which later becomes substituted by free-floating species (Fig. 8.10). Quick decomposition allows a rapid recycling of the nutrients between the communities (Sect. 9.2.3)

In most populations, the subterranean biomass is low. Subterranean root biomass of *Echinochloa polystachya* reaches about 5% of total biomass (Junk 1990). Exceptions are species with rhizomes. *Cynodon dactylon* produces rhizomes just below the soil surface. At the end of the terrestrial phase these can constitute about half of the total biomass. Another perennial species with a large proportion of subterranean biomass is

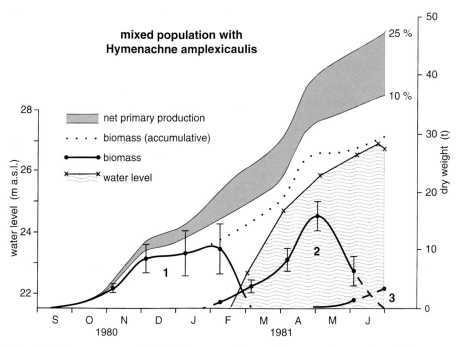

Fig. 8.10. Biomass, accumulative biomass, and net primary production (t ha^{-1} dry wt.) calculated for mixed populations of gramineans, cyperaceans and *Aeschynomene sensitiva* (*1*) during the terrestrial phase, *Hymenachne amplexicaulis* (*2*) and floating plants (*3*) during the aquatic phase in Lago Camaleão, assuming monthly loss rates of 10 and 25%. (Junk and Piedade 1993c)

Montrichardia arborescens. It produces rhizomes of up to 10 cm in diameter penetrating to a depth of 1 m below the soil surface.

8.5.3 Net Primary Production and Nutrient Supply

The high primary production rates require a large nutrient supply in the water and in the sediment. As shown in Chapter 3, várzea soils are rich in nutrients easily available for plant growth. The hydrochemical data also show a sufficient supply of dissolved nutrients, but they exhibit spatial and temporal variations as well (Sect. 4.4). The growth periodicity observed for the floating species can be related to differences in the nutrient supply during the hydrological cycle. When the water level is low, there is often an input of nutrients from surface runoff, seepage from groundwater, and the exchange with the sediment. As the water level rises, there is a continuous input of nutrients by the river water and from the decaying terrestrial vegetation in the newly flooded areas. A shortage of nutrients may occur mainly at falling water level when there is no input from the main channel and little input from surface runoff from the drying shores of the lake because of lack of rain. Therefore the growth maxima of free-floating plants occur predominantly during the periods of low, rising, and high water levels.

It is difficult to determine which nutrients are limiting for plant growth in the aquatic phase. Considering the permanent presence of relatively high concentrations of nutrient elements in the waters of Lago Camaleão, only nitrate could be a limiting factor for primary production of herbaceous plants. As shown by Schlüter and Furch (1987), the root coloration of *Eichhornia crassipes* can be taken as an indicator of nitrogen supply. When there is a lack of nitrogen compounds the roots are colored deep blue due to anthocyanin deposition. When the supply of nitrogen is good they are of a brownish color. The plants growing in Lago Camaleão sometimes had blue and sometimes brownish roots. Enclosure experiments with phytoplankton employing nitrogen and phosphorus in Amazonian várzea lakes indicated that nitrogen is more frequently a limiting factor for primary production than phosphorus (Zaret et al. 1981; Setaro and Melack 1984). However, many macrophyte species accumulate nutrients in their tissues. Adult *Eichhornia crassipes* plants grown in a greenhouse showed signs of nitrogen deficiency other than the root coloration after 30 days (Schlüter 1985). This leaves the question open as to what degree short-term shortage of nitrogen is a limiting factor for growth under natural conditions.

Long-term shortage of nutrients in acid black water inhibits or strongly reduces the occurrence and growth of most aquatic and semiaquatic spe-

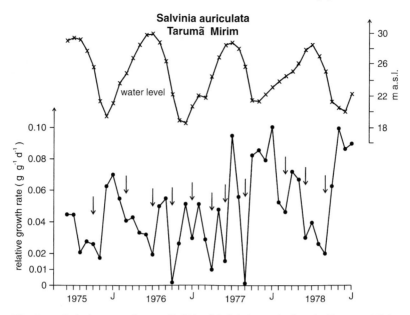

Fig. 8.11. Relative growth rate (RGR) of *Salvinia auriculata* in Taruma Mirim, a black-water habitat. The *arrows* indicate the introduction of fresh plant material from white-water habitats, because of the deterioration of the old plants

cies. *Salvinia auriculata*, *Eichhornia crassipes* and *Pistia stratiotes* were exposed in floating wire-mesh boxes in the mouth bay of Tarumã Mirim, strongly influenced by the black water of Rio Negro. The pH varied between 4 and 5 and specific conductance ranged between 6 and $8\,\mu S\,cm^{-1}$ at $20\,°C$. *Salvinia auriculata* grew well for several weeks, reaching for short periods a RGR similar to that in white water. However, after 1–2 months, plants began to deteriorate and strongly reduced in size and had to be substituted by new ones from the várzea (Fig. 8.11). *Eichhornia crassipes* and *Pistia stratiotes* did not grow well and from the beginning showed monthly weight losses of up to 35%.

The trials with *Eichhornia crassipes* in the black waters of Tarumã Mirim confirmed the results of Berg (1961). After extensive investigations in the Congo, he concluded that not only a nutrient shortage but also a low pH in the water can limit the growth of *Eichhornia crassipes*. In combination with a low supply of electrolytes, a low pH has a toxic effect.

8.5.4 Light Utilization and Photosynthetic Activity

Determinations of CO_2 absorption using infrared gas analysis provided valuable information on the photosynthetic rates and efficiency

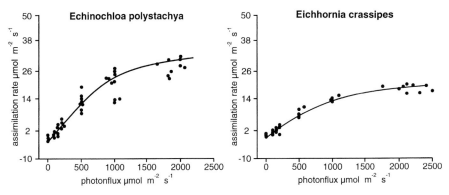

Fig. 8.12. The response of leaf CO_2 uptake to incident photon flux (light response curve) at noon ±1 h by the C_4-plant *Echinochloa polystachys* and the C_3-plant *Eichhornia crassipes* during the aquatic phase

of light utilization by the species under various environmental conditions.

The quantum flux near the equator is high. On the Ilha de Marchantaria midday values of $2500 \mu mol\, m^{-2} s^{-1}$ were recorded on cloudless days (Fig. 8.12). Light response curves indicate that most herbaceous species are "high-light" plants. With increasing light intensity, the rate of photosynthesis also increases. Saturation in C_3 plants is reached with about $1500 \mu mol\, m^{-2} s^{-1}$, in C_4 plants with about $2000 \mu mol\, m^{-2} s^{-1}$.

Most of the herbaceous plant species colonizing wetlands are C3 plants. There are, however, some C_4 plants that have colonized tropical wetlands very successfully. *Cyperus papyrus* covers large areas in African swamps, and in the Amazon várzea *Echinochloa polystachya*, *Paspalum fasciculatum* and *Paspalum repens* form large monospecific stands. C_4 plants are capable of a significantly better exploitation of the solar energy than C_3 species. Maximum rates of photosynthesis per square meter of leaf area measured for the C_3 plants *Oryza perennis*, *Hymenachne amplexicaulis*, and *Eichhornia crassipes* during the aquatic phase ranged from about 20 to $22 \mu mol\, s^{-1} m^{-2}$. For the C_4 species *Echinochloa polystachya*, *Paspalum repens*, and *Paspalum fasciculatum*, the corresponding values ranged from about 30 to $35 \mu mol\, s^{-1} m^{-2}$ (Fig. 8.12).

The primary production of a population depends not only on the rate of light conversion but also on light interception, which is regulated by the total leaf area and the spatial arrangement of the leaves. In *E. polystachya* the monthly average number of living leaves per plant varied from 6.5 to 9 (n = 20). The leaf area index (LAI) ranged from 1.5 to 4.75. This leaf area and the stems were sufficient to intercept about 71.6–89.3% of the incoming solar energy per month.

Table 8.7. Leaf area index (LAI) in different stands of aquatic and semiaquatic herbaceous plants of the várzea

Date	Species	Locality	LAI
1986	*Echinochloa polystachya*	Marchantaria	1.50–4.75[a]
13.9.88	*Paspalum fasciculatum* (young)	Marchantaria	3.87 (1.05)
7.11.88	*Paspalum fasciculatum* (well-developed)	Marchantaria	4.92 (0.55)
11.4.89	*Paspalum fasciculatum* (senescent)	Marchantaria	3.82 (0.61)
19.4.89	*Hymenachne amplexicaulis* (well-developed)	Marchantaria	3.80 (0.54)
19.4.89	*Paspalum repens* (well-developed)	Marchantaria	3.72 (0.42)
25.4.89	*Paspalum repens* (senescent)	Marchantaria	1.84 (0.22)
25.4.89	*Oryza perennis* (low density)	Marchantaria	2.98 (0.57)
27.4.89	*Eichhornia crassipes* (well-developed)	Marchantaria	2.58 (0.18)

Harvested area, $1 m^2$; n, = 10; ( ), standard deviation; [a] range of monthly samples (n = 10) during 1 year.

The LAI of *Echinochloa polystachya* is similar to that of other herbaceous species in the várzea (Table 8.7), but it is low in comparison with crop plantations. For example, the LAI of rice crops varies between 7 and 8 (Zelitch 1971). The comparatively low LAI of *Echinochloa polystachya* is explained by the flood pulse. While the water level is rising rapidly the plant rooted in the sediment continually loses the leaves on the lower parts of the stem, but these are continuously replaced by new leaves developed higher up. Since the photosynthetic rates of leaves decrease with age, the low average age of the leaves contributes to the high light-use efficiency of the photosynthetically active surfaces. During the aquatic phase young leaves of the upper part of the plant showed a 10% higher rate than older leaves of the lower part. During the terrestrial phase, when the leaves remain active for a longer period, this difference between young and old leaves reached about 30% (Fig. 8.13).

Monteith (1978) assumed that the maximum efficiency of C_4 plants in energy conversion is about $2 g MJ^{-1}$. In contrast the corresponding efficiency of C_3 plants is about $1.4 g MJ^{-1}$. Based on the data on primary production, *Echinochloa polystachya* had an average photosynthetic efficiency of $2.3 g MJ^{-1}$. The value for the biomass was about $2.1 g MJ^{-1}$ (Piedade et al. 1994; Fig. 8.14). Assuming the average energy content of the dry substance

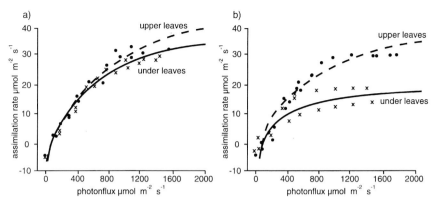

Fig. 8.13a,b. Light response curve at noon ±1h in the upper and in the lower canopy of *Echinochloa polystachya* at the peak of the aquatic phase (**a**) and during the terrestrial phase (**b**). (Piedade et al. 1994)

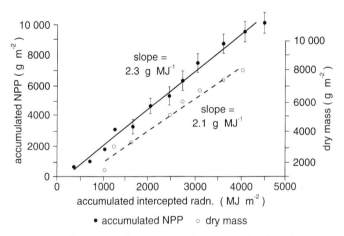

● accumulated NPP   ○ dry mass

Fig. 8.14. The accumulated net primary production (mean + SD) of *Echinochloa polystachya* plotted against the accumulated quantity of solar radiation that had been intercepted by the canopy. The best-fit line, as determined by least squares regression, show an efficiency of conversion of intercepted solar radiation into plant mass of $2.3\,g\,MJ^{-1}$. The biomass plotted against the intercepted radiation gives an efficiency of $2.1\,g\,MJ^{-1}$. (Piedade et al. 1991)

to be $20\,MJ\,kg^{-1}$, a very high utilization rate of the incoming solar energy amounting to about 4% is reached.

The C_4 photosynthetic pathway is very efficient under drought stress (Teeri and Stowe 1976). To produce 1 kg of organic matter under identical conditions, C_4 plants require 0.3–$0.4\,m^3$ of water whereas C_3 plants need 0.6–$0.8\,m^3$ (Medina et al. 1976). Therefore the number of C_4 plants in savannas is large (Teeri and Stowe 1976; Stowe and Teeri 1978; Pearcy and

Ehleringer 1984). In wetlands C_3 plants with large leaf areas, such as *Eichhornia crassipes*, can even be considered better adapted than C_4 plants because they may even benefit from high evapotranspiration rates for a more efficient cooling of the leaves and an increased uptake of dissolved nutrients (Thompson 1985). Therefore it may be surprising to find several C_4 plants in the floodplains of the Amazon.

A closer examination of the habitat conditions and preferences of the species on the flood plain provides the following explanation for the successful establishment of C_4 species in the várzea. The perennial C_4 grasses *Paspalum fasciculatum*, *Echinochloa polystachya*, and *Paspalum repens*, are the only herbaceous species that are able to colonize in large numbers habitats exposed to strong current and waves, e.g., the river banks and large lakes. The advantage of the C_4 species during the aquatic phase results from the high productivity which allows the plants to settle in habitats that cannot be permanently colonized by the less productive C_3 species. The competition between the C_4 grasses is reduced because they have different habitat requirements and occupy different ranges on the flood gradient. *Paspalum repens* can change completely to a floating mode of existence and is by far the most successful floating plant because of its high productivity, in spite of the presence of highly productive floating C_3 species such as *Eichhornia crassipes*, *Salvinia auriculata*, and *Pistia stratiotes*. At low-lying sites that are settled by *Echinochloa polystachya*, the high productivity of the species is just sufficient to keep the leaves above the water so that the plants are not damaged or carried away by the current and waves.

During the terrestrial phase, a greater water-use efficiency provides a selective advantage during very dry periods when C_3 plants suffer drought stress. *Paspalum fasciculatum* and *Echinochloa polystachya* also cover areas on the upper part of the flood gradient which can be colonized as well by trees from the floodplain forest. The great productivity of the tall grasses during the terrestrial phase inhibits the establishment of seedlings of the pioneer tree species.

Another competitive advantage may be the better nitrogen-use efficiency. The photosynthetic rate per unit nitrogen of a leaf is higher in C_4 plants than C_3 plants (Brown 1978), because C_4 plants achieve high photosynthetic rates with low concentrations of the photosynthetic enzyme RuBP carboxylase/oxygenase where more than half of the soluble protein of a leaf is located. In várzea lakes nitrogen is considered limiting for phytoplankton growth (Sect. 4.4). Várzea soils show relatively low nitrogen levels (Chap. 3). The low nitrogen content in the submersed stems of *Echinochloa polystachya*, in comparison with the high content in the photosynthetically active emergent parts, may be interpreted as a strategy to optimize the use of nitrogen. Melendez (1978) found a marked response of

E. polystachya to nitrogen under cultivation in a flooded savanna in Venezuela. Experiments in the várzea also show responses to nitrogen application (Piedade and Junk, unpubl.).

8.6 Discussion and Conclusions

With 64 families, 182 genera, and 388 species, excluding epiphytes, the herbaceous vegetation in the central part of the várzea near Manaus is astoundingly rich in species. Beck (1983) found about 400 plant species, including trees, in the flooded savannas of the Bolivian lowlands. The flora of the savannas in Humaitá encompasses 314 species of the open grasslands, 151 species of the sparsely scattered patches of shrub growth, and 67 ruderal species (Janssen 1986). Seidenschwarz (1986) reported 245 species of herbaceous plants from the open areas of the river banks along the Rio Yuyapichis and Rio Pachitea and on the nearby areas of newly cultivated fields and roadsides in the tropical lowlands of Peru. Scheßl and Prado (pers. comm.) indicate about 600 herbaceous species near Poconé for the Pantanal of Mato Grosso, floodplain of the upper Rio Paraguay.

The species spectrum ranges from ruderal plants to species with very specific habitat requirements. An important reason for this diversity is the predictability of the flooding events and the monomodal flood pulse. The species can settle along the flood gradient according to their requirements and adaptations to the specific habitat conditions. Among such adapted species are the aquatic and semiaquatic grasses *Oryza* spp., *Luziola spruceana*, *Echinochloa polystachya*, *Hymenachne amplexicaulis*, *Paspalum repens*, and *Paspalum fasciculatum*. These species depend on a regular monomodal flood pulse and have not been reported from habitats along the upper reaches of the rivers, where the floodplains are subjected to irregular, rapid pulses. The same is true for the floating aquatic macrophytes, e.g., *Eichhornia crassipes*, *Pistia stratiotes*, and *Salvinia auriculata*, which require long flood periods.

In spite of the regularity of the flood pulse, losses due to the fluctuations between the aquatic and terrestrial phase are large. Therefore, even well-adapted strongly competitive species which grow very rapidly, e.g., *Eichhornia crassipes* and the C_4 grasses *Echinochloa polystachya*, *Paspalum repens* and *P. fasciculatum*, are unable to completely displace the smaller, less competitive species. Before competition excludes other species, all populations are reduced again to a minimum by the terrestrial or aquatic phase and begin to reestablish themselves again during the following growth season. According to Odum (1959), plant communities in floodplains are maintained at a low seral stage because periodic flooding and drying result in a continuous setback in community development (pulse stability). The relatively large number of species corresponds to the intermediate disturbance hypothesis (Connell 1978), which postulates that highest diversity is maintained at intermediate scales of disturbance. It also corresponds to the flood pulse concept (Junk et al. 1989), which postulates high species diversity in predictably pulsing systems.

Small irregularities in the flood pulse and in the distribution of rainfall during the terrestrial phase change habitat conditions and assure advantages to different species each year. During years of low and slowly rising water levels semiaquatic species rooted in the ground, e.g., *Luziola spruceana*, *Oryza perennis*, and *Sagittaria sprucei* are favored, while free-floating species, e.g., *Eichhornia crassipes* and *Scirpus cubensis* occur in smaller numbers. When the water level does not fall below the 21-m level or rises quickly, only small populations of the first species group will develop while free-floating species become predominant. The fluctuations in the occurrence of the plant species control the populations of phytophagous insects, e.g., the grasshopper *Paulinia acuminata*, which feeds on *Salvinia* spp. and *Pistia stratiotes*. During periods of very low water level, its populations are strongly reduced because of lack of food and their population density may be much smaller in the subsequent flood period (Vieira and Adis 1992).

The great majority of the terrestrial species of the várzea, 273 (or 71%) of the total, prefer disturbed locations on the levees where flood stress is least. Levees are normally covered by the floodplain forest which reduces light penetration to the forest floor. Only 25 herbaceous plant species (7% of the total in the region) are able to inhabit this habitat. Clearing the forest allows a great number of low flood-tolerant species to colonize the habitats. Because most of the species that prefer low-lying habitats also occur occasionally in frequently disturbed areas at higher elevations, the total number of species is even greater.

Many of these species are ruderal plants and weeds, which are not confined to the várzea (Chase 1944; Holm et al. 1977, 1979; Aranha et al. 1982; Leitão Filho et al. 1982; Lorenzi 1982). Among them are many species introduced from other areas (neophytes) (Table 8.8). An increase in the number of ruderal species after the clearing of the forest was also observed by Seidenschwarz (1986) on river banks in the Peruvian lowlands. An additional increase in the várzea is expected to occur because of a further clearing of the floodplain forest on the levees and the introduction of new weeds by agriculture. First results point to a strong impact of grazing pressure from cattle and buffalo on species diversity in natural pastures. Ohly (1987) reported 105 herbaceous species in a natural pasture used for

Table 8.8. Herbaceous plants of the várzea that are considered worldwide, troublesome weeds, and their regions of origin

Name	Origin	Distribution
Mostly terrestrial		
Cyperus rotundus	India	Mostly Tropics and Subtropics, worldwide
Cyperus esculentus	?	Mostly Tropics and Subtropics, worldwide
Fimbristylis miliacea	Trop. America	Humid Tropics, worldwide
Fimbristylis dichotoma	Trop. America	Humid Tropics, worldwide
Cynodon dactylon	Africa or Indo-Malaysia	Mostly Tropics and Subtropics, worldwide
Eleusine indica	probable S.E. Asia	Mostly Tropics and Subtropics, worldwide
Paspalum conjugatum	Trop. America	Humid Tropics, worldwide
Panicum maximum	Africa	Tropics and Subtropics, worldwide
Brachiaria mutica (*Panicum purpurascens*)	Trop. Africa	Humid Tropics and Subtropics, worldwide
Pennisetum purpureum	Trop. Africa	Tropics and Subtropics, worldwide
Portulaca oleracea	North Africa or Europe	Worldwide, excepting northern high latitudes
Digitaria sanguinalis	Europe	Worldwide, excepting northern high latitudes
Amaranthus spinosus	Trop. America	Tropics and Subtropics, worldwide
Heliotropium indicum	Paleo-Tropics	Tropics, worldwide
Lantana camara	Trop. America	Tropics and Subtropics, worldwide
Sida acuta	Central America	Tropics, worldwide
Solanum nigrum	Europe	Worldwide
Sphenoclea zeylanica	Trop. Africa	Tropics, worldwide (excepting Australia?)
Mostly aquatic		
Ceratophyllum demersum	Unknown	Worldwide
Pistia stratiotes	Unknown	Tropics and Subtropics, worldwide
Salvinia auriculata	Trop. S. America	Tropics, worldwide
Eichhornia crassipes	Trop. S. America	Tropics, worldwide
Leersia hexandra	Trop. America	Tropics and Subtropics, worldwide

buffalo ranching; 10 years later the species number had decreased to about 55 species (Conserva and Piedade, unpubl.).

The increase in areas covered by herbaceous plants and the use of fire for pasture management and weed control lead to frequent and uncontrolled burning of the herbaceous vegetation. Most survival strategies of flood-plain species against flooding fail to provide protection against fire stress. This explains the susceptibility of many flood-adapted species, e.g., *Echinochloa polystachya* and *Paspalum fasciculatum*, to repeated burning. This points to the following tendencies in herbaceous plant community development; total species number tends to increase because of immigration or introduction of neophytes. Increased fire stress and grazing pressure reduce species number in natural pastures and favor the growth of generalized ruderal plants.

The determination of the primary production of aquatic and terrestrial herbaceous plants in the Amazon floodplain by the harvesting method is difficult, because decomposition of dead material in the hot and humid climate is very quick (Sect. 9.2.2) and dead material cannot be sampled quantitatively during the aquatic phase. Therefore, the harvesting method tends to underestimate primary production. High productivity results from good nutrient supply. In the várzea of the Rio Amazonas nutrient content in water and soils is high and there is an annual input of nutrients into the system by the flood. In the aquatic phase, nitrogen is considered a limiting factor for phytoplankton. Whether this is also correct for herbaceous plants still remains an open question. Enclosure experiments showed a positive response of *Eichhornia crassipes* to NH_4 in floodplain lakes of the Rio Paraná with levels of $0.1-0.7\,\mu mol\,l^{-1}$ inorganic nitrogen (Garignan and Neiff 1994). Mean inorganic nitrogen content in the Rio Amazonas reaches about $11\,\mu mol\,l^{-1}$ and in várzea lakes it varies between 0.5 to $18\,\mu mol\,l^{-1}$ (mainly NH_4). In Ria lakes, macrophytes grow mostly near the lake mouth stripping nitrogen from the inflowing river water (Doyle 1991). In black water habitats, e.g., the Rio Negro, low nutrient levels together with a low pH limit the occurrence and primary production of aquatic macrophytes.

During the terrestrial phase lack of rain at the beginning of the terrestrial phase strongly reduces seedling establishment on mud flats and diminishes primary production of species with a superficial root system. More efficient water use during periodic drought stress may be one of the competitive advantages of the C_4 grasses *Paspalum fasciculatum* and *Echinochloa polystachya* in the várzea.

Often, total annual net primary production (NPP) is the sum of the production of two or three plant communities which follow one another: a community growing during the terrestrial phase including individuals of semiaquatic or aquatic plants; a community of semiaquatic and aquatic grasses rooted in the ground, which substitutes the terrestrial community when the river floods the area; and a community of free-floating aquatic macrophytes. Under favorable conditions cumulative annual NPP can reach about $50\,t\,ha^{-1}$ and is two to three times greater than maximum biomass because of rapid decomposition of the dead material. Highest production and biomass values are reported for the perennial C_4 grasses *Paspalum fasciculatum* and *Echinochloa polystachya* with up to 50 and $80\,t\,ha^{-1}\,year^{-1}$ biomass and 60 and $100\,t\,ha^{-1}\,year^{-1}$ NPP.

Acknowledgments. We are very grateful to Dr. Ernesto Medina, Instituto Venezolano de Investigaciones Cientificas, Caracas, Dr. Brij Gopal, Jawaharlal Nehru University, and Dr. Rosemary Lowe-McConnell, Sussex, for critical comments on the manuscript and the correction of the English.

9 The Chemical Composition, Food Value, and Decomposition of Herbaceous Plants, Leaves, and Leaf Litter of Floodplain Forests

KARIN FURCH and WOLFGANG J. JUNK

9.1 Introduction

Structure and function of an ecosystem are determined to a large extent by its trophic dynamics (Lindeman 1942). Organic material produced by primary producers is processed by a suite of consumer and decomposer populations, the species composition and size of which depends upon the amount, quality and availability of the organic material (Boyd and Goodyear 1971). In Amazonian floodplains large amounts of non-woody biomass are available for primary consumers. Herbaceous plant communities produce up to $100\,t\,ha^{-1}\,year^{-1}$ dry matter in the várzea. The floodplain forest of the várzea provides up to $13.6\,t\,ha^{-1}\,year^{-1}$ leaf litter and that of the igapó up to $6.7\,t\,ha^{-1}\,year^{-1}$ (Chaps. 8, 11). However, only a very small fraction is utilized by herbivorous animals.

Fish feed preferentially on algae, fruit and seeds (Sect. 20.2). Birds also concentrate mainly on fruit and seeds (Chap. 22). The capybara (*Hydrochoerus hydrochaeris*) and the manatee (*Trichechus inunguis*) are the only herbivorous mammals which earlier occurred in large numbers in the Amazon floodplain (Chap. 21). Other important primary consumers were the large river turtles (*Podocnemis* spp.). Aquatic invertebrates play a minor role as primary consumers of aquatic macrophytes (Sect. 13.4). Similarly, the number of terrestrial invertebrates feeding directly on the plant material is relatively small. Most conspicuous are the grasshoppers, e.g., *Paulinia acuminata* and *Cornops aquaticum*.

Because of the relatively low uptake by herbivorous animals, most of the plant material produced on the floodplain enters into the detritus food webs. The large amounts of organic material accumulate large amounts of bioelements that are released during decomposition and modify the chemical conditions of the environment. The biogeochemical cycles of the floodplain depend on the amount and quality of the organic material and its fate.

First we present the chemical compositions and the nutritive values of the herbaceous plants, the tree leaves, and the leaf litter to examine their

Ecological Studies, Vol. 126
Junk (ed) The Central Amazon Floodplain
© Springer-Verlag Berlin Heidelberg 1997

suitability for the herbivores and their relationship with decomposition. Because no data exist on the nutritional requirements of native animals we assess the nutritive value on the basis of nutrient requirements of cattle given by the US National Research Council (NRC 1984). We then present studies of the decomposition of both herbaceous plants and tree leaf litter in water and on land, because the environmental conditions differ strongly in the terrestrial and aquatic phase as indicated by the flood pulse concept (Junk et al. 1989; Sect. 1.3).

9.2 Herbaceous Plants

9.2.1 Chemical Compositions and Nutritive Values of the Plants

Herbaceous plants include species growing mainly during either the terrestrial phase or the aquatic phase (Sect. 8.8.3). A distinction between the two categories is important because they utilize nutrients from different sources (sediments and water, respectively) and are also processed by different consumer and decomposer organisms. Some species grow in both phases and are included in both categories.

The chemical compositions of herbaceous plants growing in the várzea during the terrestrial phase are given in Table 9.1 (modified from Ohly 1987). The data show that the chemical composition of grasses (including sedges) and non-grass species differ significantly. The average nitrogen concentration of grasses and sedges is 1.50% compared to 2.88% of non-grass herbs. The grasses and sedges also have lower concentrations of phosphorus and calcium than the non-grass species, but the concentrations of sulfur, potassium, chlorides, and silica are higher than in non-grass species. Iron concentrations vary over a wide range. Highest values of 716 mg kg^{-1} were recorded in *Croton trinitatis* (Ohly 1987), whereas the copper and zinc concentrations were generally low.

The aquatic macrophytes have lower concentrations of P, Ca, and Mg, and higher concentrations of K and Na than terrestrial herbaceous plants (Table 9.2). Nitrogen concentrations are similar. The water contents of aquatic macrophytes, which vary from 69.7% in *Leersia hexandra* to 95.9% in *Utricularia foliosa* (Howard-Williams and Junk 1977), are high and lower their food value.

A comparison of these data with those reported elsewhere for aquatic macrophytes (Boyd 1970a; Cowgill 1974; Easley and Shirley 1974) and for land forages (NRC 1971; NRC 1984) shows that the concentrations of nitrogen, phosphorus, potassium, and magnesium lie within the same range,

Table 9.1. Mean contents and ranges of elements, H_2O, ash, raw protein, total fat, acid detergent fiber and acid detergent lignin, total energy and metabolizable energy of herbaceous plants growing during the terrestrial phase in natural pastures of the várzea (according to Ohly 1987) and requirements and tolerance limits of cattle according to NRC (1984)

Components		Grasses and sedges		Herbs		Cattle	
		Mean	Range	Mean	Range	Required range	Upper tolerance limit
N	% dry wt.	1.50	0.44–2.59	2.88	2.11–4.22	–	–
P	% dry wt.	0.24	0.14–0.34	0.46	0.23–0.79	0.19–0.24	1.0
K	% dry wt.	2.16	1.36–3.21	1.70	1.19–2.35	0.5–0.7	3.0
Ca	% dry wt.	0.51	0.27–0.91	1.51	1.02–2.22	0.28–0.45	2.0
Mg	% dry wt.	0.29	0.13–0.48	0.41	0.26–0.55	0.05–0.25	0.4
Na	% dry wt.	0.0129	0.003–0.032	0.015	0.003–0.023	0.06–0.1	–
S	% dry wt.	0.76	0.16–1.34	0.39	0.19–0.62	0.08–0.15	0.4
Si	% dry wt.	2.74	0.47–5.75	1.51	0.51–2.96	–	–
Cl	% dry wt.	0.13	0.05–0.28	0.06	0.01–0.15	0.1–0.2	–
Fe	mg kg^{-1} dry wt.	233.43	40.75–704.20	159.53	51.90–716.15	50.0–100.0	–
Zn	mg kg^{-1} dry wt.	4.93	1.77–10.19	5.41	2.24–10.42	20.0–40.0	–
Cu	mg kg^{-1} dry wt.	2.94	0.96–4.16	3.29	2.63–3.94	4.0–10.0	–
H_2O content	%	77.73	65.02–85.58	73.94	73.46–77.52	–	–
Organic matter	% dry wt.	86.94	79.45–92.15	87.28	80.15–93.79	–	–
Digestible organic substance	% dry wt.	51.91	39.31–58.94	55.47	49.31–65.84	–	–
Ash	% dry wt.	13.06	7.85–20.55	12.72	6.22–19.85	–	–
Raw protein	% dry wt.	9.37	2.75–16.19	18.04	13.19–26.37	8.8–11.4	–
Fat	% dry wt.	1.34	0.4–2.1	2.84	1.8–4.1	–	–
Acid detergent fiber	% dry wt.	36.18	29.56–41.94	31.48	18.71–42.74	–	–
Acid detergent lignin	% dry wt.	8.00	4.40–16.60	12.98	6.10–22.87	–	–
Total energy	MJ kg^{-1} dry wt.	15.23	13.70–16.84	16.14	14.76–17.59	–	–
Metabolizable energy	MJ kg^{-1} dry wt.	6.78	4.49–8.09	7.30	6.20–8.87	8.87–9.6	–

Seventeen grasses and sedges: *Cynodon dactylon, Digitaria horizontalis, Echinochloa polystachya* (2), *Eleusine indica, Eriochloa punctata, Hymenachne amplexicaulis, Panicum laxum, Paspalum conjugatum, P. fasciculatum* (5), *P. repens, Scleria microcarpa, Setaria geniculata.*
Seven non-grasses: *Croton trinitatis, Dalechampia tilifolia, Rhynchosia minima, Sida acuta, Teramnus volubilis* (2), *Wedelia paludosa.*
Raw protein is calculated by multiplying N-content by the factor 6.25.

but the sodium and calcium concentrations are lower in plants from the várzea. The polyphenol content is usually very low when compared with that of tree leaves from the terra firme rainforest, with some exceptions such as *Ludwigia natans, Neptunia oleracea,* and *Victoria amazonica* (Anderson 1973; Howard-Williams 1974).

Further comparison with the nutritional requirements of cattle (Table 9.1) shows that the mean crude protein content of the grasses and sedges (9.37%) is lower than the requirement of at least 12% for cattle (Boyd 1968),

Table 9.2. Mean contents and ranges of elements, H₂O, ash, raw protein, polyphenols, cell wall and total energy of grasses and sedges and non-grasses growing during the aquatic phase in the várzea. (According to Howard-Williams and Junk 1977)

Components		Grasses and sedges		Herbs	
		Mean	Range	Mean	Range
N	% dry wt.	1.74	1.20–3.40	2.26	0.85–3.56
P	% dry wt.	0.14	0.10–0.16	0.23	0.08–0.47
K	% dry wt.	2.61	1.66–3.33	3.58	1.78–5.69
Ca	% dry wt.	0.22	0.15–0.29	1.24	0.21–4.28
Mg	% dry wt.	0.16	0.09–0.23	0.36	0.11–0.79
Na	% dry wt.	0.03	0.02–0.07	0.55	0.03–1.66
Si	% dry wt.	2.62	1.17–3.72	0.99	0.13–2.84
H₂O content	%	81.0	69.7–86.1	91.3	81.4–95.9
Organic matter	% dry wt.	88.5	86.2–91.9	83.6	71.3–92.1
Ash	% dry wt.	11.53	8.1–13.8	16.36	7.9–28.7
Raw protein	% dry wt.	10.87	7.5–21.3	14.09	5.3–22.2
Polyphenols	Units g⁻¹ dry wt.	1.87	0.6–4.2	2.67	0.4–10.9
Total energy	MJ kg⁻¹ dry wt.	16.77	15.95–17.96	16.30	13.23–18.71

Eight grasses and sedges: *Echinochloa polystachya, Eleocharis variegata, Hymenachne amplexicaulis, Leersia hexandra, Oryza perennis, Paspalum repens, Rynchospora gigantea, Scirpus cubensis.*

Sixteen non-grasses: *Aeschynomene rudis, Azolla microphylla, Ceratophyllum demersum, Ceratopteris pteridoides, Eichhornia crassipes, Limnobium stoloniferum, Ludwigia natans, Marsilia polycarpa, Neptunia oleracea, Phyllanthus fluitans, Pistia stratiotes, Pontederia* sp., *Reussia rotundifolia, Salvinia auriculata, Utricularia foliosa, Victoria amazonica.*

Raw protein is calculated by multiplying N-content by the factor 6.25.

whereas the non-grasses have more crude protein (mean values up to 18%).

Grasses and sedges meet the requirements of cattle for calcium, phosphorus, and potassium. Magnesium concentrations are often higher than required by cattle, and those of *Paspalum fasciculatum, Eleusine indica, Paspalum conjugatum, Wedelia paludosa, Sida acuta* and *Teramnus volubilis* exceed the tolerance limit of 0.4%. Whereas the sodium concentrations do not meet the requirements, sulphur concentrations exceed the tolerance limit. Zinc and copper concentrations are also below the required levels whereas chlorine and iron concentrations are adequate.

Total energy content of herbs varies little within a range from 13.7 to 17.6 MJ kg⁻¹. Of this, the metabolizable energy content of grasses and sedges lies in the range of 4.5 to 8.1 MJ kg⁻¹ and that of non-grass species ranges from 6.2 to 8.9 MJ kg⁻¹. These values are similar to those reported for other aquatic macrophytes and terrestrial herbs (Boyd 1970b; Easley and

Shirley 1974) but somewhat lower than required for cattle ranching (8.9–9.6 MJ kg^{-1}).

Large differences also occur in the chemical compositions of different parts of the same plant. In *Echinochloa polystachya*, the leaves have the highest nitrogen concentration (1.89%) whereas the submerged sections of the stem which account for 85% of the total biomass (Piedade et al. 1992), have the lowest concentration (0.26%). These differences cause large variations in the nutritive values when calculated on the basis of whole plants.

The data on the chemical compositions of terrestrial plants in the igapó (Table 9.3) show lower element concentrations in grasses and sedges than in non-grass species. Compared with the concentrations in terrestrial plants of the várzea, the herbaceous plants in the igapó have considerably lower concentrations of P and Ca, slightly lower Mg and N concentrations, but higher sodium concentrations (Table 9.3). Aquatic macrophytes are sparse on the Negro River floodplain and there are no data for their composition.

In spite of the relatively low energy contents, low contents of Na, Zn, and Cu, and high contents of S and Mg, the herbaceous plants of the várzea are valuable forage for cattle and probably for large herbivores in general.

Table 9.3. Mean contents and ranges of elements, H_2O, ash, and raw protein of grasses and sedges and non-grasses growing during the terrestrial phase in the igapó

Components		Grasses and sedges		Herbs	
		Mean	Range	Mean	Range
N	% dry wt.	1.35	0 .82–1.95	2.26	1.12–3.54
P	% dry wt.	0.11	0.05–0.23	0.17	0.05–0.24
K	% dry wt.	1.72	0.36–3.24	2.22	0.8–5.28
Ca	% dry wt.	0.21	0.05–0.53	0.98	0.40–1.59
Mg	% dry wt.	0.23	0.07–0.39	0.28	0.17–0.46
Na	% dry wt.	0.02	<0.01–0.17	0.09	<0.01–0.18
H_2O content	%	73.5	66.7–83.2	72.6	67.8–77.6
Organic matter	% dry wt.	92.89	88.23–96.62	92.47	87.33–96.05
Ash	% dry wt.	7.11	3.38–11.77	7.53	3.95–12.67
Raw protein	% dry wt.	8.44	5.13–12.19	14.13	7.00–22.13

Twelve grasses and sedges: *Andropogon selloanus, Bulbostylis* sp., *Cyperus diffusus, C. ligularis, C. surinamensis, Homolepis aturensis, Oryza perennis, Panicum chloroticum, P. pilosum, Paspalum amazonicum, P. conjugatum, Scirpus cubensis.*
Seven non-grasses: *Alternanthera tenella, Borreria verticillata* (2), *B. capitata, Diodia hyssopifolia, Dracontium* sp., *Montrichardia arborescens.*
Raw protein is calculated by multiplying N-content by the factor 6.25.

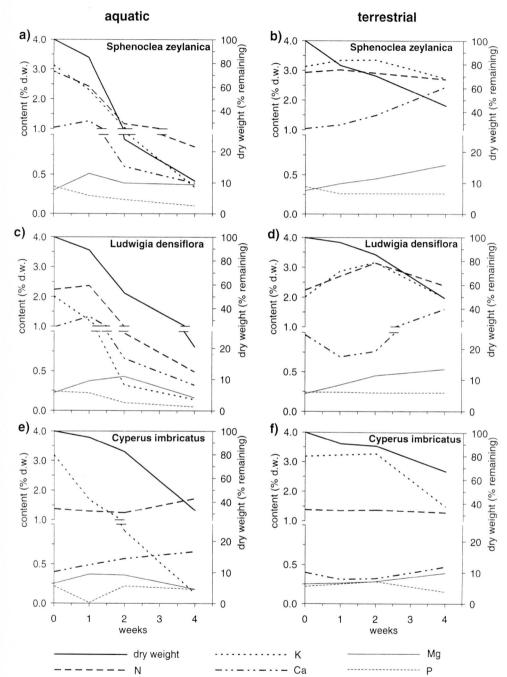

Fig. 9.1a–f. Dry weight loss and content of K, Mg, Ca, N and P in the herbaceous plant species *Sphenoclea zeylanica*, *Ludwigia densiflora*, and *Cyperus imbricatus* during decomposition under terrestrial and aquatic conditions in the Amazon floodplain. (According to Junk and Furch 1991)

9.2.2 Decomposition During the Terrestrial and the Aquatic Phases

High temperatures of about 30 °C and high humidity in tropical environments accelerate decomposition of organic material. Litterbag experiments of Howard-Williams and Junk (1976) showed that decomposition of herbaceous plants in the aquatic phase is very rapid. After 14 days about 50% of the exposed material was lost from litterbags of 1 mm mesh size. After six months, the material from five of the six species tested had lost from 80 to 90% of their initial weight. Only *Salvinia auriculata* showed a lower weight loss (about 50%). After 338 days only 3–9% of the exposed material remained. Losses of K, Na, Mg, and P were quicker than weight loss and about 50 to 95% of the initial concentrations were lost after 14 days. The amounts of N and Ca may decrease but often increase again during decomposition. These results show that the most important chemical changes in plant material decomposing in water occur at the beginning of the decomposition process.

There are considerable differences in the decomposition dynamics on land and in water (Junk and Furch 1991). In water, after 4 weeks the weight loss of *Cyperus* amounted to about 67%, of *Ludwigia* 80% and of *Sphenoclea* 90% (Fig. 9.1). At the beginning of decomposition, nutrient loss from the organic material in water was much faster than on land. After 2 weeks *Spenoclea* and *Ludwigia* had lost 92%, and *Cyperus* 76% of their initial amounts of K. After 4 weeks only 1% of the initial K amount was present in the remaining biomass of the three species. After 4 weeks, the sequence of nutrient loss was for *Ludwigia* Mg (85%) < Ca (93%) < N,P (96%) < K (99%), for *Spenoclea* Mg (87%) < Ca (96%) < N,P (97%) < K (99%) and for *Cyperus* Ca (46%) < N (59%) < P (73%) < Mg (76%) < K (99%). The highest decomposition rate was observed for *Spenoclea* showing the highest N content of 2.4%, and the lowest rate was observed for *Cyperus* with an initial N content of 1.4%. Under aquatic conditions the nutrient content of the decomposing material generally decreased, except for *Cyperus* biomass, which showed a strong decrease only in its K content (Fig. 9.1). In *Ludwigia* and *Spenoclea*, both containing large initial amounts of N, N content decreased. In *Cyperus*, with a low initial N content, N was accumulated (Fig. 9.1).

Nutrient dynamics during decomposition on land are more complex. Dry weight loss was two to four times slower than in water: after four weeks 31% was lost from *Cyperus*, 51% from *Ludwigia*, and 55% from *Sphenoclea*. Nutrient losses after four weeks were observed for N, P, and K only and were smaller than in water. They varied between 52 and 69% of the initial amounts of K, 53 and 67% of P and 37 and 58% of N. Losses of Ca and Mg were smaller than biomass losses and led to an increase in the remaining plant material.

The differences between decomposition on land and in water can be attributed in part to different leaching rates. Air-exposed plant material without contact with water did not show weight and element loss over a period of 32 days. The loss started when the material was inundated, and increased with the frequency and duration of the contact with water (Fig. 9.2). Weight loss is explained by the leaching of easily soluble organic substances (Sect. 9.2.3) and considerable amounts of nutrients (Fig. 9.2c). After 30 min exposure time, the K content had already decreased to a third and the P content to half of its initial amount.

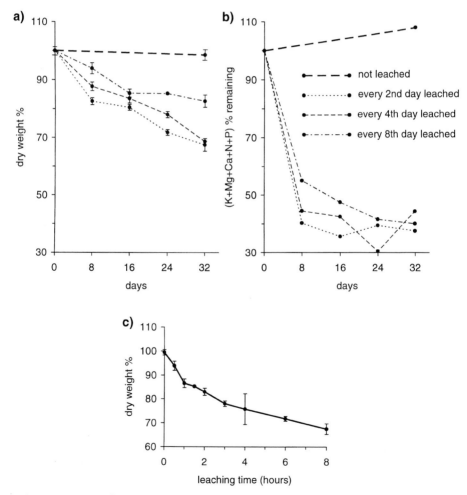

Fig. 9.2a–c. Loss of dry weight (**a,c**) and contents of K, Mg, Ca, N, and P (**b**) of *Paspalum repens* in a leaching experiment. Litterbags with sun-dried material were put into a tank filled with rainwater for 30 min every 2, 4, and 8 days, respectively, for a period of 32 days (n = 3). Chemical analyses were made from mixed samples

Microclimatic conditions are important for the dynamics of decomposition processes on land. Increased humidity and local rainfall will accelerate leaching of soluble organic substances and minerals. Because of a greater climatic and microclimatic variability in the terrestrial phase, a greater variability of the loss of soluble organic substances and inorganic nutrients from the plant material can be expected on land than in water.

9.2.3 Release of Dissolved Substances into the Water

At the beginning of decomposition easily decomposable soluble organic compounds are released in large quantities. In container experiments (Furch and Junk 1992), the high decomposition rates were reflected by an abrupt decrease in the oxygen concentration below the detection limit (Fig. 9.3). CO_2 concentration reached a characteristic peak at the beginning of decomposition as a result of the high oxidation rate. After about 2 weeks, the CO_2 content decreased again while the HCO_3 content increased and remained constant for a long period of time (Fig. 9.4a).

With the establishment of a carbonate buffer system, the pH value increased until the neutral point was reached. This increase in pH was

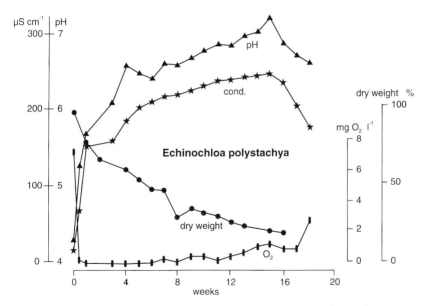

Fig. 9.3. Dry weight loss (*dry weight*) and changes in *pH*, specific conductance at 25 °C (*cond.*) and O_2 content in the water during decomposition of fresh biomass of *Echinochloa polystachya* (4 kg fresh biomass, corresponding to 1.044 kg dry wt., exposed in 700 l water). (According to Furch and Junk 1992)

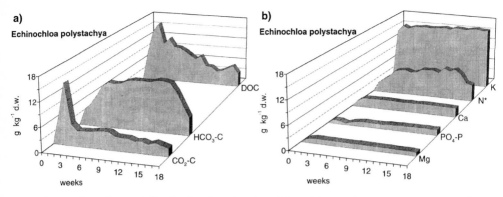

Fig. 9.4a,b. Release of dissolved carbon fractions (**a**) and the nutrients Mg, Ca, K, N*, and PO$_4$-P (**b**) into water during decomposition of fresh biomass of *Echinochloa polystachya*. (4 kg fresh biomass, corresponding to 1.044 kg dry wt., exposed in 700 l water. N* = NO$_3$-N + NO$_2$-N + NH$_4$-N). (According to Furch and Junk 1992 and unpubl. data)

accompanied by changes in the HCO$_3$ content of the water and coincided with an increase in the content of other major solutes. The specific conductance increased to about ten times its initial value (Fig. 9.3). After 2 weeks, the concentrations of dissolved organic compounds decreased. After 12 weeks, the oxygen that entered the tank water by diffusion and thermal circulation was no longer completely consumed because the remaining organic material was difficult to decompose (Fig. 9.3).

In addition to carbon, the release of mineral nutrients contributed greatly to changes in the chemical composition of the water (Fig. 9.4b). After 2 weeks of decomposition, when the plant material of *Echinochloa polystachya* had lost about 30% of its initial dry weight, the following portions of its initial nutrient amounts were found in dissolved form in the water: 85% K, 85% Mg, 78% Ca, 42% P (as PO$_4$-P), 19% N (as NH$_4$-N), and 8% C (sum of DOC, HCO$_3$-C and CO$_2$-C). Because of the large amounts of released minerals, ion-poor acid groundwater used for the experiment changed very quickly to ion-rich well-buffered water with K, NH$_4$, and HCO$_3$ being the dominant ions (Fig. 9.3).

The changes in the nutrient contents of the exposed material are given in Fig. 9.5. It shows that the biomass rapidly becomes impoverished in nutrient elements except for nitrogen during the initial period of decomposition. Later, the biomass loss rate became higher than the nutrient loss rate and a considerable increase in N, P, and Ca contents occurred. The coarse detritus left at the end of the experiment was impoverished in K, Ca, and Mg, but showed similar contents of P and N compared with the material at the beginning of the experiment. These results correspond to the findings of Howard-Williams and Junk (1976) in field experiments.

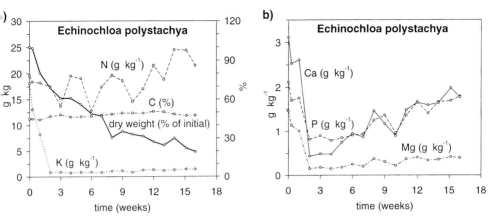

Fig. 9.5a,b. Dry weight loss and changes in the content of the bioelements N, K, and C (a) and Mg, Ca, and P (b) in the biomass of *Echinochloa polystachya* during decomposition of 1.044 kg dry material in 700 l water. (According to Furch and Junk 1992)

The differences between the amounts of nutrients lost from the detritus and those detected in the water are explained by losses of fine detritus particles from the litter bags, evolution of gaseous compounds, and assimilation of consumer organisms that are not fixed to the substrate, such as bacteria and fungi but also larvae of Culicidae.

9.3 Leaf Litter

9.3.1 Consistency and Chemical Composition of Tree Leaves and Leaf Litter

Leaves from the várzea forest are on average larger and less sclerophyllic than those from the igapó forest. Klinge et al. (1983) reported that about 60% of the leaves of the igapó belong to the microphyllic and notophyllic category (leaf area of 2.2–20.2 cm^2 and 20.2–45 cm^2, respectively) compared with 30% of the leaves from the várzea forest. The specific leaf area (leaf area/gram dry weight) corresponds to 140 cm^2 g^{-1} for the várzea forest and 97 cm^2 g^{-1} for the igapó forest. Values reported by Meyer (1991) vary between 120 cm^2 g^{-1} for the várzea forest and 85 cm^2 g^{-1} for the igapó forest. Sobrado and Medina (1980) and Medina (1981, 1984) consider the vegetation of the igapó to be sclerophyllous because of the small leaf size, low specific leaf area, low values of nitrogen and phosphorus, and large accumulation of structural carbohydrates in the leaves. Loveless (1961,

Table 9.4. Mean contents of elements of fresh leaves and leaf litter from várzea, igapó, and terra firme

Components		Várzea				Igapó				Terra firme		
		Fresh leaves		Leaf litter		Fresh leaves		Leaf litter		Fresh leaves[d]		Leaf litter[d]
		Mean	SD	Mean	SD	Mean	SD	Mean	SD	Mean	SD	Mean
N	% dry wt.	2.230[a]	(0.531)	1.350[a]	(0.040)	1.560[a]	(0.404)	1.170[a]	(0.090)	1.800	–	1.536
		2.587[b]		2.100[c]		1.731[b]		1.430[c]				
P	% dry wt.	0.156[a]	(0.043)	0.102[a]	(0.006)	0.084[a]	(0.026)	0.045[a]	(0.002)	0.054	(0.015)	0.036
		0.176[b]		0.078[c]		0.062[b]		0.018[c]				
K	% dry wt.	0.900[a]	(0.684)	0.433[a]	(0.070)	0.640[a]	(0.262)	0.468[a]	(0.050)	0.499	(0.230)	0.232
		1.318[b]		0.490[c]		0.633[b]		0.300[c]				
Ca	% dry wt.	1.375[a]	(0.831)	1.858[a]	(0.360)	0.420[a]	(0.204)	0.577[a]	(0.060)	0.426	(0.348)	0.321
		1.914[b]		1.540[c]		0.251[b]		0.500[c]				
Mg	% dry wt.	0.284[a]	(0.185)	0.302[a]	(0.080)	0.156[a]	(0.058)	0.168[a]	(0.060)	0.290	(0.140)	0.232
		0.398[b]		0.320[c]		0.122[b]		0.128[c]				
Na	% dry wt.	0.006[a]	(0.027)	0.018[a]	(0.003)	0.004[a]	(0.044)	0.007[a]	(0.001)	0.115	(0.083)	0.089
		0.020[b]		0.009[c]		0.025[b]		0.006[c]				
Sum	% dry wt.	4.951[a]	–	4.063[a]	–	2.864[a]	–	2.435[a]	–	3.184	–	2.446
		6.413[b]		4.537[c]		2.824[b]		2.382[c]				

[a] Furch et al. (1989).
[b] Klinge et al. (1983).
[c] Adis et al. (1979).
[d] Furch and Klinge (1989).

1962) regards concentrations of P lower than 0.3% as an indication of sclerophyllous conditions.

The data about the chemical composition of leaves and leaf litter show pronounced differences among species from the nutrient-rich Amazon River floodplain and the nutrient-poor Negro River floodplain (Irmler and Furch 1980; Klinge et al. 1983, 1984; Furch et al. 1988, 1989; Furch and Klinge 1989). The leaves of the igapó trees contained only about 30% of the total amount of the nutrients K, Ca, Mg, and P when compared with tree leaves in the várzea (Furch and Klinge 1989). Differences were marked for calcium and were relatively small for nitrogen (Table 9.4). This corresponds to the availability of nutrients in the soil and in the water of the respective ecosystems (Chaps. 3, 4).

Compared with leaf litter, fresh leaves of várzea and igapó trees had higher contents of K, N, and P, and lower contents of Na, Ca, and Mg. Considering the sum of the six elements, fresh leaf litter of the várzea contained about 18–30% less nutrients than fresh leaves. The difference between leaf litter and fresh leaves of the igapó was about 15–16% (Table 9.4). The concentrations of nutrient elements in leaves and leaf litter of the terra firme were similar to those of the igapó except for Na, which was higher in the leaves from the terra firme (Table 9.4).

9.3.2 Decomposition in Water and on Land

Decomposition in Water. Experiments in tanks with leaves of the várzea trees *Cecropia latiloba*, *Salix humboldtiana* and *Pseudobombax munguba*, and the igapó tree *Aldina latifolia* showed that the decomposition of leaves of the várzea species followed a pattern that is similar to herbaceous plants (Fig. 9.6, Table 9.5). After 120 days, up to 86% of the initial biomass and 89% of the nutrients (sum of K, Ca, Mg, N, and P) were lost from the litter bags (Furch et al. 1989). The decomposition of *Aldina* occurred more slowly. Only about 10% of the plant material and about 40% of the nutrients were lost over a 120-day period (Table 9.5).

Between 44 and 59% of the initial amounts of total nutrients in the leaves of várzea trees were detected as inorganic ions in the water compared to 37% of total nutrients in the *Aldina* leaves. This difference results mainly from a comparatively small release rate of N from *Aldina* leaves. The low nutrient concentrations and the low specific leaf area of most igapó species suggest that decomposition dynamics will be similar to that of *Aldina*, whereas the higher nutrient concentrations and higher specific leaf area of many várzea species indicate that decomposition dynamics will be similar to that of *Cecropia*, *Salix*, and *Pseudobombax*.

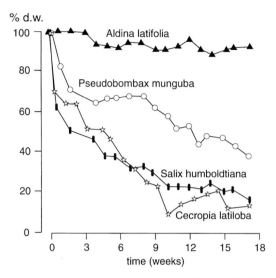

Fig. 9.6. Dry weight loss from fresh leaf material of the várzea tree species *Cecropia latiloba, Salix humboldtiana, Pseudobombax munguba* and the igapó tree species *Aldina latifolia* during decomposition in water (4.0–4.4 kg fresh leaf material was exposed to 700 l water). (According to Furch et al. 1989)

Decomposition on Land. A comparative study of the leaf litter exposed in várzea and igapó on land showed that decomposition is rather slow (Irmler and Furch 1980). After 100 days a weight loss of about 20% in the várzea and 10% in the igapó was recorded in spite of more than 500 mm rainfall during the period that they were exposed. Irmler and Furch (1980) calculated according to Olson (1963) a 95% decomposition time of 6.3 years for the igapó and 2.8 years for the várzea. Experiments with dried fresh leaf material of the igapó tree species *Eschweilera coriacea* and *Buchenavia ochroprumna* showed a greater weight loss of 40–55% corresponding to 1.0–1.7 years for 95% decomposition. Also, element losses from fresh *Eschweilera* and *Buchenavia* leaves were considerably higher than those from fresh leaf litter (K = 85–95%, Na = 75–90%, Mg = 50–75%, Ca = 20–45%, P = 35–65%, N = 60–90% and energy = 35–50% (Irmler and Furch 1980)).

An advanced stage of decomposition of leaves is also indicated by a dark color in water, melanosis (Ball 1973). Meyer (1991) found that the percentage of dark leaves in the litter layer of the várzea forest was much higher than in the litter layer of the igapó forest. Decomposition in the várzea was about twice as rapid as in the igapó. A 95% decomposition of the leaf litter was reached after 1.9 to 2.3 years in the igapó and after 0.9 to 1.4 years in the várzea. Decomposition rates in the igapó were lower on low-lying areas than on high-lying ones. Meyer (1991) relates the low decomposition rates

Table 9.5. Maximum weight loss (% dry wt.), initial content of K, Mg, Ca, N, and P (g kg^{-1} dry wt.) in leaves of the várzea species *Cecropia latiloba*, *Salix humboldtiana*, and *Pseudobombax munguba* and the igapó species *Aldina latifolia*. (According to Furch et al. 1989)

			C. latiloba	S. humboldt.	P. munguba	A. latifolia
Weight loss		%	86.0	82.2	58.5	11.9
K	Initial	g kg^{-1}	12.22	14.16	13.64	8.98
	Rel. (K$_r$)		11.10 (0.91)	9.20 (0.65)	10.94 (0.80)	8.67 (0.97)
	Loss (K$_l$)		12.07 (0.99)	14.01 (0.99)	13.24 (0.97)	8.65 (0.99)
Mg	Initial	g kg^{-1}	3.41	2.26	8.14	1.01
	Rel. (K$_r$)		2.67 (0.78)	1.30 (0.58)	4.54 (0.56)	0.65 (0.64)
	Loss (K$_l$)		3.25 (0.95)	2.17 (0.96)	7.57 (0.93)	0.66 (0.65)
Ca	Initial	g kg^{-1}	10.10	14.40	18.18	1.90
	Rel. (K$_r$)		8.60 (0.86)	7.07 (0.49)	6.68 (0.37)	0.92 (0.48[a])
	Loss (K$_l$)		8.90 (0.89)	12.76 (0.89)	14.19 (0.78)	0.50 (0.26[a])
N	Initial	g kg^{-1}	29.45	23.60	18.97	22.10
N[b]	Rel. (K$_r$)		10.42 (0.35)	6.40 (0.27)	3.95 (0.21)	2.08 (0.09)
N	Loss (K$_l$)		24.87 (0.84)	17.79 (0.75)	9.90 (0.52)	3.61 (0.16)
P	Initial	g kg^{-1}	2.16	1.55	2.08	1.28
	Rel. (K$_r$)		1.10 (0.51)	0.91 (0.59)	1.51 (0.73)	0.80 (0.63)
	Loss (K$_l$)		1.97 (0.91)	1.46 (0.94)	1.69 (0.81)	0.83 (0.65)
Sum	Initial	g kg^{-1}	57.25	55.97	61.01	35.27
	Rel. (K$_r$)		33.89 (0.59)	24.88 (0.44)	27.62 (0.45)	13.12 (0.37)
	Loss (K$_l$)		51.06 (0.89)	48.19 (0.86)	46.59 (0.76)	14.45 (0.41)

Maximum amounts of these elements (g) released into the water (Rel.) and lost (Loss) from 1 kg dry wt. of leaves are given for a 4-month period of decomposition in water.
K$_r$, Ratio of the maximum amount released to the initial amount; K$_l$, ratio of the maximum amount lost to the initial amount.
Values for K$_r$ and K$_l$ are given in parentheses.
Sodium was not considered due to its extremely low proportion in the leaf biomass.
[a] The strange relationship between the released and lost amounts of Ca is explained by methodological problems due to the extremely low Ca concentrations in leaves of *Aldina*.
[b] NO_3-N + NO_2-N + NH_4-N.

indicated by Irmler and Furch (1980) to differences in the thickness of the investigated litter layers.

9.3.3 The Impact of Terrestrial Invertebrates on Leaf Litter Decomposition

The small weight loss of *Aldina latifolia* leaves in water indicates that leaching of organic compounds is of minor importance for decomposition of sclerophyllous leaves. Feeding and fragmentation by invertebrates may play a major role. The impact of feeding activities of terrestrial macroinvertebrates on leaf litter decomposition was tested by exposing soft leaves of *Buchenavia ochroprumna* and hard leaves of *Eschweilera coriacea* in litterbags and without litterbags. Leaves exposed in litterbags (0.5 mm mesh size) could not be attacked by the fauna. Without litterbags the soft leaves of *Buchenavia* were immediately attacked by the macrofauna whereas the hard leaves from *Eschweilera* were attacked only after 2 weeks. After 100 days, weight loss of leaves exposed to macroinvertebrate attack was 5 to 9% greater and the loss of leaf area 56% greater than in leaves protected by litterbags (Irmler and Furch 1980).

Cockroaches represent an important animal group in the litter layer of the igapó. At Tarumã Mirím eight species occur which make up about 20 to 30% of the total soil macrofauna biomass ($1.4 \, \mathrm{g \, m^{-2}}$ dry weight; Irmler 1976). *Epilampra irmleri*, on average corresponded to 19% of the cockroach population. This species feeds preferentially on leaf litter and supplements its diet by saprophagy on carcasses and coprophagy to satisfy the mineral demand. The species consumed about $18.8 \, \mathrm{g \, m^{-2} \, year^{-1}}$ biomass, of which about 2 g were assumed to be animal remnants. This consumption corresponds to 5.6% of a leaf litter production of $300 \, \mathrm{g \, m^{-2}}$ during low water periods (Irmler 1979; Irmler and Furch 1979). The chemical analysis of litter and faeces showed an increase of mineral content in the faeces. Faeces supply to litter may accelerate the decomposition process, especially in the nutrient-poor igapó (Irmler and Furch 1979).

The increased participation of the soil fauna in litter decomposition in the igapó forest leads to a stronger incorporation of organic material into the upper mineral soil layers than in the várzea forest (Meyer 1991).

In a litter turnover model for the igapó, Irmler (1979) estimated a litter consumption of $50 \, \mathrm{g \, m^{-2}}$ and $40 \, \mathrm{g \, m^{-2}}$ dry material by macrosaprophages and microsaprophages, respectively, during the terrestrial phase of 7 months. This corresponds to about 13.3% of the annual litter fall of $675 \, \mathrm{g \, m^{-2}}$.

9.3.4 The Impact of Aquatic Invertebrates on Leaf Litter Decomposition

There are no detailed studies available on the impact of aquatic invertebrates on leaf litter decomposition in the aquatic phase in the várzea and igapó. This is in part because benthic animals are scarce near the bottom and inside the packages of leaf litter because of hypoxia. Fresh leaves of *Symmeria paniculata* exposed at different water depths, in order to study the colonization of the submersed canopy of the floodplain forest with aquatic invertebrates, showed the presence of an abundant and diversified invertebrate fauna (Irmler and Junk 1982; Sect. 13.4.2). However, only a few feeding marks by aquatic invertebrates were observed after 4 weeks of exposure. Most animals used the leaves as a substratum and for shelter and not as a food item; grazing on epiphyte growth and filtering of phytoplankton from the water may have been more important.

Detailed studies of benthic communities were made in black-water forest streams by Walker (1985, 1988); Henderson and Walker (1986); Walker et al. (1991). These studies show that the food chain is closely related to fungi that colonize the leaves. A few animal species (mainly small chironomids and nematodes) penetrate the epidermis and mine inside the leaves to feed on the mesophyll and associated fungi and bacteria. Powerful shredders, such as *Gammarus* spp., which quickly transform large amounts of leaf litter into fine particulate material in European streams, are absent in Amazonian streams and river floodplains.

9.4 Discussion and Conclusions

The contents of nutrient elements in plant tissue exhibit large variations among species, different parts of the plants, life-cycle stages, and habitats. Mean values and ranges of nutrients accumulated in the plant material show that concentrations in herbaceous plants and in the leaves of the floodplain forest correspond to the nutrient status of the respective habitats. Total nutrient content is higher in the vegetation of the várzea than in that of the igapó corresponding to a higher nutrient status of soils and water of the várzea. The nutritional value of herbaceous plants of the várzea is high in spite of the fact that the grasses show some deficiencies for cattle ranching, such as low concentrations of P, Ca, Mg, Na, Cu, Zn, low digestable energy content and unfavorably high water content during the aquatic phase (NRC 1971).

In pre-Columbian times herbaceous plant communities of the várzea sustained large populations of capybaras, manatees and turtles (Chap. 21). However, in comparison with tropical African and Asian floodplains, the number of large herbivorous species is rather small. Very few large herbivorous animals immigrate periodically from the surrounding terra firme into the floodplain because there are only a few species and their population size is small. For these animals the large flood amplitude may be a limiting factor for a better use of the food resources in the floodplain. The large flood amplitude and the periodicity of the food also hinders the use of the natural pastures for cattle and buffalo ranching (Chap. 1).

Most of the plant material enters into the detritus food webs. Decomposition of herbaceous plant material on land and in water is very rapid because of the high nutrient content and the low content of structural carbohydrates. In water at the beginning of decomposition, leaching of soluble organic and inorganic substances leads to a rapid weight loss. The decomposition of sclerophyleous leaf litter is slower because of greater amounts of structural carbohydrates and a lower nutrient content. In the igapó, decomposition of sclerophyleous leaf litter is accelerated because of its fragmentation by terrestrial soil invertebrates. During the terrestrial phase reduced leaching of soluble substances makes the organic material attractive for detritus feeding animals. Irmler and Furch (1980) postulated that the periodical flooding in the igapó may slow down litter decomposition because of the periodical elimination of terrestrial soil invertebrates.

The formation of sclerophyllous leaves by trees is often found in dry and wet habitats of a low nutrient status (Loveless 1961, 1962; Beadle 1966; Medina 1984). According to Vitousek (1984) and Jordan and Uhl (1978), sclerophylly favors nutrient conservation in the system because nutrient loss during decomposition is low. Meyer (1991) shows that about 70% of the annual leaf fall in the igapó is produced during the aquatic phase, but most of it is shed at the end of the flood period. Sixty-six to 80% of the decomposition occurs during the terrestrial phase when losses by leaching are small. A dense superficial fine-root system in combination with mycor- rhiza allows the trees a direct uptake of the nutrients from the litter. In the várzea about 55% of the annual litter fall occurs during the aquatic phase. Losses by leaching are large, 34 to 51% of the annual litter fall is decom- posed during the terrestrial phase. The fine-root system is not so dense and is concentrated in the upper mineral soil.

The chemical composition and digestibility of the detritus formed in water is quite homogeneous regardless of initial differences in consistency and chemistry (Furch et al. 1989; Furch and Junk 1992). Boyd and Goodyear (1971) and Polinisi and Boyd (1972) emphasized that the quality of vegetable biomass in the food web depends on its nutritional value. The physical and chemical heterogeneity of the plant material at the start of the decomposition process favors selective utilization by specialized organ- isms. The relatively homogeneous detritus produced from the herbaceous plants and from the leaf litter provides a generally uniform source of food for detritivorous organisms. The large number of detritivorous animals, including many Amazon fish species, suggests that detritus is an essential factor in the food webs of Amazonian river floodplains. However, the question as to what extent detritus of macrophytes directly enters into higher levels of the food webs is still controversely discussed (Sects. 10.4, 20.2, 20.7).

Acknowledgments. We are very grateful to Dr. Ernesto Medina, Instituto Venezolano de Investigaciones Cientificas, Caracas, Dr. Brij Gopal, Jawaharlal Nehru University, Dr. John Melack, University of California, and Dr. Rosemary Lowe-McConnell, Sussex, for critical comments on the manuscript and the correction of the English.

10 Phytoplankton and Periphyton

Rainer Putz and Wolfgang J. Junk

10.1 Introduction

Most studies of Amazonian algae concentrate on phytoplankton. Taxonomy and aspects of community structure are discussed by Grönblad (1945); Förster (1969, 1974); Martins (1980); Scott et al. (1965); Uherkovich (1981); Uherkovich and Schmidt (1974); Uherkovich and Rai (1979); Thomasson (1971), and many others. In 1973 Schmidt published the first comprehensive study on primary production of phytoplankton in a várzea lake. In the following years many researchers focussed on the subject and primary production of phytoplankton in central Amazonian floodplain lakes became rather well known (Schmidt 1976; Fisher 1979; Fisher and Parsley 1979; Forsberg 1984; Rai and Hill 1984; Forsberg et al. 1991; Rodrigues 1994; and many others).

Periphyton was neglected for many years. Research started in the 1960s when Hustedt (1965) began some taxonomic studies on periphytic diatoms. Uherkovich and Franken (1980) studied species composition of periphyton communities in seven forest streams near Manaus. In 1984 Rai and Hill published the first quantitative study on the primary production of periphyton in Lago Cristalino.[1] During the last few years several studies have been published on periphyton communities associated with aquatic macrophytes and the floodplain forest (Engle and Melack 1990, 1993; Doyle 1991; Alves 1993; Putz 1997; and others), which have increased our knowledge considerably.

---

[1] A general description of the study area is given in Sect. 2.2.

Ecological Studies, Vol. 126
Junk (ed) The Central Amazon Floodplain
© Springer-Verlag Berlin Heidelberg 1997

10.2 Phytoplankton

10.2.1 Species Number and Community Structure

According to Uherkovich's review (1984) of phytoplankton taxonomy in Amazonia, a great number of the algal taxa that occur in the Amazon basin are cosmopolitan. Most algal species are eurythermic, so a large distribution is likely. In total, 319 algal taxa were described, belonging to Desmidiaceae (313), Diatomeae (43) and other groups (33). The large number of Desmidiacean taxa is expected given that Amazonian waters are frequently acidic with a low dissolved mineral content and are oligotrophic. The high diversity in desmids does not correspond with an abundance of individuals. Numerical dominance of desmids was found in few samples. High species numbers were noted for genera of diatoms which favor acidic and oligotrophic waters. The greatest abundance was recorded for species from other genera (e.g., *Melosira granulata* var. *angustissima*).

The taxonomic studies give little information about numerical abundance and changes along environmental gradients. Studies of phytoplankton communities of specific water bodies that include hydrochemical data are scarce. The first comprehensive study of the phytoplankton community in a várzea lake was made in Lago Castanho by Uherkovich and Schmidt (1974). In 15 samples, covering a period from August 1967 to October 1969, the authors described 209 taxa. Maximum taxon numbers are cited for the falling- and low-water periods. The phytoplankton community is described as a community of *Melosira granulata* var. *angustissima*, *Sphaerocystis schroeteri*, and *Closterium kuetzingii* with occasional blooms of Cyanophyceae (*Anabaena hassalii, Aphanizomenon flos aquae, Oscillatoria limosa*).

A recent study in Lago Camaleão conducted from September 1987 to March 1989 by Rodrigues (1994) cites 262 taxa (Table 10.1). A much larger number of Euglenophyta was found, compared to the number found in Lago Castanho. Conforti (1993) describes 90 taxa of the genus *Trachelomonas*, 40 of them show a widespread or cosmopolitan distribution. The frequent occurrence of Euglenophyta is related to the abundance of dissolved organic compounds in Lago Camaleão (Sect. 4.4).

Samples from the phytoplankton community of the Negro River showed a dominance of *Melosira granulata* var. *angustissima*, *Tabellaria fenestrata*, Chlorococcales (*Dictyosphaerium pulchellum, Kirchneriella lunaris*), and Conjugatophyceae (*Closterium kuetzingii, Staurastrum quadrinotatum*). Dominant species in the Tapajós river phytoplankton were *Melosira granulata*, *Melosira granulata* var. *angustissima*,

Table 10.1. Phytoplankton species found in two Amazonian várzea lakes Lago Castanho (Uherkovich and Schmidt 1974) and Lago Camaleão. (Rodrigues 1994)

Author	Uherkovich and Schmidt	Rodrigues
Classes	No. of species	
Cyanophyta	19	9
Euglenophyta	58	185
Pyrrhophyta	1	–
Chlorophyta	108	38
Chrysophyta		
Chrysophyceae	9	3
Bacillariophyceae	14	13
Dinophyceae	–	1
Zygnemaphyceae	–	11
Cryptophyceae	–	2
Total:	209	262

Cyanophyceae (*Microcystis aeruginosa, Oscillatoria limosa*), Chloro-coccales (*Pediastrum duplex, Treubaria crassispina*) and Conjuga-tophyceae (*Staurastrum quadrinotatum, S. hystrix* var. *brasiliensis, Hyalotheca dissiliens, Mougeotia* spp.) (Thomasson 1971; Uherkovich 1976, 1981; Uherkovich and Rai 1979).

10.2.2 Biomass and Primary Production

Schmidt (1973) studied primary production in the várzea lake Lago Castanho using the ^{14}C technique. He described a distinct productivity pattern that was dependent on the water level. Light penetration was the main factor controlling productivity. The euphotic layer varied from 0.5 m at low water to 6 m at high water. For the low water period, October and November, the production per unit volume reached its maximum of $2.15\,g\,C\,m^{-3}\,day^{-1}$. During the period of rising water, which lasted from January until May, production per unit volume was reduced because of the inflow of turbid Amazon water which diluted the phytoplankton popula-tion and resulted in a narrow euphotic layer of only about 2 m. During this period, the lowest production values of $0.32\,g\,C\,m^{-3}\,day^{-1}$ were recorded. With increasing transparency at high and falling water levels from the middle of May until September, the production per unit area reached maximum values of $1.5\,g\,C\,m^{2}\,day^{-1}$. The annual mean value of net primary production was about $0.8\,g\,C\,m^{-2}\,day^{-1}$, corresponding to $3\,t\,C\,ha^{-1}\,year^{-1}$.

Gross primary production was about 25–40% higher, corresponding to about $3.9\,t\,C\,ha^{-1}\,year^{-1}$ (Fig. 10.1).

A mean algal C content of $0.8\,g\,m^{-3}$ was calculated from a mean chlorophyll content of $52\,mg\,m^{-3}$, corresponding to about 5–10% of the total dissolved and suspended C content ($10–15\,g\,m^{-3}$) in the water. Mean algal biomass calculated from the chlorophyll data reached $1.9\,g\,C\,m^{-2}$ corresponding to $3.8\,g\,m^{-2}$ dry material. From this algal biomass and the average gross primary production of $1.1\,g\,C\,m^{-2}\,day^{-1}$, Schmidt (1973) calculated a turnover rate of 1.7 days.

Studies of other authors indicate values of similar orders of magnitude and confirm the general production pattern. Fisher (1979) indicates during low and rising water an average net primary production of 2.2 g (0.82–3.5 g) $C\,m^{-2}\,day^{-1}$ for Lago Janauacá. The lowest values are characteristic of the rising water period. Melack and Fisher (1990) report a net primary production of $0.68 \pm 0.35\,g\,C\,m^{-2}\,day^{-1}$ and a gross primary production of $0.92 \pm 0.22\,g\,C\,m^{-2}\,day^{-1}$ for Lago Calado.

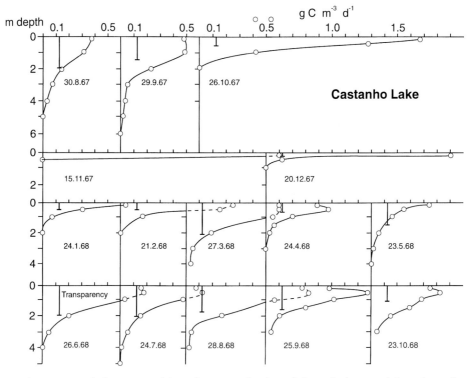

Fig. 10.1. Vertical patterns of the primary production of phytoplankton and the values of Secchi transparency in Lago Castanho from August 1967 to October 1968. (According to Schmidt 1973)

Rodrigues (1994) found highest production per unit volume ($3.1\,\mathrm{g\,C\,m^{-3}}$ $\mathrm{day^{-1}}$) and highest content of chl-a ($211\,\mathrm{mg\,m^{-3}}$) in Lago Camaleão in the surface layer during low water. During this period the lake was turbid because of suspended sediments. However, the negative impact of low transparency was counteracted by shallow water depth and additional nutrient input from the sediment, birds, and decomposing aquatic macrophytes. At rising and high water, phytoplankton concentration was low because of dilution of the plankton community by inflowing river water. The amount of inorganic suspended matter resulted in a narrow euphotic layer of 1–2 m. At falling water, transparency rose and primary production occurred to a depth of about 3 m. Mean net primary production reached $0.46\,\mathrm{g\,C\,m^{-2}\,day^{-1}}$, corresponding to $1.7\,\mathrm{t\,C\,ha^{-1}\,year^{-1}}$.

The effect of nutrient input from the river is shown by Fisher and Parsley (1979) in Lago Janauacá. They found a bloom of blue-green algae in the mixing zone of Amazon river water with lake water, which corresponded to a strong depletion of nitrate and phosphate. This poses the question: to what extent are nutrients limiting for algal production? Schmidt (1973) did not find strong responses by phytoplankton in experiments with additional inputs of nitrogen and phosphorous. Forsberg (1984) indicates phosphorous limiting factor in the black-water lake Lago Cristalino and nitrogen in

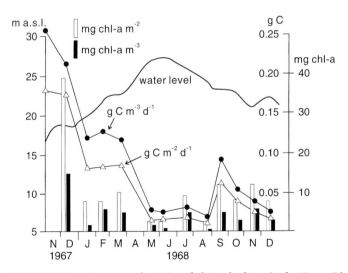

Fig. 10.2. Net primary productivity of phytoplankton in the Negro River near Ponta Negra upstream of Manaus per unit volume at optimum depth (*black points*) and per unit area (*triangles*) compared with chl-a content per unit volume at depth of maximum productivity (*black columns*), the chl-a content per unit area of the trophogenic zone (*white columns*), and the height of the river gauge (*water level*) during the investigation period. (According to Schmidt 1976)

the white-water lake Lago Jacaretinga. Setaro and Melack (1984) and Melack and Fisher (1990) considered phosphorus, during rising and high water, and nitrogen during falling and low water as limiting factors in Lago Calado. Rodrigues (1994) considered light and possibly nitrogen as limiting factors for phytoplankton production in Lago Camaleão.

Low nutrient levels and low light penetration into the water limit phytoplankton production in black-water lakes (Schmidt 1976) (Fig. 10.2). Net primary production in a bay of the Negro River varied per unit volume between 0.019–$0.266\,g\,C\,m^{-3}\,day^{-1}$ and per surface area between 0.011–$0.181\,g\,C\,m^{-2}\,day^{-1}$. The chl-a content varied between 1–$15\,mg\,m^{-3}$ and between 3–$40\,mg\,m^{-2}$ in the trophic zone. Annual net primary production was estimated to reach $230\,kg\,C\,ha^{-1}\,year^{-1}$. Highest values were reached at low water and at the beginning of rising water. This maximum is associated with an increase of nutrients entering the river at the beginning of the rainy season (Schmidt 1976). Changes in light penetration were of minor importance for the observed changes in primary production because there was little variation in transparency and the euphotic zone had a depth of about 2 m. A similar productivity pattern with higher values is given by Rai and Hill (1984) for the black-water lake Lago Cristalino.

10.3 Periphyton

10.3.1 Species Number and Community Structure

The investigation of the periphyton communities in nine habitats of seven rainforest streams near Manaus showed a total of 329 algal taxa belonging to the Cyanophyta (37), Euglenophyta (7), Pyrrophyta (2), Chrysophyceae-Xanthophyceae (6), Bacillariophyceae (146), Chlorophyceae (23), Conjugatophyceae (105), and Rhodophyta (3) (Uherkovich and Franken 1980). The streams could be classified according to the taxonomic composition of the periphyton community into streams with more than 75% Diatomeae, streams with 55–65% Diatomeae and 25–30% Conjugatophyceae and streams with 15–40% Diatomeae, 35–50% Conjugatophyceae, and a relatively high percentage (15–20%) Cyanophyta, which occur in most streams in very low numbers. The authors explain the increase of Conjugatophyceae and the decrease of Diatomeae as being due to increasing light intensity in the habitats and an increase in Cyanophyta as being due to eutrophication.

A few species were found in at least five of the nine investigated habitats (*Closterium navicula, Actinella brasiliensis, Eunotia diodon, E. lunaris* var.

genuina, E. pectinalis var. minor, E. pectinalis var. minutissima, Frustulia rhomboides, F. rhomboides var. saxonica, F. rhomboides var. saxonica f undulata). The authors categorize the studied streams as Eunotia–Frustulia streams.

Very few detailed studies are available about species numbers and community structures of periphytic algae in the Amazonian floodplains. In a recent study of the primary production of periphyton using exposed roughened cellulose-acetate foils, Putz (1997) found that attached algae embedded in gelatinous layers were characteristic at the regularly sampled black-water sites. Motile and sessile diatom species dominated black-water periphytic algae and represented up to 95% of the total attached algal community. Seasonal changes in black-water periphyton community composition were not observed. The dominant diatoms in the black-water habitats were Actinella brasiliensis and Eunotia cf. lunaris. Green algae and cyanobacteria were of minor importance at black-water sites. Compared to foil periphyton, attached algal communities on flooded forest leaves were found to be of slightly different species composition with a higher proportion of green algae (Spirogyra sp.) and cyanobacteria (Microcystis sp.).

Depth profile analysis revealed differences in periphyton composition within the 2-m euphotic zone. Inside the floodplain forest, diatoms, representing 95% of the total algal population were dominant at a depth of 0.3 m (Putz 1997). Chains of green algae and small amounts of detritus and fungi were embedded in a mucous layer. Deepest foil samples from 1.9 m had well-developed fungal matrices, a few chains of Spirogyra sp., and a large number of conspicuously long specimens of the diatom Eunotia cf. lunaris. The material contained high ciliate and nematode population densities and substantial amounts of detritus. Comparison to a site located at the margin of this forest revealed slight differences: a mucous layer of the uppermost (0.3 m depth) foil substrata coated diatoms (50%), filamentous green algae (40%) and cyanobacteria (10%). Periphyton from 1.9 m often consisted almost solely of diatoms.

Periphyton community structures in white-water habitats were of different quality. During high water, mature periphyton communities on foils exposed in the floating meadow root zone were dominated by green algae, contributing up to 80% of total periphytic algae. During rising and falling water phases their contribution was still about 50%. Diatom contribution varied between 20 and 30%, but reached about 90% during bloom periods. Cyanobacteria usually contributed no more than 20%, but blooms of Oscillatoria sp. occurred at low water level.

Low levels of chl-a point to little importance of attached algal populations in the floating vegetation in the mainstem of the Solimões (Table 10.3). Periphyton consisted primarily of diatoms with a restricted species composition. A small (7 μm) motile *Navicula* sp. was found to be the characteristic diatom species. In isolated habitats with Solimões water, characterized by low turbidity and low suspended sediments even at high water levels, diatoms were the principal component (70–90%) of the attached algal populations. Green algae and diatoms dominated the periphytic community structure with contributions up to 40% each at low water level in October and November, whereas cyanobacteria reached up to 20%. Cyanobacteria in general seemed to be promoted by low water levels at all sites.

Assemblages in mixed waters revealed a seasonality according to riverine water levels and ratios of the different waters. In Lago Janauarí during low water, with a maximum black-water proportion from November to February, periphytic assemblages mainly consisted of cyanobacteria (principally of the genera *Oscillatoria* and *Pseudoanabaena*) contributing up to 80%. Green algae were most common during rising and falling water levels, March to May and August to October, and contributed up to 90% during a *Spirogyra* sp. bloom in September 1993. Diatom contribution was secondary, except for a short-term bloom in July 1993 with a rapid decline afterwards. This diatom bloom was accompanied by a large number of undetermined ciliates that were feeding selectively on the diatoms.

10.3.2 Biomass and Primary Production

The first quantitative studies were conducted by Rai and Hill (1984) with exposed strips of cellulose-acetate foil in Lago Cristalino, a black-water lake with clear-water influence from the terra firme. Primary production was determined using the ^{14}C technique. They obtained net primary production values ranging between 0.01 and 4.6 g C m^{-2} day^{-1} and calculated a mean productivity of 1.5 g C m^{-2} day^{-1} for attached algae populations after 15 days of exposure.

Doyle (1991) measured primary production and nitrogen fixation of periphyton in the roots of aquatic macrophytes by inserting whole root clusters into plexiglass chambers with recirculated water. Primary production was measured by determining concentrations of oxygen. Production values were derived from periphyton on the roots collected under a water surface area of 0.2 m^2. This was extrapolated to 1 m^2. In the floating roots of the grasses *Paspalum repens* and *Echinochloa polystachya* Doyle found a periphyton production of 1.2 g C m^{-2} day^{-1}, corresponding to 310 g C m^{-2} for the annual growth period of the macrophytes. These values are similar to the values indicated for phytoplankton production in várzea lakes. They are high, because on average about 66% of the total photosynthetic active radiation was intercepted by the canopy of the grasses inside the macrophyte stands. Within the root zone, less than 2% of the radiation penetrated to a depth of 10 cm (Doyle 1991).

In Lago Calado, periphyton contributed about 120 t of carbon to the carbon budget of the lake which was about 25% of the amount produced by phytoplankton. Macrophyte growth during the study period was relatively low, covering only about 0.43 km^2 corresponding to 14% of the 3 km^2 area of the southern basin of the lake. The extrapolation of the area covered by aquatic macrophytes to periphyton production is somewhat problematic because inside dense macrophyte mats growth conditions for periphyton deteriorate considerably. Even so, Doyle's study demonstrates the importance of periphyton associated with aquatic macrophytes.

Engle and Melack (1990, 1993) conducted studies in Lago Calado on periphyton grown on floating meadow roots enclosed by plastic bags with a semicontinuous throughflow of lake and river water to determine the impact of Amazon water on the periphyton. The biomass data, expressed as chl-a content, ranged from 18 to 121 mg m^{-2}. Engle and Melack (1993) calculated periphyton chl-a values in two ways. They used values obtained directly from samples of a known area of floating meadows; these values ranged between 20 and 42 mg m^{-2}. Their calculation of chl-a values as mg per gram root dry weight resulted in values ranging between 0.93 and 2.16 mg. These data represent actual production data of periphyton per water surface area. A direct comparison with other studies using exposed foils is problematic because the surface area of macrophyte roots covered by periphyton per unit area of water surface may be greater than one.

Biomass of periphyton on macrophyte roots (Doyle 1991) increased from 20.5 g dry weight in January to 286.5 g m^{-2} in July. Engle and Melack (1993) reported values up to 320 g m^{-2} dry weight. These values are very high and may be overestimated because of methodological problems. The separation of periphyton from roots by manual washing includes root fragments, fine macrophyte detritus, and perizoic animals, which cannot be separated from the periphyton. Assuming a carbon: chl-a ratio of 30 for microalgae, Doyle (1991) calculated a 23% contribution of periphytic carbon biomass to total algal carbon.

Alves (1993) studied periphyton growth on leaves of the floodplain forest of Lago Calado using an incubation chamber for oxygen measurements. He reported strongly varying light extinction rates caused by the canopy of the floodplain forest with values reaching up to 90%. Light availability was higher in the dry season than in the rainy season. Whereas periphyton communities of the floating meadows showed adaptations to high light intensities (Doyle 1991), those of the floodplain forest were adapted to low light intensities. Compensation light intensities were significantly higher in the dry season compared to the rainy season ($19 \pm 11 \, \mu mol \, m^{-2} s^{-1}$ and $7 \pm 2 \, \mu mol \, m^{-2} s^{-1}$, respectively). In spots of high irradiation, light inhibition of photosynthesis was observed. According to Alves (1993), these observations suggest a shift in the physiological state of the periphyton from a light-adapted (rainy season) to a less light-adapted (dry season) community. There was no difference in photosynthetic activity on the depth scale. Mean chl-a values varied between 5 to $21 \, mg \, m^{-2}$ (mean value $11 \, mg \, m^{-2}$) during the dry season and between 4 to $12 \, mg \, m^{-2}$ (mean value $8 \, mg \, m^{-2}$) during the rainy season.

Another approach to the study of periphyton assemblages was conducted by Putz (1997) using roughened cellulose-acetate foils with defined surfaces for quantitative and qualitative investigations. Data were derived from 15 black- and white-water sites, with six sites sampled at 2-week intervals during 14 months of investigation. Nine sites were sampled once or twice for comparative purposes. Comparison of foil periphyton with leaf periphyton showed little differences in respect to chl-a content and ash-free dry weight (Table 10.2).

Table 10.2. Comparison of chl-a ash-free dry weight for three substrates of leaf and foil periphyton after 2 weeks of inundation (leaves) and exposure time (foils)

	Substrate	Chl-a[a] $(mg \, m^{-2})$	AFDW $(g \, m^{-2})$
24 June 1993	Leaf	9.24 ± 0.9	0.91
24 June 1993	Foil	8.84 ± 1.1	1.03
7 April 1994	Leaf	0.84 ± 0.2	0.2
7 April 1994	Foil	1.17 ± 0.2	0.19
19 May 1994	Leaf	9.68 ± 1.8	1.08
19 May 1994	Foil	10.22 ± 2.1	1.19

[a] Values for chl-a, $\pm SD$, $n = 3$.

Table 10.3. Electric conductivity, pH values, chl-a content, ash-free dry weight, and productivity of four regularly sampled sites, 1992–1994

	pH	Conductivity ($\mu S \, cm^{-1} s^{-1}$)	Chl-a ($mg \, m^{-2}$)	AFDW ($g \, m^{-2}$)	Productivity ($mg \, C \, m^{-2} h^{-1}$)
Solimões	6.71 (6.4–7.0)	72.2 (62.4–78.3)	1.3 (0–0.7.1)	1.1 (0.25–2.5)	6.9 (0.52–17.3)
Floating meadows	6.65 (6.0–7.3)	91.8 (69.6–385.2)	8.5 (0.6–46)	4.18 (1.1–10.6)	38.7 (0.46–139.8)
Mixed water	6.07 (5.5–6.6)	32.2 (11.4–48.5)	25.3 (9.4–41.6)	6.83 (2–19.8)	86.5 (17.5–258.1)
Igapó forest margin	4.51 (4.1–4.9)	16.8 (14.3–18.6)	4.0 (0.5–36.8)	0.31 (0.1–1.43)	23.5 (4.8–54.2)
Igapó forest	4.49 (3.8–4.8)	17.2 (10.2–20.9)	3.5 (0.3–11.4)	0.39 (0.19–1.2)	25.4 (2.2–77.8)

Data derived from foils after 2 weeks of exposure. Figures indicate means, minimum, and maximum values for the respective parameters.

Comparing data from a depth of 0.3 m, there were site specific differences of chl-a, ash-free dry weight (AFDW) and primary production (Table 10.3). In the Solimões mainstem very little chl-a and AFDW were detected. In the várzea, a significantly higher level was reached than in black-water igapó. In general, highest chl-a levels were observed during high water in June and July and at the end of April during a period of rapidly rising water levels. Variations in chl-a contribution to periphytic AFDW turned out to be independent of water levels. Mixed-water attached algae chl-a content peaked at 41.6 mg m^{-2} during the high-water phase. At that time chl-a contribution to an AFDW of 9.4 g m^{-2} was low, reaching only 0.4%. The highest biomass of the entire investigation was found to be a AFDW of 19.8 g m^{-2} with a corresponding chl-a content of 39.1 mg m^{-2} (0.2% AFDW).

Mean chl-a contents of floating meadow root zone periphyton were considerably lower varying between 0.6 mg m^{-2} (0.1% of 1.1 g m^{-2} AFDW) and 46 mg m^{-2} (0.4% of 10.6 g m^{-2} AFDW) (Table 10.3). Maximum values were measured during high water, in June and July. Solimões mainstem periphyton chl-a oscillated between 0.0 mg m^{-2} and 7.1 mg m^{-2} (Table 10.3).

Black-water biomass data were found to be considerably lower (Table 10.3); elevated chl-a was observed during high water. Attached algae chl-a content reached up to 38.9 mg m^{-2}; chl-a maximum was due to a diatom bloom, resulting in a 2.6% contribution to AFDW, which was the highest obtained during the investigation. Lowest igapó periphyton chl-a content was 0.3 mg m^{-2}.

Solimões periphyton productivity derived from ^{14}C applications was very low (Table 10.3). Mean productivity was 6.9 mg C m^{-2} h^{-1}. Periphyton productivity on foils exposed within floating meadows peaked at 139.9 mg C m^{-2} h^{-1} (Table 10.3), correlating with the highest assimilation rate of 13.84 mg C mg chl-a^{-1} h^{-1}. Minimum productivity was 0.46 mg C m^{-2} h^{-1} and minimum assimilation rate was 0.16 mg C mg chl-a^{-1} h^{-1}. Mean periphyton productivity only reached 38.7 mg C m^{-2} h^{-1}. Mixed water attached algal populations had the highest productivity, varying between 17.53 mg C m^{-2} h^{-1} and 258.4 mg C m^{-2} h^{-1}. Overall mean productivity was 86.5 mg C m^{-2} h^{-1}. Considering the large and complex surface structure in the root zone of floating meadows, values obtained from foil experiments can only be used as an index of primary production in this habitat.

Black-water periphyton productivity was found to be lower than white-water productivity, except for Solimões mainstem. This is true for all other occasionally sampled black-water sites as well, with a periphyton productivity differing only marginally from mean values derived from three regularly sampled black-water sites. Attached algal productivity was highest within an extensive igapó inundation forest with a mean productivity of 25.4 mg C m^{-2} h^{-1} (Table 10.3). Maximum and minimum productivity were

Table 10.4. Chlorophyll-a, ash free dry weight and productivity of the depth profiles measured at two sites; Igapó forest and Igapó forest margin

| Site | 27.5.1993–24.6.1993 | | | | | | 23.3.1994–20.4.1994 | | | | | |
| | Igapó forest | | | Igapó forest margin | | | Igapó forest | | | Igapó forest margin | | |
	chl-a (mg m^{-2})	AFDW (g m^{-2})	Prod. (mgC m^{-2} h^{-1})	chl-a (mg m^{-2})	AFDW (g m^{-2})	Prod. (mgC m^{-2} h^{-1})	chl-a (mg m^{-2})	AFDW (g m^{-2})	Prod. (mgC m^{-2} h^{-1})	chl-a (mg m^{-2})	AFDW (g m^{-2})	Prod. (mgC m^{-2} h^{-1})
0.3 m	15.1 +4.3	n.d.	64.68 +9.9	38.3 +1.1	n.d.	11.93 +2.1	3.97 +0.3	0.57	53.37 +7.8	4.13 +0.7	0.31	74.9 +8.8
0.7 m	9.8 +0.5	n.d.	48.4 +9.4	28 +9.8	n.d.	13.33 +5.2	5.41 +0.3	0.46	22.6 +4.9	3.6 +0.5	0.29	32.34 +7.6
1.1 m	5.5 +2.4	n.d.	30.14 +6.1	29.4 +2.2	n.d.	12.01 +4.1	3.98 +0.1	0.67	23.41 +1.4	3.9 +1.2	0.26	34.8 +8.2
1.5 m	8.8 +3.4	n.d.	14.59 +7.3	13.4 +1.7	n.d.	9.37 +1.8	1.02 +0.2	0.46	8.25 +1.1	1.25 +0.2	0.32	9.13 +1.6
1.9 m	1.5 +0.3	n.d.	[a]	1.3 +0.3	n.d.	1.02 +0.2	0.2 +0.1	0.36	1.2 +0.2	0.92 +0.2	0.25	4.65 +0.8

All means from n = 3, SD.
n.d., No data.
[a] Without any measurable net primary production.

$77.82\,\mathrm{mg\,C\,m^{-2}\,h^{-1}}$ and $2.21\,\mathrm{mg\,C\,m^{-2}\,h^{-1}}$. Mean productivity was slightly lower, reaching $23.5\,\mathrm{mg\,C\,m^{-2}\,h^{-1}}$ at the igapó forest margin. Maximum productivity was obtained in May 1994 reaching $56.16\,\mathrm{mg\,C\,m^{-2}\,h^{-1}}$, whereas the minimum was measured at $4.82\,\mathrm{mg\,C\,m^{-2}\,h^{-1}}$.

In 1993 and 1994, black-water periphyton biomass and productivity were measured at five depths from 0.3 m to 1.9 m during high water. After 4 weeks of exposure, periphytic chl-a and productivity could be measured to 1.9 m. Both parameters remained at the same level to a depth of 1.1 m (Table 10.4). These data correspond to Alves (1993) who found no variation in periphyton production in the upper surface layer. Beyond 1.5 m both chl-a and productivity declined. In 1994, AFDW was uniform within the trophic zone at both sites. Periphytic chl-a content and productivity within the black-water igapó forest was similar to levels in mixed water forests during the high water period (May–August) after exposure of four weeks. Therefore, it may be that after 2 weeks of exposure mature periphyton was not established in black water, and that primary production in black-water habitats may be underestimated.

10.4 Discussion and Conclusions

Algal biomass and primary production in Amazonian floodplains is strongly influenced by the floodpulse (Junk et al. 1989; Sect. 1.3). All authors mention light as a limiting factor at least periodically for phytoplankton and periphyton production in várzea lakes (Sect. 4.2). Some planktonic algae can adjust their position in the water column according to light intensity. Periphyton growth depends on the position of the available substratum. In the floodplain forest light interception by the canopy of up to 90% may limit algal primary production. Considering the narrow euphotic layer of about 2 m extent and the seasonal rise and fall of the water level, most surfaces are available in the euphotic layer for the growth of periphytic algae during about 3 months of the year. The short availability of surfaces is in part compensated for by a surface:area index greater than 1. It corresponds to the sum of the surface areas of submerged leaves, branches, stems and the soil per unit area of water surface and may reach up to 8 in areas covered by a dense vegetation of shrubs.

A more favorable habitat for periphyton is offered by the root bunches of free-floating macrophytes, because the plants follow the water level fluctuations. Intense algal growth is observed on the roots of singular plants on the margins of the stands. Inside the stands, light intensity is reduced. According to Piedade et al. (1991), light interception of the canopy of an *Echinochloa polystachya* community varied between 72 and 89% throughout a 1-year cycle. Doyle (1991) indicates a mean light interception of 65%. In dense stands of free-floating macrophytes, light penetration into water and algal growth is reduced to the upper few centimeters. In very dense floating islands algal production is negligible, because plants and detritus cover the water surface completely (Sect. 8.2.3).

The depth of the euphotic layer in open water depends on the amount of suspended matter and dissolved colored organic substances. High load of suspended matter and related low transparency of the water strongly reduce algal production in the main channel of the Amazon river. Phytoplankton production can be neglected, algal periphyton production is low, reaching a mean value of about $84 \, mg \, C \, m^{-2} \, day^{-1}$.

The following general pattern can be considered typical for Amazon floodplain lakes. When the river invades the floodplain at rising water, transparency is slightly increased in comparison with the main channel. Phytoplankton density and primary production are low because of dilution of the existing plankton community and a narrow euphotic layer. During this period, periphyton growth in the root zone of aquatic macrophytes is favored because floating plants and attached algae accompany the rise of the water level in the euphotic layer and the periphyton can take up nutrients from the river water (Engle and Melack 1993). When suspensoids sink, transparency increases and primary production per unit area reaches maximum values at falling water level. At low water, when lakes reach a depth less than 2–3 m, transparency decreases due to wind- and fish-induced mixing, which stir the surface sediment layer. High levels of nutrients and low water depths result in high phytoplankton densities and in maximum values of primary production per unit volume. In black water, light penetration is attenuated by large amounts of colored humic substances, which allow phytoplankton and periphyton primary production in the surface layer to a maximum depth of about 2 m (Schmidt 1976; Rai and Hill 1984; Putz 1997).

In addition to light, nutrients may become limiting for algal production. The low nutrient status of blackwater (Sect. 4.3.2) is reflected by a low phytoplankton production, which reached $230 \, g \, C \, ha^{-1} \, year^{-1}$ in the Negro River representing about 10% of the production in most white-water lakes. Periphyton production in black-water lakes reached about 30% of the values in várzea lakes (Putz 1997). Maximum phytoplankton and periphyton production is reached in mixed-water lakes, which combine high transparency with adequate nutrient supply. Considering a mean surface area index of 2.5 (1 for the soil and 1.5 for leaves, stems, and branches), an exposure time of 90 days, and mean production rates of $86 \, mg \, C \, m^{-2} \, h^{-1}$ for mixed-water forests and $25 \, mg \, C \, m^{-2} \, h^{-1}$ for igapó forests, the respective production of periphytic algae corresponds to 2.32 t C ha^{-1} and 0.67 t C ha^{-1}, respectively, per flood period. The hypothesis that the black-water floodplains are less productive than the white-water floodplains (Sioli 1955, 1968; Fittkau et al. 1975; Junk 1984) is confirmed for phytoplankton and periphyton. The recorded mean algal productivity of Amazonian black-water habitats corresponds to oligotrophic and mesotrophic lakes in temperate zones; that of várzea lakes corresponds to temperate eutrophic lakes.

Forsberg (1984) shows that nitrogen and phosphorus are limiting nutrients for phytoplankton growth in a várzea lake and in an igapó lake, respectively. Setaro and Melack (1984) and Melack and Fisher (1990) indicate phosphorus limitation at rising and high water and nitrogen limitation at falling and low water in Lago Calado. In várzea lakes, major sources for nitrogen and phosphorous are water and sediments from the Amazon River (Sects. 4.3.1, 4.4). Aquatic macrophytes and associated periphyton grow in large quantities near the entrances of the lakes where they take up nutrients from the inflowing river water (Doyle 1991; Engle and Melack 1993). A quantification of the uptake of nutrients by periphyton is still problematic, because the in situ experiments do not differentiate between the nutrient uptake by periphyton and the uptake by the macrophytes themselves.

Differences in the behaviors of phytoplankton and periphyton populations can be related to specific morphological peculiarities of the lakes, which result in a change in the pattern of transparency due to changes of water- and sediment-exchange with the river and the impact of wind on the lake surface. The influence of the river on lakes isolated during most of the year will be much smaller than in lakes permanently connected with the river. Nutrient input into the euphotic layer also occurs from deeper water layers because of vertical circulation. Differences in phytoplankton production in várzea lakes are therefore attributed also to the circulation pattern, which is in part correlated with lake size and shape (Melack 1984; Sect. 4.4). In Ria Lakes, affluents from the terra firme are important sources of nitrogen (Sects. 4.4, 4.5). In these lakes, periphyton on floodplain forest leaves perform a filtering function for nitrate from terra firme affluents similar to that performed by aquatic macrophytes near the mouth of the lakes where nutrients are introduced by the Amazon river (Alves 1993).

The presence of large quantities of free-floating and rooted emergent aquatic macrophytes or trees in floodplain forests reduces light intensity within the water by shadowing the surface, but offer surfaces for periphyton growth. Competition for nutrients between algae and aquatic macrophytes may be of importance in lakes with large amounts of aquatic macrophytes as indicated for the Pantanal of Mato Grosso (Heckman 1994; Prado et al. 1994).

The importance of planktonic and periphytic algae for food webs in Amazonian floodplains is not yet quantified. Junk (1973) points to periphyton and phytoplankton as a main food source for perizoon living in the root zone of aquatic macrophytes. "Aufwuchs" diatoms in particular are subjected to a severe invertebrate feeding pressure. Putz (1992) reported high ciliate densities associated with the "*Aufwuchs*" in a River Rhine floodplain, consuming diatoms highly selectively according to species and size. This phenomenon could be frequently observed in the diatom dominated igapó periphyton communities as well (Putz 1997).

Turnover rate of periphytic assemblages is high. Doyle (1991) indicates a turnover time of 4.7 days for Lago Calado. Werner (1977) records cell division rates for diatoms of up to 2.3 days^{-1}. McIntire (1973) carried out an area-based simulation model for periphyton dynamics in a laboratory stream. The attached diatom standing crop of $10\,\mathrm{g\,m^{-2}}$ AFDW was found to be sufficient to support a primary consumer biomass of $150\,\mathrm{g\,m^{-2}}$.

Algal biomass is an important food source in the food chain. Araujo-Lima et al. (1986), deduced from an analysis of five adult detritivorous fish species, that they fed on carbon originating in phytoplankton and periphyton. The nutritional value of periphyton is high with an average $C:N:P$ ratio of $171:18:1$ compared to $260:28:1$ in phytoplankton (Smith and Fisher, unpubl.). However, according to a calculation by Bayley (1989) phytoplankton and periphyton production in the várzea is not sufficient to sustain the high fish production (Sects. 20.2, 20.7). Algae may have a key function as a food source for juvenile fishes, directly or indirectly via aquatic invertebrates. We suggest that the relatively high primary production of periphyton, which is concentrated in a biofilm of 1–5 mm in diameter on submerged leaves and branches of the floodplain forest, is of major importance for the food webs of black-water floodplains.

Acknowledgments. We are very grateful to Prof. Dr. Schwoerbel, University of Konstanz, Dr. John Melack, University of California, and Dr. Hakumat Rai, Max-Planck-Institute for Limnology, Plön, for critical comments on the manuscript and the correction of the English.

11 The Forest Ecosystem of the Floodplains

MARTIN WORBES

11.1 Introduction

The Amazon floodplain forests are of special interest for botanical research due to the special physiognomy, the easy attainability of their margins and their exceptional environmental conditions. They were described in terms of their dominant and conspicuous tree species by Huber (1910), Ducke and Black (1953), Takeuchi (1962), and Hueck (1966). The hydrochemical differentiation of the Amazonian floodplains (Sioli 1954) into white-water (várzea) and black-water (igapó) inundation areas was confirmed at the botanical level by Prance (1979, 1989) and from a taxonomic point of view by Kubitzki (1989). Detailed forest inventories were carried out by Pires and Koury (1959), Keel and Prance (1979), Revilla (1981, 1991), Worbes (1983), and Campbell et al. (1986). Worbes et al. (1992) defined different forest communities of the várzea as successional stages.

Prance (1989) considered edaphic and hydrological aspects in his classification of inundation forests in tropical South America. Ranked by order of increasing length of inundation these classifications are the gallery forest, the floodplains, and the inundation forests (várzea and igapó). In this chapter várzea and igapó are combined and classified as floodplains according to the definition of Junk et al. (1989; Sect. 1.2).

11.2 Floristics and Forest Structure

The following description of floristics and forest structure is based on 19 ha of forest inventories in the igapó and the várzea in Central Amazonia. Detailed information of methods, site conditions, and complete species lists of these inventories were given in Revilla (1981, 1990), Piedade (1985), Worbes (1986), Rankin-De Merona (1988), Ferreira (1991), Worbes et al. (1992). Study sites are described in Section 2.2.

Ecological Studies, Vol. 126
Junk (ed) The Central Amazon Floodplain
© Springer-Verlag Berlin Heidelberg 1997

11.2.1 Species Diversity

The description of diversity and the discussion of possible reasons for diversity patterns play an important role in the vegetation analysis of tropical forests (Gentry 1982; Prance 1989). There are several underlying problems with the estimation of diversity, i.e., how to choose plot size, dbh size (diameter at breast height) of the selected trees and the selection of stands with a homogenous vegetation structure. A widely accepted method for comparing diversity at different sites is to construct species-area curves.

Within plots with relatively homogenous site conditions and a homogenous vegetation structure the area-species curve flattens out for small plot sizes of 0.22 ha in an igapó stand (Keel and Prance 1979). No flattening point is visible when several igapó stands are considered (Fig. 11.1). The species-area curve of várzea stands at the Ilha de Marchantaria flattens out at about 0.5 ha with 40 species. Within a 12 km area on the same island in an additional 1.21 ha study site only ten new species occur (Fig. 11.1).

In the most diverse record in the várzea, 106 species were found in 1 ha at the Costa de Marrecão. In 10 ha at the same site and 5 ha at the Costa de Barroso the species-area curves flatten at between 2 and 3 ha at different

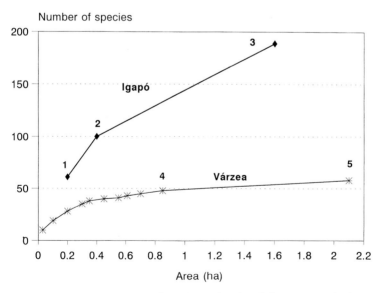

Fig. 11.1. Species area curves from igapó (*1-3*) and from várzea (*4,5*). New species were added to the number of species of the previous stand. Stand *1*, Tarumã Mirím (Worbes 1983); stand *2*, white sand, Negro River (Keel and Prance 1979); stand *3*, Praia grande, Negro River (Revilla 1981); stand *4*, from Ilha de Marchantaria, Lago Camaleão; stand *5*, another stand from Ilha de Marchantaria, Central Lake

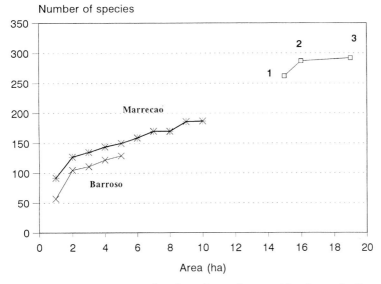

Fig. 11.2. Species area curves for várzea forests from ten 1 ha plots at the Costa de Marrecão, five 1 ha plots at the Costa de Barroso. *1*, Combination of Costa de Marrecão and C. de Barroso; *2*, additional stand at the Ilha de Carreiro; *3*, additional stands at the Ilha de Marchantaria

levels of species numbers, respectively (Fig. 11.2). The species inventory of the two sites differs by approximately 60 species and shows more than 250 species in total. Additional stands at the Ilha de Marchantaria and at the Ilha de Careiro increase the total number of species slightly (Fig. 11.2).

Keel and Prance (1979) and Worbes (1983, 1986) counted in small igapó plots of 0.22 and 0.21 ha 54 and 61 species, respectively. This is almost twice the number of species found on plots of comparable size in the várzea. A total of 292 species are found on 19 ha of várzea forests in the vicinity of Manaus between the Costa de Marrecão (near the city of Manacapurú) and the Ilha de Carreiro (near Manaus). About 200 species were counted on 1.4 ha of investigated igapó forests.

There is not much available information on the neotropical inundation forests outside of Central Amazonia. Most information is on várzea forests. There are 53 species ha^{-1} in a tidal várzea forest at the Guamá River, Pará (Pires and Koury 1959), 109 species ha^{-1} in a várzea forest at the Tefé River (Ayres 1993) and 149 species ha^{-1} at the Napo River, Ecuador (Balslev et al. 1987). Tuomisto (1993) sampled in gallery forests of the white water rivers in the Peruvian foothills of the Andes and found 306 species in an undefined area. On non-flooded sites (terra firme) in the vicinity of Manaus, figures vary between 65 and 176 species ha^{-1} (Klinge and Rodrigues 1968; Prance et al. 1976). Gentry (cited in Balslev et al. 1987) counted between 179 and 285 species of trees per hectare in Amazonian Peru. Investigations in floodplain forests and adjacent terra firme stands at the Napo River, Ecuador (Balslev et al. 1987) and at the Xingú River, Brazil (Campbell et al. 1986) show in both cases a higher diversity in the non-flooded areas.

The data suggest that species diversity in igapó forests is higher than in várzea forests. On a local scale, diversity in floodplain forests is lower than in regional neighbouring terra firme forests. Diversity tends to increase in the Amazon basin from east to west in both flood-plains and adjacent terra firme sites.

11.2.2 Floristic Relationships Between Amazonian Várzea, Igapó and Terra Firme

Lists of common tree species of igapó and várzea differ considerably depending on the view point of the respective author. Under aspects of taxonomy and species distribution, (Kubitzki 1989), economic interest (Klenke and Ohly 1993) and general descriptions (Prance 1989) the most often cited species are *Ceiba pentandra, Hura crepitans, Nectandra amazonum*, and *Cecropia* spp. for the várzea, and *Aldina latifolia, Swartzia* ssp., *Parkia discolor* for the igapó. Abundant species of the igapó are listed in Table 11.1, and abundant species of the várzea are listed in Table 11.2.

The different species compositions in várzea, igapó and the terra firme are traced back to differences in the hydrological regime and, in the case of the floodplain ecosystems, to water chemistry (Kubitzki 1989; Prance 1989). However, the more important factor for plant growth is the soil type. Considerable differences in soil type and soil chemistry occur between várzea and igapó and within the igapó between the sites Tarumã Mirím and Anavilhanas. The content of total phosphorus increases from 13 ppm in the Anavilhanas to 130 ppm at the Tarumã Mirím and to 530 ppm on the Ilha de Marchantaria in the várzea. In other tropical forests the differentiation of vegetation is traced back to differences in P contents of the soil (Tilman 1982).

Nevertheless some tolerant species, i.e., *Alchornea castaneaefolia, Tabebuia barbata, Piranhea trifoliata, Triplaris surinamensis, Macrolobium acaciifolium* and others are common in both floodplain forests. Moreover a number of species especially from várzea forests also occur on the terra firme near Manaus, e.g., *Mabea caudata, Heisteria spruceana, Minquartia guianensis, Pithecolobium jupumba*, and *Vatairea guianensis*. Representatives from all leading families of the terra firme (Gentry 1990) are also found in the floodplain forests (Fig. 11.3). Studies on terra

Table 11.1. Distribution of abundant and frequent species in the igapó

Stand No:	1	2	3	4	5	6	7	8	9	10	11
Species											
A *Borreria capitata*	30										
Dalbergia inundata	5										
Pithecellobium adiantifolium	15	19	9								
Myrciaria dubia	12	19	6								
Eugenia chachoeriensis		9	9								
E. chrysobalanoides		19	34								
Schistostemon macrophyllum		10									
B *Couepia paraensis*	9				6	8	+	+	+		
Leopoldinia pulchra	+	+	6				+	+	16		
Anacampta rupicola	+		9				+	5		5	
Licania apetala	+				5		+		15		
Himatanthus attenuatus		+	+		+		+	+	+	+	+
C *Malouetia furfuracea*					17	+		+	+		+
Mollia speciosa			+		12			+	+		
Duroia velutina			+		6					+	
Neoxythece elegans					6	+		7	+	+	
Quiina rhytidopus					11		+				
Eschweilera tenuifolia					7		+				
Myrciaria floribunda					8	+		+			
Acmanthera latifolia					10		11	+			
Hevea spruceana					7		7	+		+	+
Alchornea schomburgkiana					+	53					
Macrolobium acaciifolium						11					
Mabea nitida					+	6	+	+		+	+
Maprounea guianensis					+		5	+			
D *Swartzia argentea*					+		+	+	+	+	
Swartzia polyphylla					+		+	+	+	+	
Tabebuia barbata					+		+	+	+	+	
Psychotria lupulina					+		5	5		+	
Caraipa grandiflora					+		+	6		+	
Eschweilera parvifolia					+		+	5		+	
Virola elongata					+		+	7		5	
Ferdinandusa rudgeoides							+	5	+	+	
Parkia discolor							+	+	+		
Aldina latifolia							+	+		5	
Penthaclethra macroloba										10	
Astrocaryum jauari					5		+	+		+	66
Calophyllum brasiliense											5

Figures give rel. density (%) in the respective stand; +, density below 5%.
Stands 1–3, white sand; stands 4–11 clay.
Stands 1–4, long-lasting inundation (>260 days/year); stand 5, 200–260 days/year; stands 6 and 7, 150–200 days/year; stands 8 and 9, 75–150 days/year; stands 10 and 11, 25–75 days/year.
Stands 2 and 3 from Rio Negro near Manaus (Keel and Prance 1979).
Stands 1 and 9 from Rio Negro, Praia Grande (Revilla 1981).
Stands 5, 7, 8, and 10 from Tarumã Mirím (Ferreira 1991).
Stands 4, 6, and 11 from Rio Negro, Anavilhanas (Piedade 1985).
Species group A, low-level community on white sand; species group B, tolerant to different site conditions; species group C, low-level community on clay; species group D, mid-level and high-level community on clay.

Table 11.2. Várzea species of low-level shrub communities (Junk 1989), mid-level forest communities (Worbes et al. 1992) and 12 important species from a 10-ha record of a high-level forest community at the Costa de Marreacão. (Revilla 1991)

Low-level shrub community	Mid-level community	High-level forest community
Coccoloba ovata	*Casearia aculaeta*	*Ceiba pentandra*
Eugenia inundata	*Cecropia latiloba*	*Hevea spruceana*
Ruprechtia ternifolia	*Crescentia amazonica*	*Licania heteromorpha*
Symmeria paniculata	*Ilex inundata*	*Malouetia furfuracea*
	Labatia glomerata	*Manilkara amazonica*
	Laetia corymbulosa	*Olmedioperebea sclerophylla*
	Macrolobium acaciifolium	*Piranhea trifoliata*
	Nectandra amazonum	*Pseudobombax munguba*
	Pseudobombax munguba	*Pterocarpus amazonum*
	Tabebuia barbata	*Rollinea exsucca*
	Vatairea guianensis	*Tapura amazonica*
	Vitex cymosa	*Virola surinamensis*

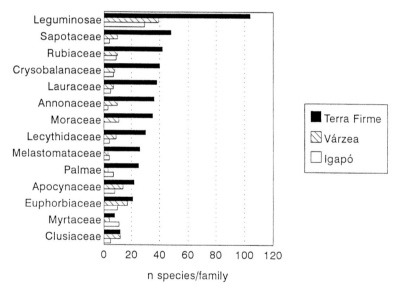

Fig. 11.3. Number of tree species in 12 species-rich families in the Amazonian terra firme (Gentry 1990) in várzea and in igapó

firme and adjacent floodplain forests in the wider Amazon region show concurrences of 18% (Balslev et al. 1987) and 45% (Campbell et al. 1986) at the species level. Taxonomic and chorological investigations (Kubitzki 1989; Prance 1989) show that in general the flora of the igapó is closely related to the flora of the nutrient-poor savannas. On the other hand, close connections are visible between the flora of the várzea and the flora of more fertile sites on the terra firme.

Many of the tree species from Central Amazonian inundation forests have a wide geographic distribution. Quite a few of them, e.g., *Hura crepitans, Casearia aculeata, Unonopsis guatteroides* or *Pseudobombax munguba*, are distributed throughout the Neotropis (Prance and Schaller 1982; Prance 1989). There is, however, a recognizable differentiation between those species distributed throughout all of northern South America and those that are found in the eastern or in the western Hylaea (Kubitzki 1989).

11.2.3 Plant Communities and Site Conditions in the Igapó

Most of the vegetational records from the igapó were carried out with regard to the flood gradient. Keel and Prance (1979) and Revilla (1981) give relative positions of recorded subplots or communities with respect to elevation. Piedade (1985), Worbes (1986) and Ferreira (1991) used planimetry to find the actual water level of the Negro River sites. This information together with the daily water level record in the harbour of Manaus makes it possible to calculate the length of inundation in any year or for longer periods of time (Worbes 1985; Junk 1989; Sects. 2.3, 8.2.1).

Besides the flood regime the soil type has been taken into consideration for all records. Two different soil types are distinguishable at the lower Negro River and its small tributaries. The soil is nutrient-poor white sand near the margins of the low level river. At some distance from the low-level shore line the soil is at least 50% clay (Worbes 1986).

We can differentiate between plant communities according to different site conditions (Table 11.1). There is a distinct difference between the species composition on sand and the species composition found in the clay. On clay soils a further differentiation can be made between a low-level community exposed to more than 150 days of annual inundation, a mid-level community with annual inundations between 75 and 150 days and a high-level community which is inundated for less than 3 months. Transitions between these communities are gradual. It is remarkable that a number of species, e.g., *Couepia paraensis, Leopoldinia pulchra, Anacampta rupicola* and others, are tolerant of long-lasting inundation, very poor soil conditions of the white sand as well as to poor light saturation at the forest floor of the high-level forest community.

According to changes in species composition the structure and the physiognomy of the vegetation change along the flood gradient. On both soil types at sites with a long-lasting mean annual inundation of about 250

Fig. 11.4. a Tarumã Mirím (igapó) at low water level with the investigated stand at the left-hand side. **b** *Salix humboldtiana* stand on a sand bar downstream of the Ilha de Marchantaria (várzea). **c** *Piranhea trifoliata* trunk with secondary branches on the Ilha de Careiro (várzea)

Fig. 11.4c

days, the vegetation is composed of relatively few shrubs and small trees of a maximum height of 10 m. This vegetation type turns gradually into a well-structured and diverse forest at inundation levels of less than 200 days year^{-1}. At the Tarumã Mirím River site (Fig. 11.4a) the mid-level community is dominated by mighty buttressed trees of *Aldina latifolia* and associated species with a maximum height of 30 m. At the forest floor the herb *Ischnosyphon ovatus* is conspicuous beneath a high number of shrub species. The appearance of *Selaginella stellata* at the forest floor indicates the transition to the never-flooded terra firme (Adis 1984).

11.2.4 Plant Communities and Site Conditions in the Várzea

Zonation in the Várzea. The influence of the length of the inundation period on species composition of the várzea forests is often mentioned (Hueck 1966; Junk 1989) (Sect. 8.2). At low-lying sites at the 19.5-m level, which were exposed to inundations 300 days year^{-1} on average and frequently experienced in three consecutive years, Junk (1989) found stands of dead shrub communities. At slightly elevated sites (between 20 and 21 m) shrub species may persist. The lower tree line in the várzea is located where average annual inundation is 230 days (22 m). Among others the tree species *Piranhea trifoliata* and *Vitex cymosa* are often found in very low lying locations. In contrast, *Ceiba pentandra* and other species colonize less frequently in shortly flooded areas (Hueck 1966). A species list of low lying shrub communities and a high lying community from a 10-ha record at the Costa de Marrecão is given in Table 11.2.

Successional Differentiation in Várzea Forests. Different várzea stands at the same flood level can be dominated by different tree species. Pure stands of *Cecropia* ssp. or *Salix humboldtiana* are frequently distributed at the river margins, the latter also dominate young sand bars (Fig. 11.4b). On sand bars the trees are exposed to high annual sedimentation rates of 50 cm or more. *Salix humboldtiana* tolerates high sedimentation stress by annually forming new root layers at the soil surface to avoid hypoxic conditions in the root zone. The typical tree of the medium sized Marchantaria island is *Pseudobombax munguba*. On the large Careiro island and at the Costa de Marrecão *Pterocarpus amazonum* and *Piranhea trifoliata* are conspicuous and very frequently found (Fig. 11.4c).

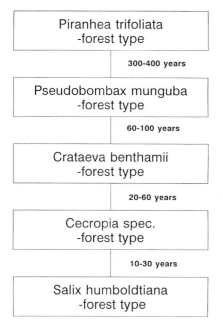

Fig. 11.5. Successional pattern of várzea forest types. Approximate time spans of development of the stages are indicated between the *boxes*. (After Worbes et al. 1992)

Worbes et al. (1992) used an analysis of the species inventories of nine stands located on similar elevations and therefore under equal flood stress to show that different groups of tree species dominate at different sites (Fig. 11.5). Age datings of the dominant trees by means of tree-ring analysis, as described below, showed considerably different age structures for the investigated stands (Fig. 11.6a). The dominating species groups can be differentiated by features such as life span, growth rates and wood density. The

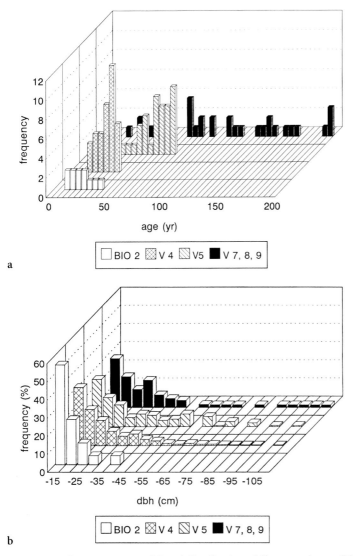

a

b

Fig. 11.6a,b. Age structure (a) and distribution of diameter classes (b) of trees in four forest types in the várzea. (*Bio 2, V4, V5* Young forest stands on the Ilha de Marchantaria. *V7–V9* Climax forest stands at the Costa de Marrecão and on the Ilha de Careiro

young stands are dominated by trees with characteristics that are typical of pioneers (Swaine and Whitmore 1988), i.e., low life expectancy, fast growth and low wood density (*Salix humboldtiana, Cecropia* ssp., *Pseudobombax munguba, Macrolobium acaciifolium*). The oldest stands are dominated by trees with converse features (*Piranhea trifoliata, Manilkara* sp., *Tabebuia barbata*).

The presence of tree species with opposing life-history strategies supported by the differences in age structures of the stands allows us to define the investigated stands as a successional time series (Fig. 11.5). The features of the defined successional stages in the várzea also occur in Central American forests (Budowski 1961) as well as in the European floodplain forests (Carbiener et al. 1988).

The initial stage of primary succession in floodplain habitats is dominated by high growing grass communities (Ellenberg 1988) (Sect. 8.2.3). The following pioneer forest community is dominated by representatives of the genus *Salix* (Worbes et al. 1992). Pioneer stages are followed by the "early secondary" stage (Budowski 1961), which is characterized by a species-poor formation dominated by *Cecropia* sp. (examples in Martius 1989). The *Crataeva benthamii* community (stand 4) is analogous to the young "late secondary" stage described by Budowski (1961). This stage soon undergoes a transition to the *Pseudobombax* forest type (stand 5). The *Pseudobombax munguba* community represents the beginning of a dynamic stage between the "late secondary" and the climax community (Hallé et al. 1978). By this time the fast growing dominant species *P. munguba* has reached its maximum age and is eventually replaced by slower growing species from the lower forest canopy. The *Piranhea trifoliata* community (stands 6, 7, 8) represents the typical climax forest. This community is dominated by dense-wood emergents with multisecular rotation and subdominants with a shorter life span.

In general, the successional sequence from pioneer to climax forest is characterized by a decrease in radial growth rates and an increase of wood density of the dominant trees. The division into "light-wood" and "dense-wood" associations of European floodplain forests (*Populion albae* vs. *Ulmenion* with *Quercus*) was originally explained by different degrees of flood tolerance among species. Recently this diversification is interpreted as a consequence of successional development (Carbiener et al. 1988; Ellenberg 1988).

Structural Changes During Dynamic Development. The shift in species composition from pioneers to species of climax forests during a successional time sequence is accompanied by basic structural changes (Fig. 11.6b). Along the age gradient the diversity increases continuously from monospecific stands to stands with about 100 species ha^{-1} (Crow 1980). Tree density declines from a maximum of some 1000 stems ha^{-1} at the age of about 10 years to 500–700 stems ha^{-1}. Biomass increases in the same order. A similar development for the relationship of density to biomass is observed in temperate zones and is known as a self-thinning process (White 1981). The distribution of diameter classes changes rapidly from a high percentage of small diameters in the youngest stand to J-shape distribution observable in any forest community (Fig. 11.7). The differences between the stands that are older than 40 years are very small. Comparison with the age structure of the same stands shows that the age of the stands cannot be determined from diameter development.

According to other tropical forests, the várzea forest appears well structured into different crown layers. The frequency polygon diagrams of tree heights in an 80-year-old stand on the Ilha de Marchantaria shows five crown layers (Fig. 11.8). The layers are composed of trees of different species. Shrubs and trees shorter than 10 m in height are *Psidium acutangulum*, *Trichilia singularis*, *Pithecolobium inaequale* and *Pseudoxandra polyphleba*, from 10 to 20 m *Crataeva benthamii*, *Ilex inundata*, *Sorocea duckei* and *Vitex cymosa* occur. *Laetia corymbulosa*, *Triplaris surinamensis* and *Eschweilera* sp. can grow to a height of 25 m and

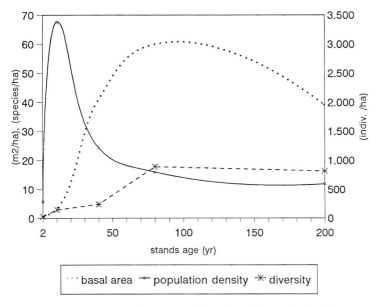

Fig. 11.7. Development of basal area, diversity and tree density along a successional time series in the várzea. Forest stands of different ages are treated as a successional sequence. (Worbes et al. 1992)

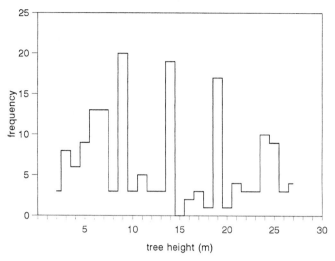

Fig. 11.8. Frequency polygon plot of tree heights measured in an 80-year-old stand on the Ilha de Marchantaria. Five crown layers are indicated by the peaks of the frequency curve

Pseudobombax munguba dominates the crown layer with a maximum height of 30 m.

11.3 Responses to the Flood Pulse

11.3.1 Leaf-Fall Behavior

The casual observer easily recognizes that in the várzea many conspicuous tree species, e.g., *Vitex cymosa, Pseudobombax munguba* and others, are defoliated at the end of the submersion phase (Fig. 11.9). The igapó forest seems to be composed more of evergreens (Worbes 1983). A more detailed analysis shows that evergreen and deciduous species exist in both habitats. Species may shed their leaves at the beginning or at the end of the submersion phase. In the igapó a maximum of leafless species is observed during the end of the aquatic phase (Fig. 11.10). In the várzea the maximum leaf fall is observed during the aquatic phase with a smaller peak in November (cf. Fig. 11.21). In the várzea, the outbreak of new leaves may occur when the trees are still flooded (*Crataeva benthamii, Vitex cymosa*; Fig. 11.9) or at the beginning of the rainy season in November or December (*Ilex inundata*). Two distinct peaks of leaf flush were observed in igapó stands at Praia Grande (Fig. 11.10). One is in August–September when the water runs

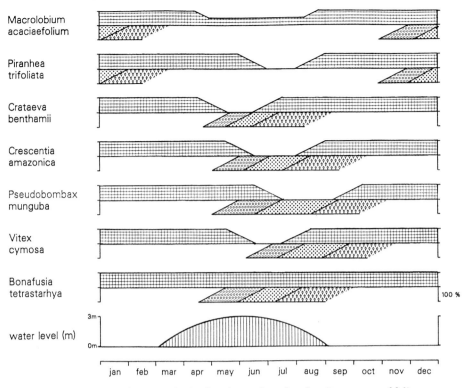

Fig. 11.9. Phenology of tree species in the várzea. *Cross hatches*, Percentage of foliage; *open circles*, flowering; *points*, fruiting; *triangles*, fruit fall. *Below*, Water level in 1981 in the stand V5 at the Ilha de Marchantaria

out of the forests. The second is in November and December at the beginning of the rainy season. The peaks of leaf fall and new flushes in November–December may be a reaction to the rapidly drying soil during the months between the end of the aquatic phase and the beginning of the rainy season (Worbes 1986).

11.3.2 Periodical Growth of Roots, Wood and Shoots

The electrical resistance of the cambium can be interpreted as the conductivity of the cambium and therefore as cambial activity (Shigo and Shortle 1985). Cambial activity of trees in the várzea measured throughout one year shows lower activity values during the aquatic phase than during the non-flooded period (Fig. 11.11). A small peak of high values was observed during the end of the aquatic phase, probably due to the beginning of flowering or the outbreak of new leaves.

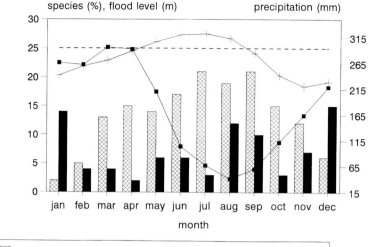

Fig. 11.10. Percentage of completely defoliated tree species and percentage of species with new flushes from 95 investigated species in total at the Praia Grande, Rio Negro (igapó) adopted from Revilla (1981). Monthly precipitation (*right y-axis*) and flood level of the Rio Negro are included. The mean elevation at the stand is 25 m a.s.l. and marked with a *dotted line*

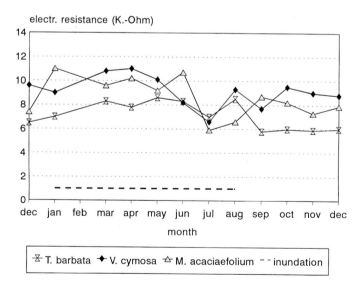

Fig. 11.11. Cambial resistance in *Tabebuia barbata*, Bignoniaceae (n = 1), *Vitex cymosa*, Verbenaceae (n = 2) and *Macrolobium acaciifolium*, Caesalpiniaceae (n = 3) from the Ilha de Marchantaria, várzea from December 1986 until December 1987. High values of resistance indicate low cambial activity and vice versa. The *dotted line* marks the aquatic phase at the investigation site

The low cambial conductivity in trees lasted throughout several months of the aquatic phase, indicating a cambial dormancy. The dormancy is also reflected by the existence of annual rings in the wood of most species of the inundation forests (Worbes 1985, 1986). Different species were marked by cambial woundings (Mariaux 1967) some weeks before the beginning of the aquatic phase and were felled later (Fig. 11.12). Wood production stopped shortly after water covered the roots of the tree. This proves the existence of annual rings in most trees of várzea and igapó. Annual growth periodicity was proven independently by radiocarbon datings of individual growth zones using the nuclear weapon effect (Worbes and Junk 1989). The increase of ^{14}C in the atmospheric CO_2 in the 1950s and early 1960s had worldwide effects with an equal increase in the radiocarbon concentration of the wood of trees. The annual variations of ^{14}C concentration in the growth zones serve as an artificial marker for age dating. Measurements of the ring widths of species with distinct growth zones (Worbes 1985) as well as density estimations of tree-ring series (Worbes et al. 1995) show that the

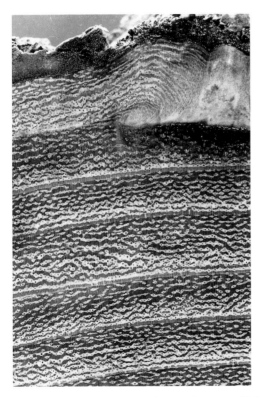

Fig. 11.12. Annual rings and scars from cambial woundings in the wood of *Tabebuia barbata*, Bignoniaceae from the várzea

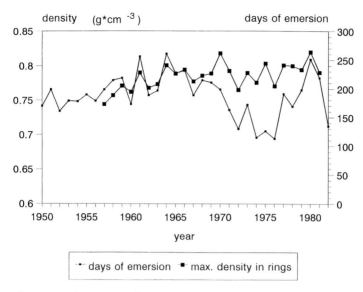

Fig. 11.13. Time series of maximum wood density in annual rings from várzea trees and time series of duration of annual terrestrial phases. The mean density curve is calculated from mean tree curves (n = 2) of four species. (Worbes et al. 1995)

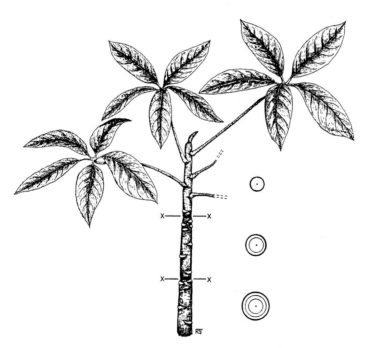

Fig. 11.14. Annual shoot growth in *Pseudobombax munguba*, Bombacaceae, várzea. Distinction between annual shoots is marked with x–x. Cross section with annual rings of the respective shoots is indicated at the *right-hand side*

widths of the rings and the maximum density are dependent on the length of the terrestrial phase (Fig. 11.13). From the investigation of the occurrence of distinctiveness of growth zones of 82 tree species from the igapó and the várzea we found 67 species have distinct growth zones allowing us to determine the age of individual trees. Fourteen species showed indistinct rings; in *Aldina latifolia* no definite ring structure could be detected.

In congruence with the annual periodicity of radial growth an annual shoot extension can be expected. This congruence is well known for the temperate zones (Rauh 1939). For tropical trees only two positive examples were given by Hallé and Martin (1968) and Gil (1989). Periodical shoot extension can easily be proven by comparison of the number of rings and the number of shoot segments limited by characteristic scars (Fig. 11.14). Six of seven investigated tree species from igapó and várzea show an annual rhythm of shoot extensions (*Pseudobombax munguba*, *Psidium acutangulum*, *Salix humboldtiana*, *Macrolobium acaciifolium*, *Cecropia latiloba* and *Zanthoxylum compactum*). *Bonafusia tetrastarhya*, a shrub species in the understory, has very indistinct growth zones. The existence of annual growth periodicity and the presence of annual rings in woody plants allows us to use tree-ring patterns as a tool for ecological analysis ranging from age determinations to the investigation of growth rates and growth behaviour under different ecological conditions, as shown in the preceeding sections.

The growth of roots was directly observed with a mini-rhizotron by Meyer (1991). In the várzea, root production increased continuously from October to the following February. During the dry October 1987, the igapó root production was considerably lower than in the other months of the terrestrial phase. A high proportion of living roots at the end of the aquatic phase indicates that roots can survive the anoxic conditions during this period of unfavourable conditions.

11.3.3 Morphological and Physiological Adaptations of the Roots

Adaptations and reactions of herbaceous and woody plants to anaerobic soil conditions are summarized by Crawford (1982, 1989). Most investigations were carried out under laboratory conditions and only for short time periods of hours or days with species from temperate zones. Studies on tropical species and on the effects of very long-lasting floodings up to 300 days year^{-1} under natural site conditions are rather poor.

When the water floods a site the pore volume in the soil is rapidly filled with water and the gas exchange between soil and atmosphere is hindered. The remaining oxygen in the soil is consumed by the respiration of roots

Respiration

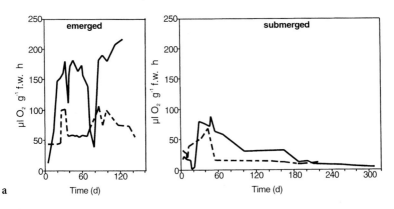

Ethanol

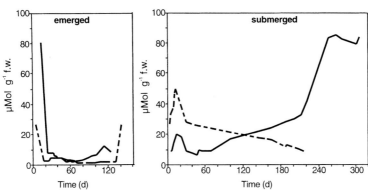

Dis. Carbohydrates

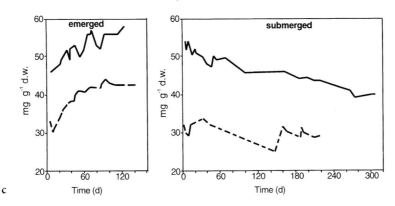

— Macrolobium ac.   – – Astrocaryum jauari

Fig. 11.15. Root respiration (**a**) and concentrations of ethanol (**b**), soluble carbohydrates (**c**), malate (**d**) and lactate (**e**) in the roots of *Astrocaryum jauari*, Palmae and *Macrolobium acaciifolium*, Caesalpiniaceae from the várzea during the terrestrial phase (*emerged*) and inundation phase (*submerged*). (After Schlüter 1989)

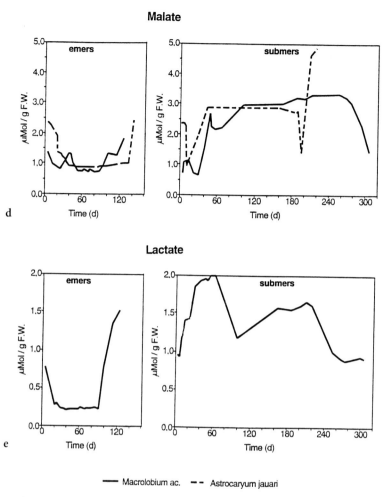

Fig. 11.15d,e

and microorganisms within hours or days (Ponnamperuma 1984). Flooding levels of 5–8 m above the forest floor and root lengths up to 30 m in adult trees preclude the possibility of gas exchange processes between stem and root tips by means of air transport systems of cellular dimensions. Investigations of the xylem from approximately 100 dicotyledonous tree species did not give indications of any aerenchyma (Worbes 1986). Usually trees of the inundation forests cannot support the O_2 demand for the respiration of their roots during the aquatic phase via lenticells in the stems and air conducting systems or via pneumatophores, as can mangroves (Gessner 1959), *Taxodium distichum* and others.

Nevertheless some observations point to morphological and anatomical stress avoiding systems, e.g., aerenchyma tissues were found in the roots of two-year-old seedlings of the palm species *Astrocaryum jauari* (Schlüter 1989). The connection to an above-ground air transport system remained unclear. In experiments with shallowly flooded seedlings of *Cecropia latiloba* (Junk pers. comm.) and in seedlings of *Macrolobium acaciifolium* on natural sites at the beginning of the aquatic phase (Schlüter 1989), the development of lenticells was observed. Two-and 3-year-old *Salix humboldtiana* trees form a dense layer of adventitious roots on the stem up to a height of 2 m. Under the regular high-rising flooding at the natural sites in várzea and igapó, anoxia-avoiding anatomical and morphological systems (Crawford 1989) are useful only during a short period of rising or falling water level. To tolerate anoxia trees must possess physiological mechanisms to survive the long periods of high flooding. At the beginning of the aquatic phase root respiration decreases to very low values (Fig. 11.15), according to studies of root physiology of the palm species *Astrocaryum jauari* (Schlüter et al. 1993) and the dicotyledonous species *Macrolobium acaciifolium* (Caesalpiniaceae) (Schlüter 1989) in igapó and várzea. After a few more days respiration increases suddenly and reaches its maximum level after 50 days of inundation. During this time, energy metabolism is switched to anaerobic pathways, indicated by the increasing concentrations of ethanol, lactate, alanin and malate in the roots. Whereas production of ethanol, alanin and lactate results in a direct energy gain, the function of malate synthesis is probably as a sink for superfluous CO_2 (Schlüter 1989). The decrease in respiration after 50 days of inundation indicates that the anaerobic pathway of energy gain is only a provisional solution. Moreover in *Macrolobium acaciifolium* after 200 days the concentration of ethanol, which is supposed to be a cell poison (Crawford 1977), increased rapidly. Little information is available on the removal of harmful end products of metabolism. In *Macrolobium acaciifolium* and other species the parenchyma tissue at the ring boundaries produced in the inundation period is filled with undetermined secondary plant substances. These substances may play a role in removal of cell poisons (Worbes 1986).

Carbohydrates as the basic source for the energy gain are transferred into the roots during the terrestrial phase (Fig. 11.15) and will be dissimilated during the aquatic phase. A considerable concentration of soluble carbohydrates in the roots of both investigated species at the end of inundation points to a reduced metabolism during inundation limited by the concentration of toxic end-products of the anaerobic metabolic pathways.

11.3.4 Reproduction

Reproductive behaviour of most species is oriented towards the aquatic phase (Revilla 1981; Worbes 1986; Ziburski 1990, 1991). The majority of species flower and fruit during or at the end of the inundation. Seeds with hydrochor distribution are supplied with special aerated tissues and are able to float for several days (*Pseudobombox munguba*) or months (*Aldina latifolia*). Fruits with high specific weights (e.g., *Astrocaryum jauari*) sink to the bottom and have a high tolerance for anoxic conditions. Moreover submersion breaks dormancy in this species and supports the germination at the beginning of the following terrestrial phase. Few species possess anemochor dispersal like *Triplaris surinamensis*, fruiting during the terrestrial phase, and *Salix humboldtiana* fruiting the entire year. Other dispersal agents are birds, mammals and iguanas.

Fruits are essential in the diet of the Amazonian ichthyofauna (Sect. 20.2). At least 100 fish species eat fruits from várzea and igapó trees (Goulding 1983). The role of ichthyocory for the dispersal of seeds is con-

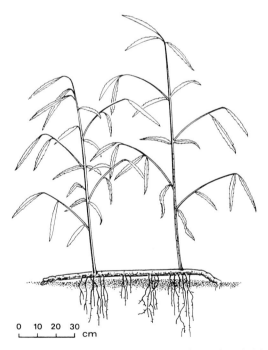

0   10   20   30
⌐___⌐___⌐___⌐ cm

Fig. 11.16. Vegetative propagation of *Salix humboldtiana*, Salicaceae through parts of broken stems at a sand bank downstream of the Ilha de Marchantaria

troversial. Gottsberger (1978) and Goulding (1983) propose that flowering plant groups are dependent on fishes for seed dispersal and that there is a coevolutionary relationship between fish and tree species in the floodplains. According to Ziburski (1991), only a few species with heavy fruits or with fruits which must to be opened for the germination of the seeds (e.g., *Crescentia amazonica*) are dependent on ichthyochory. In general, ichthyochory seems to be less important because seed dispersal by more than one agent is very frequent.

Vegetative propagation plays an important role in those species which form monospecific stands. The lower branches of *Eugenia inundata* often become rooted. In young stands of *Salix humboldtiana* the lateral shoots of fallen trees become transformed into several orthotropic axes with independent newly formed root systems (Fig. 11.16). Vegetative propagation is of advantage in environments where establishment of seedlings is impeded by sedimentation (*Salix*) or anoxic soil conditions (*Eugenia*).

Many species are able to develop stump sprouts, e.g., *Crataeva benthamii, Piranhea trifoliata, Tabebuia barbata* and *Triplaris surinamensis*, indicated by stem disks with more than one center (Fig. 11.17). Stem discs with signs of stump sprouting were frequently found in forests at the margins of Ilha de Marchantaria, indicating an anthropogenic deforestation.

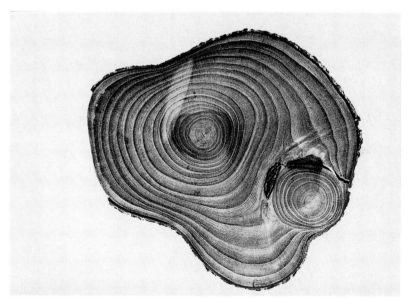

Fig. 11.17. Stem disk of the lower part of a *Tabebuia barbata*, Bignonoaceae tree from the margins of the Ilha de Marchantaria. The two centers indicate overwhelmed stump sprouts

11.3.5 Zonation

In the igapó, trees can apparently persist at lower sites (21 m) than in the várzea (Ferreira 1991). Schlüter (1989) suggests that there is a higher photon flux in the black water than in the white water. The leaves of some evergreen species in the igapó are photosynthetically active at a low level during submersion and that might be an advantage in tolerating flooding. Better oxygen conditions in black water habitats can also favour the occurrence of trees at lower sites (Sects. 4.4.3 and 20.5).

On an igapó site at Tarumã Mirím Ferreira (1991) found distinct variations in the distribution of tree species dependent on the elevation of sites and consequently along a flood level gradient. Besides generalistic species, such as *Malouetia fufuracea*, which occur along the entire flood gradient, other species show a centred distribution to a certain flood level (*Hevea spruceana* in the igapó at 23 m a.s.l). Few species are restricted to extremely high (*Mora paraensis*, *Penthacletra macroloba*) or low elevations (*Symmeria paniculata*) exclusively.

Distribution patterns can be explained at least partially with reproduction biology and flood tolerance of the seedlings. Along the flood gradient in the igapó, *Swartzia polyphylla* has a low, *Mora paraensis* a high, and *Aldina latifolia* a mid-level distribution centre. Mortality of seedlings exposed to three months of flooding at 27 m a.s.l. was highest in *Mora paraensis* and lowest in *Swartzia polyphylla* (Ziburski 1990). A similar correlation between differences in distribution centre and mortality of seedlings can be made in the várzea for *Crataeva benthamii*, *Vitex cymosa* and *Pterocarpus amzonum* (Fig. 11.18).

A discussion on zonation and specific differences in flood tolerance must incorporate the fact that many tree species occupy a wide range along the flood gradient and that a large number of species from the várzea can also be found on terra firme sites with a definite dry season of several months (Kubitzki 1989). Generally an inundation results in reduced root growth and a reduced uptake of water and nutrients (Lee 1977, 1978). The water potential in the leaves is lowered. For many tree species inundation results ultimately in leaf fall because an insufficient amount of water can reach the crown (Crawford 1982).

Tree species which grow equally well on both very dry and periodically flooded sites are well known in the temperate zones (Ellenberg 1988) where survival is additionally a function of species tolerance to periodically low winter temperatures (e.g., *Quercus* spp.). One can assume that those widespread species found at all sites in tropical South America are not primarily adapted to flooding, but generally tolerant to seasonally poor growing conditions (Worbes 1985; Junk 1989; Adis 1992a).

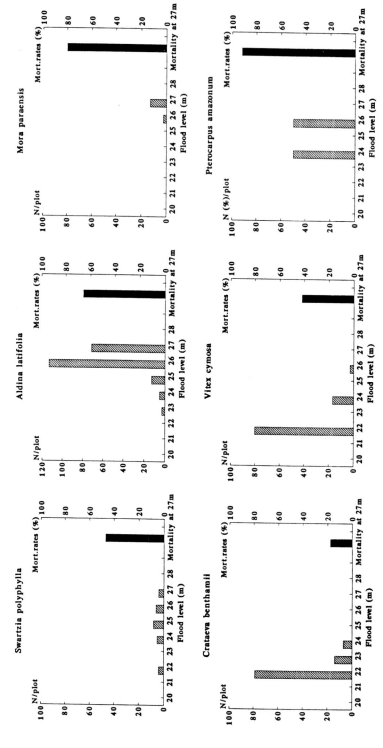

Fig. 11.18. Distribution of adult tree species (*left y-axis*) along the flood gradient (*x-axis*) and mortality of seedlings (*right y-axis*) during the submersion phase 1988 at 27 m a.s.l. of flood level. Species from igapó (*above*), from várzea (*below*). Data after Ziburski (1990) and Ferreira (1991)

11.4 Biomass and Primary Production

Net primary production (NPP) of a forest is defined as the sum of the increase of wood biomass, the total below- and aboveground litter production and the losses by herbivory. In all forest ecosystems values of belowground production and losses by herbivory are difficult to evaluate and rarely measured. In tropical forests the estimation of NPP is generally subjected to further specific difficulties, which were discussed in detail in Jordan (1983) and Whitmore (1984). Above all, the estimation of biomass increment is problematic because of the generally assumed absence of tree rings (Whitmore 1993), which in temperate zones are the basis for estimates of the wood increment. The majority of NPP calculations for tropical forests are based on leaf-fall measurements (Lieth and Whittaker 1975). In these cases NPP is estimated by multiplying leaf fall by a conversion factor of 2.5 (Murphy 1975).

Using tree-ring analyses of trees from várzea forests as described above we can calculate long-term biomass increase and, with measurements of the coarse (Martius 1989) and fine litter (Adis et al. 1979; Adis, unpubl. data), give a reliable figure of the aboveground NPP in the investigated várzea stands. In the igapó, the amount of fine litter fall and radial wood growth of the trees serves as an indicator for the NPP. All weight values in this section refer to dry weight.

11.4.1 Litter Production

Adis et al. (1979) measured litter production, including leaves, twigs, flowers and fruits, in the igapó at Tarumã Mirím and in four stands on the Ilha de Marchantaria using litter samples of $1\,m^2$ area over 1 year. In the same forests Meyer (1991) repeated the measurements during the emersion phase and found similar values. Within the várzea the litter fall varied from $7.8\,t\,ha^{-1}\,year^{-1}$, in an early successional 12-year-old *Cecropia latiloba* stand, to $13.6\,t\,ha^{-1}\,year^{-1}$ in a mixed forest stand with an approximate age of 60 years (Table 11.3). The mean of four estimations is $10.3\,t\,ha^{-1}\,year^{-1}$. The litter fall in the igapó is considerably lower with $6.7\,t\,ha^{-1}\,year^{-1}$. At both sites leaves constitute about 80% of the total amount of litter fall.

A distinct seasonality of leaf fall is observable in várzea and igapó. At both sites the maximum leaf fall is at the end of the submersion phase from May until July (Fig. 11.19). In the várzea a second (lower) peak is found in November during the dry season, when the soils are partially

Table 11.3. Fractions of fine litter, total fine litter (n, 4 plots) and total nutrient storage of the uppermost 30 cm soil in selected Amazonian forest stands

Forest	Igapó[a]	TF Manaus[a]	TF Belém[b]	Várzea[c]
Litter:				
Leaves (%)	79	81	81	82
Twigs (%)	15	13	13	9.7
Fruits (%)	7	6	6	8.3
Total ($t ha^{-1} a^{-1}$)	6.7	7.9	9.9	10.3
Soil nutrients:				
N ($kg ha^{-1}$)	4290[d]	4263[e]	–	3 350[d]
P ($kg ha^{-1}$)	255[d]	71[e]	–	1 455[d]
K ($kg ha^{-1}$)	2220[d]	58[e]	–	20 860[d]
Na ($kg ha^{-1}$)	–	35[e]	–	1 290[d]

TF, Terra firme.
[a] Franken et al. (1979).
[b] Klinge (1977).
[c] Adis (unpubl.).
[d] Worbes (1986).
[e] Klinge (1975).

dried out to the wilting point for some weeks (Worbes 1986). The igapó trees in contrast to várzea trees have smaller and more sclerophyllous leaves (Klinge et al. 1983), which are less sensitive to water stress.

The annual litter fall in the várzea is within the range of the world-wide mean value of $9.8 t ha^{-1}$, which is taken from different tropical lowland forests given by Medina and Klinge (1983). Compared to other Amazonian forest stands várzea forests produce the highest and igapó forests the lowest amount of litter fall. The differences between the inundation forests in the amount of litter correlate with differences in the nutrient content of the soils (Table 11.3; Vitousek 1984). Igapó soils are poor and várzea soils are comparably rich in phosphorus (see Chap. 3). However, the terra firme soil is even poorer even though litter fall is higher than in the igapó, where the vegetation period is restricted by the floodings.

The amount of coarse litter fall (dead wood) was measured by Martius (1989) in different stands in the várzea on Ilha de Marchantaria. In the stand (V4) he measured $6.6 t ha^{-1} year^{-1}$ and in another stand at the Central Lake (V5) $11.4 t ha^{-1} year^{-1}$ of coarse litter (Sect. 12.2).

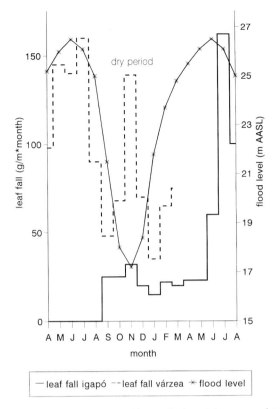

Fig. 11.19. Monthly leaf fall in the igapó in 1976 and 1977 from Adis et al. (1979), monthly leaf fall in the várzea in 1981 and 1982 adopted from Adis (unpubl. data) and pattern of flood level variation of Negro River

11.4.2 Root Production

Meyer (1991) measured the root production in várzea and igapó stands with minirhizotrons (Taylor 1987) during the emersion phase down to a depth of 40 cm. The lengths and the densities of fine roots (<1.0 mm) were measured at the same sites on soil slides on a millimetre scale down to a depth of 80 mm. In both stands fine roots are concentrated in the humus soil profile directly beyond the litter layer. Depending on the duration of flooding on different sites the density of fine roots in the igapó was 1.4 to 2 times higher than in the várzea. The amount of fine roots increased with increasing length of inundation period from 19.9 to 35.7 km 10^3 ha^{-1} in the igapó and from 14.1 to 18 km 10^3 ha^{-1} in the várzea (Fig. 11.20).

A comparison of the lengths of fine roots in different Amazonian forest stands (Klinge 1973a) is shown in Fig. 11.20. The nutrient contents of

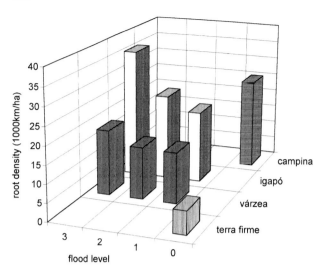

Fig. 11.20. Root density in km ha^{-1} in different Amazonian forest stands. The campina and the terra firme stands are never flooded (*flood level 0*) data from Klinge (1973). Length of inundation in the igapó: *flood level 1*, 80 days; *flood level 2*, 150 days; *flood level 3*, 175 days. Length of inundation in the várzea: *flood level 1*, 110 days; *flood level 2*, 180 days; *flood level 3*, 220 days

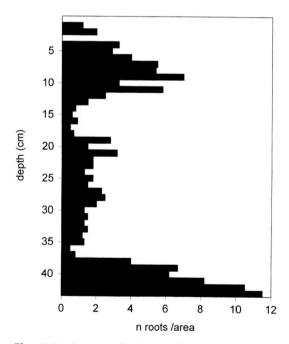

Fig. 11.21. Root production at different soil depths in the várzea measured with a minirhizotron. Unit of root production is number of newly formed roots per 4 cm^{-2} over 5 months

the soils are lowest in Campina on podsol, and increase from igapó, to terra firme to várzea (Klinge 1977; Worbes 1986). With respect to the increase of root biomass, this order is changed due to the influence of the inundation.

In general root production is higher in the várzea than in the igapó (Figs. 11.21, 11.22). High rates of production were observed in the upper horizon at both sites. In the várzea a second maximum occurred at a depth of 37 cm, where a sharp transition from clay to sand was observed (Worbes 1986).

The results show that root density and production depend strongly on nutrient content of the soil, soil type, and length of inundation. In the sandy soil horizons, with higher pore volumes (Worbes 1986), fine roots find better physical growth conditions than in denser clay soils. Stress factors, e.g., low nutrient supply and flooding, may stimulate the production of fine roots. A prolonged inundation means a short vegetation period for the trees. The activity of the roots is restricted during the submersion phase. At least at the beginning of the flood the reduced water and

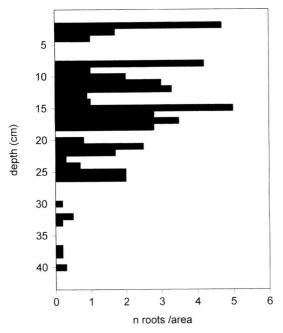

Fig. 11.22. Root production in different soil depths in the igapó measured with a minirhizotron. Unit of root production is number of newly formed roots per $4\,cm^{-2}$ over 5 months

nutrient transport may be compensated by the production of a higher root biomass.

The inverse relationship between nutrient supply and root biomass production is observed in tropical forests as well as in forests of the temperate zones (Klinge and Herrera 1978; Keyes and Grier 1981; Scherfose 1990). The high root biomass on poor soils, as in the igapó forest, is explained by the need to optimize exploitation of very limited nutrient resources (Medina et al. 1977). The lower rates of root production in combination with a higher root density in the igapó indicates a longer life span for the roots in the igapó (Meyer 1991). Klinge et al. (1983) assumed a longer life span for the leaves of igapó trees. Both findings indicate a highly efficient use of the low nutrient resources in the igapó (Nadelhoffer et al. 1985).

11.4.3 Wood Increment

The existence of annual rings in trees from inundation forests (Worbes 1984; Worbes and Junk 1989) allows us to apply dendrochronological methods for the estimation of wood increment. With this method we can make increment measurements over arbitrary time periods. In tropical forests the standard estimations of wood increments are for short periods of time and do not take into account growth changes caused by climatic variations or dynamic developments (Kira et al. 1967). The following calculations are based on the increment, i.e., each tree-ring width for every year of the entire life span of each tree. That is the longest possible measurement period.

Measurement of annual tree-ring widths over the complete life span of 203 trees show mean figures of 3.5 mm year^{-1} in the várzea and 1.7 mm year^{-1} in the igapó (Table. 11.4). The influence of light on tree growth is shown in Fig. 11.23. Upper-storey trees have considerably higher increments than under-storey trees, which is typical of any forest community. A more detailed examination of the results from the várzea according to stands of different successional development shows a decrease in radial increment as stand age increases from 9.9 mm year^{-1} in a 20-year-old stand to 1.7 mm year^{-1} in a stand that is more than 200 years old. In most cases these values exceed those of other tropical forest investigations and were evaluated by repeated dbh measurements (Table 11.4). The values from igapó and from the two oldest várzea stands are of the same order of magnitude as those from a lowland humid forest in Venezuela (Veillon 1985). One reason for the high radial increment rates in the inundation forest is that these measurements include the growth of the trees during the

Table 11.4. Mean radial increments with extreme values in brackets of trees in different tropical forests

	Period of measurement (years)	Radial increments ($mm\,a^{-1}$)
Repeated dbh measurements		
Natural Forest Sarawak[a] (n = 187)	1–4	0.7 (0.5–1.6)
Natural Forest[a] (1–2 years after Expl.) (n = 44)		2.0 (1.5–2.4)
Natural Forest[a] (3–4 years after Expl.) (n = 44)		1.1 (0.8–1.5)
Costa Rica, moist forest,[b] (n = 2010; BHD ≥ 10 cm)	13	1.3 (0.2–4.6)
Costa Rica, moist forest,[b] canopy		2.1 (0.5–4.6)
Costa Rica, moist forest,[b] subcanopy		0.9 (0.3–2.8)
Costa Rica, moist forest,[b] understory		0.5 (0.3–1.4)
Venezuela, woodland[c] (BHD ≥ 10 cm)	24	0.5
Venezuela, dry forest[c]		1.25
Venezuela, semidry forest[c]		1.75
Venezuela, moist forest[c]		2.25
Venezuela, Estado Barinas, seasonal forest[c]		1.8–2.55
Malaysia, Diptero.-forest,[d] (n = 3909; BHD ≥ 10 cm)	34	1.3 (0.4–2.5)
Malaysia, Diptero.-forest,[d] emergents		2.0 (1.5–2.5)
Malaysia, Diptero.-forest,[d] main canopy		1.1 (0.7–1.5)
Malaysia, Diptero.-forest,[d] understory		0.7 (0.4–1.1)
Panama, moist forest[e] (n = 3590; BHD ≥ 2.5 cm)	10	1.3 (0.1–5.8)
Tree-ring measurements ≥5 cm[f]		
Brasil, Igapó (n = 45)		1.7 (0.8–3.1)
Brasil, Várzea (n = 158)		3.5 (0.7–12.3)
Venezuela, seasonal forest[f] (n = 56)		4.3 (0.9–12.5)
Brasil, Várzea: stand age 20 years (n = 9)		9.9 (4.7–12.3)
Brasil, Várzea: stand age 45 years (n = 49)		4.8 (2.4–10.1)
Brasil, Várzea: stand age 80 years (n = 80)		2.7 (0.7–7.6)
Brasil, Várzea: stand age >200 years (n = 20)		1.7 (0.7–2.6)

Own estimations on the base of tree ring analysis, others on the base of repeated dbh measurements.
[a] Primack et al. (1985), only Moraceae.
[b] Lieberman et al. (1985).
[c] Veillon (1985), 62 investigation areas.
[d] Manokaran and Kochummen (1987), only the 20 most important species.
[e] Lang and Knight (1983).
[f] Worbes (1994).

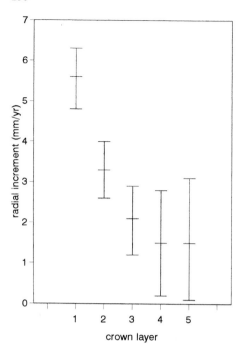

Fig. 11.23. Radial wood increment in mm (means and deviation) from 113 trees of the várzea (stand V5) in five crown layers from upper story (*1*) to understory (*5*). Further details of structural composition of this stand are given in Chapter 11

early life history, when the radial wood increments in young forest communities are higher than in older and differentiated stands. The assessment of data on radial growth rates should always take into consideration the successional stage and the structure of a forest stand.

Rates of radial increment are only relative figures with no direct relation to biomass production on a given area. Aboveground wood production was estimated by dividing the aboveground biomass by the ages of the stands (Worbes 1994). Aboveground biomass was measured by harvesting in the stands BIO1 and BIO2 (Klinge et al. 1996). In the other stands biomass for the stem of each tree was estimated from tree height, diameter at breast height and density of the wood. Ten percent of the result was added for the biomass of the branches and twigs (Ellenberg et al. 1986).

The calculation of wood production of individual trees (Worbes 1994) shows great variations between individuals of different crown layers but little variation between different species growing in the same stratum (Table 11.5). The mean wood production of the pioneer species *Pseudobombax munguba* ($12\,kg\,tree^{-1}\,year^{-1}$) is equal to the mean production of *Piranhea trifoliata*, a species of the mature forest. Together with

Table 11.5. Individual annual stem wood production with deviations and extreme values (min.–max.prod.) of different tree species from the várzea, with dimensions of the sample trees: height, radial increment, wood density, and number of samples (n)

	Height (m)	Radial incr. (mm a^{-1})	Wood density (g cm^{-3} a^{-1})	Min. and max. prod. (kg a^{-1} tree^{-1})	n	Mean production (kg a^{-1} tree^{-1})
Cecropia latiloba	14–15	8.8–10.6	0.42–0.46	5.2–8.4	2	6.8 ± 2.3
Crataeva benthamii	4–16	2.9–5.3	0.39–0.48	0.2–3.8	4	1.9 ± 1.6
Elaeoluma glabrescens	20	2.7–3.4	0.56–0.59	7.6–10.8	2	9.2 ± 2.3
Eschweilera albiflora	17	1.5	0.75		1	5.0
Eschweilera spec.	20–25	3.3–4.8	0.50–0.55	15.1–19.7	2	17.4 ± 3.3
Ilex inundata	15–20	1.9–3.3	0.43		1	4.9 ± 0.9
Laetia corymbulosa	18–24	2.2–2.9	0.60–0.64	4.0–13.5	4	7.2 ± 4.4
Luchea spec.	22	3.9	0.57		1	15.3
Luehea spec.	18	2.5–3.3	0.37–0.42		1	3.0
Macrolobium acaciifolium	15–25	3.4–6.5	0.42–0.50	4.8–24.2	7	11.3 ± 6.8
Mouriri guianensis	17	2.7	0.82		1	6.6
Myrciaria amazonica	17–18	1.5	0.68–0.74		1	2.1
Nectandra amazonum	14–15	2.4–4.9	0.39–0.47	2.3–5.6	2	4.0 ± 2.3
Piranhea trifoliata	15–32	1.2–4.5	0.83–0.93	3.1–19.3	8	12.0 ± 5.5
Pithecolobium inaequale	10	1.8	0.62		1	1.6
Pseudobombax munguba	6–28	4.0–10.0	0.20–0.24	3.9–23.5	3	12.2 ± 10.1
Pseudoxandra polyphleba	8	1.9	0.51		1	0.7
Psidium acutangulum	4–18	0.7–1.6	0.80–0.85	0.1–4.0	2	2.1 ± 2.8
Salix humboldtiana	7–10	4.7–12.3	0.42–0.43	1.8–8.5	6	4.1 ± 2.4
Sorocea duckei	15	2.0	0.58		1	2.6
Tabebuia barbata	8–25	1.1–4.8	0.55–0.85	1.1–20.6	16	6.0 ± 5.4
Trichilia singularis	6	1.6	0.51		1	0.6
Triplaris surinamensis	26	4.2	0.62		1	14.1
Vitex cymosa	10–16	3.1–4.6	0.58–0.62	1.4–11.3	3	5.7 ± 5.1

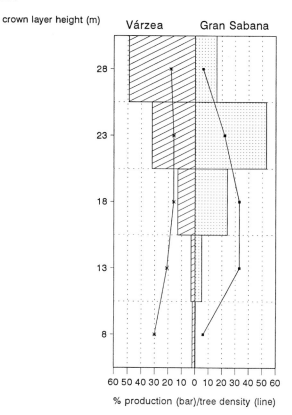

crown layer height (m)

Fig. 11.24. Percentage of wood production and percentage of trees in different crown layers in the várzea stand V5. For comparison, on the *right-hand side* the corresponding figures for a disturbed stand of the Gran Sabana, Venezuela are documented. The Gran Sabana stand is characterized by crown damage to the trees so that even the third crown layer contributes a considerable proportion to total wood production. (Worbes 1994a,b)

structural data of tree density in all layers of stand (V5) on the Ilha de Marchantaria, values of individual wood production of trees from all crown layers allow the estimation of area-related figures of production. Separate estimations for five crown layers show that about 50% of the total wood production is caused by the individuals from the upper storey which comprise only 20% of the individuals in the stand (Fig. 11.24). On the other hand, 50% of the individuals grow in the two understorys (up to 15 m height) but contribute less than 10% to the total amount of wood production. This is the result of decreasing radiation from the top to the forest floor as it is decribed by Kira and Yoda (1989) for a forest stand in Thailand.

   The results of the area-related calculations document, in contrast to the behavior of the radial growth, an increase in biomass production with

Table 11.6. Aboveground biomass, total aboveground annual wood production in different forest types of the várzea (Worbes 1994a,b) and figures from literature (with extreme values). Detailed stand descriptions in Worbes et al. (1992)

	Biomass (t ha^{-1})	Biomass increments (t a^{-1} ha^{-1})
Várzea:		
Pionier (2 years old)	3	1.5
Pionier (4 years old)	14	3.4
Pionier (12 years old, BIO1)	98	8.1
Young secondary (44 years old, BIO2)	258	7.3
Late secondary (80 years old, V5):		
Gap	118	2.2
Most productive plot	470	8.6
Mean of 7 subplots	279	*7.2*
Worldwide tropics:		
Plantations[a] (15 means)		11.9 (4.9–27.5)
Natural forests[a] (10 means)		*7.3* (4.9–12.4)

[a] Jordan (1983).

increasing age of the stands (Table 11.6). A two-year-old monospecific pioneer stand of *Salix humboldtiana* produces only 1.5 t ha^{-1} year^{-1} wood. The wood increment increases to 8.1 t ha^{-1} year^{-1} in a 12-year-old *Cecropia latiloba* stand. In the 80-year-old *Pseudobombax munguba* stand the amount of wood production decreases slightly to 7.2 t. This stand is a patchwork of relatively unproductive gaps (2.2 t ha^{-1} year^{-1}) and highly productive plots with well-structured and differentiated tree stands (8.6 t ha^{-1} year^{-1}).

These figures match closely the worldwide mean from ten investigations (Jordan 1983) of 7.3 t ha^{-1} year^{-1}. This result confirms earlier findings (Jordan 1983; Medina and Klinge 1983) that wood production in tropical forests even under good nutrient conditions does not much exceed the wood production of temperate forests.

11.4.4 Total Net Primary Production

The measurements of the aboveground wood production and of the coarse and fine litter production as given above were summarized to estimate the total aboveground primary production. For herbivory and for belowground production, estimates from the literature were added. Total NPP was calculated for a 40-year-old early secondary stand dominated by *Crataeva benthamii* (V4) and an 80-year-old late secondary stand domi-

Table 11.7. Aboveground NPP without losses from herbivory in the investigated várzea stands, in a rain forest in Thailand (Kira et al. 1964) and in a beech forest in the temperate zone. (Ellenberg et al. 1986)

Site	Fine litter	Dead wood	Wood increment	NPP
Várzea (V4, 40 years)	9.9	6.6	7.3	23.8
Várzea (V5, 80 years)	13.6	11.4	8.6	33.6
Rainforest, Thailand	12.0	13.3	3.1	28.4
Beech for., Germ. (120 years)	3.6	0.4	6.7	10.7

The age of the rain forest in Thailand was unknown.
Data for fine litter from Adis (unpubl. data).
All figures in $t\,ha^{-1}\,year^{-1}$.

nated by *Pseudobombax munguba* (V5), which are described in Chapter 11. The total aboveground NPP, without losses due to herbivory, is $23.8\,t\,ha^{-1}\,year^{-1}$ in stand V4 and $33.6\,t\,ha^{-1}\,year^{-1}$ in the most productive plot of stand V5.

Grazing losses of the leaves vary between 2% in an oligotrophic tropical forest (Jordan and Uhl 1978) and 3–8% in tropical forests on fertile soil (Odum 1970). Estimations of root production in temperate forests vary from 10% (Ellenberg et al. 1986) to 170% (Reichle et al. 1973) of the aboveground wood production. Jordan and Escalante (1980) give a figure of 33% for a forest in San Carlos de Rio Negro, Venezuela. In the inundation forest it is very likely that the investment for root production is even higher than in non-flooded forests as pointed out above.

Adding a mean figure for grazing of 5% of the leaf production and a cautious estimation of 30% of the wood increment as root production, the total NPP of well-structured várzea forests varies between $27\,t\,ha^{-1}\,year^{-1}$ in stand V4 and $36.9\,t\,ha^{-1}\,year^{-1}$ in stand V5. Both figures considerably exceed the estimates of net primary production of other tropical forest stands (Table 11.7).

11.5 Discussion and Conclusions

The analysis of Neotropical forest vegetation, due to its great diversity, necessarily focuses on taxonomic inventories and descriptions. This work, based on the investigations of Martius et al. (1840–1869), is recently being edited for *Neotropical Flora* by Prance. The enormous quantity of taxonomical information led to theories on the evolution of neotropical plant species and in consequence to chronological explanations of distribution patterns of the neotropical vegetation (Prance 1982; Kubitzki 1989).

A large-scale attempt is being made to classify the neotropical vegetation based on structural and physionomical differences of vegetation types linked with climatic, and partially with edaphic, variations (Holdrige 1966; Walter and Breckle 1991). Information on the phytosociological level of tropical forests and the linkage between the occurrence of plant communities and site conditions is relatively scarce. A great number of forestry inventories and the detailed knowledge of site factors of the vegetation of the Amazonian inundation forests allows us to make a phytosociological analysis on the basis of deterministic explanations.

Flood Pulse

The most obvious growth factor in the inundation forests is the annual flooding defined by Junk et al. (1989) as flood pulse (Sect. 1.3). The predictable flood patterns lead to a distinct seasonality, which determines many growth patterns in the floodplains. To begin with, woody species rooting in the annually flooded soil have a periodic growth rhythm with a cambial dormancy during the aquatic phase. This is indicated by low cambial activity during flooding, the existence of annual rings in the wood, the existence of annual shoot growth and increased leaf fall during the aquatic phase. The concurrence between time series of the annually varying length of the flood-free period and tree-ring patterns shows that the main vegetation period is the terrestrial phase. This explains why reproductive behavior of tree and shrub species is oriented towards the flooding. Trees use the vegetation period for as long as possible for flowering, fruiting and the maturation of fruits. Many tree species of the várzea and the igapó start flowering during the aquatic phase, just as trees of the temperate zones flower as early as possible in the spring and trees of the tropical terra firme start flowering at the end of a dry phase (Alvim and Alvim 1978; Franco 1979). This example supports the view that flood pulse triggers all growth processes, as do low winter temperatures in the temperate zones and dry phases in tropical and subtropical regions.

As in habitats periodically influenced by other unfavorable climatic conditions, certain adaptations are required to tolerate the varying growth conditions. Since the maturity of fruits occurs during the inundation phase, fruits must have the ability to swim or to tolerate anoxic conditions on the bottom of the water body. Comparing fruits of closely related species from inundation forests and terra firme, Ziburski (1991) concluded that special buoyant tissues in fruits were developed secondarily.

The most important condition for woody species in the floodplains is the ability of the individual to tolerate the annual flooding. The flood tolerance is composed of two strategies. Cambial dormancy during the aquatic phase connected with deciduousness or sclerophylic leaves in evergreen species reduces water consumption as far as possible. The shift of root metabolism from respiration to anoxic pathways enables living roots to survive until the beginning of the following terrestrial phase.

Different metabolic pathways result in differing levels of flood tolerance between species. In contrast to *Astrocaryum jauari*, Palmae, the dicotyledonous species *Macrolobium acaciifolium* is able to use the production of lactate and alanin as an energy supply during the aquatic phase. *Astrocaryum jauari* is restricted to sites of higher elevation and shorter inundations than *Macrolobium acaciifolium* (Piedade 1985; Ferreira 1992). Stress-avoiding systems (Crawford 1989) such as aerenchymas, lenticells and adventitious roots occur in some species but seem to be of minor importance under the long-lasting and high-rising floods in várzea and igapó.

Slopes on river banks cause a rapid change in duration of flooding and change site conditions over short distances. This, along with the aforementioned differences in flood tolerance of tree species, leads to a clear zonation of the forest vegetation.

Soil Conditions

Plant communities in the Amazonian floodplains differ as the nutrient conditions of the soils differ. The concentration of nitrogen in the investigated soils is almost always the same, but there are great differences in P content (Chap. 3). Phosphorus seems to play an important role as a differentiating factor (Tilman 1982). This explains differences in species composition between várzea and igapó sites (cf. Schmidt 1973, 1976) as well as differences within the igapó between nutrient-poor white sand soil and the somewhat richer clay soil.

Stand Dynamics

In tropical vegetation analyses an often neglected differentiating factor is the influence of the dynamic stage of a forest community. The basic problem is the assumed impossibility of dating the exact age of trees and forest stands, primarily due to the assumed absence of tree-rings in tropical trees (Whitmore 1990). Various indirect or time-consuming methods were tested, e.g., interviews with inhabitants (Uhl 1980) and estimation of mortality rates (Lieberman et al. 1985), which led to vague information. Pure assumptions result in different datings with variations of 100% for the same stand depending on the author (Foster and Brokaw 1983; Lang and Knight 1983). Exact age dating by dendrochronological methods matches the definition of successional stages of forest communities (Worbes et al. 1992). The change in species composition over time incorporates change in structural elements, i.e., increasing biomass, decreasing population density and decreasing individual growth rates.

Growth Factors and Diversity

The three differentiating factors in vegetation analysis of tropical plant communities are climate (in this case flood pulse), soil conditions and successional development. These factors lead to a better understanding of varying species composition, as discussed above, and of structural features, e.g., patterns of species richness.

The trends in the Amazonian floodplain forest are diversity, which increases in order from young to old stands, from the eutrophic várzea to more oligotrophic stands in the igapó and with decreasing flood stress (Ayres 1993). These findings can be explained using Grime's model of stress, competition and disturbance (Grime 1979) and can be linked with observations from non-flooded tropical forest communities (Gentry 1982; Ashton 1989). In Grime's model species diversity is low for sites with either high or low degrees of stress and disturbance. A high degree of disturbance and stress commonly occurs at the river margins of the Amazon. On sites often exposed to erosion and sedimentation, only monospecific (*Salix humboldtiana*) or tree communities with low species diversity are able to persist. A low degree of disturbance occurs in floodplain forests in the center of great islands and outside the main river channel where 106 species ha^{-1} were documented (Revilla 1991).

On non-flooded sites in Amazonia species richness increases from east to west, from 87 species ha^{-1} near Belém (Black et al. 1950) to over 179 species ha^{-1} near Manaus (Prance et al. 1976). Gentry (1982) explained species richness generally increases as precipitation increases. However, the increase in annual precipitation is correlated with a decrease in seasonality.

The flood pulse and periodic droughts can be interpreted as climatic stress factors, which are correlated with a depression of diversity. The effects of both factors on species richness are linked in Fig. 11.25. Drought occurs on sites with a deficit in precipitation, flood pulse occurs on sites with superfluous water. At the transition between the two habitat types

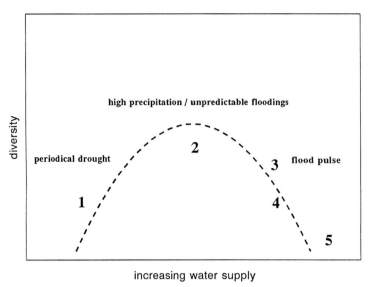

Fig. 11.25. Model of relative correlation between species diversity and water regime. The figures stand for forest inventories: *1* Terra firme, Belém (Black et al. 1950); *2* terra firme and floodplain, Napo River (Balslev et al. 1987); *3* Igapó, this work; *4* Várzea, this work; *5* low-lying shrub communities in the várzea. (Junk 1989)

either a high amount of precipitation throughout the year (at the left angle of the curve) or short and unpredictable floodings (at the right angle of the curve) are typical. Balslev et al. (1987) investigated a terra firme and an adjacent floodplain forest at the Napo River in Ecuador that fit the previously described conditions. Both forests show extremely high species diversity in comparison with terra firme sites with lower precipitation (Gentry 1982) or similar floodplain sites with a long-lasting flood pulse as in Central Amazonia.

The differences in species diversity of sites in igapó and várzea exposed to equal flood stress show the influence of nutrient supply in the soil. According to Tilman (1982), in tropical forests in Malaysia diversity is highest under moderate conditions, in the Amazonian floodplains these sites are in the igapó with clay soils. The above examples show that nutrient and climatic conditions of the site, as well as dynamic processes of the vegetation, must be considered for a comprehensive interpretation of the patterns of species composition and species diversity.

Biomass Production

Total primary production of the várzea forest is high when compared with other tropical forests. The growth period of many species is concentrated in a terrestrial phase of only about 200–250 days and often includes a period of drought stress of about four weeks. The impact of flood stress on tree growth in the várzea is compensated for by the input of nutrients during the floods. In the igapó, there is no nutrient input from the nutrient- and sediment-poor water of the Negro River. The igapó forest is very unproductive, as indicated by the low figures for litter fall and radial wood growth compared with figures from forest stands in San Carlos de Rio Negro (Medina and Klinge 1983).

The comparatively low wood production in comparison to the total NPP may be a disappointing result for those who expected to find a great potential for sustainable use and forestry in the tropical forest. The reason that tropical trees do not invest a greater proportion of the relatively high NPP in wood production is probably the high respiration loss from the sprout axis. A tree's stem and branches are not only dead organic matter, as they are often defined (Begon et al. 1990), but also a scaffold for the leaves. Their function in water transport and storage assimilation must be maintained through annual reproduction of the respiring cambial tissues. Total stem respiration in a tropical forest may be as much as $19 \, t \, ha^{-1} \, year^{-1}$ (Medina and Klinge 1983). Therefore the costs for the maintenance of the wood biomass increases with weight. On the other hand the investment in "unproductive" wood tissue is the stabilizing factor of forest communities that are in competition with other highly productive plant communities, such as the perennial grass communities of the várzea. The conflict between developing a stabilized system and keeping down the costs for this system generally restrict the investment in wood production.

Acknowledgments. I am in debt to Prof. Dr. J. Bauch, Prof. Dr. H. Ellenberg, Prof. Dr. F.H. Schweingruber and C. Martius for valuable comments on ealier versions of the manuscript. The indefatigable discussions with W.J. Junk and the immovable willingness of G. Lemke to prepare new versions helped fundamentally to finish the manuscript.

12 Decomposition of Wood

Christopher Martius

12.1 Introduction

Decomposition includes two important basic processes: mechanical fragmentation of substrate and chemical mineralization of organic materials. In the várzea, these processes take place both in the aquatic and the terrestrial phases. In both phases, mineralization is accomplished mainly by bacteria and fungi. Wood fragmentation in the terrestrial phase is mainly due to wood-feeding termites (Chap. 18) and Coleoptera (Cerambycidae and Buprestidae; Irmler, unpubl.), which are also able to mineralize wood components to a certain extent (Sect. 12.5). In the aquatic phase, larvae of the mayfly *Asthenopus curtus* are important wood fragmenters (Sect. 12.4).

Wood generally decays more slowly than leaves. This is true also in the várzea. Here, wood fragments are submitted to several intermittent aquatic and terrestrial decay phases until complete mineralization, which may result in markedly different decay histories of the samples, influenced by the relative duration of both phases. It is also important whether decay starts in the terrestrial or aquatic phases. The analysis of studies on wood decomposition is further complicated by the use of different methods and time intervals, ranging from 5 to 20 months. For a description of sampling sites, see Sect. 2.2.

12.2 Wood Litter Production and Distribution

The total annual input of wood litter to the forest floor of the várzea is 6.0 t ha^{-1}, based on 1 years sampling of dead wood falling to the ground on two areas of 75 m^2 each (Martius 1980). The "real" production must be higher, as some wood decays in the canopy. Small- to intermediate-sized litter (<10 cm diameter) represents the bulk of the input (62%) and represents a far steadier flux than large wood (>10 cm diameter). Monthly varia-

Ecological Studies, Vol. 126
Junk (ed) The Central Amazon floodplain
© Springer-Verlag Berlin Heidelberg 1997

tion is mainly due to differences in large wood litter input and is apparently independent of the periodic flood pattern. However, flooding redistributes dead wood; low-density wood floats and is deposited at the high-water margin after the withdrawal of the flood. In 1985, an additional input of $1.2\,t\,wood\,ha^{-1}$ was registered at the high water level. The sources of this material are adjacent lower areas of the forest and distant areas upstream (Martius 1989).

Dead wood can be classified according to size (twigs, branches, logs), and quantified on the basis of either volume or weight. It consists of the wood found in the litter layer on the forest floor and the "standing dead wood" (Swift 1977). The latter includes standing dead boles (trees) and dead branches which are still attached to living trees.

Dead wood stocks were assessed on one secondary forest site of $1425\,m^2$ in the várzea (Lago Cobra). The total volume of standing boles ($27.3\,m^3$ ha^{-1}) was about 80% that of the wood litter on the forest floor ($33.9\,m^3$ ha^{-1}; Martius 1989). The volume of standing dead branches, which was not assessed, can be assumed to be equivalent to that of standing dead trees (Ovington and Madgwick 1959; Christensen 1977); consequently, a total dead wood volume of about $90\,m^3\,ha^{-1}$ can be assumed. The distribution of dead wood size classes was analyzed by Martius (1989). For dead wood on the ground and standing dead wood, the total volume of fine wood litter (3 to 10 cm diameter) is much lower (12 and 46%, respectively) than that of coarse matter (10 to 100 cm diameter; the volume of very fine material <3 cm diameter was not assessed). In contrast to leaf litter, which is more or less evenly distributed, the distribution of wood litter is clumped, particularly when standing dead wood is taken into account.

Table 12.1 shows that the dead wood mass ("standing crop") depends on the age of the stands which in turn reflects the size and biomass of the trees. The Lago Central stand is older than that at Lago Cobra. The Lago Cobra

Table 12.1. Dead wood mass ($t\,ha^{-1}$) in the litter layer of different várzea forest stands on Marchantaria Island

Site	Height (m.a.s.l.)	Wood of the litter layer ($t\,ha^{-1}$)
Lago Cobra	23–25	3.57[a]
	25–26	10.38[b]
	≈26–27	5.90[a]
Lago Central	23–25	11.38

Lago Cobra data: values marked[a,b] are significantly different at $P = 5\%$.

data also show that the distribution of dead wood differed along the flood gradient. The average for that site is $6.62 \, t \, ha^{-1}$, but the dead wood stock in the litter layer of the intermediate zone (25–26 m above sea level) was higher than that in the lower or higher areas of the site because it received additional wood input from the flood (Martius 1989).

The amount of dead wood can be higher in transitional phases between early and later successional stages. This was observed in a stand dominated by 40-year-old dying trees of *Cecropia* sp., which, at the time of the study, were being replaced by a more diverse tree community which had already become established on the site. A similar rise in dead wood was observed where *Pseudobombax munguba*, which dominates a later successional stage, died (Worbes, pers. comm.). Both *Cecropia* and *P. munguba* have wood of very low density that is easily consumed by *Nasutitermes* species (Bustamante 1993). During these transitional stages much of this low-density dead wood is available to wood-feeding termites and their populations temporarily rise to large numbers (Sect. 18.3).

The dead wood mass of the várzea ($3.6–11.4 \, t \, ha^{-1}$; Table 12.1) is generally lower than that found in non-flooded terra firme forests ($10.5–21.2 \, t \, ha^{-1}$; Klinge et al. 1975; Fearnside unpubl.; Martius and Bandeira unpubl.). It is less than half of the $25 \, t \, ha^{-1}$ previously assumed for the várzea (Richey 1982; Junk 1985).

12.3 Wood Mineralization in the Terrestrial Phase: Decay Rates of Boles

Wood decomposition in the terrestrial phase is influenced by many factors including wood density, moisture and nitrogen content (Cowling and Merrill 1966; Merrill and Cowling 1966; La Fage and Nutting 1978; Anderson and Swift 1983), toxic metabolites, chemical and physical bark properties (Käärik 1974; Janzen 1985), and stochastic factors which affect the succession of decomposers in a given piece of wood (Martius 1989; Bustamante 1993).

Boles of 11 várzea tree species with densities varying between 0.235 and $0.716 \, g \, cm^{-3}$ (average $0.495 \pm 0.002 \, g \, cm^{-3}$; Martius 1992b) were exposed to up to 20 months of decomposition on Marchantaria island (Martius 1989). Microbial decay was measured as loss of mass per volume (density) with time; mass loss due to fragmentation was not assessed.

Wood decay is exponentially related to time in analogy to leaf litter decay rates (Olson 1963), according to the relationship:

$$D_t = D_0 \cdot e^{-kt},$$

with D_t and D_0 being the specific density of wood ($g\,cm^{-3}$) at time 0 (start of the experiment) and time t (in months), respectively. From this equation, the decomposition coefficient k_e is derived. The average k_e for the 11 tree species studied was 0.344, but the variation between the species was very high (0.049–1.000). Only low-density wood species (*Cecropia* sp., *Pseudobombax munguba*) decayed completely within a few months; the others persisted over more than 20 months. At a density of ≈0.15–0.20 g cm^{-3} the wood physically "dissolves". The species–specific decay coefficients are correlated with wood density; however, as wood moisture is significantly and negatively linked to density, it is not possible to separate both factors. Other elements play a role in determining the decay rates of boles. Some tree species (*Crataeva benthamii, Gustavia augusta*) contain organics which act as deterrents to decomposer organisms such as termites. Their decay rates are lower than their wood density would suggest, and the palatability of species such as *Pseudobombax munguba* and *Cupania cinerea* is reduced when soaked in aqueous extracts of *C. benthamii* and *G. augusta* (Martius 1989).

Bark, which is generally rich in feeding deterrents (Janzen 1985), only slightly reduced the attack of decomposers on the dead boles, and sometimes even favored their presence in the wood by forming cavities in which arthropods could hide. Bark did not protect the boles against the infiltration of water: water content was elevated after the flood, a factor which enhanced the subsequent decay in some tree species, but reduced it in others.

The concentrations of nitrogen, determined by C/N analysis, increased in the wood of *Cecropia latiloba*, but nitrogen dynamics in the wood of most of the other species showed no clear trend, and the nitrogen concentration in the wood had no measurable influence on the decay coefficient k_e.

Qualitative observations show that the succession of decomposers on wood in the várzea seems to follow some general rules: dry-wood termites attack the standing wood and have never been found in wood on the ground. Larvae of Cerambycidae are found only in very fresh dead wood (standing or lying). *Nasutitermes* species readily attack fresh wood of *Pseudobombax munguba, Salix humboldtiana* or *Cecropia latiloba*, but most other wood species are consumed only in advanced states of decay, and fungi probably make part of the obligate diet of these termites (Bustamante 1993). Rhinotermitidae are found in wood samples which are in a much more advanced state of decay. Nevertheless, several factors such as individual decay history, differential attack by insects, the partial influence of the flood (Sect. 12.4) and the formation of decay compartments (Shigo and Marx 1977; Shevenell and Shortle 1986) contribute to a mosaic of widely differentiated states of decay within a given piece of wood.

The decomposition coefficient found for microbial decay of várzea wood is in agreement with values for the decay of boles in Panamanian and Costa Rican rainforests (k_e = 0.461 and 0.354, calculated from data in Lang and Knight 1979 and Leberman et al. 1985, respectively). Only in a Puerto Rican rainforest was bole decay much slower (k_e = 0.115; data from Odum 1970). Thus, the terrestrial decay of boles in the várzea does not differ from that of most other tropical regions.

12.4 Aquatic Wood Decay and the Role of the Flood Pulse

Figure 12.1 shows the weight loss of wood blocks of two várzea tree species which were exposed for different time intervals to aquatic decay in the suspensoid-rich water of the Amazon River. This technique allowed the evaluation of the impact of leaching and microbial activity in the water without the influence of attack by *Asthenopus curtus* larvae, and permitted the assessment of subsequent terrestrial decay on the várzea forest floor (Martius 1989). In the first weeks of submersion (triangles in Fig. 12.1) a rapid weight loss was observed in comparison to terrestrial decay (small squares in Fig. 12.1). Over longer periods (6 months), aquatic decay was much slower than terrestrial decomposition (differences are significant at $P \leq 0.01$). The terrestrial decomposer guild is much more diverse than the aquatic guild, the former consists of several species of termites and beetles, whereas few insect species attack wood in the water (e.g., the *Asthenopus curtus* larvae). Fewer bacteria are capable of degrading cellulose anaerobically underwater (Schlegel 1976) and physical decay processes are slow (Day 1983; Holt 1983; Harmon et al. 1986). Total weight loss of *Laetia corymbulosa* and *Cupania cinerea* blocks in the water was 12.9 and 7.1% in 6 months, respectively, and the blocks subject to terrestrial decay lost 15.9 and 9.7%, on average, of their initial weight (Martius 1989).

Terrestrial decay varied with period of previous submersal. If wood was immersed in the river water for 1–4 weeks (large squares and lozenges in Fig. 12.1), the decay rate in the subsequent terrestrial phase was significantly higher than that of wood which had not been submerged; it was also higher than that of wood that had been flooded for longer periods (6 months; triangles in Fig. 12.1 and Fig. 12.2). White-rot attack was visibly enhanced on these blocks. A possible explanation is that short immersion increases the water content of the wood blocks which enhances attack by fungi in the subsequent terrestrial phase. Wood blocks submerged for a

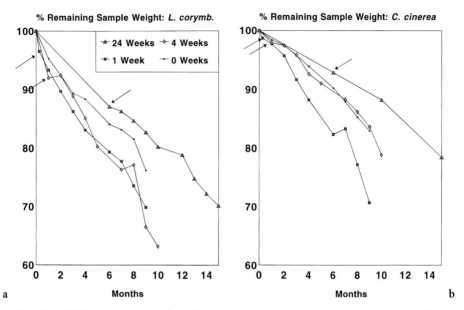

Fig. 12.1. Differences in wood decay in an intermittent aquatic/terrestrial várzea environment. Sets of wood blocks (4 × 4 × 4 cm) of **a** *Laetia corymbulosa* and **b** *Cupania cinerea* were exposed to terrestrial decay on the várzea forest ground after exposure to aquatic decay for 0, 1, 4 and 24 weeks (four series). The terrestrial phase lasted 9 months; every month a set of wood blocks of each series was sampled, dried and weighed. The graph shows the percentage dry weight loss of the wood blocks in the aquatic and terrestrial decay phase related to their previous weight. *Arrows* indicate start of the terrestrial phase. Each *point* represents the average of a set of ten blocks

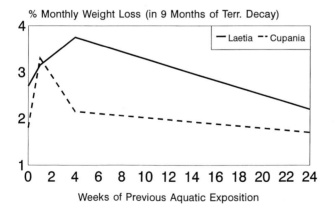

Fig. 12.2. Differences in wood decay in an intermittent aquatic/terrestrial várzea environment: Monthly average weight loss of wood blocks during the terrestrial phase (9 months) of the decay experiment as a function of the duration of the previous total aquatic exposure of the samples (*x-axis*, duration of aquatic decay phase in weeks)

longer period are far more depleted of nutrients and therefore less attractive to decomposer organisms.

In the presence of the mayfly *Asthenopus curtus*, wood destruction in the aquatic phase is strongly accelerated (Braga 1979; Sect. 13.5). The larvae are filter feeders and mine in any hard substrate (generally wood). The tunnels, with a diameter of 3–5 mm, can reach a length of 60–85 mm, with their deepest point at 7–26 mm below the wood surface (Sattler 1976). This makes *A. curtus* larvae very efficient wood destroyers during the aquatic phase. The direct effect of the fragmentation by the larvae is increased by secondary attack of fungi and bacteria, which gain better access to the internal parts of the logs. The abundance of larvae is related to high amounts of phytoplankton and fine detritus, low amounts of inorganic suspensoids, and the availability of oxygen (Sect. 13.5). In 37 out of 40 living trees, Braga (1979) found small numbers of *A. curtus* tunnels in or below the bark. A much stronger attack occurs in dead wood and is linked to wood density. Wood blocks ($12 \times 8 \times 8$ cm) of three tree species with a density of 0.27–0.36 g cm^{-3} that were exposed floating near the surface in Lago Janauacá for five months lost approximately 21% of their original weight. Larva density reached 740 individuals per block after 4 months of exposure, further colonization being limited only by the density of tunnels in the surface layer of the wood blocks (Braga 1979).

Available evidence therefore suggests that oscillation of wetting and drying in the seasonally flooded forests significantly enhances the decay rate of wood in comparison to that in nonflooded and in permanently flooded sites.

12.5 Carbon Cycling Through Wood-Feeding Termite Populations

Figure 12.3 shows a model for the quantitative role of termite populations in wood decomposition in the várzea. We used the conventional terms of production ecology (MR = C + NU, C = P + R + F; Petrusewicz and Macfadyen 1970) and quantitative data (dry weight biomass units) for one of the wood-feeding species, *Nasutitermes macrocephalus*.

Wood litter, including imports by the flood, is produced at a rate of 7.2 t ha^{-1} (Sect. 12.2). Wood removed from the litter stock of 10.4 t ha^{-1} by the wood feeders (MR) is only partly consumed (C); a small fraction is occasionally lost (NU1), and about 10% are wood particles directly incorporated into the nests of the termites (NU2). In the following section, "C" refers to consumption, whereas "carbon" has always been spelled out.

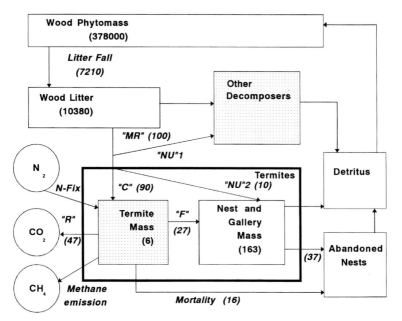

Fig. 12.3. Model of biomass cycling through a population of wood-feeding *Nasutitermes macrocephalus* in the várzea. *Squares* and *circles* Pools of solid and gaseous matter; *arrows* fluxes. *Italic letters* indicate fluxes; *MR* removed material; *C* consumed material; *NU1* material not ingested and not used by the consumers; *NU2* material not ingested but used for nest construction; *R* respiration; *N-fix* nitrogen fixation from the atmosphere; *F* feces. *Numbers in brackets* give the size of the pools (kg dry weight ha^{-1}) and fluxes (kg dry weight ha^{-1} year^{-1})

One of the differences between termites and other decomposers is the fact that termite feces are used to construct nests and are not dissipated in the ecosystem. By this process, carbon and other nutrients which are concentrated in the nests often accumulate near trees ("selective nutrient allocation", Salick et al. 1983; cf. Sects. 18.2, 18.3).

Another difference is the symbiosis of termites with their gut-inhabiting microfauna. There has been a controversy over whether termites depend completely or only partially on the microbes to degrade cellulose and lignin (Eutick et al. 1978; Odelson and Breznak 1985; Rouland et al. 1989a,b; Pasti et al. 1990). Potts and Hewitt (1973) and Schulz et al. (1986) showed that termites produce their own cellulases; termites are even able to degrade lignin to some extent (French 1975; French and Bland 1975; Butler and Buckerfield 1979; Cookson 1987; 1988). Breakdown of wood in the termite hindgut occurs both aerobically and anaerobically; the latter process produces methane at low rates which yields a CH_4/CO_2 ratio of generally <1% (Heyer 1990; Martius 1994a; Sect. 7.5). Gut symbionts can also fix atmospheric nitrogen (Breznak et al. 1973; Bandeira 1978, 1983; Sylvester-Bradley et al. 1978, 1983). Thus, xylophagous termites and beetles eliminate carbon from the system as CO_2 and CH_4, at the same time providing a small input of nitrogen.

For *N. macrocephalus*, Martius (1989) determined the population size in the field (termite biomass = 5.6 kg ha^{-1}), and the consumption rate in laboratory assessments (49 mg wood g termite^{-1} day^{-1}). Ecological efficiencies (C/MR = 90%, Martius 1989; A/C = 70%, Wood 1978; P/A = 25%, Wood and Sands 1978) allow the calculation of the components and fluxes in the model that were not measured. A *N. macrocephalus* population of 5.6 kg ha^{-1} produces a calculated biomass of 16.4 kg ha^{-1} year^{-1}, which corresponds well with the production-to-biomass (P/B) ratio of 3/1 postulated for termites (Wood and Sands 1978). It is estimated that almost half (47.3%) of MR (52.5% of the consumed carbon) is expired to the atmosphere as CO_2, a result in agreement with Butler and Buckerfield (1979) who found that 57% of the wood carbon consumed by a population of *N. exitiosus* in Australia was respired.

Both C and NU2 enter the "termite compartment", which consists of the termites themselves and their nests (Fig. 12.3). The input of 100 kg ha^{-1} year^{-1} is balanced by an output of 16 kg ha^{-1} year^{-1} through mortality (which has three components: natural mortality, alate flight, and death due to predation), plus 37 kg ha^{-1} year^{-1} from decaying nests, and a calculated respiration of 47 kg ha^{-1} year^{-1}. Smaller fluxes as methane emissions (calculated at 0.1 kg ha^{-1} year^{-1} on the basis of Martius et al. 1993) and nitrogen fixation (determined to 0.04–0.25 kg N$_2$ ha^{-1} year^{-1} on the basis of data from Bandeira 1983, cf. Martius 1994a) do not contribute significantly to the mass budget.

The proportion of the woody litter removed by *N. macrocephalus* (100 kg ha^{-1} year^{-1}) and *N. corniger* (289 kg ha^{-1} year^{-1}) (Martius 1989) correspond to 1.4 and 4.0%, respectively, of the annual wood litter production. The whole wood-feeding termite community in the várzea comprising dry-wood termites, Rhinotermitidae, and Nasutitermitinae probably turns over up to 20% of the annual wood litter production. This is a high rate when compared to the impact of termites in other ecosystems (Martius 1994b).

Dead wood has been considered one of the large nutrient pools of ecosystems which helps to buffer external fluctuations (Swift 1977). A similar role has been postulated for the large and persistent termite nests in Australian and African savannas which were seen as pools where nutrients are protected from rapid mineralization (Lee and Wood 1971; Wood 1976; Wood and Sands 1978; Grassé 1984). However, recent studies in Australia show that decay rates in termite nests are, in fact, enhanced, due to higher microbial biomass in the nests than in soil (Holt 1987; Coventry et al. 1988).

In the várzea, dead wood is an insignificant pool, because it is poor in nutrients and corresponds to only 2.7% of the living wood biomass. *Nasutitermes* nests, which accumulate only small masses and have a short life time span (Sect. 18.3), represent an even smaller stock of 641 kg ha⁻¹, only 6% of the dead wood mass. In the várzea foodplain forest, as in semiarid Australia, the effect of the activity of termites is to accelerate dead wood decay rather than to build up nutrient stocks.

12.6 Discussion and Conclusions

Wood decay is probably one of the least-studied major processes in ecosystems in general, because it generally takes periods of time which exceed those covered by most science funding programs, and because the methodical aproach is difficult. Unfortunately, the very different approaches used in the studies on wood decomposition in the várzea severely hamper a quantitative comparison of the results. However, some general conclusions can be drawn.

Fragmentation, due mainly to termites, and, to much lesser extent, wood-feeding beetles, is high in the terrestrial phase. In the aquatic phase, we must distinguish between the open river channel with its suspensoid-rich water and the várzea lakes where mineral particles have sedimented. In the channel, fragmentation is low, but in the lakes the wood-colonizing *Asthenopus curtus* occur in high densities and wood fragmentation is high. Microbial mineralization in general is far slower than fragmentation. The interference of fragmentation and mineralization was not studied here, but it is well known that fragmentation due to wood-feeding insects greatly enhances the possibility for microbial attack and thus accelerates wood mineralization.

In the terrestrial phase, mineralization of large boles occurred largely at the same rates as in other tropical regions that are not subject to aquatic decay. Aquatic decay is initially fast, probably due to leaching of soluble wood constituents, but in the long term, mineralization is much slower than in the terrestrial phase because anaerobic processes dominate. The model experiment with small blocks, however (Sect. 12.4), showed that the intermittent aquatic and terrestrial decay can substantially accelerate wood decomposition in floodplains when compared to decay processes that occur under exclusively aquatic or exclusively terrestrial conditions. The flood pulse accelerates wood decay and thus contributes to faster nutrient cycling.

Part IV
Animal Life in the Floodplain

13 Aquatic Invertebrates

Wolfgang J. Junk and Barbara A. Robertson

13.1 Introduction

The aquatic invertebrates of Amazonian floodplains can be divided into three large communities: the zooplankton, the benthos and the perizoon. Most studies deal with zooplankton, concentrating on taxonomic aspects and on community structure during the hydrological cycle.

Very few studies have been conducted on the benthos and the perizoon. There are several reasons for this lacuna: collecting quantitative samples is difficult in the coarse litter layer, in the canopy of the inundated floodplain forest, and in the floating vegetation. Animal distribution can be patchy because it depends on substratum structure, which is often very heterogeneous. The separation of the animals from sediments and plant material is time consuming, making quantitative statements difficult. Furthermore, there are large taxonomic deficiencies. New species are to be expected, principally in the perizoon. Many of the adult aquatic insects are described, but not the larvae. Satisfactory ecological research at the community level requires the collaboration of taxonomists to solve the main taxonomical problems. Therefore our level of knowledge is rather limited.

Details about the physico-chemical conditions in central Amazonian floodplain lakes and the investigated localities are given in Chapters 2 and 4.

13.2 The Zooplankton

A review of the zooplankton of Amazonian floodplain lakes is given by Robertson and Hardy (1984). The authors indicate a total number of 250 species of rotifers, 20 limnetic Cladocera excepting Macrothricidae and Chydoridae, and about 40 Copepoda, the majority being Calanoid copepods. The number of rotifers has since increased to about 300

Ecological Studies, Vol. 126
Junk (ed) The Central Amazon Floodplain
© Springer-Verlag Berlin Heidelberg 1997

species (Koste unpubl.). The following species are considered numerical dominants:

1. Pelagic Rotifera: *Brachionus dolabratus, B. falcatus, B. patulus, B. zahniseri gessneri, B. zahniseri reductus, Epiphanes macrourus, Hexarthra intermedia braziliensis, Keratella americana, K. cochlearis, Lecane (M.) bulla, Lepadella cristata, Polyarthra vulgaris, Polyarthra sp., Ptygura pedunculata, P. melicerta socialis, Testudinella patina triloba.*
2. Pelagic Cladocera: *Bosminopsis deitersi, Ceriodaphnia cornuta, Daphnia gessneri, Moina minuta, M. reticulata,* often a species of *Bosmina* and *Diaphanosoma* and sometimes *Holopedium amazonicum.*
3. Pelagic Copepoda: Calanoida; *Notodiaptomus amazonicus, N. coniferoides, Dactylodiaptomus pearsi*; Cyclopoida; *Mesocyclops longisetus, M. leukcarti, Thermocyclops minutus, Oitona amazonica.*

The occurrence of most planktonic species is not restricted by water-chemical parameters (Chap. 4). An exception is *Bosminopsis negrensis*, which apparently occurs only in acidic, electrolyte-poor black water. Differences in species composition of plankton communities between black-, clear- and white-water lakes seem to exist mainly at the level of dominant species associations (Brandorff 1978; Schaden 1978). In acidic and electrolyte-poor waters Bosminidae seem to dominate, whereas in water with a higher pH and higher electrolyte content Daphnidae are more frequent. With respect to standing stock, available data do not show significant differences between the different water types, suggesting that zooplankton are also using nutrient sources other than phytoplankton (Sect. 10.2). The data on the stomach contents of fishes in black- and white-water lakes suggest that zooplankton may occur in lower quantities in black water because it represents a very small fraction of the stomach contents (Sect. 20.2).

Quantitative studies in various lakes point to the dominance of copepods in the zooplankton communities, mainly due to the presence of young nauplii and copepodite stages. According to Carvalho (1981), nauplii were responsible for 60% of the total standing stock in Lago Manaquiri, while adult copepods made up less than 1%. The average relationship between numbers of Copepoda, Rotifera, and Cladocera was estimated to be about 60, 30, and 10%, respectively. In Lago Castanho and Lago Jacaretinga crustaceans, mainly young copepods, were dominant (Brandorff 1977; Brandorff and Andrade 1978). Brandorff (1977) found ten species of Cladocera and eight species of Copepods, and a maximum number of about 900 individuals l^{-1} in Lago Castanho (Fig. 13.1). He related the maximum, found at rising water, to additional nutrient input from inflowing

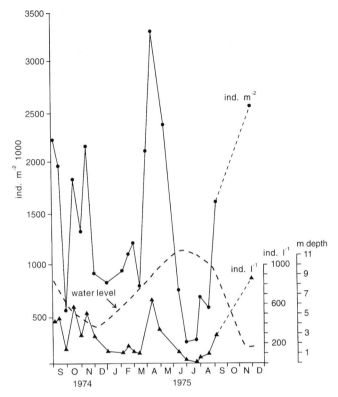

Fig. 13.1. Water depth and standing stock (individuals $m^{-2} l^{-1}$) of crustacean zooplankton in Lago Castanho. (Brandorff 1977)

river water and decomposing plant material, which favored phytoplankton development. The maximum at falling water is related to nutrient input from the surrounding floodplain forest. Low crustacean density at high water is attributed to a restricted oxygenated water layer, about 3 m (Fig. 13.2; Sect. 4.4.3) and to low phytoplankton production due to nutrient deficiency. At low water, numbers of crustaceans were intermediate in spite of a bloom of *Microcystis*. Brandorff postulated that *Microcystis* is not an adequate food for crustaceans and the low egg number implied a low nutritional status.

Large amounts of inorganic suspended particles can negatively influence zooplankton density. They reduce phytoplankton production by decreasing the euphotic layer and increase the nondigestable fraction of the suspended material. In Lago Jacaretinga, Brandorff and Andrade (1978) observed that planktonic crustaceans, present in relatively high numbers during the low water period (about 200 individuals l^{-1}), suffered a severe decline and almost disappeared when the inflow of Amazon River water

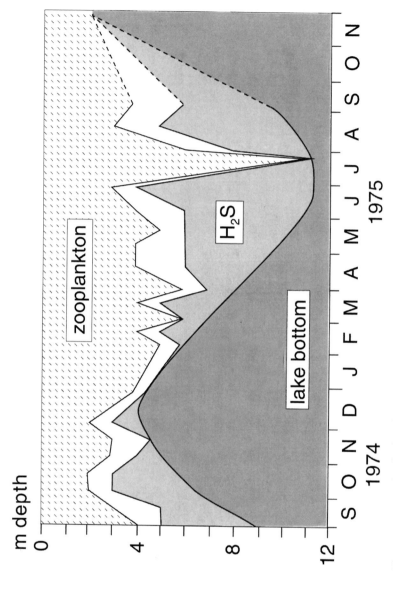

Fig. 13.2. Water depth, distribution of H_2S and occurrence of crustacean zooplankton in Lago Castanho. In the intermediate layer zooplankton was absent and H_2S was sometimes present. In July 1975 a short mixing of the whole water column occurred. (Brandorff 1977)

increased inorganic suspensoids. However, transparency alone is not sufficient as an indicator of the abundance of zooplankton. Frequently, at low
water, transparency is low because sediments have been stirred up by wind
and fishes. Phytoplankton density in the shallow water can be very high
(Schmidt 1973; Sect. 10.2.2) and there should be enough food available to
support large amounts of zooplankton and perizoon (Junk 1973).

A study by Fisher et al. (1983) indicates diel changes in the distribution
pattern of zooplankton in Lago Calado at high water. There was a slight
tendency by adult *Daphnia gessneri* to occupy the deepest part of the
oxygenated layer during daytime at a depth of about 3–4 m and to migrate
to a depth of about 1 m during the nighttime. This migration to regions
with low light intensity is explained as a strategy for avoiding predation by
visual planktivores during daytime.

Zooplankton studies conducted between January 1981 and October 1982
in Lago Camaleão indicate a coenosis that is different from other várzea
lakes in that it is dominated by rotifers (Fig. 13.3; Hardy et al. 1984); 175
species were identified (Koste and Robertson 1983; Koste et al. 1983, 1984).
The rotifer assemblage included numerous nonplanktonic forms, which
points to the close interaction with aquatic macrophytes.

Crustaceans were rare but relatively diverse. There were 14 species of
Cladocera and seven species of Copepoda (Reid 1989):

1. Cladocera, Bosminidae; *Bosminia* sp., *Bosminopsis deitersi, Bosminopsis*
   sp.

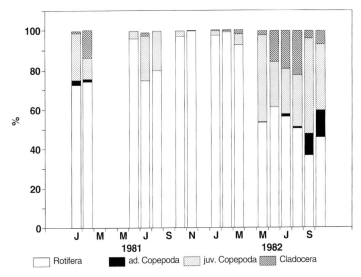

Fig. 13.3. Relative abundance (%) of Rotifera, Copepoda, and Cladocera in Lago Camaleão.
(Hardy et al. 1984)

2. Daphnidae; *Ceriodaphnia cornuta, C. reticulata.*
3. Moinidae; *Moina minuta, M. reticulata, Moinodaphnia macleayi.*
4. Sididae; *Diaphanosoma spinulosum, Diaphanosoma* spp., *Latonopsis fasciculata.*
5. Macrothricidae; *Macrotrix* sp., *Illiocryptus spinifer.*
6. Chydoridae. Copepoda, Calanoidea; *Notodiaptomus amazonicus, N. coniferoides, N. kieferi, Dactylodiaptomus pearsi.*
7. Cyclopoida: *Mesocyclops* sp. *Thermocyclops decipiens, T. minutus.*

Species richness of rotifers increased as the water level rose. At first, planktonic forms were abundant and forms associated with the periphyton were absent. In April, at intermediate water level, when large amounts of terrestrial vegetation were decomposing and oxygen concentrations were very low, semiplanktonic forms were found as well as a "rotten-mud" assemblage consisting of species such as *Brachyonus patulus patulus, B. patulus macracanthus, Dicranophorus claviger, D. braziliensis, Lepadella rhomboides, Mytilina acantophora. M. bisulcata, M. unguipes, Rotatoria neptunia, R. rotatoria R. tartigrada* and *Testudinella patia.* During high water when aquatic macrophytes were abundant, the semiplanktonic forms, the swimming-creeping forms of the periphyton, and the sessile species reached their maxima.

Maximum species numbers were recorded during the high water period. In July 1981 138 taxa were found and in July 1982 64–82 taxa were identified. Lower values in 1982 were related to very high water levels which

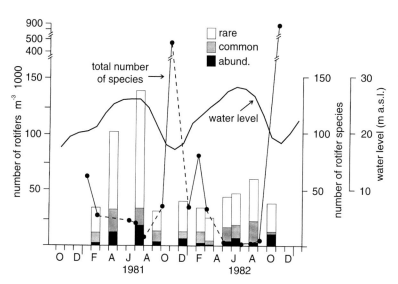

Fig. 13.4. Standing stock (individuals m^{-3}) of rotifers and numbers of species assigned to the categories rare, common and abundant in Lago Camaleão. (Hardy et al. 1984)

swept part of the zooplankton out of the lake. Minimum values were found at low water and at the beginning of rising water, when suspensoid-rich river water enters the lake. In 1981 and in 1982, 33 and 34 species were recorded, respectively (Fig. 13.4).

During falling and low water periods diversity decreased markedly. The taxocenosis was composed of residual populations of planktonic rotifers (*Brachionus zahniseri gessneri, Filinia longiseta, F. saltator, Keratella americana, K. lenzi, Polyarthra vulgaris, Trichocerca similis*) and semiplanktonic forms (*Brachionus bidentata, B. falcatus, B. quadridentata, B. caudatus, Epiphanes clavulata, E. macrourus*), but mostly by the inhabitants of the decomposing aquatic macrophytes (*Brachionus patulus, Cephalodella* spp., *Lecane* spp. *Lepadella rhomboides, Rotatoria rotatoria, Testudinella patina, T. mucronata haueriensis*). At low water the "rotten-mud" assemblage was established again.

Rotifers are not only the most diverse but also the most abundant group of the three main groups of zooplankton in Lago Camaleão, contributing between one third and more than 99% to the total number. Maximum values of about 900 individuals l^{-1} were found (Fig. 13.4; Hardy et al. 1984). The crustacean fauna was strongly reduced. Only during the latter part of 1982 did crustaceans participate in the zooplankton assemblage. It is interesting to note that the initial presence of crustaceans was due almost entirely to calanoid copepod nauplii.

Standing stocks of rotifers in Lago Camaleão reached two peaks during the year. The greatest occurred during the low water period, a smaller one occurred at the beginning of the rising water period. After this numbers declined markedly. The same pattern was observed for two consecutive years. In 1981, the peak high water level did not flood the entire island. The low zooplankton density during this time is related to strong hypoxia in the lake. During the peak high water level of 1982 the entire island was flooded and zooplankton was swept out of the lake.

The impact of the flood pulse is suported by studies about ergasiloid parasitic copepods. Much of what is known, particularly the taxonomy, is based on the study of females. This is because only the adult females attach themselves to the fish hosts while males and the young remain free living components of the plankton throughout their lives. Thus in order to obtain a more complete concept of their morphology, development, and biological and ecological relationships it is necessary to examine not only the fish hosts but plankton samples. Plankton samples collected during the peak dry season in Lago Camaleão revealed rare males and free-living ovigerous females of various species of parasitic copepods, particularly *Vaigamus retrobarbatus* (Thatcher and Robertson 1984). These animals were not found in plankton samples collected at other times of the year. We postulate that the peak dry season, when plankton and fish are concentrated in the remaining water bodies, is the time of year when the parasites most succesfully complete their life cycles. The probability that males and females will encounter each other and their hosts is much higher than in the high water period when density of parasites and hosts in the water is much lower.

13.3 The Benthos

Expeditionary research in the 1950s and 1960s included several surveys of benthic animal numbers. Braun (1952) studied some clear- and black-water lakes in the mouth region of Rio Tapajos near Santarém and found a maximum of 12600 benthic animals m^{-2} during low water at Lago Jurucui. During high water the number decreased to 3000 animals because of unsuitable oxygen conditions. In other lakes the numbers of benthic animals varied between 400 to 9000 specimens m^{-2}. Marlier (1965, 1967) reported 90 to 1500 individuals m^{-2} and a biomass of $0.245 g m^{-2}$ dry weight for Lago Redondo, a várzea lake near Manaus. He found there was very low colonization of the bottom of the black-water lake at Rio Preto da Eva. In Lago Jari, another black-water lake on an affluent of the lower Rio Purús, he found 670 individuals m^{-2}.

The first detailed studies were performed by Reiss (1974, 1976a,b, 1977) on permanently aquatic habitats, and Irmler (1975, 1976, 1979) on the aquatic terrestrial transition zone in the floodplain forest. Animal densities reached a maximum of about 5000 and 8000 individuals m^{-2} at low water and 50 and 250 individuals m^{-2} at high water in the white-water lakes Lago Calado and Lago Jacaretinga (Reiss 1976b). There are no data on biomass. In Lago Tupé, a black-water lake, animal density reached a maximum of only 1700 individuals m^{-2}, which corresponds to a biomass of 0.23–$0.24 g m^{-2}$ wet weight. Mean total abundance in 1971/1972 was 721 animals m^{-2} corresponding to a wet weight of $0.136 g m^{-2}$. In white-water lakes Lago Muratú and Lago dos Passarinhos, mean annual animal density was about the same order of magnitude with 736 and 810 individuals m^{-2}. The corresponding biomass of 2.65 and $2.06 g m^{-2}$ wet weight was considerably higher, because the animals in Lago Tupé (Ostracoda, Chaoboridae, Acari) were much smaller than those found in Lago Muratú and Lago dos Passarinhos (large and heavy larvae of *Chironomus gigas* and *Campsurus notatus*) (Reiss 1976b, 1977).

The main factor limiting the occurrence of the macrobenthos is the low oxygen concentration and the occurrence of H_2S in the water layer near the bottom of the lakes as shown for Lago Jacaretinga (Fig. 13.5). Samples from the hypoxic Lago Camaleão showed no benthos at all during most of the year. During a period of very high water in 1982 when oxygen-rich water from the Amazon River flooded the whole island, small numbers of benthic chironomids and ephemeropterans were observed. These disappeared when the water level decreased and the water near the lake bottom became anoxic (Junk, pers. observ.).

The benthos of the black-water lake Lago Tupé shows little variation in species number and biomass during the year (Reiss 1977). Oxygen concentration is usually less than $0.5\,mg\,l^{-1}$ throughout the year with a slight increase to about 1 mg at low water. A small increment in numbers and biomass is observed at low and rising water, with a minimum in August at falling water level.

The distribution of benthic species as a function of environmental conditions is discussed by Reiss (1976a). To construct their tubes *Chironomus gigas* and *Campsurus notatus* prefer sediments with high silt contents available only in white-water and mixed-water lakes. The larvae of *Chironomus paragigas* are restricted to black-water lakes. They have strongly elongated anal papillae for osmoregulation and long tubuli which are considered an adaptation to the electrolyte-poor and acidic black water. Reiss (1974) considered the closely related species *Ch. gigas* and *Ch. paragigas* to be

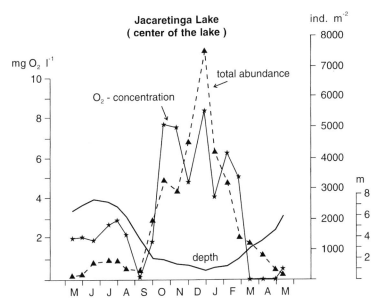

Fig. 13.5. Water level, oxygen content near the bottom, and total abundance of the macrobenthos in the center of Lago Jacaretinga. (Reis 1976a, modified)

stenoecious and indicator organisms for specific habitat conditions. In experiments, *Campsurus notatus* larvae developed well in white water, but died in black water when reaching a length of 7–8 mm, probably because of unsuitable hydrochemical conditions (Nolte 1986). The absence of Bivalvia in black water is probably due to the low concentrations of calcium and the low pH.

Periodic desiccation or anoxia near the bottom during the aquatic phase force benthic animals to migrate vertically or horizontally into permanently aquatic habitats or habitats with a better oxygen supply. *Goeldichironomus pictus* and *G. natans* colonize lake sediments at low water, but migrate into the floating vegetation with rising water (Reiss 1976a). Studies of the benthos in the aquatic–terrestrial transition zone made by Irmler (1975, 1976) reinforced the results of Reiss. *Campsurus notatus, Brasilocaenis irmleri* and *Laevapex aguadae* are able to avoid unfavorable oxygen concentrations by migration. Others, e.g., *aff. Aphylla* and *Eupera* make use of short periods of favorable conditions in specific habitats and are only found for a short period of time each year. The age

Table 13.1. Characteristic species of the benthic communities of different floodplain forests during the aquatic phase. (Irmler 1975)

	Low water	Rising water	High water	Falling water
White water		*aff. Aphylla* Culicidae *Chironomus latistylus* -----*Brasilocaenis irmleri* ----- -----*Laevapex aguadae* -------	*Eupera simoni* *E. bahiensis* ---------*Campsurus	*Chironomus gigas* Culicidae notatus*---------
Mixed water		*aff. Aphylla* -----*Brasilocaenis irmleri* ----- -----*Laevapex aguadae* ------- -----*Opistocysta flagellum*----- -----*Chironomus latistylus* ----	*Eupera simoni* *E. bahiensis*	*Chironomus gigas* *Pisidium sterkianum* *Campsurus notatus*
Black water	Gomphidae *Tanytarsus* *Gundlachia bakeri*	Tubificidae *Chironomus latistylus* *Brasilocaenis irmleri* -------------------*Naididae* -------------------*Ostracoda* ------------- *Euryrhynchus	---------Ostracoda ---------Chaoboridae *Gundlachia bakeri* *Polypedium sp* ---------------------- ---------------------- burchelli*--------------

distribution of the larvae of *Chironomus gigas* indicates that young larvae are able to migrate, whereas large larvae are resident.

In spite of the often inhospitable conditions in the water near the bottom, some species are predominantly benthic, e.g., *Chironomus gigas* and *Campsurus notatus* in white-water lakes, the bivalvia *Pisidium sterkianum* in mixed-water lakes and *Chironomus paragigas* and some *Tanytarsus* species in black-water lakes. Different types of benthic communities in floodplain forests are characterized in Table 13.1 (Irmler 1975). Considering the whole flood period, species diversity and biomass are greatest in the areas with the longest aquatic phase.

13.4 The Perizoon

In addition to the sediment, the floodplains of the Amazon and its large tributaries offer a great variety of surfaces that can be colonized by animals. These animal communities are called perizoon. The main habitats for the perizoon are rooted and free-floating herbaceous plants and the inundated floodplain forest. Both cover thousands of square kilometers. In black-water and clear-water rivers, herbaceous plant communities are much less common than in white-water rivers, as shown in Sect. 8.4.

13.4.1 The Fauna of the Aquatic Macrophytes

Junk (1973) considered the fauna of the floating vegetation of the várzea to be the richest aquatic invertebrate community in the Amazon. Most aquatic animal groups have representatives in the floating vegetation. Rheophilic forms like Plecoptera are missing, specific benthic groups, e.g., Tubificidae and Bivalvia are very rare. There are no data about numbers of species because little taxonomic work has been done. Some groups such as Hydracarina seem to be very rich in species. Viets (1954), working with Hydracarina from different Amazonian rivers and their floodplains, studied 1033 specimens and found 104 species; 64 were described as new. Koste (1974) reported 152 species of rotifers from floating root bunches collected during low water in a lake at the mouth of Rio Tapajós.

Numerically the most abundant are Copepoda, Ostracoda, Cladocera, and Diptera, which sometimes reach population densities of more than $100\,000$ individuals m^{-2} in *Paspalum repens* stands. Other groups, e.g., Conchostraca (*Cyclestheria hislopi*), Ephemeroptera and Trichoptera reach up to $10\,000$ individuals m^{-2}. Mollusks of the genus *Biomphalaria* and Hydracarina may reach up to 1000 specimen m^{-2}.

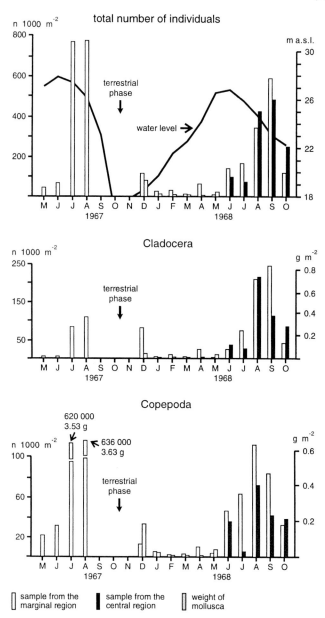

Fig. 13.6. Specific conductance, transparency, number and dry weight of aquatic inverte-brates in the root systems of floating aquatic macrophytes at the mouth of Lago Manacapurú. At the beginning of the aquatic phase suspensoid-rich white water dominates the habitat keeping animal numbers low. With rising water level black water, indicated by low conductivity values, becomes dominant; number of animals remains low. At falling water level transparent mixed water, indicated by rising specific conductance, from Lago

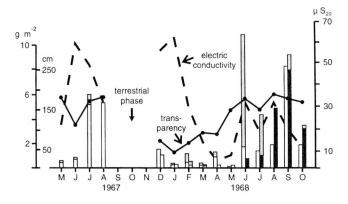

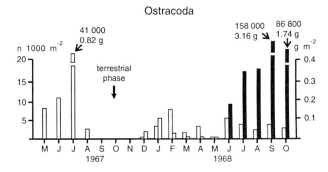

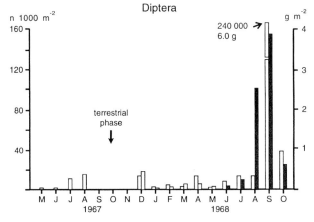

Manacapurú with phytoplankton and fine organic detritus dominates the habitat and favors the development of a very abundant invertebrate fauna (Junk 1973). *Total number* includes all aquatic invertebrates >0.2 mm. *Total dry weight* excludes large molluscs (Ampullariidae) and decapods

The perizoon community is strongly influenced by the amount of inorganic suspended matter, the oxygen concentration in the root zone and the availability of food (Junk 1973; Engle 1993). Flowing white water is rich in inorganic suspended matter and oxygen, and poor in phytoplankton and fine organic detritus. Perizoic organisms suffer from strong currents and the high loads of inorganic particles that cover the surface of the submerged parts of the plants, reduce growth of periphyton and hinder filter feeders and grazers in food uptake (Engle 1993). Total numbers of animals and biomass are rather small reaching up to $100\,000$ individuals m^{-2} in *Paspalum repens* stands, corresponding to 0.3–$4.2\,g\,m^{-2}$ dry weight. Numbers of animals increase from the edges of the stands to the inner parts, because currents and the amounts of suspended inorganic particles are reduced, and oxygen levels remain high (Junk 1973).

In várzea lakes the amounts of suspended inorganic particles are lower than in the river channels during most of the year. Food availability increases as growth of periphyton, bacteria and fungi increase, and when phytoplankton is washed into the roots from the open water. In experiments conducted to study decomposition of aquatic macrophytes culicids developed in large numbers. They fed on the bacteria decomposing the large amounts of organic substances leached from the macrophyte material (Walker 1986; Junk, unpubl. data). Experiments with an artificial substratum indicate that macrophyte material plays a minor role as food for the perizoon. An artificial substratum exposed in wire-mesh boxes was colonized by aquatic invertebrates in Lago Calado in an abundance pattern similar to aquatic macrophytes (Junk 1973). When exposed in the black water of Tarumã Mirim, only about 20% of the individuals and 10% of the biomass were attained compared to experiments in white water (Junk, unpubl.).

In várzea lakes, the number of animals is high or very high reaching up to $800\,000$ individuals m^{-2} on the edges of the stands. Towards the inner parts, population density and biomass decrease. In very dense stands anoxia prevails and H_2S is present and the number and biomass of aquatic invertebrates is low. Short life cycles and high reproduction rates of the dominant animal groups allow a quick reaction to changing environmental conditions. Population densities and compositions can vary strongly in monthly intervals, in different parts of a lake, and between lakes (Fig. 13.6).

13.4.2 The Fauna of the Submerged Canopy of the Floodplain Forest

There exists only one study on perizoon of the submerged canopy of the floodplain forest (Irmler and Junk 1982). Leaves were exposed to different water depths in small plastic cups in a black-water floodplain forest at Tarumã Mirim and a white-water floodplain forest at Lago Castanho and were sampled at monthly intervals. A distinct vertical zonation of the fauna was found. Naididae with the genera *Nais, Dero* and *Aulophorus* and larvae of the chironomid *Chironomus latistylus* were very abundant in deep water layers. *Macrobrachium amazonicum, Tenagobia melini, Hebetancylus moricandi, Asthenopus curtus* and *Cyclestheria hislopi* preferred the upper water layers.

Diversity and biomass changed with water depth. A reduction in species diversity, indicated by the index of Shannon and Weaver, from 2.5 at the 0.5 m layer to about 1.0 at the 8 m water layer was observed. Annual average biomass decreased logarithmically from the water surface to the bottom because of the preference of the large and heavy species, *Macrobrachium amazonicum* and *Asthenopus curtus*, for the upper water layer. Seasonality did not influence species diversity, but had a strong impact on total biomass. In the white-water habitat, the maximum total biomass was found at low water; in the black-water habitat maximum total biomass occurred at high and decreasing water levels.

There was no similarity between the fauna of the submerged canopy of the floodplain forest and the zooplankton, a rather weak similarity to the benthos and a strong similarity to the fauna in the floating vegetation. Benthic animals were mainly found in the container samples exposed near the bottom. The experiments allowed the collection of mobile species but offered also more shelter for invertebrates than the majority of the natural habitats in the canopy of the floodplain forest. Therefore the results cannot be used for calculating the biomass of perizoon.

13.5 Key Aquatic Invertebrates

Some of the numerous aquatic invertebrates in the Amazon floodplain have special importance in the ecosystem due to their large biomass. A few of them have been studied in more detail and can be used as examples for specific survival strategies.

Macrobrachium amazonicum, Fam. Palaemonidae. Fifteen freshwater shrimp species are described from central Amazonian waters (Magalhães and Walker 1986). Most species occur in black-water and clear-water rivers or streams; some are also found in white-water rivers, e.g., *Macrobrachium jelskii, Palaemonetes ivonicus* and *Euryrhynchus amazoniensis*. *Macrobrachium amazonicum* is restricted to white-water habitats. It does not tolerate pH values lower than 4.5–5 (Magalhães 1984) as frequently found in black-water and some clear-water streams and rivers. The species tolerates prolonged periods of hypoxia (Favareto et al. 1976) and elevated temperature (Guest and Durocher 1979). It grows up to 10–15 cm in length depending upon its habitat. Because of its abundance, it is an important food item for nearly all predatory and omnivorous fish species in the várzea, at least in certain periods of their life. In the states of Pará and Amapá it is exploited for human consumption and contributes locally to about 25% of the catch of inland fisheries. About 80% of the macrocrustacean biomass in the várzea is made up by *M. amazonicum* (Ordinez-Collart 1988).

All Amazonian freshwater shrimps are omnivorous, feeding on detritus, aquatic invertebrates (mainly chironomids), and periphyton. They are benthic but also occur in floating vegetation and floodplain forests. *M. amazonicum* produces up to 2300 small eggs (1.05–0.77 mm in size) and has 10–11 planktonic larval stages (Magalhães 1985). All other palaemonids occuring in the Amazon basin produce a much lower number of eggs, normally less then 100, which are considerably bigger than those of *M. amazonicum* and have only one to three larval stages. *M. amazonicum* larvae hatch as planktonic palaemonid zooea larvae. The larvae of all other species hatch in a more advanced stage, are about twice as long (5–6 mm) as those of *M. amazonicum* and are not planktonic. The larvae of the *Euryrhynchus* spp. have only one larval stage which shows a remarkable resemblance to the adults.

Magalhães and Walker (1986) suggest that abbreviated metamorphosis with direct development arose in plankton-poor environments. In black- and clear-water rivers, a few individuals that hatch at an advanced larval stage (K strategy), hide in litter banks, can exploit food sources similar to the adult food sources and can maintain a relatively small population. In nutrient- and plankton-rich environments the production of numerous small planktonic larvae with many zooea stages compensates for higher larval mortality of *M. amazonicum* (r strategy). This strategy is reinforced by the sex ratio, which shows a strong tendency towards females, between 78–88% on average. Reproduction occurs throughout the year and allows *M. amazonicum* to maintain a large population in spite of high losses during the annual dry period and strong predation pressure.

Pomacea lineata, Fam. Ampullariidae. From the Amazon region 34 species of the family Ampullariidae have been reported (Merck 1994). The status of the taxonomy is not satisfactory and the collection of specimens does not cover the whole area. Snails of the family Ampullariidae have a gill and a lung and can therefore take up oxygen from air and water. In ephemeral water bodies they are able to survive the terrestrial phase by estivation.

They feed on plant material and detritus. Because of their large biomass and sometimes large numbers they play an important role in the food webs. Several fish species, caimans, and birds feed on the snails. The snail-kite *Rostrhamus sociabilis* is specialized to feeding on ampullariids.

According to Merck (1994), adult *Pomacea lineata* reach about 10 cm and are very common in white-water rivers and floodplains. The water chemistry is suitable for mollusks (Sect. 4.3). The pH values of the water are near pH 7. There is sufficient calcium to allow the formation of the calcified shells (containing 35–39% Ca) and calcified egg masses. The nutrient-rich water provides an abundant food supply.

To cope with environmental conditions, *P. lineata* has an r strategy. Egg masses containing up to 2200 small calcified eggs (1.5–2 mm in diameter) are deposited out of the water on trees and herbaceous plants and resist low air humidity and solar irradiation, even during the dry season, without major losses. Egg deposition occurs every few days and can be observed easily in captivity even under adverse conditions. Even during the dry season, the animals reproduce in permanent waterbodies. Hatching occurs after 2–3 weeks.

In the igapó of the Rio Negro, *Pomacea lineata* is replaced by *Pomacea papyracea*, which is similar in size but occurs in much smaller numbers. Living conditions are adverse for mollusks because of the low calcium content in the water of about 0.3 mg l^{-1} and the low pH values, between 4 and 4.5. *P. papyracea* builds a very thin shell mostly of conchiolin with 13–18% content of calcium. Eggs are not calcified, are large (4–5 mm in diameter) and are covered by an elastic membrane. They suffer desiccation and increased mortality at low air humidities. Egg masses contain 250–1000 polyembryonic eggs (1–4 embryos of different sizes). Reproduction is restricted to high water periods, when air humidity in the floodplain forest is near saturation point. Egg deposition seems to be much less frequent and could not be observed in captivity.

Asthenopus curtus, Fam. Polymitarcyidae. *Asthenopus curtus* is a large mayfly that colonizes all types of waterbodies in Amazonia, although in varying densities. Using its very strong mandibles the larvae mine U-shaped tunnels in hard substrata, particularly dead wood. A middle wall is formed and the inner surface of the tunnel is coated with a silky secretion. Three tufts of hairs on each foreleg, one on each mandible, and a transverse row of hairs on the front of the head form a filter that removes phytoplankton and detritus from the flow of water through the tunnel. This flow is created by the movement of the strong lateral external gills (Sattler 1967). Their mining behavior and abundance make the larvae of *Asthenopus curtus* very efficient destroyers of wood in the floodplain (Sect. 12.4).

Adult females reach about 13 mm in length, about twice as big as males. The larvae reach a body length of 15–17 mm and a dry weight of about 10–13 mg. According to Braga (1979), emergence of the adults and subsequent reproduction occurs every day throughout the year. However, mass emergence is often observed in sultry weather with rain or at the beginning of a thunderstorm. The emergence of the adults is synchronized, which tends to occur in the evening. Egg masses are deposited at night and float near the water surface until they come in contact with a substratum. Numbers of eggs per female vary between 3600 and 10 500. Parthenogenesis occurs but the viability of unfertilized eggs (5.3%) is much smaller than that of fertilized eggs (about 71%). Most larvae hatch after 7–13 days; however, some eggs can hatch as late as 3 months after egg deposition.

Asthenopus curtus is another example of a successful r-strategist. The simultaneous emergence of many thousands of individuals increases the chance that the females will escape predation and allows for rapid mating. Adults have a life span of only a few hours. Larval development requires about 3 months. However, this strategy is only successful in nutrient-rich white-water floodplains, where food supply is large. In nutrient-poor black-water and clear-water floodplains, *Asthenopus curtus* occurs only in small numbers or is absent.

13.6 Discussion and Conclusions

While the species composition and abundance of aquatic invertebrates varies from lake to lake all authors agree on the existence of an annual cycle mainly related to turbidity, nutrients, and oxygen levels caused by the flood regime. Standing stocks of zooplankton in várzea lakes tend to be high at falling and low water and low at rising and high water, when river water rich in suspended solids dominates the lakes. Most invertebrates react with horizontal and vertical migrations and the displacement of habitats during the flood cycle. When the benthic habitat becomes anoxic during high water levels, many benthic animals colonize the floating vegetation or the floodplain forest. When the floating vegetation and the forest dry up at falling water level, many species move to benthic habitats. In small lakes with abundant macrophyte growth, zooplankton samples can show large numbers of rotifers, which are characteristic of the benthic or perizoic communities. The exchange of species between habitats results periodically in a greater diversity than expected.

Following the terrestrial phase, recolonization starts from different resting stages and also occurs by drift from permanent water bodies and by immigration via air. Local survival is guaranteed by resting eggs for Cladocera, Ostracoda, Conchostraca, diaptomid Copepoda, and Rotifera. Cyclopoid and harpacticid Copepoda survive as adults or copepodites in the dry mud. Bryozoa build sessile and floating statoblasts, Spongillidae produce gemmulae, and ampullariid snails and bivalves estivate. Many aquatic insects survive as larvae in neighboring wet habitats or as adults on land. In experiments, egg deposition by aquatic insects in water containers began within a few hours after exposure and the first chironomid larvae were observed after 24 h; larvae of culicids and mayflies were observed after 2 days. Notonectidae, Corixidae and Veliidae immigrated during the first days and quickly started to reproduce. Chaoborids were found after 4–10 weeks (Nolte 1986). Interconnection between a large set of different permanently aquatic habitats is in the long run a basic requirement for the great species diversity of aquatic invertebrates in Amazonian floodplains.

Unstable habitats favor colonization by animals with short life cycles and high reproductive rates (r-strategists) because they can make efficient use of the changing amounts of food and habitats, and can quickly replace population losses. The chironomid *Chironomus strenzkei* completes its life cycle in 10–12 days (Fittkau 1968) and culicids in 4–7 days (Nolte 1986). Semelparous species (chironomids and mayflies) are frequent. Cladocera, Ostracoda, Conchostraca, Rotifera, some harpacticoid Copepoda, some Hydrachnellae and some insects (*Asthenopus curtus*, *Caenis cuniana*) are parthenogenetic. Naididae propagate asexually by fission of the body. Monthly sampling intervals are probably too long to describe the dynamics of the development of the populations in Amazonian floodplain lakes.

An r strategy is very efficient in the nutrient-rich várzea lakes, where phytoplankton, periphyton, terrestrial and aquatic herbaceous plants and easily decomposable detritus and associated bacteria and fungi are available in large quantities. Food availability in black-water habitats is less and of lower quality, therefore aquatic invertebrates occur in much smaller numbers. Sometimes várzea species are substituted by better-adapted relatives, e.g., *Macrobrachium amazonicum* by other palaemonid shrimps, *Pomacea lineata* by *P. papyracea* and *Chironomus gigas* by *Ch. paragigas*; however, the black-water species are not able to build up large populations. Highest values are recorded in mixed-water areas with low sediment contents, high primary production and high oxygen content.

The largest diversity and biomass values of aquatic invertebrates are found in the aquatic macrophyte communities of white-water and mixed-water lakes. Macrophytes seem to be important as food for some herbivorous and detrivorous species, e.g., snails of the Family Ampullariidae. Many species use the floating plants primarily as a substratum and shelter and feed on other resources, e.g., periphyton or bacteria, phytoplankton and detritus filtered from the open water.

Acknowledgments. We are very grateful to Prof. Dr. Schwoerbel, University of Konstanz, Dr. John Melack, University of California, and Prof. Dr. Zwick, Max-Planck-Institute for Limnology, Plön/Schlitz, for critical comments on the manuscript and the correction of the English.

14 Terrestrial Invertebrates: Survival Strategies, Group Spectrum, Dominance and Activity Patterns

JOACHIM ADIS

14.1 Introduction

Terrestrial invertebrates in periodically flooded ecosystems require special "survival strategies". Development of these strategies is determined by the kind of flooding. This is defined by the number of floods (frequency), their height (maximum elongation or amplitude) and the duration of the flooding in the annual cycle (period of oscillation). These characteristics represent the "flood pulse" (Sect. 1.3; Junk et al. 1989). This is the primary control mechanism or primary ecofactor (Schaefer and Tischler 1983) in the system.

In Central Amazonia, the flood pulse is monomodal (Sect. 2.3). The floodplain forests and their adjacent shores, located between rivers and the floodplain forests, are covered by several meters of flood water for 5–7 months each year (March/April–August/September), depending on the terrain elevation and the height of the annual flood (cf. Adis 1981, 1992a; Junk 1989). This natural process is believed to have been repeating itself for several million years (Adis 1984; Sects. 2.1, 2.7). Through obvious long-term seasonal adaptations to this ecosystem, the animals and plants are able to adjust to regular, periodic flooding (cf. Irmler 1981; Worbes 1986a, 1989; Junk 1989; Junk et al. 1989). Along many other rivers, such as the Rhine, the frequency, amplitude and period of oscillation of flood pulses are variable, thus unpredictable, and catastrophic conditions often occur (Dister 1985, 1988).

In this chapter survival strategies of terrestrial invertebrates are outlined which have been determined through ecological studies of Central Amazonian floodplains in the surroundings of Manaus over the last 20 years (cf. Beck 1983; Adis 1992a). Furthermore, data on the group spectrum, dominance and activity patterns of terrestrial arthropods from white-water (várzea) floodplain forests located on the lower Solimões River, and from black-water (igapó) floodplain forests adjacent to the lower Negro River are presented. Sampling sites are described in Chapter 2.

Ecological Studies, Vol. 126
Junk (ed) The Central Amazon Floodplain
© Springer-Verlag Berlin Heidelberg 1997

14.2 Survival Strategies

Survival strategies are adaptations of organisms to unfavorable external conditions which increase their ability to survive (Tischler 1984). Strategies are genetically determined behavior patterns, which arise through natural selection (Southwood 1977, 1988). This does not imply that there is a conscious action by the organisms, and the terms "pattern" or "option" can be substituted for strategy (Chapleau et al. 1988).

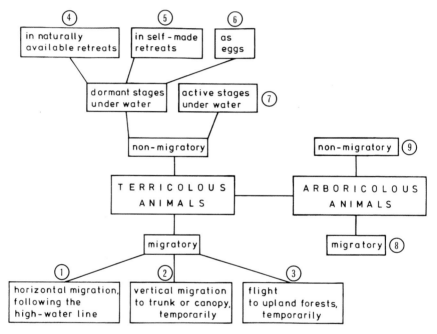

Fig. 14.1. Survival strategies of terrestrial invertebrates from floodplains along the Solimões River and Negro River in Central Amazonia. Examples of species studied (according to Adis 1992b):

1 a) Oligochaeta: Glossoscolecidae: *Tairona tipema* (Adis and Righi 1989)
   b) Coleoptera: Staphylinidae: *Lathrobium* sp. (Irmler 1979a)
2 a) Oligochaeta: Glossoscolecidae: *Andiorrhinus venezuelanus tarumanis* (Adis and Righi 1989)
   b) Pseudoscorpiones: Chtoniidae: *Tyrannochthonius amazonicus, Tyrannochthonius migrans* (Adis and Mahnert 1985)
   c) Opiliones: Cosmetidae: *Eucynortula lata*; Gonyleptidae: *Auranus parvus* (*Stygnidius inflatus* in Friebe and Adis 1983)
   d) Araneae: Gnaphosidae: *Camillina taruma, Tricongius amazonicus*; Ochyroceratidae: *Theotima minutissima*; Oonopidae: *Xyccarph* sp. (Höfer 1990b; Platnick and Höfer 1990)

The terrestrial invertebrate fauna of the Central Amazonian floodplain forests and their adjacent shore regions comprises terricolous and arboricolous animals. The ground or "soil" animals (according to Adis 1981) are mainly nocturnal inhabitants of the organic layer and the upper inorganic soil layer (e.g., Adis and Ribeiro 1989). The arboreal or "tree" animals live on trunks and in the canopy region of trees (e.g., Adis et al. 1984). Both groups include migrants and nonmigrants (Fig. 14.1). Nonmigrants complete their life cycles in only one habitat, either on the ground or in the trees. However small-scale movements within the habitat and temporary occurrences in other habitats are possible. Migrants reproduce primarily in only one of the habitats, but their life cycles include stages that change habitat or biotope.

◄───

Fig. 14.1 (*Continued*)

e) Symphyla: Scutigerellidae: *Hanseniella arborea* (Adis and Scheller 1984; Adis et al. 1996b)

f) Diploda: Paradoxomatidae: *Mestosoma hylaeicum*; Fuhrmannodesmidae: *Cutervodesmus adisi* (Adis 1992c; Adis et al. 1995a)

g) Archaeognatha: Meinertellidae: *Neomachilellus scandens* (Adis and Sturm 1987a,b; Wolf and Adis 1992)

h) Isoptera: Termitidae: *Anoplotermes* spp. (Martius 1989, 1990)

i) Coleoptera: Cicindelidae: *Pentacomia egregia* (Paarmann et al. 1982)

j) Coleoptera: Carabidae: *Scarites* sp. (Adis et al. 1990)

k) Hymenoptera: Formicidae: *Acromyrmex lundi carli* (Adis 1982a)

3 a) Coleoptera: Carabidae: *Polyderis nympha* (Adis et al. 1986)

4 a) Symphyla: Scolopendrellidae: *Ribautiella amazonica* (inside roots; Adis 1992b; Adis et al. 1996b)

b) Coleoptera: Oedemeridae: *Sisenopiras gounellei* (inside logs; Adis and Arnett 1987)

5 a) Diplura: Japygidae: *Grassjapyx* sp. (silken cocoon; Adis et al. 1989a)

b) Homoptera: Pseudococcidae, Cicadidae (wax protection; Adis and Messner 1991; Messner and Adis 1992a)

6 a) Acari: Eremobelbidae: *Eremobelba foliata*; Hypochthoniidae: *Parhypochthonius* sp. (Beck 1972)

b) Chilopoda: Henicopidae: *Lamyctes adisi* (Adis 1992b; Zalesskaja 1994)

c) Collembola: Arthropleona & Symphypleona (Adis and Messner 1991)

7 a) Acari: Haplozetidae: *Rostrozetes foveolatus* (Beck 1969; Messner et al. 1992)

b) Diplopoda: Pyrgodesmidae: *Gonographis adisi* (Adis 1986; Messner and Adis 1988)

8 a) Pseudoscorpiones: Miratemnidae: *Brasilatemnus browni*; Olpiidae: *Pachyolpium irmgardae* (Adis and Mahnert 1985; Adis et al. 1988)

9 a) Araneae: Pholcidae: *Blechrocelis* sp. (Höfer 1990b); Pisauridae: *Trechalea amazonica* (*T. manauensis* in Carico et al. 1985; cf. Carico 1993)

b) Diplopoda: Pseudonannolenidae: *Epinannolene arborea* (Adis 1984)

c) Archaeognatha: Meinertellidae: *Meinertellus adisi*; *Neomachilellus adisi* (Adis and Sturm 1987b)

d) Isoptera: Termitidae: *Nasutitermes* spp., *Termes medioculatus* (Martius 1989, 1992a)

e) Coleoptera: Carabidae: *Agra* spp., *Euchila* spp., *Miotachys* spp., *Moirainpa amazona* (*Tachyina* sp. in Adis 1982b)

f) Hymenoptera: Formicidae: *Cephalothes atratus*, *Daceton armigerum* (Adis 1981)

Terricolous Animals: Migrants. Three migration modes are distinguished which represent reactions of animals on the ground to seasonal flooding (Fig. 14.1).

1. Horizontal migration: This kind of migration is undertaken by animals that move in the direction of the terra firme (flood-free uplands) in front of the advancing water line during periods of rising water. They return to the floodplain forest with the receding waters. Species of animals using this strategy are given in category 1 of Fig. 14.1.
2. Vertical migration: Animals undertaking this kind of migration temporarily climb tree trunks and spend the period of flooding there or in the canopy. Characteristics are:
   - All or only some of the active life stages of the species begin to climb tree trunks either just before the flooding or several weeks earlier, at the beginning of the rainy season.
   - Main reproduction occurs on the forest floor during the terrestrial phase.
   - Reproductive cycle and duration of life stages are synchronized with the periodic fluctuations in water level.

   Species of animals using this strategy are given in category 2 of Fig. 14.1.
3. Flight: Animals capable of flight undertake this kind of migration by temporarily flying to neighboring forests in the terra firme that are not flooded. Characteristics are:
   - Departure from the floodplain forest usually occurs several weeks before flooding, but after the beginning of the rainy season.
   - Main reproduction takes place on the forest floor during the terrestrial phase.
   - Duration of the terricolous developmental stages and timing of the entire reproductive cycle are synchronized with the flood pulse.

   Species of animals using this strategy are given in category 3 of Fig. 14.1.

Terricolous Animals: Nonmigrants. Individual stages of animals on the ground that do not migrate remain for 5–7 months each year underwater. A distinction is made between species that are active underwater and those that remain dormant during the period of flooding (Fig. 14.1).

1. Dormancy underwater: The group of animals that do not leave their habitat and spend the period of flooding in dormancy on or in the ground of the floodplain forest:
   - In naturally available retreats.
   - In self-made retreats.
   - As eggs.

Species of animals using this strategy are given in categories 4–6 of Fig. 14.1.

2. Activity underwater: Those animals that remain active on the floor of the floodplain forest and on those parts of the tree trunks that are underwater. Species of animals using this strategy are given in category 7 of Fig. 14.1.

Arboricolous Animals: Migrants. In this group are animals that live mainly on the trunks and in the canopies of trees where they have their main reproduction. Characteristics are:

– During the terrestrial phase, life stages that live on the ground appear as well; there they have a secondary reproduction.
– Either all or only some of the stages in the life cycle take part in the downward and upward migration on tree trunks.

Species of animals using this strategy are given in category 8 of Fig. 14.1.

Arboricolous Animals: Nonmigrants. This group comprises animals that live and reproduce exclusively on the trunks and in the canopies of trees where they undertake only small-scale movements. Characteristics are:

– During the terrestrial phase, stages of the arboricolous animals appear only temporarily on the ground, if at all.

Species of animals using this strategy are given in category 9 of Fig. 14.1.

14.3 Group Spectrum, Dominance and Activity Patterns

14.3.1 Arthropods of Várzea Forests

The reaction of terrestrial arthropods to the annual inundation (submersion period) of a várzea island forest at Lago Camaleão (Ilha de Marchantaria; lower Solimões River) was studied in 1981–1982. Their activity density on the ground surface and abundance in the soil (emersion period) was determined by means of ground photoeclectors (emergence traps; cf. Adis and Schubart 1984) and soil extraction (cf. Adis 1987). Their vertical up- and downward migration (both periods) on trunks of the common várzea tree species *Pseudobombax munguba* (Bombacaceae; cf. Prance 1979; Adis 1981; Worbes 1983, 1986a) was monitored using arboreal photoeclectors (trunk traps; Funke 1971; Adis 1981). In addition, arthropods from the canopy of four different várzea tree species (emersion period) were obtained with the fogging technique (cf. Adis et al. 1984;

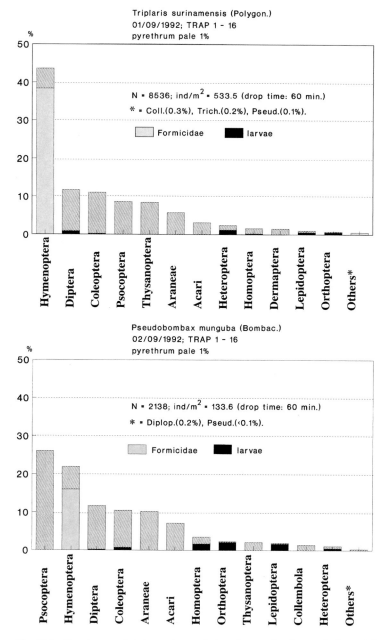

Fig. 14.2. Dominance, total number and abundance of arthropods fogged from the canopy of four tree species in a várzea island forest on Ilha de Marchantaria (emersion period) in 1992

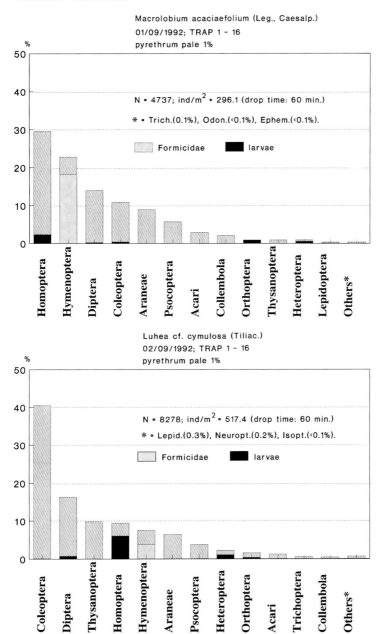

Fig. 14.2 (*Continued*)

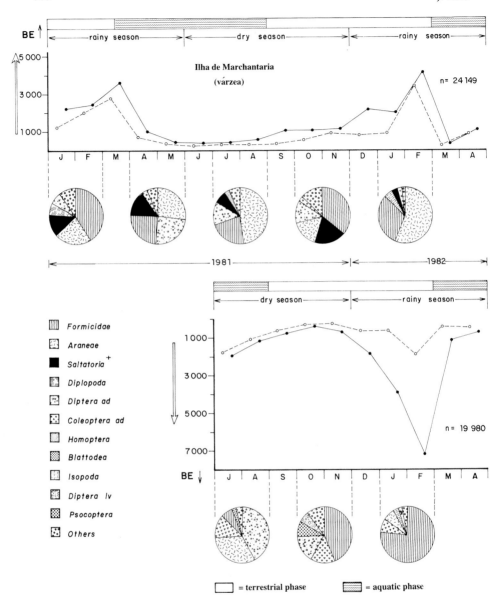

Fig. 14.3. Activity density, trunk ascents (↑) and trunk descents (↓) of arthropods and dominance of common taxa on *Pseudobombax munguba* (Bombacaceae), one arboreal photoeclector (*BE*) on each tree, in a várzea island forest on Ilha de Marchantaria between January 1981 and April 1982. Acari and Collembola are not included; [+]Caelifera and Ensifera; *black dots*, total catch; *circles*, total catch without Formicidae

Höfer et al. 1994b), using natural pyrethrum 1% (pyrethrum pale) diluted in diesel oil.

The majority of terrestrial arthropods in the várzea island forest (Acari and Collembola excluded) were found to live in the trunk/canopy region (Adis 1981; Adis and Schubart 1984).

In the canopy, adult Coleoptera (21%), Formicidae (20%) and adult Diptera (13%) represented more than half of the total number of arthropods obtained by canopy fogging (n = 23 689) after the flood had receded from the forest. However, dominance of groups varied according to the tree species sampled (Fig. 14.2).

The trunk fauna was dominated by Collembola and Acari (≤95% of the total Arthropoda; Table 14.1). During the terrestrial phase (emersion period), numerous animals came from the canopy/trunk region to the forest floor for nutrition (esp. Formicidae) or oviposition (esp. adult Coleoptera and Diptera). More than 60% of arthropods caught in the lower tree–trunk region (n = 44 129, Collembola and Acari omitted; Fig. 14.3) were thus represented by Formicidae (45%), adult Diptera (10%) and Coleoptera (6%). In addition, Araneae were also numerous (23% of the total catch). However, the dominance of groups not only changed throughout the year but also within the same season of different years. For example, during the rainy season (January/February) Formicidae (41%) dominated the total catch of trunk-ascending arthropods in 1981 but Araneae (56%) prevailed in 1982. Average biomass (dry weight) month^{-1} of ascending arthropods (Acari and Collembola excluded) was 5.7 g trunk^{-1} (Saltatoria 23%, Formicidae 22%, Araneae 21%, adult Coleoptera 16%) and 10.8 g trunk^{-1} for descending arthropods (adult Coleoptera 61%, Formicidae 26%).

On the forest floor, Acari (49–58%) and Collembola (11–24%) dominated in the litter and upper soil layer (0–3.5 cm). Up to 13 400 ind. m^{-2} were extracted from soil samples during the rainy season (from December onwards) and represented ≤60% of the total Arthropoda (cf. Table 16.2, Chap. 16; and Fig. 1 in Adis and Schubart 1984).

Excluding these two groups, Formicidae (32%) and Diplopoda (22%, mainly Polydesmida) dominated. Catches of ground photoeclectors were six times higher in the várzea island forest (6000 ind. m^{-2}) when compared to a 2.5 km distant and more exposed riparian forest (Fig. 14.4). The four arthropod groups which dominated on tree trunks of the várzea island forest were also the most abundant in ground photoeclectors (Fig. 14.4). The high number of Diptera was due to mycetophagous Sciaridae. These represented about 75% of all emerging Diptera from the soil during the rainy season. Staphylinidae, Pselaphidae and Ptiliidae amounted to 37% of all Coleoptera caught. They were even more numerous in soil samples and on tree trunks (77 and 54%, respectively).

Table 14.1. Mean abundance (ind. m^{-2} month^{-1} with sample standard deviation) and percentage (%) of Collembola, Acari and other Arthropoda caught per month during trunk ascents (↑) and descents (↓) in arboreal photoeclectors (BE) in a várzea and igapó forest

Arboreal photoeclectors	Trunk ascents (BE↑)		Trunk descents (BE↓)		Trunk ascents and descents (BE↑↓)	
Várzea	(ind. trunk^{-1} month^{-1})	(%)	(ind. trunk^{-1} month^{-1})	(%)	(ind. trunk^{-1} month^{-1})	(%)
Collembola	45760 (± 60202)	93.5	2277 (± 2146)	46.9	32557 (± 45137)	91.8
Acari	1675 (± 1545)	3.4	579 (± 291)	11.9	1194 (± 808)	3.4
Other Arthropoda	1509 (± 1163)	3.1	1998 (± 2111)	41.2	1711 (± 1426)	4.8
Total Arthropoda	48944 (± 61831)	100.0	4854 (± 2792)	100.0	35462 (± 46163)	100.0
Igapó	(ind. trunk^{-1} month^{-1})	(%)	(ind. trunk^{-1} month^{-1})	(%)	(ind. trunk^{-1} month^{-1})	(%)
Collembola	6097 (± 6950)	75.3	127 (± 98)	6.3	4128 (± 5195)	68.5
Acari	652 (± 1135)	8.0	373 (± 935)	18.5	515 (± 961)	8.6
Other Arthropoda	1350 (± 1086)	16.7	1516 (± 2160)	75.2	1379 (± 1147)	22.9
Total Arthropoda	8099 (± 7549)	100.0	2016 (± 2953)	100.0	6022 (± 5570)	100.0

Traps were mounted on common tree species. In the várzea island forest (Ilha de Marchantaria) on *Pseudobomax munguba* (Bombacaceae) 1 BE↑, Jan. 1981–April 1982; 1 BE↓, July 1981–April 1982. In the igapó forest (Tarumã Mirim) on *Aldina latifolia* (Leguminosae) 1 BE↑, Jan. 1976–May 1977; 1 BE↓, July 1976–May 1977.

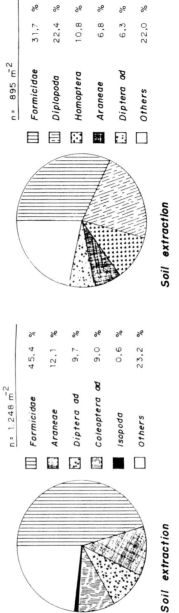

Island Forest (MA)

n = 6.056 m⁻² *

⊡	Diptera ad	61,4 %
▥	Coleoptera ad	12,3 %
☰	Formicidae	10,0 %
▦	Araneae	5,9 %
⊡	Hymenoptera *	4,7 %
☐	Others	5,7 %

Ground photo-eclector

n = 895 m⁻²

☰	Formicidae	31,7 %
⊟	Diplopoda	22,4 %
⊡	Homoptera	10,8 %
▦	Araneae	6,8 %
⊡	Diptera ad	6,3 %
☐	Others	22,0 %

Soil extraction

Riverine Forest (IC)

n = 1.009 m⁻²

⊡	Diptera ad	67,9 %
▥	Coleoptera ad	19,5 %
▦	Araneae	4,3 %
⊡	Hymenoptera *	3,5 %
⊡	Lepidoptera ad	3,2 %
☐	Others	1,6 %

Ground photo-eclector

n = 1.248 m⁻²

☰	Formicidae	45,4 %
▦	Araneae	12,1 %
✦	Diptera ad	9,7 %
▦	Coleoptera ad	9,0 %
■	Isopoda	0,6 %
☐	Others	23,2 %

Soil extraction

Fig. 14.4. Abundance or activity density of arthropods and dominance of the five most common taxa on the floor of a várzea riverine forest, Ilha de Curari, (IC) and a várzea island forest, Ilha de Marchantaria, (MA) during the rainy season. Data from IC (Adis and Schubart 1984) and MA were gathered between January and March 1976 and 1982, respectively. Acari and Collembola are not included; *without Formicidae

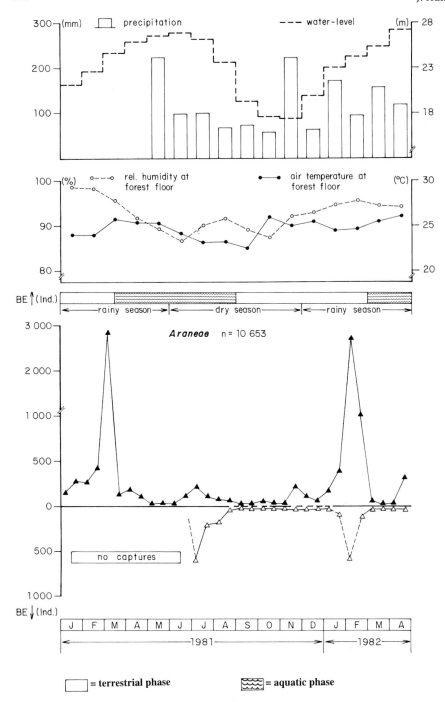

Fig. 14.5. Trunk ascents (↑) and descents (↓) of Araneae on *Pseudobombax munguba* (Bombacaceae), one arboral photoeclector (*BE*) on each tree, in a várzea island forest on Ilha de Marchantaria between January 1981 and April 1982. No correlation was found between trunk ascents and the climatic factors measured at the study site

Nonflying "terricolous migrants" were low in abundance and species (e.g., Pseudoscorpiones, Chilopoda) or totally absent (e.g., Symphyla, Uropygi) when compared to igapó forests (cf. Adis 1977, 1992a; Adis and Mahnert 1990). Conditions for a rich soil fauna are considered unfavorable in the very thin litter layer of várzea forests (Adis 1981). Leaves either glue together, due to annually deposited river sediments, and/or are partly lost during inundation, mostly due to a strong current. Primary decomposers were represented mainly by Diplopoda (esp. Polydesmida; cf. Golovatch 1992), Isopoda (esp. Philosciidae) as well as Isoptera (cf. Martius 1994a,b; Chap. 18) and predators were mainly represented by Araneae (esp. Gnaphosidae, Ochyroceratidae, Symphytognathidae; cf. Höfer 1990b; Chap. 19) as well as Opiliones (esp. Laniatores; cf. Friebe and Adis 1983). Together with leaf-cutting ants (cf. Adis 1982b) they passed forest inundation of 6–7 months duration (March/April–September) in the lower trunk region, mostly under loose bark, in crevices, inside freshwater sponges, abandoned termite nests (cf. Martius et al. 1994) and hollow trees, and also on floating logs. They were not collected from the upper canopy (Fig. 14.2; cf. Adis et al. 1984). Opiliones, Isopoda and some species of Araneae were shown to escape flooding only shortly before actual inundation of their habitat (Fig. 14.5; cf. Adis 1984, 1992a). Duration of trunk ascent thus corresponded with the velocity of the rising water level. For example, a constant increase of the Solimões River in January/February 1981 resulted in a uniform trunk ascent of Isopoda throughout a 4-week period (Fig. 14.6). However, due to the rapidly rising water level in early February 1982, their trunk ascent was concluded within 2 weeks.

14.3.2 Arthropods of Igapó Forests

Arthropods of an igapó forest at Tarumã Mirim were studied between 1975 and 1988, together with various collaborators (cf. Adis 1981, 1992a; Chaps. 16, 17, 19). The main study site was in the upper igapó (Adis 1984) which was annually inundated for 5–7 months to a height of 3–4 m. In this chapter we present data from: (1) the forest floor (soil extraction; Adis 1977); (2) one arboreal photocelector mounted on the common tree species *Aldina latifolia* var. *latifolia* (Leguminosae; Adis 1981; Worbes 1983) and (3) the upper canopy (fogging technique with pyrethrum (3%) diluted in kerosene; Adis et al. 1984).

The arthropod community in the canopy of the igapó resembles that of várzea forests: ants (43%) dominated and together with adult Diptera (15%) and beetles (5%), they made up more than half of the total catch obtained with the canopy fogging technique (Adis et al. 1984). However, most of the canopy guild represented different species when compared to

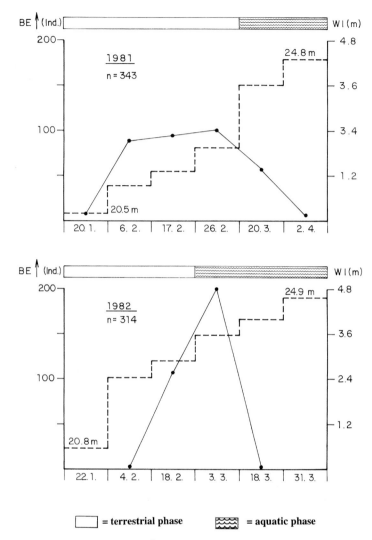

Fig. 14.6. Trunk ascents (↑; *continuous line*) of *Prosekia* albamaculata (Philosciidae, Isopoda; Lima 1996) on *Pseudobombax munguba*, one arboreal photoeclector (*BE*), in a várzea island forest on Ilha de Marchantaria during rising water level. *Dotted line*, Water level (*WL*) of the lower Solimões River in 1981 and 1982

the canopy inhabitants of the várzea (see Adis et al. 1984 for Formicidae; Erwin 1983, 1988 for Coleoptera; Chap. 19 for Araneae; Martius 1989, 1994b, and Chap. 18 for Isoptera). The first fogging data also indicated that arthropod density in the canopy of the igapó (62 ind. m^{-2}) may be twice as high as that found in the várzea (32 ind. m^{-2}) but less than half that found in a primary upland forest (161 ind. m^{-2}; cf. Adis et al. 1984).

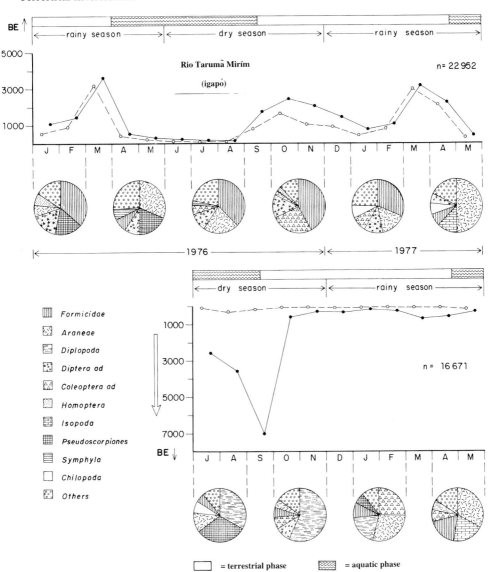

Fig. 14.7. Activity density, trunk ascents (↑) and trunk descents (↓) of arthropods and common taxa on *Aldina latifolia* (Leguminosae), one arboreal photoeclector (*BE*) on each tree, in the igapó forest at Tarumã Mirim between January 1976 and May 1977. Acari and Collembola are not included; *black dots*, total catch; *circles*, total catch without Formicidae

On the trunks the Collembola dominated, but their catch numbers in the igapó were lower by far (4000 ind. trunk^{-1} month^{-1}) when compared with the várzea (32500 ind. trunk^{-1} month^{-1}; Table 14.1). Activity densities of other arthropods on the trunks (Collembola and Acari omitted) were comparable to catch numbers obtained in the várzea (Figs. 14.3, 14.7). However, in the igapó trunk ascents and trunk descents were characterized by "migrating terricolous animals" (cf. Fig. 14.1) which passed the aquatic phase in the trunk/canopy region (cf. Araneae, Pseudoscorpiones, Symphyla and Chilopoda in Fig. 14.7).

During the terrestrial phase, some ants from the canopy visited the forest floor throughout the season with less precipitation ("dry season"); however, this was to a much lesser extent than compared with the várzea (cf. Figs. 14.3, 14.7).

In the litter and upper soil layer (0–3.5 cm) the Acari dominated. They accounted for 59–64% of the total catch (Table 16.2, Chap. 16). Together with the Collembola, they represented ≤85% of the extracted soil arthropods during the rainy season and their total abundance was twice as high (≤22 300 ind. m^{-2}) when compared to the várzea (≤13 400 ind. m^{-2}). If these two groups are disregarded, the nonflying terricolous migrating arthropods (Fig. 14.1) dominated, but their abundance changed throughout the terrestrial phase. For example, during the season with less precipitation (in November), Diplopoda (240 ind. m^{-2}), Pseudoscorpiones (220 ind. m^{-2}) and Araneae (110 ind. m^{-2}) accounted for 44% of the total catch. During the rainy season (in January) they represented only 30% of the total number of Arthropoda (although their abundance was higher: Diplopoda 370 ind. m^{-2}, Pseudoscorpiones 430 ind. m^{-2}, Araneae 500 ind. m^{-2}), by virtue of flying taxa, which were more common (Coleoptera: 1060 ind. m^{-2}, Homoptera: 590 ind. m^{-2}).

In the lower igapó (Adis 1984) where the aquatic phase lasted longer than 7 months/year on average, the nonflying terricolous migrating taxa were absent. Even the flood-resistant animal groups (cf. Fig. 14.1) were lower in numbers (Acari, Collembola, Coccoidea) or not found at all (Japygidae, Henicopidae). Flying insects, in particular Coleoptera (70%) as well as Diptera (5%) and Thysanoptera (2%), represented most of the arthropod guild in the upper litter and soil layers (total catch 7500 ind. m^{-2} or 1100 ind. m^{-2} without Acari and Collembola; Adis, Wantzen and Franklin, unpubl.).

14.4 Discussion and Conclusions

According to the results obtained at the species level, the flood pulse can be regarded as the original determinant of the upward and downward migrations of terrestrial invertebrates on tree trunks. However, it is still the primary control mechanism or ecofactor only among certain species [e.g., *A. v. tarumanis* (Oligochaeta), *E. latus*, *A. parvus* (Opiliones), *C. taruma*, *Tr. amazonicus* (Araneae); Fig. 14.1]. Most of the invertebrates have apparently become sensitive to secondary, mainly abiotic, ecofactors, which are no longer directly related to the flood pulse. The migration of animals from the ground to tree trunks and the flight to upland forests is triggered mainly by the rainy season (December–May; Ribeiro and Adis 1984), which begins 3–4 months before the flooding, and by the changes in the edaphic and climatic factors it causes. Examples are the increasing wetness of the soil and the relative humidity of the air [*Ty. amazonicus*, *Ty. migrans* (Pseudoscorpiones)], as well as the decreasing maximum air temperature and the decreasing difference between the maximum and minimum air temperatures near the forest flood [*P. nympha* (Carabidae, Coleoptera); cf. Fig. 14.1]. Vertical migrations of terricolous invertebrates are observed in the floodplain forests of Central Amazonia but not in the terra firme forests there (cf. Fig. 69 in Adis 1992a).

Vertical migrations were not only associated with changes in local climatic conditions ("dry" versus rainy season; cf. Ribeiro and Adis 1984) but also with macroclimatic influences such as the El-Niño Southern Oscillation (Adis and Latif 1996; cf. Ropelewski and Halpert 1987; Richey et al. 1989):

1. In years without El Niño events (e.g., 1975/1976, 1983/1984), trunk ascents of most arthropod species were correlated mainly with changes in the edaphic and climatic factors (e.g., the increasing relative humidity or the decreasing temperature of the air 3–4 months before forest inundation at the beginning of the rainy season (Adis 1981, 1992a).

2. In a "weak" El Niño year (1976/1977), an additional positive correlation (bifolded linear correlation-test) between weekly captures of continuously ascending species and total precipitation per week was found for Symphyla (*Hanseniella arborea*, $P = 5\%$: migrating adults), Diplopoda (*Cutervodesmus adisi*, $P = 5\%$: migrating juvenile stages) as well as for Pseudoscorpiones (*Tyrannochthonius amazonicus*, $P = 1\%$, and *T. migrans*, $P = 5\%$: migrating tritonymphs; cf. Adis and Mahnert 1985; Adis et al. 1996a,b).

3. In a "strong" El Niño year (1982/1983) none of these correlations were found. The rainy season and the flooding of the forest were delayed by 2 months in relation to years without El Niño events. The time lag observed in the trunk ascending-species given above was positively correlated instead with the water level of the Negro River, i.e., the beginning of forest inundation ($P = 1\%$ for *T. amazonicus* and $P = 5\%$ for all

other species). In some species where trunk ascents were normally represented by adults, juvenile stages now occurred as well (e.g., in *Xyccarph* sp., Araneae; Oonopidae; Höfer 1990b).

The relatively high current of the lower Solimões River causes a permanent removal of floodplain sediments and their temporary deposit downstream (cf. Sioli 1956, 1984; Irion 1976a, 1984a; Meade et al. 1979b, 1985; Adis 1981; Salo et al. 1986; Sect. 2.5). From the ecological point of view várzea floodplains adjacent to the lower Solimões River thus represent a permanent naturally disturbed ecosystem (Adis 1992a). From the entomological point of view terrestrial invertebrates inhabiting these várzea floodplains are therefore expected to represent mainly a pioneer fauna (opportunists and generalists) with high reproduction rates, short life spans (r-strategy) and great mobilities (Adis 1981, 1982b, 1992c; Smith and Adis 1984; Adis et al. 1990, 1993; Chap. 15).

Our data confirmed that arthropods from várzea island forests and riparian forests respond to the flood pulse with great mobility. This is also true for those species which inhabit the grass belt of the treeless bank margin, but which pass the flood period in the trunk/canopy region of the floodplain forest (cf. Adis 1982b, 1992a,c; Adis et al. 1990). However, due to the unfavorable living conditions on the forest floor, most of the nonflying "terricolous migrants" are absent. Moreover, only two examples can be given for "terricolous nonmigrants" that pass the submersion period underwater in a dormant stage: adult Collembola (*Onchiurus* sp.; Adis and Messner 1991) and larvae of Cicadidae (Tibicinae; Messner and Adis 1992a). Further studies have to show if vertical migrations of terrestrial invertebrates are more pronounced in those várzea floodplain forests that are more distant from the riverbed of the lower Solimões River, as their forest floor is subject to a lower water current and therefore is less disturbed.

Floodplain forests along the lower Negro River have been periodically flooded for at least 1 million years (cf. Adis and Schubart 1984; Chap. 2). Their invertebrate fauna comprises many endemic species (cf. Adis and Mahnert 1990). However, it also includes components that must have undertaken secondary immigrations from the shore region adjacent to the floodplain forest and from the higher terra firme region. Three taxa from the terra firme forests have not succeeded in colonizing Central Amazonian floodplain forests: the Palpigradi, Ricinulei and the Protura (cf. Adis et al. 1989c). It has been shown that the development of specific reproduction cycles by the terricolous and the arboricolous animals is related to the transition from the terra firme to the floodplain forest (Adis et al. 1988). The acquisition of an annual periodicity primarily as a reaction to a 6-

month flood period can be viewed as a basic requirement for the colonization of the temperate zones. An example of this is provided by the family
Carabidae (Coleoptera), the evolutionary center of which is located in the
floodplains along tropical rivers (Erwin 1979, 1981; Erwin and Adis 1982).
Many species of animals in the floodplain forests differ ecologically and
phenologically as well as genetically to such a great degree from those in
neighboring upland biotopes that they must be regarded as new species
and subspecies (Adis and Sturm 1987b; Adis et al. 1988; Wolf and Adis
1992). Thus, the floodplain forests of Amazonia contribute greatly to the
diversity of neotropical invertebrates (cf. Adis 1990), both by virtue of the
endemic species found (e.g., in Pseudscorpiones and Diplopoda) and due
to possible speciation processes observed in immigrants as a response to
the flood pulse over long periods of time (taxon pulse; Erwin and Adis
1982). However, phenotypic plasticity (morphological, physiological and
ecological states in response to environmental conditions, according to
West-Eberhard 1989) will have to be studied for each species (cf. Tomiuk et
al. 1996; Walker 1992a,b). The characteristic terrestrial invertebrate fauna
provides evidence that its ecosystem has become fully independent when
compared with the shore region and the terra firme region.

The adaptations of terrestrial invertebrates to the floodplain forests of
Central Amazonia are generally of an ethological and/or physiological nature. Morphological adaptations are exceptional (e.g., Messner and Adis
1988; Chap. 15). Amid a multitude of habitats and resources, survival
probabilities of terrestrial invertebrates in the floodplain forest depend on
their ecological valence, i.e., the diversity of their survival strategies.

Acknowledgments. I am grateful to all colleagues who in one way or another contributed to this paper, especially to many taxonomists and to the
scientists of the Tropical Ecology Working Group at the Max-Planck-
Institute for Limnology (MPI) in Plön (FRG.) as well as of the National
Institute for Amazonian Research (INPA) in Manaus (Brazil). Sincere
thanks are expressed to our technical staff in Manaus and Plön for the
valuable help received in the field and laboratory. Dr. Nigel Stork (The
Natural History Museum, London), Dr. Helen Read (Burnham Beeches,
Slough, UK), Dr. Christopher Martius (Univ. of Göttingen, FRG.), and
Privatdozent Dr. Charles Heckman (MPI Plön, FRG.) are specially thanked
for valuable comments and help received regarding manuscript format.
Canopy fogging was part of a biodiversity programme funded by the Ger-
man Research Foundation (DFG) and the German Agency of Technical
Cooperation (GTZ).

15 Adaptations to Life Under Water: Tiger Beetles and Millipedes

Joachim Adis and Benjamin Messner

15.1 Introduction

An adaptation is an attribute of an organism that enhances survival and reproduction in the biotope which it inhabits (cf. Schaefer 1992). In Amazonian floodplains, many terrestrial invertebrates show adaptations which enable them to live in a periodically flooded ecosystem. Two animal groups have been selected to exemplify different survival strategies, the tiger beetles and the millipedes. The tiger beetle *Megacephala* (*Tetracha*) *sobrina punctata* Laporte from várzea floodplains is amphibious and its adaptations are of an ecoethological, ecophysiological, and morphological nature. Many terricolous millipedes of Central Amazonian floodplain forests, including Polydesmida, migrate from the soil into the trunk/canopy where they pass the aquatic phase of 5–7-months duration (Adis 1992a; Adis et al. 1996a). Some polydesmidan species, however, show a "flooding ability" which ranges from a few hours or days (submersion tolerance) up to several weeks or months (submersion resistance); this is possible due to plastron respiration (Messner 1988) under water.

15.2 Results and Discussion

15.2.1 Tiger Beetles (Coleoptera: Carabidae: Cicindelinae)

Ecoethological Adaptations. Amphibious behavior in tiger beetles was previously only known from *Oxycheila polita* Bates, a neotropical species living in riverside floodplains of Costa Rica. When approached by predators, adults actively jump or fly into the river and are carried downstream. After some seconds, they rise passively (dorsal side up), open their elytra and hind wings just below the water surface, and emerge from the water in full flight. Beetles were also observed to occasionally enter the river water for a few seconds while they feed on sessile aquatic insects on rock surfaces along the water's edge. Maximum duration of activity during diving experiments in the laboratory was 5.8 min in aerated water and 3.6 min in nonaerated water. Maximum tolerance to submersion was not tested (Cummins 1992).

In Central Amazonia, cicindelid species of nonflooded upland areas and of várzea floodplains survive 2–18 h when being submerged in nonaerated water (submersion at $20\,^{\circ}C$ (4.0–5.1 mg $O_2 l^{-1}$) in strainers of metal gauze; width of mesh 0.6 mm). Species tested from the upland areas were *Odontocheila cayennensis erythropŭs* Chaudoir, *O. margineguttata rugatula* Bates *O. luridipes* Dejean, and *Cenothyla varians* (gory); species tested from the várzea floodplains were *Megacephala (Phaenoxantha) klugi* Chaudoir, *M. (Tetracha) spinosa* Brullé, *Odontocheila confusa* (Dejean), *Pentacomia (Pentacomia) egregia* (Chaudoir) and *P. (Poecilochila) lacordairei* Gory (Adis et al., unpubl.).

An exception is *Megacephala s. punctata* from várzea floodplains which has a flood tolerance of 24–30 h. This mostly "running-active" tiger beetle (cf. Adis 1982b) is capable of flight, nocturnal and univoltine. During the low-water period (about September–March) adults are found in dry cracks of desiccated lakes as well as under flotsam washed ashore (esp. aquatic macrophytes) on riverbanks and sandbanks of the Solimões River (Adis et al. 1993; cf. Pearson 1984; Rodriguez et al. 1994). When disturbed during the day near the water's edge, they run into the water where they may hide submerged for several hours.

Larval period is from September to March. Larvae and pupae are not flood tolerant as reported for other species of tiger beetles (cf. Wilson 1974; Pearson 1988). In low riverine areas, their development is threatened in years with a fast-rising water level (Chap. 14, Fig. 14.6). Adults of the new generation hatch from January onwards. During the period of rising water they migrate to the higher elevated floodplain forest (Adis 1982b), move into the trunk and canopy region, or inhabit driftwood, where they hide or live submerged during the day.

Ecophysiological Adaptations. In order to renew the air, diving animals walk backwards along the driftwood towards the water surface until the tip of the abdomen is above the water's surface. The terminal abdominal segments are then moved downwards and the air under the elytra can be exchanged (complete, emersed ventilation). The new air is stored in the subelytral cavity and serves as a physical gill under water (cf. Schaefer 1992). This respiration procedure only takes a few seconds (Adis et al. 1993). Maximum time per diving event was 2 h. No correlation was found

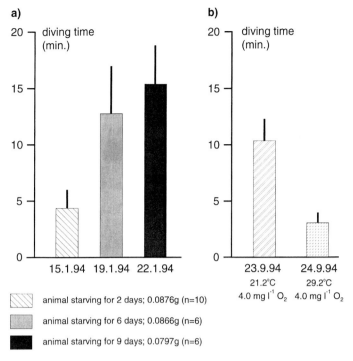

Fig. 15.1a,b. Diving behavior in a male of *Megacephala sobrina punctata* related to **a** the state of nutrition in laboratory experiments (Manaus; water temperature 21 °C, 4.2 mg $O_2 l^{-1}$; ANOVA: $p < 0.05$), **b** the water temperature in laboratory experiments (Plön; ANOVA: $p < 0.05$)

between the period of diving and breathing. Frequency and duration of diving varies from animal to animal (interindividual variability; Siepe 1989). The diving time increases during starvation but decreases with rising water temperature (Fig. 15.1).

During the diving event of adult *Megacephala s. punctata*, additional uptake of oxygen from the surrounding waters occurs: the diffusion area is enlarged through active ventilation of the air under the elytra by means of an abdominal bubble under water (partial submersed ventilation). This bubble (Fig. 15.2) is increased or reduced in size by vertical up and down movements of the abdomen (cf. Adis et al. 1993). The hydrophobic hairs on the hind wings, which face the subelytral plastron cavity (see above), serve as air-retaining structures. According to Messner (1988) and Messner and Adis (1992b), this type of plastron respiration (facilitated diffusion), enables extension of the diving time under water.

Beetles which lost their abdominal bubble in laboratory experiments under water had to surface after about 5 min, beetles with an exceptionally

Fig. 15.2. Abdominal bubble in a diving female of *Megacephala sobrina punctata*. (Photo courtesy of K. Hirschel and J. Adis)

large bubble (more than one-half the size of their abdomen) showed longer diving times (up to 2 h at 21 °C; 4.2 mg $O_2 l^{-1}$.

Morphological Adaptations. In comparison with species of tiger beetles from nonflooded upland forests in Central Amazonia (e.g., *Odontocheila c. erythropa*, *O. m. rugatula*; cf. Adis et al. 1993), from shore areas of the North and Baltic Sea as well as from river banks in Germany (*Cicindela maritima*, *C. hybrida*; cf. Adis et al. 1993), from wet rocks near waterfalls in southern Brazil (*Oxycheila tristis* Fabricius (Fig. 15.3a; cf. Mandl 1981), and from riverside floodplains in Costa Rica (*Oxycheila polita* Bates (Fig. 15.3b; Cummins 1992), the following morphological adaptations were found in adult *Megacephala s. punctata*:

– bulgy uprising spiracles with reticulate, microtrichia-covered lateral areas (cf. Fig. 15.3c)
– a two to four times higher density of hair on the hind wings which face the abdomen and cover the spiracles (cf. Fig. 4 in Adis et al. 1993)
– more elevated lateral margins of the abdomen

Similar structures were also found in *Megacephala (Tetracha) spinosa* Brullé (cf. Fig. 15.3d) from the same habitat in várzea floodplains. In this species, however, diving ability was not observed and flood resistance (8–10 h) was relatively low. Apparently, additional ecophysiological adaptations are needed to make use of morphological adaptations for diving.

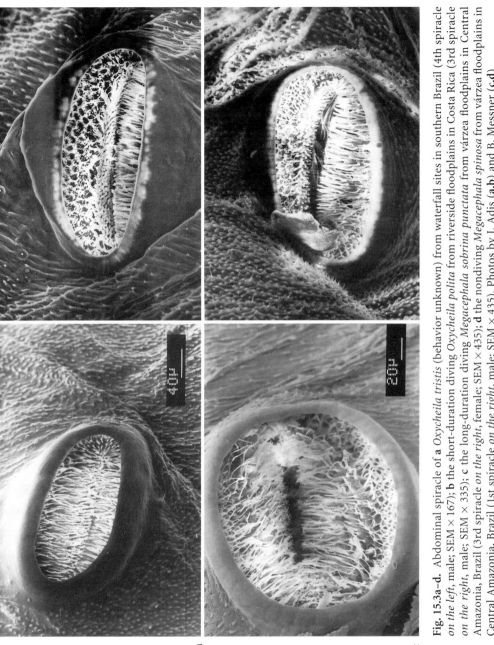

Fig. 15.3a–d. Abdominal spiracle of **a** *Oxycheila tristis* (behavior unknown) from waterfall sites in southern Brazil (4th spiracle on the left, male; SEM × 167); **b** the short-duration diving *Oxycheila polita* from riverside floodplains in Costa Rica (3rd spiracle on the right, male; SEM × 335); **c** the long-duration diving *Megacephala sobrina punctata* from várzea floodplains in Central Amazonia, Brazil (3rd spiracle on the right, female; SEM × 435; **d** the nondiving *Megacephala spinosa* from várzea floodplains in Central Amazonia, Brazil (1st spiracle on the right, male; SEM × 435). Photos by J. Adis (**a,b**) and B. Messner (**c,d**)

15.2.2 Millipedes (Diplopoda: Polydesmida)

Plastron Structures. Plastron structures hold a thin layer of air, depending on the size of the arthropod, into which oxygen from the surrounding water is added by means of diffusion to the same extent as oxygen is withdrawn through the connected tracheal system by breathing. In addition, this plastron can be actively ventilated. In flowing waters, small bubbles of atmospheric air in the water can be captured and added to the plastron (cf. Messner and Adis 1992b, 1994).

The plastron is of an appropriate size so that the animals may actively move underwater without expenditure of energy, i.e., preventing buoyancy. The efficiency of plastron structures depends on the type and is species-specific. In the Arthropoda plastron respiration has evolved polyphyletically, apparently from hydrophobic morphological structures that above all protect terrestrial animals from getting wet during rainfall or during temporal inundation.

Based on their origin, plastron retaining structures in aquatic arthropods and in terrestrial arthropods with flooding ability can be classified into three groups (cf. Messner and Adis 1992b):

1. Cuticular structures, such as hydrophobic trichia, scales, microtrichia, pustula and papillae as well as thoracical gills, have been found in several species of insects, mites and recently in polydesmidans (cf. Messner and Adis 1994; Messner et al. 1996).
2. Secretions of the cuticula are deposited:
   - as rapidly drying hydrophobic layers shortly after or during molting on the entire or on part of the cuticula (cerotegument) in many oribatids and as a secretion layer in a polydesmid (*Gonographis*) from Central Amazonia (cf. Messner and Adis 1988; Messner et al. 1992)
   - as waxes which are segregated in the same way to make specific cuticular surfaces hydrophobic (e.g., in Pseudococcidae and larvae of Cicadae, Homoptera; cf. Messner and Adis 1988; Adis and Messner 1991; Messner et al. 1992).
3. Secretions of glands are found as a hydrophobic silken cocoon in submersed living larvae as well as pupae of a parasitic wasp (*Agriotypus*, Agriotypidae) and a microlepidopteran (*Acentria*, Pyralidae) or they occur as hardening, air-containing egg foam in leaf beetles (*Galeruca*, Chrysomelidae; cf. Messner et al. 1981, 1987).

Plastron of Spiracles. A relationship between the morphological, plastron-retaining structures of spiracles and flooding ability has been found by means of comparative investigations in seven species of polydesmidans

Table 15.1. Characteristics of Polydesmida (Diplopoda) with plastron-retaining spiracles

Species	Family	Occurrence/habitat	Body length in adults (mm)	Flooding ability	Form of atrium wall	Structure of microtrichia	References
Aphelidesmus sp.	Platyrhacidae	Central Amazon/floodplain forests[a]	40	Several hours[b]	Elevated	Setiform, non- or bifurcate	Messner et al. (1996)
Pycnotropis epiclysmus	Platyrhacidae	Central Amazon/floodplain forests[a]	50	<3h (<6h)[c1]	Elevated	Setiform, polyfurcate	This chapter, Hoffman (1995)
Mestosoma hylaeicum Jeekel	Paradoxomatidae	Central Amazon/riverine floodplains	32	<26h (<7 days)[c2]	Elevated	Setiform, polyfurcate	Adis (1992c), Messner and Adis (1994), Messner et al. (1996)
Selminosoma chapmani Hoffman	Paradoxomatidae	New Guinea/stream-banks in caves[d]	23	?; Enters pools[b,d]	Elevated	Setiform, polyfurcate	Chapman (1976), Hoffman (1977/78)
Oxydus gracilis (Koch)	Paradoxomatidae	Cosmoplite/riverine wetlands	25	5–7 days[c]	Elevated	Setiform, polyfurcate	Causey (1943), Hoffman (1977/78), Resh et al. (1990 Stauder (1990), Tadler and Thaler (1993)
Polydesmus denticulatus Koch	Polydesmidae	Europe/wetlands esp. riverine floodplains[a,e]	17	(50 < 75 days)[c3,c4]	Deepened	Piston-like	Steinmetzger (1982), Tadler and Thaler (1993), Tarasevich (1992), Voigtländer and Dunger (1992), Zulka (1989, 1991, 1996)
Polydesmus inconstans Latzel	Polydesmidae	Europe/wetlands[a]	12	?	Deepened	Piston-like	Steinmetzger (1982), Tadler and Thaler (1993), Voigtländer and Dunger (1992), Zulka (1991)
Polydesmus (Brachydesmus) superus (Latzel)	Polydesmidae	Europe/wetlands, esp. riverine floodplains[a]	12	(>60 days)[c4]	Deepened	Piston-like	Tadler and Thaler (1993), Zulka (1991)

[a] In nonflooded habitats as well.

[b] Field observations only.

[c] Submergence tests in nonaerated (aerated) water at (1) 22 °C, (2) 15, 20, 25 °C, (3) 4 °C, (4) 9 °C. (1) & (2): animals submerged individually in strainers of metal gauze; width of mesh 0.6 mm; (3) & (4): see Zulka (1991, 1996).

[d] Eyeless trogobolite, inhabiting the damp floor near subterranean streams and entering large, shallow pools to presumably feed on the microflora.

[e] Advanced juvenile stages (V, VI) and subadults hibernate underwater.

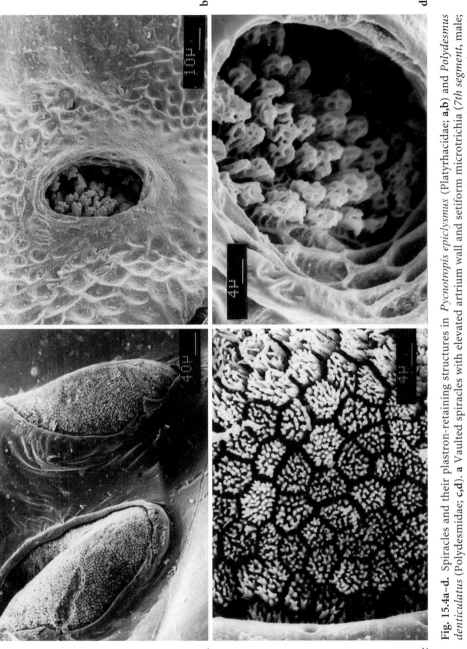

Fig. 15.4a–d. Spiracles and their plastron-retaining structures in *Pycnotropis epiclysmus* (Platyrhacidae; **a,b**) and *Polydesmus denticulatus* (Polydesmidae; **c,d**). **a** Vaulted spiracles with elevated artrium wall and setiform microtrichia (*7th segment*, male; SEM × 167). **b** Network of setiform microtrichia covering the atrium wall (*7th segment*, male; SEM × 1675). **c** Vaulted spiracle with deepened atrium wall (*4th segment*, male; SEM × 670). **d** Atrium wall consisting of a cuticular lattice which carries piston-

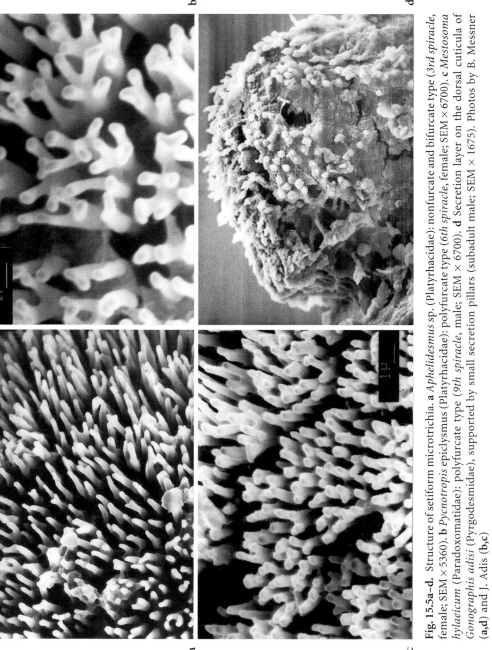

Fig. 15.5a–d. Structure of setiform microtrichia. **a** *Aphelidesmus* sp. (Platyrhacidae): nonfurcate and bifurcate type (*3rd spiracle, female*; SEM × 5360). **b** *Pycnotropis epiclysmus* (Platyrhacidae): polyfurcate type (*6th spiracle, female*; SEM × 6700). **c** *Mestosoma hylaeicum* (Paradoxomatidae): polyfurcate type (*9th spiracle, male*; SEM × 6700). **d** Secretion layer on the dorsal cuticula of *Gonographis adisi* (Pyrgodesmidae), supported by small secretion pillars (subadult male; SEM × 1675). Photos by B. Messner (**a,d**) and J. Adis (**b,c**)

which inhabit wetland habitats of different climate zones (Table 15.1, Fig. 15.6). All species show hemispherically vaulted spiracles (Fig. 15.4a,c) with hydrophobic, plastron-retaining microtrichia; these enable respiration under water. The atrium wall of the spiracles is either elevated (Fig. 15.4a) and covered by a network of setiform microtrichia (Fig. 15.4b) or it is deepened (Fig. 15.4c) and consists of a cuticular lattice (Eisenbeis and Wichard 1985; Hopkin and Read 1992) which carries piston-like microtrichia (Fig. 15.4d). In both cases the spiracle opening is narrowed to a nonclosable slit.

Simple plastron structures are nonfurcate or bifurcate setiform microtrichia (Fig. 15.5a). More complex structures are polyfurcate microtrichia (Fig. 15.5b). The most complex structures are piston-like microtrichia (Fig. 15.4d). Polydesmidan species from tropical and temperate regions with setiform microtrichia showed a submersion tolerance from a few hours up to several days, whereas species with piston-like microtrichia had a sub-

Fig. 15.6. Aggregation of adult *Pycnotropis* epiclysmus (Platyrhacidae) on the bark of a tree trunk near the water line, passing the 5–7 months of inundation in a mixed-water floodplain forest. (Photo by J. Adis)

mersion resistance of several weeks, probably months (Table 15.1; cf. Messner et al. 1995 for details).

Plastron of Secretions. The highest flood resistance was found in the small polydesmid *Gonographis adisi* Hoffman (Pyrgodesmidae; Adis 1986, 1992a; Messner and Adis 1988); it represents a terricolous nonmigrant with active stages under water (Adis, Chap. 14). Advanced juvenile stages (V, VI) and subadults (0.3–0.5 cm in length) pass the 5–7 months of inundation under loose bark of submerged tree trunks (mainly of *Aldina latifolia*, Legum.) in black-water floodplain forests along the lower Negro River. The nocturnal animals graze on algae, mostly Cyanophyceae and Chrysophyceae. They show uptake of dissolved oxygen greater than $10 \mu l \, mg^{-1}$ dry weight h^{-1}, which indicates a relatively high metabolic rate (cf. Dwarakanath and Job 1974; Hopkin and Read 1992). Some subadults become adults during the following terrestrial period and reproduce. Most of their progeny reach the subadult stage before the next flooding period and undergo inundation along with the remaining subadults from the preceding generation. Survival of immatures in laboratory cultures was 11 months, i.e., 4–6 months more than they would need to encounter terrestrial conditions on the forest floor. The entire cuticula of submersed living immatures is covered by a white secretion layer, visible as a netting of fine white lines between the tuberculi (cf. Messner and Adis 1988). This secretion layer extends from the tergites into the coxal region of the sternites where it covers the spiracles. It is hydrophilous on the outside but hydrophobic on its inner surface, and has a rough surface with irregular cracks and crevices. The entire layer is supported by small secretion pillars (Fig. 15.5d) and forms a cavity of 6.4–9.6 μm in depth on the tergites and about 5.0 μm on the sternites. A thin air casing, held in this cavity and invisible from the outside, enables plastron respiration. Submerged living immatures are found in the upper water layers of the black-water forests, where the concentration of dissolved oxygen in the water is high enough to enable plastron respiration (cf. Adis and Messner 1991). In adults the secretion layer hardly surpasses the sternal region, i.e., it has no connection with the spiracles. For this reason adults cannot withstand flooding and die during the beginning of the aquatic phase. In addition, submersion is made difficult due to the uncovered part of their hydrophobic cuticula. *Gonographis adisi* does not occur in other Central Amazonian forest types and is considered endemic. Up to now, it represents the only millipede which is adapted to living submerged for almost a full year.

Acknowledgments. We thank all participants of the postgraduate course 'Entomological Field Ecology', at INPA/Univ. Amazonas, Manaus/Brazil (given by J. Adis in 1992 and 1994) who joined the field and laboratory activities. The following colleagues kindly provided specimens for SEM studies: Mrs. Michelle Powell Cummins (Pasadena, USA), Prof. Dr. Wolfram Dunger and Dr. Karin Voigtländer (State Museum of Natural History, Görlitz, FRG), Prof. Dr. Richard L. Hoffman (Virginia Museum of Natural History, Martinsville/USA) as well as Dr. Peter Zulka (Zoological Institute, Univ. of Vienna, Austria). Special thanks for valuable comments on the draft to: Prof. Dr. David L. Pearson (Arizona State University, Tempe, USA), Prof. Dr. Henrik Enghoff (Zoological Museum, Kopenhagen, Denmark), Prof. Dr. Otto Kraus (Zoological Institute and Museum, Univ. of Hamburg, FRG) and, in particular, to Dr. Helen Read (Burnham Beeches, Slough, UK) and Prof. Dr. Richard L. Hoffman.

16 The Oribatid Mites

Elizabeth Franklin, Joachim Adis, and Steffen Woas

16.1 Introduction

The occurrence of soil fauna populations increases nutrient release by fragmentation of litter, grazing of microflora and improvement of soil structure. In floodplain forests of Central Amazonia, Acari represent up to 64% of the arthropod fauna in the litter and upper soil layers (0–3.5 cm; Table 16.2) and up to 19% of the monthly catches on tree trunks (Chap. 14). The oribatid mites are often a numerically dominant group among the soil arthropods in all soil types. These animals inhabit a variety of habitats, such as soil, rotten wood, litter, mosses, and trees. In the Central Amazon lowlands, Beck (1969, 1971, 1972, 1976) and Adis and Ribeiro (1989) have reported on the ecological significance of oribatid mites in inundation forests. In this chapter we discuss their abundance, distribution, population dynamics and submersion resistance with respect to the flood pulse (Sect. 1.3; Junk et al. 1989) in the igapó of Tarumã Mirím and in the várzea of Ilha de Marchantaria. Sampling sites are described in Chapter 2.

Adult oribatid mites have been assigned to species and their arrangement in tables and figures follows the systematic ranking suggested by Woas (1997; cf. Fig. 16.5; see also Woas 1981, 1986, 1990). A total of 78 species (Franklin 1994) were used for the comparison of catch data from soil, litter and tree trunks in the igapó and várzea forests. Of these, only 22 species have already been described in the literature. The 78 described and undescribed species represented about 75% of all adult oribatid mites collected and are referred to as "named oribatid mite species" in the text. Diversity indices, however, are based on the total number of species in catches, i.e., 78 named and additional unnamed oribatid mite species. Voucher specimens were deposited in the Systematic Entomology Collections of the National Institute for Amazonian Research (INPA) in Manaus/AM, Brazil.

Ecological Studies, Vol. 126
Junk (ed) The Central Amazon Floodplain
© Springer-Verlag Berlin Heidelberg 1997

16.2 Results and Discussion

16.2.1 Fauna of Litter and Soil

The vertical distribution of the oribatid fauna in litter and soil was monitored during the terrestrial (nonflooded) phase (1981/1982) by means of the Kempson extraction method (cf. Adis 1987). In addition, oribatid mites were obtained from litter which was sampled during both the terrestrial and aquatic phases (1989) and treated with the flotation method and/or the Kempson extraction method (Sect. 16.2.3). Figures 16.1 and 16.2 show the monthly abundance of Acari in the upper 3.5 cm of litter and soil in the várzea and igapó forests.

In the várzea forest, about 73% of the total catch (n = 6558 in 5 months) was represented by oribatid mites. Of these, about 6% were immature. The mean number of total oribatid mite species recorded month^{-1} (based on adult specimens) was 7.1 ± 2.1. *Galumna* sp. A and *Paralamellobates sp. A* were the most dominant species. In the igapó forest, about 89% of the total catch (n = 9852 in 6 months) was represented by oribatid mites. Of these, about 4% were immature. The mean number of oribatid mite species recorded per month was 14.6 ± 6.3. Samples to a soil depth of 14 cm were also studied in the igapó. About 70% of the total oribatid mite fauna (n = 12 352 in 6 months) was found in the upper 3.5 cm, 22% were from 3.5–7 cm depth, and only 8% were from 7–14 cm depth. In the lowest soil layer (10.5–14 cm), the mean number of oribatid mite species recorded month^{-1} de-

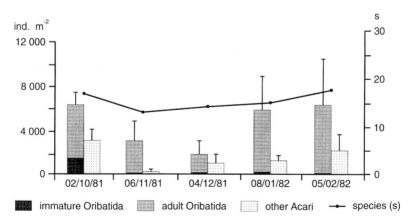

Fig. 16.1. Mean abundance with sample standard deviation of Oribatida (adults and immatures) and other Acari (individuals m^{-2}) and total number of oribatid mite species (*S*) extracted per month from the litter and upper soil layers (0–3.5 cm) between October 1981 and February 1982 (terrestrial phase) in the várzea forest on Ilha de Marchantaria

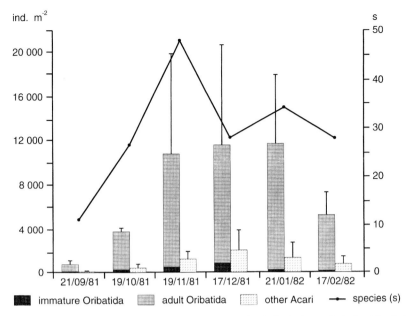

Fig. 16.2. Mean abundance with sample standard deviation of Oribatida (adults and immatures) and other Acari (individuals m^{-2}) and total number of oribatid mite species (S) extracted per month from the litter and upper soil layers (0–3.5 cm) between September 1981 and February 1982 (terrestrial phase) in the igapó forest at Tarumã Mirím

creased to 2.1 ± 1.8. *Rostrozetes foveolatus* Sellnick dominated all monthly catches at 0–3.5 cm depth and *Eremobelba* sp. B was dominant at 3.5–14 cm soil depth.

On average about 23% more Acari lived in the litter and upper soil layers (0–3.5 cm) of the igapó (8175 ind. m^{-2} month^{-1}) as compared to the várzea (6311 ind. m^{-2} month^{-1}). Oribatid mites prevailed at both sites, but the Pterogasterina (Galumnidae, Scheloribatidae, Haplozetidae) represented 54% of the named oribatid mite species in the várzea and only 31% in the igapó (Table 16.1). At least some species of the Galumnidae and Scheloribatidae are omnivorous, or even predaceous (Woas, unpubl.) and it is noteworthy that the percentage of the most-predaceous non-oribatid mites in the total catch was also higher in the várzea (27%; igapó, 11%). Altogether, the similarity of named oribatid mite species in igapó and várzea soils (0–3.5 cm) was only 14% (Sörensen index), and the relative dominance of named common species was low (Renkonen coefficient: 4%).

In the igapó, the constant increase in oribatid mite abundance and number of species within the first 3 months after the aquatic phase indicates a gradual recolonization of the litter and upper soil layers (Fig. 16.2). The maximum number of species (based on adults) was observed in No-

Table 16.1. Number of oribatid species and their share within the respective ecological group in the litter and upper soil layers (0–3.5 cm) of the várzea and igapó forests during the terrestrial phase 1981/1982 (according to Woas 1996 and unpubl.; see Fig. 16.5 and Franklin 1994 for details)

	Várzea		Igapó	
	Species	(%)	Species	(%)
Lower Oribatida	3	23.1	9	31.0
Basic Higher Oribatida	0	0.0	6	21.0
Peripheric Higher Oribatida				
– form Galumnidae	3	23.1	5	17.0
– form Scheloribatidae/Haplozetidae	4	30.7	4	13.7
– form Oppiidae	0	0.0	3	10.3
– other forms	3	23.1	2	7.0
Total	13	100.0	29	100.0

Basic Higher Oribatida: higher Oribatida that in their morphological organization are dominated by regressive characters.
Peripheric Higher Oribatida: higher Oribatida that in their morphological organization are dominated by static characters.

vember 1981 (n = 47). Immatures represented 5–8% of the total number of Oribatida caught between September and December 1981 and only 2–3% of those taken in January/February 1982. In the várzea, 17 out of 18 species were detected only 4 weeks after the aquatic phase (in early October) in the upper litter and soil layers (Fig. 16.1) and 23% of the oribatid mites were represented by immatures.

These differences may reflect not only the dominance of different ecological groups in the igapó forest as compared to the várzea island forest, but also the different abiotic situations in the two ecosystems. The várzea forest has no contact with terra firme areas. The climatic differences are more pronounced on the floor of the várzea island forest which has a more open canopy and a lower density of trees (cf. Adis 1981, 1984, 1992a; Meyer 1991; Worbes et al. 1992; Klinge et al. 1996). The soil surface in the várzea island forest is mostly bare of litter at the beginning of the terrestrial phase, due to sediment import and litter export by the Solimões River during the aquatic phase, and may therefore dry up more rapidly than the igapó soil, which carries a dense litter layer at the beginning of the terrestrial phase and experiences little sedimentation during the aquatic phase (cf. Irmler 1976; Meyer 1991; Chap. 14). This may result in a more rapid eclosion of immature oribatid mites and a faster increase of species in the upper soil layer of the várzea island forest than in the igapó forest (cf. Figs. 16.1, 16.2). This may also explain a reduction in species richness and abundance in the várzea.

The abundance of Acari in the litter and upper soil layers of nonflooded upland forests in Central Amazonia (Kempson extraction method) is higher than in floodplain forests, except in a secondary forest on yellow latosoil where the former primary forest had been cut and burnt (Table 16.2). The abundance of Oribatida in the igapó (approx. 11 000 ind. m^{-2}) is greater than that of the várzea (approx. 6000 ind. m^{-2}), and is close to the numbers obtained in a primary forest on yellow latosol during the season with less rainfall (approx. 10 000 ind. m^{-2}). Oribatida represented 66–74% of the total Acari in the várzea forest and 85–91% in the igapó forest (Table 16.2; Adis 1988). Data are comparable with those obtained by Beck (1971) from litter and soil extractions (500 cm^3) with Berlese funnels; 61–70% in várzea forests and 77–79% in igapó forests. In a várzea island forest on Ilha do Careiro (Amazonas River), about 30 km distant from Ilha de Marchantaria, oribatid mites represented 59–66% of the total Acari obtained with the Kempson extraction method (cf. Table 16.4; Adis and Ribeiro 1989) and 64–70% of Acari obtained with Berlese funnels (Beck 1971).

The number of oribatid mite species as well as their diversity was lower on the forest floor in the várzea (13–18 species; Shannon–Wiener index: H (ln) = 1.1–1.5) than in the igapó (11–47 species; H (ln) = 1.5–2.1), but higher in nonflooded primary upland forests (terra firme) of Central Amazonia (71–74 species; H (ln) = 4.5–4.7) and of Peru (130 species; H (ln) = 3.5; Ribeiro and Schubart 1989; Wunderle 1992a; Franklin 1994). Behan-Pelletier et al. (1993) found 30–51 oribatid mite species in forest habitats of Venezuela.

In general, the abundance of oribatid mites in tropical forest soils is lower than in forest soils of temperate regions. Irrespective of the many species found in tropical forests, the majority of them represent the basic higher Oribatida which are characterized by a low reproductive rate. In contrast, representatives of the periferic higher Oribatida dominate in temperate forest soils. They have a high reproductive rate which results in a high number of individuals per species. For example, Wunderle (1992b) reported that 82 oribatid mite species in the litter of a moder beech forest in SW Germany had an average population density of 61 500 adults m^{-2}.

16.2.2 Fauna of Tree Trunks

Arboreal Photoeclectors. The activity density of Acari was monitored using arboreal photoeclectors (trunk traps; cf. Funke 1971; Adis 1981; Adis and Schubart 1984) during the terrestrial and the aquatic phases on the most common tree species at the study site in the várzea forest [1981/1982:

Table 16.2. Mean abundance (individuals m⁻²) of Acari and Oribatida in the litter and upper soil layers (0-3.5 cm) obtained with the Kempson extraction method

Forest type	Study area	Sampling period	Acari			Oribatida			References
			Drier (d) season (ind.m⁻²)	Rainy (r) season (ind.m⁻²)	% of total Arthropoda (d-r)	Drier (d) season (ind.m⁻²)	Rainy (r) season (ind.m⁻²)	% of total Acari (d-r)	
Floodplain forests									
White water (várzea)	Ilha de Marchantaria (03°15'S, 58°58'W)	1981/1982	≤9500	≤8500	51-42	≤6300	≤6300	66-74	This study
	Ilha de Careiro (03°10'S, 59°44'W)	1986/1987	13400	12400	40-47	7900	8100	59-66	Adis and Ribeiro (1989)
Black water (igapó)	Rio Tarumã Mirím (03°02'S, 60°17'W)	1981/1982	≤11800	≤13500	64-59	≤10700	≤11500	91-85	This study
Upland (terra firme) forests			Drier (d) season (ind.m⁻²)	Rainy (r) season (ind.m⁻²)	% of total Arthropoda (d-r)	Drier (d) season (ind.m⁻²)	Rainy (r) season (ind.m⁻²)	% of total Acari (d-r)	
Primary – Yellow latosoil	Res. Flor. A. Ducke (02°55'S, 59°59'W)	1987	18200	(9200)	60-64	10300	(6700)	57-88	Adis et al. (unpubl.)
– White-sand soil (campinarana)	Res. Biol. INPA/SUFRAMA (02°30'S, 60°00'W)	1988	21500	32000	59-62	12800	17500	60-55	Adis et al. (1989b,d)
Secondary – Yellow latosoil									
Unburned (≥ 25 years old)	INPA Campus, Manaus (03°08'S, 60°01'W)	1985/1986	25700	29000	61-66	11700	17200	46-59	Adis et al. (1987a,b)
Burned (± 13 years old)	Distr. Agropec. SUFRAMA ZF-2 (60°60'W, 02°03'S)	1990/1991	≤13700	≤9000	54-82	≤8900	≤4800	65-53	Ribeiro (1994)

One monthly sample at each site was taken during the drier season (June–November, 550 mm of rainfall) and the rainy season (December–May, 1550 mm of rainfall) (Ribeiro and Adis 1984) in floodplain forests and nonflooded upland forests near and at Manaus between 1981 and 1991 (for methodological details, see references). The low abundance numbers obtained in March 1987 at Reserva Florestal A. Ducke (rainy season) may be based on an El Niño event (Ropelewski and Halpert 1987; Richey et al. 1989; Adis and Latif 1996).

The percentage of the total abundance of Arthropoda that are Acari and the percentage of Acari that are Oribatida are given for each season.

Pseudobombax munguba (Mart. & Zucc.) Dugand; Bombacaceae] and in the igapó forest (1976/1977: *Aldina latifolia* var. *latifolia* Spruce ex Benth.; Leguminosae: Fabaceae).

In addition to the catches with arboreal photoeclectors, Acari on tree trunks were monitored using the bark-brushing method (cf. André and Lebrun 1979). For this purpose, five trees each of two common tree species were selected in the várzea forest (*P. munguba* and *Macrolobium acaciifolium* Benth.: Leguminosae: Caesalpiniaceae) and in the igapó forest (*A. latifolia* var. *latifolia* and *Mora paraensis* Ducke: Legum.: Caesalpiniaceae). On each trunk, four $49\,cm^2$ squares were randomly selected once per month (February 1989–January 1990) about 1.5 m above the forest floor (terrestrial phase) or above the water surface (aquatic phase). Acari were collected by abrading the bark with a hairbrush.

In the várzea, the monthly density of Acari on the bark surface (bark-brushing method; Fig. 16.3) varied between 2 and 53 individuals $m^{-2} tree^{-1}$. On average, 26 Acari were collected $m^{-2} month^{-1}$ from one trunk of *Macrolobium acaciifolium* and 15 Acari $m^{-2} month^{-1}$ from *Pseudobombax munguba*. About 61% of the total catch (n = 251) was represented by Oribatida, and of these 80% (n = 120) were adult. The mean number of oribatid mite species recorded $month^{-1}$ (based on adult specimens) was 0.3 ± 0.7 on *P. munguba*, where *Scheloribates* sp. A dominated (84% of all adult oribatid mites collected) and 0.5 ± 0.9 on *M. acaciifolium*, where *Litholestes* sp. A was more common (49% of the adult Oribatida obtained). The similarity of named oribatid mite species collected from the two tree species was 40% (Sörensen index) and the relative dominance of named common species was low (Renkonen coefficient: 14%).

In the igapó, the monthly density of Acari on the bark surface (bark-brushing method; Fig. 16.4) varied between 2 and 1259 individuals $m^{-2} tree^{-1}$. On average, 322 Acari $m^{-2} month^{-1}$ were collected from one trunk of *Aldina latifolia* and 278 Acari $m^{-2} month^{-1}$ from *Mora paraensis*. About 98% of the total catch (n = 3514) was represented by Oribatida; of these 98% (n = 3419) were adult. On *A. latifolia*, the mean number of oribatid mite species recorded per month was 6.4 ± 7.1. Five species were more common and represented 64% of all adult oribatid mites collected. On *Mora paraensis*, the mean number of oribatid mite species recorded per month was 4.9 ± 5.8. Six species represented 72% of the adult oribatid mites obtained. Similarity of named oribatid mite species between the two tree species was high (Sörensen index: 93%) and the relative dominance of named common species amounted to 56% (Renkonen coefficient).

In both floodplain forests under study, the density of Acari on bark was remarkably higher during the aquatic phase than during the terrestrial phase. As in the catches in arboreal photoeclectors (Table 16.3), non-

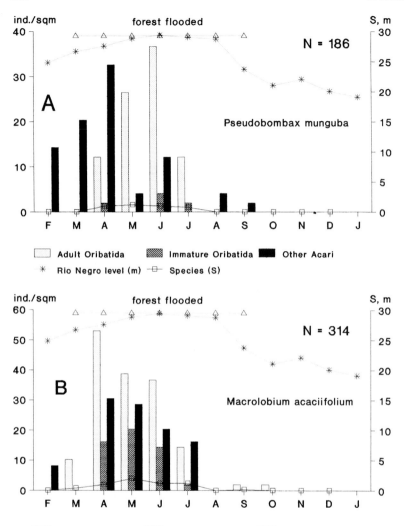

Fig. 16.3. Density of Oribatida (adults and immatures) and other Acari (individuals m^{-2}), their total catch (N m^{-2}) and mean number of oribatid mite species (S) on the bark of two common tree species in the várzea forest on Ilha de Marchantaria. Data represent average numbers of monthly collections on five tree trunks each with the bark-brushing method between January 1989 and February 1990

oribatid mites were much more common on the bark of várzea trees than on bark of trees in the igapó. However, an identification and habitat assignment of dominant species and their development stages is necessary to understand the increased catches in the arboreal traps of the two forests during the first months following the aquatic phase.

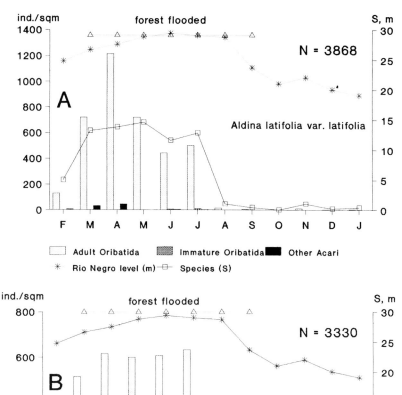

Fig. 16.4. Density of Oribatida (adults and immatures) and other Acari (individuals m⁻²), their total catch (N m⁻²) and mean number of oribatid mite species (S) on the bark of two common tree species in the igapó forest at Rio Tarumã Mirím. Data represent average numbers of monthly collections on five tree trunks each with the bark-brushing method between January 1989 and February 1990

The similarity of named oribatid mite species from the bark of the two forest types was low (Sörensen index: 9–16%) and so was the relative dominance of named common species (Renkonen coefficient: 1–12%). A preference by adult oribatid mite species for one of the tree species under study was not observed. It has been suggested that the distribution patterns

Table 16.3. Mean activity density (individuals month^{-1} trunk^{-1} with sample standard deviation) and percentage (%) of total Acari, Oribatida (with common species) and other Acari captured during trunk ascents (↑) and trunk descents (↓) in arboreal photoeclectors (BE) on common tree species in the várzea forest (Ilha de Marchantaria) and igapó forest (Tarumã Mirím) under study

Arboreal photoeclectors	Trunk ascent (BE↑)		Trunk descent (BE↓)		Trunk ascent & descent (BE↑↓)	
Várzea	Ind.month^{-1} trunk^{-1}	(%)	Ind.month^{-1} trunk^{-1}	(%)	Ind.month^{-1} trunk^{-1}	(%)
Oribatida	338 (± 594)	20.2	21 (± 15)	3.6	190 (± 297)	15.9
- *Paralamellobates* sp. A	198 (± 448)	11.8	5 (± 3)	0.9	99 (± 226)	8.3
- *Litholestes* sp. A	38 (± 69)	2.3	1 (± 3)	0.2	21 (± 35)	1.8
- *Scheloribates* sp. A	32 (± 36)	1.9	0 (± 0)		19 (± 19)	1.6
Other Acari	1337 (± 1108)	79.8	558 (± 290)	96.4	1004 (± 611)	84.1
Total Acari	1675 (± 1545)	100.0	579 (± 291)	100.0	1194 (± 808)	100.0
Igapó	Ind.month^{-1} trunk^{-1}	(%)	Ind.month^{-1} trunk^{-1}	(%)	Ind.month^{-1} trunk^{-1}	(%)
Oribatida	37 (± 45)	5.7	266 (± 698)	71.3	111 (± 288)	21.6
- *Rhysotritia* sp. A	4 (± 11)	0.6	227 (± 655)	60.9	77 (± 264)	15.0
Other Acari	615 (± 1114)	94.3	107 (± 240)	28.7	404 (± 701)	78.4
Total Acari	652 (± 1135)	100.0	373 (± 935)	100.0	515 (± 961)	100.0

Várzea – *Pseudobombax munguba*, 1 BE↑ January 1981–April 1982, 1 BE↓ July 1981–April 1982; igapó – *Aldina latifolia*, 1 BE↑ January 1976–May 1977, 1 BE↓ July 1976–May 1977.

Data on Acari are based on catches from one out of four collecting containers per arboreal photoeclector (Adis 1981). The catches of other arthropods in the container selected (Collembola and adult Diptera omitted) were nearest to the arithmetic mean calculated for the catches from all four containers of weekly or fortnightly collecting periods.

of arboricolous oribatid mites in Europe are related to the presence and structure of epiphytes (André 1979, 1984, 1985; Perez-Iñigo 1987). A low number of epiphytes on tree trunks apparently resulted in a low density of oribatid mites (Woltemade 1982). In subtropical forests and savannas of Africa, Nicolai (1986, 1989) observed that – in addition to the presence or absence of epiphytes – the diversity of oribatid mites also depended on the structure of the bark of tree species; the bark varied in temperature from the surrounding air due to the different angles of incidence of sunbeams. Few species of the named oribatid mites were found on tree trunks in the várzea island forest (cf. Fig. 16.5) in comparison with the igapó site. This may be related, for example, to a much lower presence of lichens, mosses and epiphytes (for unknown reasons) on the bark of várzea trees; a

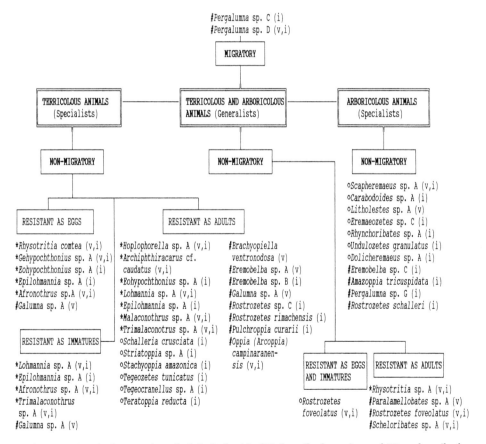

Fig. 16.5. Survival strategies of adult Oribatida (11 described species and 30 undescribed species) from várzea (*v*) and igapó (*i*) floodplain forests in Central Amazonia. See section 16.2.3 for explanation of ecological terms; *, lower Oribatida; o, basic higher Oribatida; #, peripheral higher Oribatida

smoother and less-sulcate bark structure with few fissures; a more open forest structure (*Scapheremaeus* sp. A and *Litholestes* sp. A represent species of more exposed areas, also outside forests); or the island character of the study area, which lacks connection to neighboring upland forests (cf. Adis 1981, 1992a).

16.2.3 Reaction of Oribatid Mites to the Flood Pulse

Submersion Resistance. The catch data on forest floor and tree trunks served to evaluate which oribatid mite species emigrate to the trunk/ canopy area to escape flooding. To detect the submersion resistance (cf. Chap. 15) in oribatid mites, 48 wire cages (35 × 2 × 7 cm; Franklin 1994) were filled with litter and set up on the floors of the várzea and igapó forests shortly before inundation in 1989. Half of them were removed during maximum flooding and those remaining were removed at the end of the aquatic phase. On both occasions, oribatid mites from the litter of eight cages were either extracted by using the Kempson method, obtained by means of the flotation method (using sugar; Adis et al. 1989a), or gathered from an incubation experiment lasting 2 months with manual removal of the animals every fourth day and final extraction of the litter with the Kempson method (cf. Franklin 1994). In addition, oribatid mites were extracted using the Kempson method from litter which was collected from the flooded forest floor at both times that the cages were removed. From all of these samples, the named oribatid mite species were compared with those detected in litter collected before forest inundation and were treated using the Kempson extraction and flotation methods. The named species obtained from soil and litter extractions during the terrestrial phase in 1981/1982 were also considered for purposes of comparison.

In the várzea forest, eight out of ten named oribatid mite species which were found during the terrestrial phase on the forest floor survived the period of flooding (206 days). One species was only found during the terrestrial phase and a second did not survive to the end of the aquatic phase (Table 16.4). One additional species (*Malaconothrus* sp. A) was only detected during the aquatic phase. In the igapó forest, 23 out of 39 named oribatid mite species from litter and soil of the noninundated forest floor (53%) survived the period of flooding (183 days). Seven species were found only during the terrestrial phase and 9 species did not survive until the end of the aquatic phase (Table 16.4). An additional 5 species were found only during the aquatic phase [*Cyrthermannia simplex, Suctoribates* sp. A, *Tecteremaeus cachoeirensis* (cf. Franklin and Woas 1992), *Tegeozetes tunicatus, Oppia* sp. B].

Table 16.4. Submersion resistance in named oribatid species from the soil and litter of the várzea forest (Ilha de Marchantaria) and igapó forest (Tarumã Mirim)

	No submersion resistance	Intermediate submersion resistance	High submersion resistance
Várzea	Rhysotritia comtea[a]	Up to 69 days (maximum water level) Neosuctobelba transitoria[c]	Up to 206 days (end of inundation) Rostrozetes foveolatus[c] Scheloribates sp. A[c] Oppia (Arcoppia) campinaranensis[a]
Igapó	Eremaezotes sp. A[b] Pergalumna sp. A[c] Pergalumna sp. D[c] Pergalumna sp. F[c] Pergalumna sp. G[c] Rostrozetes schalleri[c] Neosuctobelba transitoria[c]	Up to 72 days (maximum water level) Acaronychus proximus[a] Eohypochthonius sp. A[a] Epilohmannia sp. A[a] Afronothrus sp. A[a] Carabodes irmayi[b] Truncozetes mucronatus[b] Lamellobates sp. A[c] Porozetes cuspidatus[c] Suctobelba variosetosa[c]	Up to 183 days (end of inundation) Holophorella sp. A[a] Archipthiracarus cf. caudatus[a] Oribotritia sp. A[a] Rhysotritia sp. A[a] Parhypochthonius sp. A[a] Malaconothrus sp. A[a] Trimalaconothrus sp. A[a] Teleioliodes sp. A[b] Schalleria crusciata[b] Striatoppia sp. A[b] Stachyoppia amazonica[b] Tegeocranellus sp. A[b] Teratoppia reducta[b] Plateremaeus costulatus[b] Eremobelba sp. B[c] Rostrozetes foveolatus[c] Rostrozetes sp. C[c] Rostrozetes rimachensis[c] Haplozetes sp. A[c] Scheloribates sp. A[c] Scheloribates sp. C[c] Pulchroppia curarii[c] Oppia (Arcoppia) campinaranensis[a]

[a] Lower Oribatida.
[b] Basic higher Oribatida.
[c] Peripheric higher Oribatida.

Survival Strategies. Oribatid mite species which inhabit both soil and trees with a frequency >25% and a dominance >1% are considered "generalists". Oribatid mite species which occur more in litter and soil (frequency >25%, dominance >1%) than on tree trunks (frequency <25%, dominance <1%), or vice versa, are considered "specialists" (Wunderle 1992a). Terrestrial invertebrates of Central Amazonian floodplains comprise terricolous and arboricolous animals. Both groups include migrants and non-migrants (Adis 1992a; Chap. 14).

According to these premises, a first classification of the oribatid mite fauna in várzea and igapó forests is presented (Fig. 16.5). However, identification of species is mainly based on adult specimens. Immature oribatid mites mostly have singular morphological characters compared to adults (in particular representatives of the higher Oribatida) and can rarely be assigned to species without rearing. For this reason, phenology in most cases remains unknown and our ecological classification is tentative.

Of the 78 named oribatid mite species, 10 species (13%) occurred only in the várzea forest, 48 (61%) solely in the igapó forest, and 20 species (26%) in both forest types. Forty-nine species (63%) were collected only from the litter and soil, 12 species (15%) only on tree trunks and 17 species (22%) in both habitats (material from the canopy has not yet been identified). Of the 78 named oribatid mite species, 41 species (53%; Fig. 16.5) could be classified ecologically: 6 species (15%) represented generalists, 11 species (27%) arboricolous specialists, and 24 species (58%) terricolous specialists. Four of the generalist species occurred in both forest types. Of the terricolous specialists, 12 species were collected only in the igapó, 3 species solely in the várzea and 9 species occurred in both forest types. Of the arboricolous specialists, 9 species were found only in the igapó, 1 species solely in the várzea and 1 species occurred in both forest types. All 24 species of the terricolous specialists and 4 of 6 generalist species (terricolous and arboricolous animals) passed the aquatic phase underwater in the litter and soil:

- as eggs (*Gehypochthonius* sp. A and *Rhysotritia comtea*)
- as eggs and as immatures (*Afronothrus* sp. A)
- as immatures and adults (*Lohmannia* sp. A and *Trimalaconothrus* sp.)
- as eggs, immatures and adults (*Galumna* sp. A, *Eohypochthonius* sp. A *Epilohmannia* sp. A and *Rostrozetes foveolatus*)
- in the adult stage (and probably in other stages as well)

All species cited in the first four categories are considered parthenogenetic (thelytoky; Norton, pers. commun.) Seven of 9 species (78%) with submergence resistant eggs or immatures represented the lower Oribatida, and 17 of 25 species (68%) with submergence resistant adults the higher Oribatida

(Fig. 16.5). Only two species of the periferic higher Oribatida (the generalists *Pergalumna* sp. C from the igapó and *Pergalumna* sp. D from both forest types) showed a vertical migration in response to the flood pulse; adults moved from the soil to tree trunks prior to inundation (in February/March) and returned to the forest floor at the beginning of the terrestrial phase (in September/October). During the aquatic phase they were not detected underwater, despite all the methods employed (extraction, flotation, incubation).

In contrast to many terricolous invertebrates which pass the aquatic phase in the trunk/canopy region (cf. Chap. 14), vertical migrations in response to the flood pulse are of minor importance to terricolous oribatid mite specialists. These mites belong to the "nonmigrating terricolous" invertebrate group of Central Amazonian floodplains that are resistant to submergence. Representatives may either be active or dormant underwater during the aquatic phase (cf. Adis 1992a; Chap. 14). Initial morphological and ethological studies have indicated that long-term activity of submerged oribatid mite species depends mainly on the availability of sufficient dissolved oxygen in the water and on the presence of well-developed plastron-retaining cuticular structures (e.g., a cerotegument; Figs. 16.6, 16.7; Messner et al. 1992). However, most oribatid mites pass the aquatic phase in a dormant stage due to the lack of oxygen in the water near the forest floor (Adis and Messner 1991). Experiments in the laboratory under controlled conditions in environmental chambers (12 h of light at 27 °C and 12 h of darkness at 21 °C) showed that adult *Rostrozetes foveolatus* became inactive 24 h after they were submerged in water with no dissolved oxygen. Under this condition, they presumably changed to anaerobic respiration. Sømme and Conradi-Larsen (1977) maintained the boreal and circumpolar species *Carabodes labyrinthicus* and *Calyptozetes sarekensis* for 96 days under anoxic conditions at 0 °C and observed an accumulation of lactate. In *R. foveolatus*, adults returned to activity only 12–24 h after the oxygen level had been raised to 68–74%, through exchange or aeration of the water, and continued to feed on algae (presence of fecal pellets; Messner et al. 1992). Maximum submergence resistance, observed in adult oribatid mites under conditions of 61% dissolved oxygen at 20.2 °C, were 684 days in *R. foveolatus* and 671 days in *Rhynchoribates* sp. A, the latter species being more common on tree trunks than on the forest floor (cf. Fig. 16.7; Guimarães et al. 1993). Maximum submergence resistance reported for circumpolar oribatid mites was considerably lower: 226 days in the boreal *Steganacarus magnus* (Schuster 1978) and 210 days in *Hermannia subglabra* (Weigmann 1973). The facultative parthenogenetic species *Rostrozetes foveolatus* (cf. Beck 1969, 1971) is able to complete its life cycle underwater; larvae hatching from eggs of submerged females in the labora-

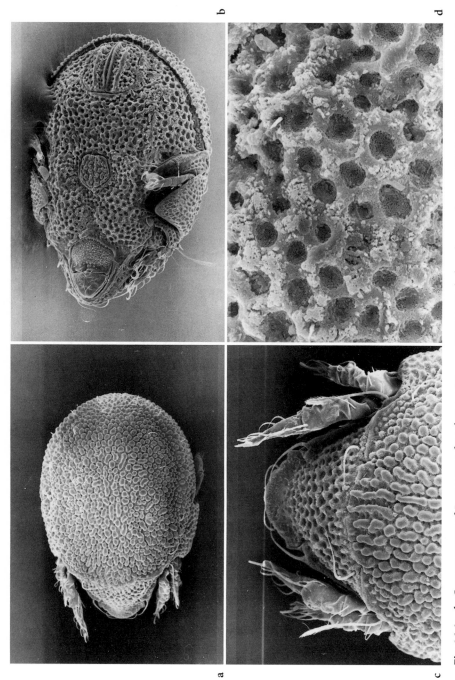

Fig. 16.6a–d. Cerotegument of *Rostrozetes foveolatus*. **a** Dorsum SEM × 123; **b** dorsal propodosoma (SEM × 245); **c** ventrum (SEM × 152); **d** ventral opisthosoma (SEM × 980)

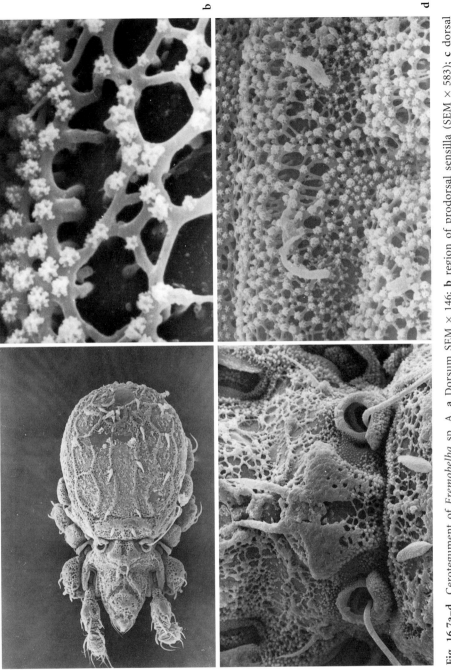

Fig. 16.7a–d. Cerotegument of *Eremobelba* sp. A. **a** Dorsum SEM × 146; **b** region of prodorsal sensilla (SEM × 583); **c** dorsal notogaster (SEM × 4485); **d** epimeral region (SEM × 1332)

tory reached the adult stage (Guimarães and Franklin, unpubl.).
Eremobelba sp. B (probably *E. foliata* in Beck 1972), which was the most
dominant terricolous species in igapó soils below 3.5 cm depth and whose
adults resisted submersion for almost 10 months in the laboratory
(Guimarães and Franklin, unpubl.), presumably has the same ecological
valence.

16.3 Faunal Differences

The main difference observed between the oribatid mite guilds in the two
floodplain sites under study was the lower number of arboricolous individuals and of species representing the basic higher Oribatida in the várzea
forest. This may be related to different abiotic and biotic conditions. For
example, the igapó forest is connected with the neighboring nonflooded
upland forest, and faunal interchange is facilitated when compared to the
várzea forest, which is situated on an island. Also, the climatic and edaphic
conditions of the two forest types differ considerably: the várzea island
forest has a more open structure (lower tree density) and less litter on the
ground than is found in the igapó forest; thus the oribatid mite fauna is
more exposed to climatic oscillations (e.g., in temperature, humidity) than
the igapó forest fauna is. Different biotic conditions, for example, a
smoother bark structure and a lower supply of food resources (lichens,
mosses and epiphytes) on tree trunks in the várzea forest are considered
important for the species composition of the oribatid mite guild.

Beck (1971, 1972, 1976) assumed that the ecological requirements of
oribatid mite species in Central Amazonian floodplain forests do not differ
substantially from those species inhabiting nonflooded upland forests. A
certain similarity in the faunal compositions of igapó and upland forests
and a difference as compared with várzea forests was also observed in
symphylans (Adis et al. 1996b), but not in other groups such as pseudo-
scorpions and millipedes (Mahnert and Adis 1985; Adis and Mahnert 1990;
Golovatch 1992, 1994; Adis and Golovatch, unpubl.): in other words, requirements depend on the taxonomic group under study. Our data on
oribatid mites show that the number of generalist species in floodplain
forests is low (cf. Fig. 16.5) when compared to Amazonian upland forests,
where generalists accounted for 40–50% of the oribatid mite fauna (Beck
1971; Wunderle 1992a). The predominance of terricolous and arboricolous
specialists may be attributed to the generally more unstable living conditions in the ecosystem caused by annual flooding.

Acknowledgments. Sincere thanks are expressed to Prof. Dr. L. Beck (Karlsruhe), Prof. Dr. G. Alberti (Heidelberg), Prof. Dr. G. Weigmann (Berlin) and to Dr. H.O.R. Schubart (Brasilia/Manaus) for constant help and valuable comments. Thanks to Prof. Dr. G.W. Krantz (Corvallis, Orgeon), Prof. Dr. R.A. Norton (Syracuse, New York) and Dr. B. Robertson (Manaus) for their comments on earlier drafts. B.Sc. T. F. Hayek, B.Sc. R.L. Guimarães, M.Sc. M. Amorim and E. de A. Silva (Manaus) are thanked for their dedicated help in the field and laboratory. The German Academic Exchange Service (DAAD) is thanked for financial support.

17 The Collembola

Ulrich Gauer

17.1 Introduction

Collembola are soil-inhabiting insects with a worldwide distribution. They are always the insect group with the greatest number of individuals. Their ecological significance results from their involvement in the decomposition of organic material and nutrient cycling processes. In spite of their key role in litter decomposition the group has not been well studied in the Neotropics. Winter (1962) published some ecological results from the Pre-Andean catchment area. In the Central Amazon lowlands, Beck (1976) reported on Collembola in inundation forests and Schaller (1969) made some observations of behavior. Sampling sites are described in Chapter 2.

17.2 Species Inventory

Most of the Collembola collected in the inundation forests of the white waters (várzea: Ilha de Marchantaria) and the black waters (igapó: Tarumã Mirim) were provisionally assigned to morphospecies. (Morphospecies are "types" which can be separated according to morphological criteria, but for which a complete taxonomic analysis has not yet been completed). At present (September 1994), we can distinguish 65 different "species" from the Amazon floodplain near Manaus: 22 are already described and 43 are morphospecies. In the várzea of Ilha de Marchantaria they belong to 26 genera representing 9 families and in igapó at Tarumã Mirím to 39 genera representing 12 families. Thirteen genera were found solely in the igapó and one genus only in the várzea.

For some genera the taxonomic analysis is already in progress. From the floodplains four new species of the genus *Isotomiella* and eleven new species of the genus *Sphaeridia* have been described (Deharveng and Oliveira 1990; Oliveira and Deharveng 1990; Bretfeld and Gauer 1994). It is evident

Ecological Studies, Vol. 126
Junk (ed) The Central Amazon Floodplain
© Springer-Verlag Berlin Heidelberg 1997

that the taxonomic analysis of Collembola in these areas, together with the even less known fauna of the terra firme, will turn up an inestimatible number of new species.

17.3 Phenology and Abundance

The development of the Collembola fauna in the igapó was investigated during three sampling periods during the terrestrial phase 1989/1990 (Fig. 17.1). The first sampling date (7 November 1989) was approximately 1 month after the floodwaters had receded from the forest soils; the second date (30 January 1990) was after the beginning of the rainy season; and the third (1 March 1990) was 1 week before the new flooding of the area.

The abundance of the Collembola decreased during this time; however, at genus level there were very different, even contrary, developments. The

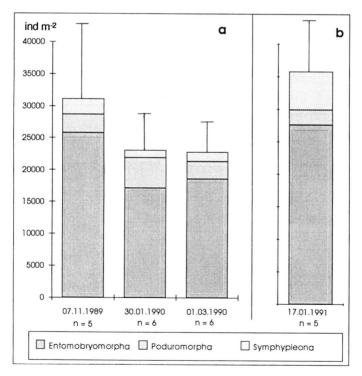

Fig. 17.1a,b. Mean abundance of Collembola in the lower igapó **a** at Tarumã Mirím and **b** in the várzea forest at Ilha de Marchantaria; samples were taken to a depth of 14 cm

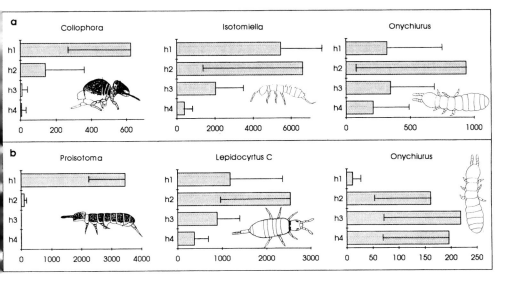

Fig. 17.2a,b. Vertical distribution of representantives of epedaphic, hemiedaphic and euedaphic Collembola in individuals m^{-2} **a** in the igapó of Tarumã Mirím (n = 17) and **b** in the várzea on Ilha de Marchantaria (n = 5). *h1*, 0–3.5 cm; *h2*, 3.5–7 cm; *h3*, 7–10.5 cm; *h4*, 10.5–14 cm

dominant genus is always *Isotomiella* with 15 551 ind. m^{-2} (n = 17; s = 6900), representing 57% of total Collembola. For all genera four patterns of population development can be distinguished:

1. Decrease (*Isotomiella, Collophora*).
2. Increase (*Cyphoderus*).
3. Increase and subsequent decrease (*Onychiurus, Mesaphorura*).
4. Decrease and subsequent increase (*Dicranocentrus, Trogolaphysa*).

These patterns are probably associated with strategies for recolonization after water levels drop (Sect. 17.4). The animals of the first group quickly, and these of third group slowly, build up their populations from dormant eggs; the second group is represented by slowly, and the fourth by quickly immigrating species.

The populations colonize different strata. The mean depth (the "center of gravity" of the populations, Usher 1970) of the Collembola in the igapó is 4.5 cm, which is also the mean depth of the dominant genus *Isotomiella*. Sixty-five percent of the genera and morphospecies prefer higher strata: 35% prefer deeper soil layers, up to 6.5 cm (*Onychiurus*) and 6.9 cm (*Neotropiella*). Figure 17.2a shows the distribution of three genera, representing one genus each of epedaphic (*Collophora*), hemiedaphic (*Isotomiella*), and euedaphic (*Onychiurus*) genera.

Displacements in vertical abundance during the dry phase occur; stratification differs between a stable situation for most groups, especially for epedaphic genera (as *Lepidocyrtus* A), and an increase in the number of animals in upper (*Isotomiella, Lepidosira, Willemgastrura, Cyphoderus*) or lower horizons (*Mesaphorura*).

The distribution of abundances throughout the flooded areas of the igapó was investigated along a 240-m transect on 15 December 1989. The boundary between lower and upper igapó is at approximately 150 m. The Collembola are divided according to their abundances into the following groups:

1. More abundant in lower igapó (*Collophora*).
2. More abundant in upper igapó (*Cyphoderus, Brachystomella*).
3. Equally abundant (*Isotomiella*).
4. Absent in the middle of the igapó (*Onychiurus, Lepidocyrtus* A).
5. More abundant in the middle of the igapó (*Pararrhopalites*).

Isotomiella was the dominant genus in all samples.

Samples from Ilha de Marchantaria (17 January 1991) registered a higher abundance than those from the igapó (Fig. 17.1b). The distribution of dominance is more balanced in the várzea than in the igapó: the dominant genus is *Trogolaphysa* with 8089 ind. m^{-2} (n = 5, s = 4458) representing 23% of the Collembola and *Isotomiella* with 7644 ind. m^{-2} (n = 5; s = 2794), 21% respectively. The mean depth of all Collembola at 4.4 cm is nearly the same as in igapó; 62% of genera and morphospecies prefer higher strata: *Isotomiella* lives at a mean depth of 5.2 cm in deeper horizons than in the igapó; however, the dominant genus *Trogolaphysa* prefers 4 cm. The deepest stratum is occupied by *Onychiurus* at 8.8 cm. Figure 17.2b shows the stratification of three typical representatives for epedaphic (*Proisotoma*), hemiedaphic (*Lepidocyrtus* C) and euedaphic (*Onychiurus*) animals on the Ilha de Marchantaria. The differences between igapó and várzea needs careful evaluation as the soil samples were not taken at the same time and seasonal effects cannot be excluded.

17.4 Adaptation to the Flood Pulse: Recolonization After Receding Floodwaters in the Igapó

The soil in the igapó is subject to periodic flooding that can reach, in the lower igapó, a height of 4–5 m and can last up to 7 months. Even at the edge of the receding waters, Collembola can be found in large numbers. In an area that had been dry for 24h, abundances can reach 34 400 animals m^{-2}. The coenosis is dominated by 2 genera: *Proisotoma*, with 16 510 individuals m^{-2} (n = 8, s = 17 143) and *Brachystomella* with 7715 ind. m^{-2} (n = 8, s = 7952). Two weeks later, both genera were present with only 451 individuals m^{-2} (n = 8, s = 208) and 671 individuals m^{-2} (n = 8, s = 208). Then, *Isotomiella* dominates the coenosis with 13 139 individuals m^{-2} (n = 8, s = 6570). This exchange of dominant genera over a period of only 2 weeks indicates a drastic change in the environmental factors in the soil.

Two strategies could be proven, which guarantee a rapid recolonization:

1. Mass migrations of animals, which follow the retreating waterline (Fig. 17.3a–d). Within 10 m of the waterfront, individuals of the genera *Brachystomella* and *Proisotoma* were especially dominant (Fig. 17.3a). At larger distances *Dicranocentrus* (70 m, Fig. 17.3) and *Lepidocyrtus* A (100 m, Fig. 17.3c) followed at lower densities, but in their own migration waves. At a distance of 130 m no distinctive migration was discernible (Fig. 17.3d). *Brachystomella* and *Proisotoma* individuals were mostly adult animals, whereas *Lepidocyrtus* A and *Dicranocentrus* were predominantly juveniles.

2. The animals hatch immediately after the floodwater has receded from eggs that had been laid in the soil before the flood. This effect was investigated using extractions of litter from flood boxes. In these extractions only animals appeared, which hatched from eggs in the flooded litter. Table 17.1 shows the numbers of individuals of the extracted genera from two boxes.

Table 17.1. Genera and number of individuals extracted from two flood boxes (36 × 20 × 8 cm) filled with litter (ca. 200 g dry weight) exposed to 179 days flooding in the lower igapó. The litter was subsequently incubated (14 days) and extraction was performed using heat (Kempson method)

Genera	Number
Isotomiella	1207
Collophora	493
Proisotoma	150
Onychiurus	126
Calvatomina	87
Mesaphorura	40
Hylaeanura	23
Megalothorax	10
Brachystomella	7
Sphyrotheca	1

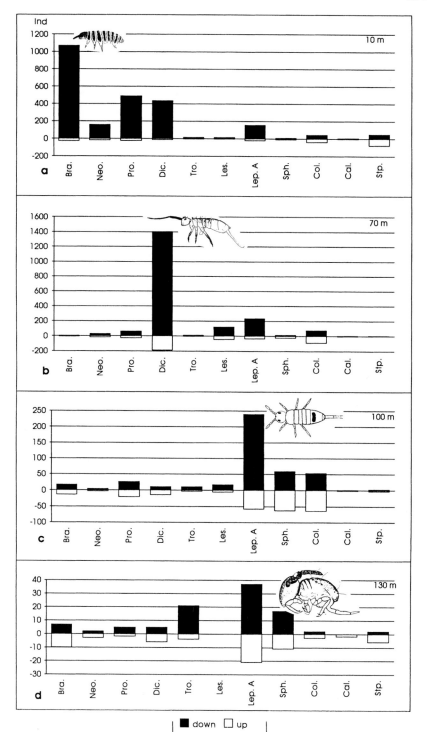

Isotomiella is the dominant genus (and has also been found dominant in the soil samples 2 weeks after the flood and during the terrestrial phase). A fast and a slow group can be differentiated: the fast group builds up a large population of adults within a few days (*Isotomiella, Collophora*), in the slower group adults appear a few months after the floodwater has receded (*Onychiurus, Mesaphorura*).

The immigrating *Proisotoma* also occurs in the flood boxes. Thus, it seems likely that these migrating Collembola also hatch from eggs and then occur at the receding water edge in numbers which grow exponentially.

17.5 Discussion and Conclusions

Soil animals can be classified as migrants and nonmigrants, and subclassified depending on the form of migration (horizontal, vertical, flying) or on the type of flood resistance (dormant or active under water; Chap. 14). The Collembola coenosis of the várzea and igapó shows a dynamic that has not been observed in these organisms before, whereby the dominant genera can also be divided into categories of "migrant" and "nonmigrant". It can be assumed that at the species level other special reactions of the population occur.

When water rises, Entomobryomorpha and Poduromorpha show small flood avoiding movements within 5 m from the waters edge. Mass concentrations of Collembola were not observed, so it can be assumed, that they become victims of the flood (Beck 1976).

An early presence on the drying soil surface is not only of importance for the competitiveness of species within interspecific competition, but also for building up populations within the Collembola coenosis and the entire soil community. Two factors provide convenient starting conditions. The soil is, in the beginning, still very damp and provides an ideal environment for physiological and reproductive processes of hygrophilous animals. How-

◄ ──

Fig. 17.3a,d. Individual densities of a few genera caught between 10 and 11 September 1990 in the igapó in pitfall traps opening in one direction only. Width of opening 10 cm, running time 24 h, liquid for catch aqueous picrid acid. In each position, two traps with openings to the water's edge captured animals migrating towards terra firme (*negative numbers*), and two traps with openings facing in the opposite direction captured animals migrating towards the water (*positive numbers*). Distances are given in meters to the water. *Bra., Brachystomella; Cal., Calvatomina; Col., Collophora; Dic., Dicranocentrus; Lep.A, Lepidocyrtus A; Les., Lepidosira; Neo., Neotropiella; Pro., Proisotoma; Sph., Shaeridia; Stp., Stenognathriopes; Tro., Trogolaphysa*

ever, this condition deteriorates after only a few weeks. The nutrient supply for the saprophagous Collembola is enriched by sedimented organic particles and algae from the retreating floodwater (Walker 1992).

Two strategies for quick recolonization of drying areas in the lower igapó can be differentiated: (1) quick development of dormant eggs and (2) mass immigration of Collembola, following the water's edge in high individual densities and covering large distances in a short time. Those species which follow these different strategies also show differences in population development during the terrestrial phase: the migrating species occur after a few weeks in very low abundances or not at all, whereas the animals hatching from eggs remain at the top of the dominance scale. Presumably, the migrating species follow the optimal environmental and nutritional conditions, which only appear for short periods in the drying soil. After a few days they become victims of increasing dryness and lack of food. For one migrating genus (*Proisotoma*), it was proven that they also develop from flooded eggs. How the individual densities of the other highly abundant migrating genera are built up cannot be conclusively deduced from the present examinations. Their eggs are possibly intolerant of long-term flooding because they were absent from the litter exposed in the flood boxes for 179 days. Possibly their eggs tolerate only a short inundation and they originate from the shorter-flooded upper igapó.

Uptill now it could not be proven whether adult or subadult Collembola survive inundation in the soil and thus provide an essential contribution to recolonization. Recolonization from trees does not seem to be of significance. All Collembola collected with arboreal eclectors belong to morphospecies that were very rare, or were not collected, on the ground (*Xenylla*, *Lepidocyrtus* F, *Seira*, *Sturmius*). The observations made by Schaller (1969) that the behavior of *Lepidosira* shows orientation toward trees, seem to be the result of fundamental arboriophily by this atmobiont genus; this can be regarded as a preadaptation for life in inundation forests.

The life cycle of the rapidly developing Collembola (*Isotomiella*, *Collophora*) is very short: after 2–3 days they have reproductive adults. For *Isotomiella* parthenogenesis can be assumed, because until now no males have been found in *Isotomiella prussianae* or *I. similis* from várzea forests (Oliveira and Deharveng 1990). This can be considered as an r-strategy.

The differences between the Collembola coenoses structure of the igapó and of the flooded forests of the várzea are considerable and cannot only be traced back to seasonal effects. Fundamental differences between the two areas exist in the flooding of the soils by flowing sediment-rich water on the Ilha de Marchantaria and the lack of any connection to the terra firme. Here, recolonization can only take place from trees or through flood-resistant eggs. This, however, can have negative effects on species diversity, since those that rely on flood avoidance through horizontal movement cannot survive. Also, species which immigrate during the terrestrial phase from the unflooded upland area cannot be found here.

Large amounts of inorganic suspensoids on the Ilha de Marchantaria cover the organic material and close the soil pores. The sample from Ilha de Marchantaria shows a greater number of Collembola than the sample from the igapó (Adis and Schubart 1984) and a deep penetration into the soil by some genera. Because of the low number of data it has to remain unresolved whether this was an exception. Considering the quick reproduction of Collembola, the higher number in the várzea could be the result of the better quality of food in comparison with the igapó.

Because of a lack of comparable data from the terra firme it cannot be said to what extent species composition in the floodplain differs from that of the terra firme, and to what extent species with special adaptations evolved in response to the flood pulse.

Acknowledgments. I am very grateful to Dr. Kenneth Christiansen, Grinnell, Iowa, USA, and Dr. Gerhard Bretfeld, Kiel, Germany, for critical comments on the manuscript and the correction of the English.

18 The Termites

CHRISTOPHER MARTIUS

18.1 Introduction

Termites (Insecta: Isoptera) are among the most abundant arthropods in the humid tropics, together with ants, springtails and mites. In tropical ecosystems they are an important element of the detritivorous food chain and process all kinds of litter: wood, leaves, and soil organic matter (Lee and Wood 1971; Wood 1976; Brian 1978; Lal 1987). In this chapter we discuss their diversity and biology in várzea floodplains. Abundance, distribution, and population dynamics are analyzed with respect to the flood pulse. A final section is dedicated to termite-nests as structural ecosystem elements. The quantitative importance of Isoptera in the process of wood decay is studied in Section 12.5.

18.2 Diversity and Biology of Floodplain-Inhabiting Termites

About 150 termite species have been described from Amazonian forests (Constantino and Cancello 1992), roughly 5% of the number of termite species known worldwide (Mill 1991); however, a great number of species from Amazonia still remain undescribed (Bandeira and Harada 1991).

In Amazonia, there are approximately 70–90 species of 3 families (Kalotermitidae, Rhinotermitidae, Termitidae) that are generally found in 1 ha of nonflooded terra firme rainforest (Martius 1994b). In contrast, only 12 termite species belonging to 2 families (Rhinotermitidae and Termitidae) were recorded on two sites of white-water influenced floodplain forest (várzea) (Table 18.1). Mill (1982) found 11 termite species in black-water igapó forest. No dry-wood termites (Kalotermitidae) were recorded in the studies cited. However, more recently, 31–61% of the alates (winged reproductives) caught in light traps over a 2-year period on

Ecological Studies, Vol. 126
Junk (ed) The Central Amazon Floodplain
© Springer-Verlag Berlin Heidelberg 1997

Table 18.1. Termites (Isoptera) recorded in várzea forests in central Amazonia (Marchantaria Island; 1.465 ha; Martius 1989) and eastern Amazonia (Marãa; 0.25 ha; Constantino 1992)

Locality Family/subfamily	Species	Nutrition	Nest type
Marchantaria Island (Amazon River)			
RHINOTERMITIDAE			
Coptotermitinae	*Coptotermes* sp. 1	Rotting wood	In dead wood
Rhinotermitinae	*Rhinotermes marginalis*	Rotting wood	In dead wood
	Rhinotermes sp. 1	Rotting wood	In dead wood
TERMITIDAE			
Termitinae	*Termes medioculatus*	SOM/dead wood	Arboreal nest
Apicotermitinae	*Anoplotermes* sp. *A*[a]	SOM	Mixed endogeic/ arboreal nest
	Anoplotermes sp. *C*[a]	SOM	Arboreal nest
	Anoplotermes sp. *D*[a]	SOM	Inquilines
Nasutitermitinae	*Nasutitermes macrocephalus*	Dead wood	Arboreal nest
	Nasutitermes ephratae	Dead wood	Arboreal nest
	Nasutitermes corniger	Dead wood	Arboreal nest
	Nasutitermes surinamensis	Dead wood	Arboreal nest
	Nasutitermes tatarendae	Dead wood	Arboreal nest
Marãa (Juruá River)			
RHINOTERMITIDAE			
Rhinotermitinae	*Rhinotermes marginalis*	Wood	Wood
TERMITIDAE			
Termitinae	*Cavitermes tuberosus*	Humus	Inquiline
	Termes hispaniolae	Wood	Arboreal nest
	Termes medioculatus	Wood	Arboreal nest/inquiline
Apicotermitinae	*Anoplotermes* sp. *C*[b]	Wood	Arboreal nest
	Anoplotermes sp. *G*[b]	Humus	Arboreal nest
Nasutitermitinae	*Coatitermes clevelandi*	Humus	Inquiline
	Ereymatermes rotundiceps	Humus	Inquiline
	Nasutitermes corniger	Wood	Arboreal nest
	Nasutitermes sp. *C*[b]	Wood	Arboreal nest
	Nasutitermes sp. *D*[b]	Wood	Wood
	Rotunditermes bragantinus	Rotting wood, roots?	Epigeal/arboral nest

Families and subfamilies according to Grassé (1986).
SOM, Soil organic matter.
[a,b] Species codes of the two assessments are not identical.

Marchantaria island belonged to this family, evidence that colonies of these termites must exist in the dead wood of the várzea forest canopy (Rebello and Martius 1994).

At present, it is difficult to assess the similarities of the termite species compositions of várzea sites. Of those species found at the two sites studied, seven and eight species, respectively, are known to science; of these, only three species are common to both sites, which are about 500 km apart (Table 18.1). This indicates that intersite (β) diversity of termites in Amazonian floodplain forests might be as high as in terra firme forests (Constantino 1992). Other studies in the várzea will probably add more species to this list.

The marked reduction of the termite diversity from terra firme to floodplains can be attributed to the flood pulse (Junk et al. 1989). The termite community of the várzea consists almost entirely of tree-dwelling species (Fig. 18.1). These include two groups: species which construct arboreal nests and those which use nests built by other species (inquilines). The periodical flooding does not permit the survival of soil-nesting termite species. Leaf litter-harvesting species, which are abundant in the terra firme (Martius 1994b), are soil dwellers, and do not occur in the várzea forest. The floodplain communities consist of wood and soil-feeding species only.

The termite diversity in the floodplain must also be affected by other factors, since far more tree-dwelling termite species exist in the terra firme forest canopy than in the várzea. These factors may include increased predation pressure and the flight range of alates. The much higher density of ants in the várzea indicates that predation by ants upon alates is much higher here than in the terra firme, which could account for the nonexist-

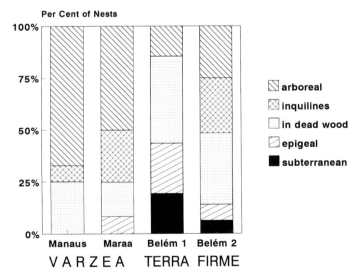

Fig. 18.1. Distribution of termite nest types in várzea and terra firme sites. *Inquilines* Termites which inhabit nests built by other species. Várzea (*left*, Marchantaria Island, Martius 1989; *right*, Marãa, Constantino 1992) and terra firme (*left*, Belém, Constantino 1992; *right*, Belém (different site), Bandeira and Macambira 1988)

ence of some highly susceptible species. Ants account for 43% of the várzea soil fauna biomass (Adis and Ribeiro 1989), but only 24% of the biomass in terra firme (Bandeira and Torres 1985). They appear to be an important mortality factor for young *Nasutitermes* nests in the várzea (Martius 1989). Comparative assessments of termite communities in flooded (11 spp.) and nonflooded forests (19 spp.) on an island in the Negro River by Mill (1982), and the light-trap catches of Rebello and Martius (1994), suggest that the distance of the river islands from the terra firme might prevent colonization by alates of many termite species which have a limited flight range.

The várzea termite communities consist of species found only in the várzea as well as some generalist terra firme species. The species unique to the várzea show a range of traits which appear to be adaptations to the life in the floodplain forest. For example, the wood-feeding *Nasutitermes macrocephalus* produces massive conglomerates of digested wood piled in its nests, which are apparently stocks of food and, as such, could help these termites to survive the flood period (Martius 1992a).

Another example is the life cycle of the still undescribed soldierless termite *Anoplotermes* sp. A, which is synchronized with the flood pulse (Fig. 18.2). The colony starts with a royal couple breeding in a soil chamber, near the base of a tree. At the beginning of the wet season – perhaps induced by increased soil moisture (Adis 1992) – the workers construct an epigeal nest of soil material attached to the side of the trunk. The top of the nest can reach a height of 6 m above the soil surface ensuring that it will emerge from the coming floodwater level. The rising water table forces the whole termite colony (including the royal couple) to migrate into the emergent part of the nest. Apparently there is enough organic matter stored in the very thick nest walls (2 to 4 cm; Martius 1994a) to ensure sufficient food supply. The queen remains active during the flood, as shown by the many egg piles found in the top of the nest, and the formation of alates continues, an energy-consuming process (Nielsen and Josens 1978). The alates swarm at the beginning of the terrestrial phase. During the flood, termites are absent from the drowned nest parts. Termites are not able to survive more than one hour underwater. The submerged nest parts are invaded and destroyed by Oligochaeta. In the terrestrial phase most nests together with their inhabitants fall to the ground. The termites generally reestablish the colony from the nest fragments and restart the cycle (this has consequences for the turnover rate of nest matter; Sect. 18.3). Apparently, this type of colony foundation ("budding"; Grassé 1984) is far more successful than the foundation of new colonies by alates (Martius, unpubl.), although it does not allow the colonization of distant places and leads to a high continuity of colony sites.

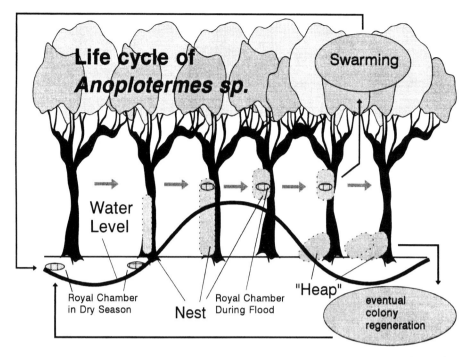

Fig. 18.2. The life cycle of the termite species *Anoplotermes* sp. *A* is synchronized with the flood pulse. The royal chamber, which is found in the soil during the terrestrial phase, is transferred to an arboreal nest constructed by the termites at the beginning of the inundation. The termites survive in the top of the nest. The flood destroys the submerged nest parts; the colony may recover, or a new colony can be started from winged reproductives, which disperse and reproduce at the beginning of the terrestrial phase

18.3 Abundance, Distribution, and Exchange Rate

The termite fauna of the várzea forest is dominated, in terms of nest density, by two groups: wood-feeding Nasutitermitinae and geophagous Apicotermitinae (soldierless termites). *Nasutitermes* populations have densities of 37–68 nests ha^{-1} (Table 18.2). The most important species is *Nasutitermes corniger* (about 32–47 nests ha^{-1}).

A large number of nests of *N. corniger* and *N. tatarendae* belong to polycalic (multinest) colonies (Grassé 1984). Generally, two to four nests belong to one colony, but in some cases as many as ten different nests can be linked together. Thus, the number of individual colonies is much lower than the number of nests in these species. Isolation of groups of workers foraging distantly from the nest due to rising water may enhance the

Table 18.2. Termite nest densities in várzea forests on Marchantaria Island

Study plot	Topographic hight (m above sea level)	Size of studied site (m²)	Species	Density (nests ha⁻¹)	Total nest volume (m³ha⁻¹)
Lago Cobra	23–26	14650	Nasutitermes corniger	47	2.00
			N. tatarendae	11	0.60
			N. macrocephalus	4	1.44
			N. surinamensis	3	1.38
			N. ephratae	3	0.41
			Nasutitermes 5 spp.	**68**	**5.83**
Lago Cobra		1050	Anoplotermes sp. A	219	16.43
Lago Cobra, 2. site		1450	Anoplotermes sp. A	62	4.65
Lago Cobra		1050	Anoplotermes sp. C	362	–
C (before flood)		6000	Anoplotermes sp. A	30–50	2.25–3.75
C (after flood)		6000	Anoplotermes sp. A	5	3.38
Lago Cobra	23–25	6825	Nasutitermes 5 spp.	43	
	25–26	4000	Nasutitermes 5 spp.	113	
	26–27	3825	Nasutitermes 5 spp.	68	
Lago Central	23–25	6000	Nasutitermes corniger	32	0.78
			N.mac., sur., tat.	4	2.05
			Nasutitermes 4 spp.	**37**	**2.83**
			Anoplotermes sp. A	3	0.25
			Anoplotermes sp. C	49	–

formation of new nests. Prolonged isolation may also allow distant nests to become independent colonies by the formation of secondary reproductives (sociotomy). Multiple colonies of up to 37 connected nests were also found in mangrove forest in Panama (Levings and Adams 1984; Adams and Levings 1987).

The distribution of termites in the várzea is restricted to the forest (<200 days of flood year⁻¹, on average; Junk 1989). The distribution of the termite species in the forest is influenced by differences in topography which determine the duration of the flooding. *N. macrocephalus* has the most extended range, occurring even on isolated *Vitex cymosa* trees outside the closed forest in areas which are flooded for 200–240 days year⁻¹.

The distribution of soil-feeding Apicotermitinae is limited to areas flooded for less than 170 days year⁻¹. Within this restricted range, the nest-building species *Anoplotermes* sp. A and C can reach densities of up to 219 and 362 nests ha⁻¹, respectively (Table 18.2). Their density, however, varies considerably, and is far lower on most parts of Marchantaria island. This distribution pattern could be influenced by soil factors, e.g., differences in particle-size distribution (Su et al. 1991) or in soil organic matter (J.A. Holt, pers. comm.) between ridges and depressions. In addition to site-dependent variations, nest densities of *Anoplotermes* sp. *A* vary considerably from one year to another (Table 18.2, site C) due to differences in flood levels (Sects 2.3 and 12.2). In 1989, the community at one site was almost extinguished and has since recovered only slowly. This makes it doubtful whether these termites are really adapted to the floodplain environment.

The distribution of wood-feeding *Nasutitermes* spp. appears to be positively correlated with the availability of dead wood. Termite nest density seems to be related to small-scale variations in the abundance of dead wood, which sometimes appears in "bands" deposited by the flood at its highest margin (Martius 1987). This is reflected in the relatively high nest density of 113 nests ha^{-1} in the intermediate zone (25–26 ma.s.l.; Table 18.2) where much dead wood is deposited by the flood (Sect. 12.2; Table 12.1). Available dead wood is more abundant in some early successional stages of the forest (Sect. 12.2). Termite nests are more numerous in younger stands (e.g., 68 vs. 37 *Nasutitermes* nests ha^{-1} on Lago Cobra, a 40-year-old stand, and Lago Central, an 80-year-old stand, respectively; Table 18.2). According to Mill (1982), the abundance of termites in floodplains is comparable to that in terra firme forest. Nest density in older várzea forests is similar to that of the terra firme (Lago Central: 89 nests ha^{-1}; Table 18.2); however, numbers of termite nests in younger várzea stands exceed by 2–3 times the 100–120 nests ha^{-1} generally found in terra firme forests (Table 18.2; Martius 1994b). Nest densities give only an approximate idea of the size of the termite population because nests are of different sizes and the size of the nests is not correlated with the biomass of the termites which inhabit them (Martius et al. 1993). The population size of one species, *Nasutitermes macrocephalus*, was estimated as 544 individuals m^{-2} from the number of nests ha^{-1} and from the individual number in subsamples from several nests (Martius 1989). This corresponds to 8.5% of the 6400 individuals m^{-2} of all macroarthropods in soil samples (Adis 1987, heat extraction). This figure is also high compared to the density of termites in other tropical forests (Table 18.3). *N. macrocephalus* had a biomass of 0.6 g m^{-2} DW whereas the termite biomass in soil cores (0.04 g m^{-2} DW; similar to that of Acari and Collembola) corresponds to only 0.4% of the soil arthropod mass (10.5 g m^{-2} DW; Adis and Ribeiro 1989). This indicates that counts from soil extraction alone tend to underestimate the relative abundance and biomass of social insects within the arthropod communities.

A *Nasutitermes* population of an initial 71 nests was monitored for 1 year (Martius 1989). The total nest number remained stable in this period, but the exchange rate was high: About 25% of the nests died annually and

Table 18.3. Density and biomass of termites in tropical forests of the world. Above, assessments of total termite communities. Below, assessments of single species, mostly from Amazonia

Forest type	Locality	Termite species	Density (ind. m^{-2})	Biomass (g DW m^{-2})	Source
Rainforest	Amazonia	All	1865	2.0–2.5[a]	Bandeira (1989) (corrected)
Rainforest	West Indies	All	4450	4.5[b]	Strickland 1944
Lowland Dipterocarp	Malaysia	All (n = 57)	3200–3800	2.2–2.6	Abe and Matsumoto (1979); Collins (1988)
Lowland Dipterocarp	Sarawak	All (n = 59)	1527	0.8[a]	Collins (1988)
Swampy alluvial	Sarawak	All (n = 31)	390	0.3[a]	Collins (1988)
Riverine	Central Africa	All	1000	3.7[a]	Maldague (1964)
Rainforest	Costa Rica	Nasutitermes costalis	87–104	0.1	Wiegert (1970)
Rainforest	Malaysia	4 spp.	1330	1.1[a]	Matsumoto (1976)
Rainforest	Amazonia	Labiotermes labralis	112–157	0.1–0.2	Ribeiro (unpubl.)
Rainforest	Amazonia	Anoplotermes banksi	22–33	<0.1	Ribeiro et al. (unpubl.)
Rainforest	Amazonia	Syntermes sp. (n = 2)	48	0.5	Martius (unpubl.)
Várzea	Amazonia	Nasutitermes macrocephalus	544	0.6	Martius (1989)

Italic letters, assessments from wetlands.
The numbers of Strickland (1994) for the West Indies have recently been questioned by J. Torres (pers. comm. 1994); they are probably too high.
[a] Recalculated from fresh weight (FW) = 3 dry weight (DW).
[b] From average biomass (1 mg/individual).

were replaced by newly founded colonies, a high rate in comparison to other sites (Table 18.4). Mortality factors included ant raids, starvation in nests on very isolated trees with limited food supply during the flood, and flooding, which drowned some nests. In general, however, *Nasutitermes* nests are found above the average high water level, and animals in low-lying nests are able to emigrate and build new nests.

Table 18.4. Exchange rate of termite populations in different ecosystems

Forest type	Locality	Species	Exchange rate (% year^{-1})	Source
Várzea forest	Marchantaria Island	*Nasutitermes* 5 spp.	17–30	Martius (1989)
Rainforest	Barro Colorado Island (Panama)	All arboreal nests	16	Lubin and Young (1977)
Rainforest	San Carlos del Rio Negro (Venezuela)	All nests[a]	11	Salick et al. (1983)
Savanna	Africa	*Trinervitermes geminatus*	11	Baroni-Urbani et al. (1978)
Not indicated	North Australia	*Tumulitermes* 2 spp.	50	Williams (1968) (cited from Lee and Wood 1971)

[a] "Nests", every record of termites within 50 m^2 (Salick et al. 1983).

18.4 Termite Nests as Structural Elements

In várzea forest, the existence of spatial niches such as abandoned termite nests may contribute to the survival of other arthropods which migrate into the tree canopy during the flood. Abandoned termite nests are frequently recolonized by the same or different termite species. Of all *Nasutitermes* nests found in the forest, 51% were partly decaying or totally abandoned (Martius 1994a). These offer a habitat for secondary colonizers of termite nests. Soon after flooding a wide variety of arthropods from the soil, particularly ants, can be found in abandoned nests. The surface areas of chambers in all abandoned nests of *N. corniger* on trees were calculated to be 197–408 m^2 ha^{-1} of forest (0.43 ± 0.06 m^2 l^{-1} of nest volume). This is small when compared to the surface area of the lower 10 m of the tree trunks ($\approx$13 000 m^2 ha^{-1}), but the nests represent well-protected, humid and thermally stable niches for which competition is apparently high (Martius 1994a; Martius et al. 1994).

Inquilines in termite nests include *termitophilous* arthropods which are particularly adapted to life in a termite colony, often by mimicking the termites' behavior or interacting with their pheromone-based communication system, and *termitariophilous* arthropods which are nonobligatory inhabitants of termite nests (Araújo 1970). The former have never been studied in the várzea and among the latter, ants play a particularly important role. One ant species, *Dolichoderus bispinosus* (Dolichoderinae) is found in 62% of the living and 69% of the dead nests of *N. corniger*, in a lestobiotic relationship (Martius 1989, 1994a). Although the ants benefit from the nest structure and eventually prey upon the termites, the chemical defense of the nasute soldiers prevents complete extinction of the termite colony, unless it is very young. Termites are forced to withdraw from nest areas conquered by the ants, but are well protected from external disturbance by the very aggressive ants. These disturbances include predatory birds, and parrots or bees searching for nest sites (Kerr et al. 1967; Camargo 1970, 1984; Koepcke 1972).

The nests of *Anoplotermes* sp. *A* provide shelter for other arthropods including other termite species (*Coptotermes, Rhinotermes, Anoplotermes* sp. *D*) and also for small vertebrates during the flood. Submerged nest parts contain large numbers of Oligochaeta which feed on the organic-rich nest material (Sect. 18.2).

Anoplotermes sp. *C* constructs thin soil coatings which are extended over the lower 12–14 m of the tree trunks. They are found on about a quarter of the trees >3 cm diameter. When abandoned, these constructions allow some soil arthropods to extend their activity range into the canopy.

18.5 Discussion and Conclusions

There are about five times less termite species in floodplain forests than in nonflooded terra firme forest (Sect. 18.2). Obviously, the flood pulse eliminates all soil-dwelling species with the exception of those which have adjusted their life cycle to the periodic flooding (e.g., *Anoplotermes* sp. A). However, the flood pulse also acts indirectly, by greatly enhancing the nutrient status of forested floodplains in comparison to the nutrient-poor terra firme. It is probably due to this that the soil fauna in floodplains is about 20 times higher than in the terra firme (soil core samples extracted by heat; Bandeira and Torres 1985; Adis and Ribeiro 1989). The termite biomass in these soil samples is equal in both habitats ($0.04 \, \mathrm{g \, m^{-2}}$, Martius 1994b), but total termite biomass in várzea sites is probably much higher than that suggested by the data, as most of the termites are absent from soil samples but are concentrated in arboreal nests. At present, we are not able to explain why the diversity of termites is so low in the várzea. The overall nutrient status of a site influences the biomass and the diversity of plants and animals. Both phenomena, the increase in termite biomass from terra firme to várzea and the marked decrease in termite species numbers, agree with theories which explain the effect of stress (lack of nutrients, in this case) on communities of organisms. However, we are far from understanding the factors which lead to the large β-diversity of termite communities in Amazonian forests in general, the forces which drive the apparent adjustment of termites to intermittently terrestrial – aquatic environments, and the causes of time-scale variations in the size of termite populations.

The distribution of species such as *Nasutitermes macrocephalus* and the still enigmatic *Anoplotermes* sp. A should be studied further in order to understand to what extent there is a real "adaptation" to floodplains. The fact that the population size of *Anoplotermes* sp. A oscillates widely in space and time indicates that this species might not really be adapted to the long-term floodings of Amazonian floodplains. Species like the multinest builder *Nasutitermes corniger* occur in mangrove forests of Panama, an aquatic environment with daily flooding patterns. The features shown by termites in floodplains (multinest colonies, accumulation of "storage food", and elaborate patterns of migration from soil to canopy as well as of nest reconstruction) are all found in other wet habitats also. Termites are probably largely tolerant to permanently wet or intermittently aquatic–terrestrial habitats, a fact which has not been previously fully acknowledged; however, at present no special adaptation to floodplains in particular can be recognized.

19 The Spider Communities

HUBERT HÖFER

19.1 Introduction

Most of the very few studies on spiders in Amazonia deal with taxonomic aspects. Only recently have comprehensive studies on spider communities begun (Höfer 1990a,b; Höfer et al. 1994a,b; Vieira and Höfer 1994). From August 1987 to May 1988 the spider fauna of the litter, ground surface, lower vegetation and lower trunk area was studied in the igapó of Tarumã Mirím and in the várzea of Ilha de Marchantaria, using circular pitfall traps (see Platnick and Höfer 1990), ground-photoeclectors and arboreal funnel traps. Comparative studies have been conducted since 1991 on terra firme in the forest reserve "Reserva Ducke" near Manaus. In 1991 and 1992 canopy fogging allowed the collection of spiders from the canopies of trees in Reserva Ducke and on Ilha de Marchantaria. Details about the sampling sites are given in Chapter 2.

19.2 Species Inventories

In the igapó of Tarumã Mirím, 210 species, belonging to 39 families, were collected (Höfer 1990a,b). The most species-rich family was Salticidae (38 spp.), followed by Araneidae (32), Corinnidae (12) and Theridiidae (12, Table 19.1). These four families made up 42% of all species (Fig. 19.1). However, these species-rich families never included dominant species. Dominant species comprising more than 10% of all individuals were: in pitfall traps *Ancylometes* sp. (Pisauridae), *Meioneta* sp. (Linyphiidae) and *Theotima* cf. *minutissima* (Ochyroceratidae); in ground-photoeclectors *Theotima* cf. *minutissima*, *Anapistula* sp. (Symphytognathidae), *Tricongius amazonicus* (Gnaphosidae), *Pseudanapis hoeferi* (Anapidae) and *Meioneta* sp.; in arboreal funnel traps *Xyccarph migrans* (Oonopidae). Webs of *Ischnothele guianensis* (Dipluridae) and of two *Blechroscelis*

Ecological Studies, Vol. 126
Junk (ed) The Central Amazon Floodplain
© Springer-Verlag Berlin Heidelberg 1997

Table 19.1. Species numbers and species identity (Soerensen indices) in eight taxonomically well-resolved families and one infraorder of spiders

Families	Igapó A	Várzea B	Identic sp. A − B = G1	Identity 2 × G1/A + B	Terra firme C	Id. sp. A − C = G2	Identity 2 × G2/A + C	Id sp. B − C = G3	Identity 2 × G3/B + C
Mygalomorphae	3	2	1	0.4	21	0	0.00	0	0.00
Anyphaenidae	4	4	1	0.25	18	0	0.00	0	0.00
Araneidae	32	21	7	0.26	75	26	0.48	8	0.17
Corinnidae	12	7	4	0.42	45	11	0.38	3	0.11
Ctenidae	7	8	7	0.90	16	7	0.61	8	0.66
Gnaphosidae	6	8	4	0.57	9	1	0.13	1	0.12
Ochyroceratidae	4	3	3	0.85	7	1	0.18	1	0.20
Theridiidae	12	10	3	0.27	60	5	0.14	4	0.11
Trechaleidae	4	4	2	0.50	3	0	0.00	0	0.00
Sum	84	67	32	0.42	254	53	0.31	28	0.17

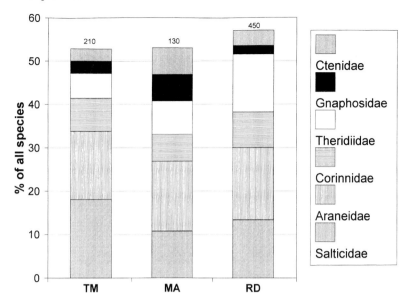

Fig. 19.1. The most species-rich families of spiders in the three study sites. *TM* Black-water floodplain forest at Tarumã Mirím; *MA* white-water floodplain forest on Ilha de Marchantaria; *RD* terra firme forest in Reserva Ducke. *Numbers* above columns are total species numbers

(Pholcidae) species were very abundant on the ground and during inundation on tree trunks.

A total of 130 species, belonging to 35 families, have been collected from the várzea of Ilha de Marchantaria. Araneidae (21), Salticidae (14) and Theridiidae (10) were the most species-rich families (Table 19.1), together making up 33% of all species (Fig. 19.1). Dominant species were *Thaumasia annulipes* (Pisauridae) and Lycosidae sp. 1 in pitfall traps and *Theotima* cf. *minutissima* and *Anapistula* sp. in arboreal funnel traps. *Ischnothele guianensis* and *Blechroscelis* spp. were also abundant on the ground and on tree trunks.

In the canopy samples 46 species of 16 families were identified as morphospecies (95% juveniles). Araneidae dominate the species list with 20 species. Dominant species in all samples was *Eustala* sp. 1 (41%), followed by *Thaumasia* sp. (19%) and another *Eustala* species (11%). It is not clear whether the small juveniles of *Thaumasia* sp. belong to the species that was abundant on the ground. None of the other abundant species of the ground were found in the canopy samples.

From the noninundated terra firme forest in "Reserva Ducke" 472 species of 53 families have been identified to date (Höfer et al. unpubl.). Part of the spider material collected during one year using the same methodology as in floodplain forests cannot be identified at the moment, but based on the number above we estimate the total number of morphospecies in all samples to exceed 500. The most species-rich families were Araneidae (75), Theridiidae (60) and Corinnidae (37) (Table 19.1). In Salticidae only 38 species were identified, but we expect the samples to include more than 100 species. The four families together make 47% (Fig. 19.1) of all morphospecies. We also collected 21 mygalomorph species of 10 families in the reserve. Dominant species did not occur in any of the sampling methods.

In the canopy samples of Reserva Ducke 80 species of 23 families were identified as morphospecies (70% juveniles). No dominant species occurred and approximately 70% of all species were singletons (represented by one specimen). Araneidae, Theridiidae and Salticidae were the most species-rich and abundant families (Höfer et al. 1994b).

19.3 Species Identity in the Two Floodplain Forests and the Terra Firme Forest

In eight families that are taxonomically well studied and the infraorder Mygalomorphae, species identity was checked by comparing species lists and by calculating Soerensen indices (Table 19.1).

Mygalomorph spiders, Anyphaenidae, Araneidae, Corinnidae, Ctenidae and Theridiidae exhibit distinctly higher species-richness in the non-inundated forest. Species identity of the two floodplain forests ranges from 0.25 (Anyphaenidae, Araneidae) to 0.9 (Ctenidae). For all these families together, species identity is 0.42. Species identity values are, with one exception (Araneidae), lower when comparing floodplain forests to the terra firme forest, and in most cases lower for the várzea–terra firme than for igapó–terra firme comparisons (Table 19.1).

Of all species represented by more than one individual in the várzea forest on Ilha de Marchantaria, at least 22 (27%) were never collected in the igapó. Fourteen of the 24 singletons (58%) did not appear in the igapó. Of eight, in at least one trap type, dominant species, in Tarumã Mirím, three species (Höfer 1990a) never appeared in Ilha de Marchantaria, and only two species were also dominant in Ilha de Marchantaria (*Theotima* cf. *minutissima*, *Meioneta* sp.). In floodplain forests a few species of very tiny web-building spiders, Orbiculariae (Anapidae, Linyphiidae, Mysmenidae, Symphytognathidae) and Ochyroceratidae, dominated the litter (Fig. 19.2). Members of these families were also collected on the ground in Reserva Ducke, but no species occured in both areas and they were never dominant.

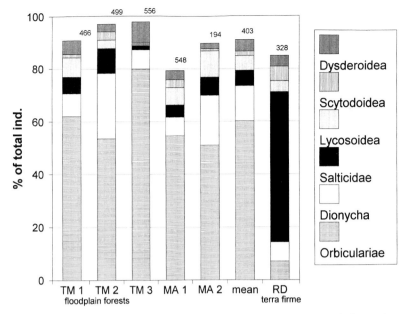

Fig. 19.2. Dominance of ecologically defined superfamilies in ground photoeclectors. *Numbers* above columns indicate mean activity density per trap (100%). Orbiculariae includes Anapidae, Linyphiidae, Mysmenidae, Ochyroceratidae, Symphytognathidae, Theridiidae, Theridiosomatidae; Dionycha includes Corinnidae and Gnaphosidae; Lycosoidea includes Ctenidae, Lycosidae, Pisauridae; Scytodoidea includes Pholcidae, Scytodidae; Dysderoidea includes Caponiidae, Oonopidae. *TM 1* 8 traps 1976/1977; *TM 2* 1 trap 1983/1984; *TM 3* 3 traps 1987/1988; *MA 1* 3 traps 1981/1982; *MA 2* 2 traps 1987/1988; *RD* 5 traps 1991/1992; *mean* mean of all samples in floodplain forests

In captures of ground-photoeclectors in Reserva Ducke, Salticidae were not only species-rich (Fig. 19.1) but also provided the three most abundant species (altogether 35%, Fig. 19.2).

The fogging samples from four different tree species on Ilha de Marchantaria (*Luhea* cf. *cymulosa, Macrolobium acaciaefolium, Pseudobombax munguba, Triplaris surinamensis*) differ greatly from the fogging samples of two trees of the one species (*Goupia glabra*) fogged in Reserva Ducke (Höfer et al. 1994b). Species identity is zero; different families dominate with respect to species number and abundance. Samples from the floodplain forest appear to represent a "canopy fauna". These are dominated by Araneidae (44% of all species, more than 70% of all individuals) and contain no representatives of ground-dwelling species. Spiders of the Araneid genus *Eustala* seem to be typical inhabitants of higher vegetation (e.g., canopy, pers. observ.). A few individuals of species of this genus also appeared in the canopy samples of Reserva Ducke, but these samples include a higher portion of non-web-building species (Anyphaenidae,

Heteropodidae, Salticidae, Thomisidae). Araneidae were poorly represented and Theridiidae was the most species-rich family (21 species).
Theridiidae and Salticidae were co-dominants in canopy samples in other
neotropical sites (Coddington, pers. comm.).

With more than 400 species the spider fauna of a terra firme site is much
more species-rich than the fauna of an inundation site of comparable size
(igapó: 210 spp., várzea island 130 spp.). Presumably, floodplain forests are
poorer in species than terra firme forests because of the impact of the
annual flood. Soil- and litter-inhabiting mygalomorphs and Zodariidae for
example seem to be groups which were not able to colonize floodplain
forests.

In the igapó Tarumã Mirím, which directly borders a terra firme forest
(Adis 1981; Höfer 1990a) we collected more species than in the study site
situated on the central part of Ilha de Marchantaria, which in most years is
completely inundated for 3–4 months. This might, at least in part, be a
result of the more intense sampling in Tarumã Mirím, but may also reflect
the more isolated situation (no connection to terra firme forests) and
harsher environment (longer-lasting inundation, deposition of river sediments) on the island.

19.4 Phenology of Abundance and Activity Density
on the Ground and on Tree Trunks

Capture rates in pitfall traps, arboreal funnel traps, and other similar traps
are denominated "activity density" (Schaefer and Tischler 1983) because
they depend at the same time on abundance and activity of the trapped
animals (Uetz and Unzicker 1975; Curtis 1980; Höfer 1990b). Activity density of spiders on the ground increased from the beginning of the terrestrial
phase in September and reached a maximum 2–3 weeks prior to inundation of the ground (Höfer 1990a,b). Activity density was high in the igapó
near the waterside, reflecting the flight of spiders from the slowly advancing water line, but was relatively low in the várzea. Density activity on tree
trunks was low (0–10 ind. day^{-1}) during the whole noninundated period
and increased abruptly (up to 250 ind. day^{-1}) a few days before inundation
of the trunk base. The day/night captures during a 3-week period revealed
a distinctly higher night activity in all traps (ratio day/night: 1.5:4).

Activity density and species-richness was lower in Ilha de Marchantaria
than in Tarumã Mirím at all times and in all trap types. This coincides with
results in other arthropod groups (Adis 1981; Chap. 14). The group of tiny
web-building spiders (Orbiculariae and Ochyroceratidae) dominates the

litter in both floodplain forests, but in Ilha de Marchantaria both groups are represented by fewer species. Three species of larger and very agile wandering spiders (Lycosidae: 2 spp. and *Thaumasia annulipes*) are more abundant on the ground in Ilha de Marchantaria than in Tarumã Mirím. The trunk ascent (capture of the arboreal funnel traps during inundation period) was much lower in Ilha de Marchantaria; the dominant species of the trunk ascent in Tarumã Mirím, *Xyccarph migrans*, does not live in Ilha de Marchantaria.

In the black-water floodplain forest on Tarumã Mirím agile spiders (large wandering spiders) avoid flood by a steady horizontal movement away from the waterline. Examples are species of Corinnidae, Ctenidae, Lycosidae, Pisauridae. At least some of them profit from the increased number of prey (insects) near the water line (*Ancylometes* sp., *Thaumasia* sp., *Trechalea amazonica*). Less-agile spiders avoid flood by climbing to vertical structures (dead wood, small plants, tree trunks). Examples are Caponiidae, Dipluridae, Gnaphosidae, Palpimanidae, Pholcidae). Many small spiders (e.g., Orbiculariae) climb upon nearby vertical structures and then start ballooning. By this means they are able to travel long distances and reach noninundated ground or tree trunks. These reactions are reflected in the increasing density activity in pitfall traps and arboreal funnel traps during the terrestrial phase.

In contrast to the black-water floodplain forest no strong increase in density activity of spiders on the ground during the terrestrial phase was detected by the pitfall traps in the white-water floodplain forest. The peak just prior to inundation might have been missed; however, horizontal migration may not play a role as in the igapó. Since there is no contact with terra firme areas at Ilha de Marchantaria, spiders cannot survive by moving away horizontally. This may explain a reduction in species-richness and abundance. Another factor responsible for a lower species-richness and abundance in the várzea can be differences in habitat conditions. Adis (1981) reported from the várzea forest (Ilha de Curari) four to six times fewer soil-living arthropods than from an igapó, due to unfavorable conditions in the litter which is covered annually by fine river sediments during the flood period. Recent observations and experimental work in Reserva Ducke clearly showed the importance of litter quantity (depth) for spider abundance (Höfer, unpubl.).

The capture of arboreal funnel traps in 1987/1988 in Ilha de Marchantaria showed no spider species migrating to the trunk/canopy area in abundances comparable to those observed in the Igapó. However, Adis (Sect. 14.3.1) observed a very high trunk ascent of spiders in Ilha de Marchantaria in 1981 and 1982 and believes that the trunk ascent is highly dependent on the velocity of the flood. My field observations suggest that at least *Blechroscelis* (2 spp.) and *Ischnothele guianensis* migrate to the trunk area in Ilha de Marchantaria, but these web-building spiders were not adequately sampled by my methods. In the Manaus area *Ischnothele guianensis* seems to live exclusively in floodplain forests, perhaps because of a general preference for brighter environments (Kovoor, pers. comm.). Coyle (1995) reports that only rarely, if ever, this Amazon basin species has been collected in undisturbed areas. Marechal (pers. comm.) studied light preferences of *I. guianensis* and observed negative phototaxis to appear only late in the development. He hypothesized different habitats for juveniles and adults, which coincide with the life cycle observed in both floodplain areas. This species seems to survive the inundation in higher strata (e.g., trunks) in subadult stages and mates towards the end of this period. Juveniles then spread out on the newly emerged floor (Höfer 1990a).

Broad belts of "floating meadows" (Junk 1970, 1984b) around Ilha de Marchantaria offer large retreat areas. For at least a few very agile species (Lycosidae spp., *Ancylometes bogotensis* and *hewitsoni*, *Ctenus* spp., *Thaumasia* cf. *velox*), we suppose the grass belt to be an important refuge habitat. For some species (*Alpaida veniliae*, *Acanthosoma pentacanthum*, *Asthenoctenus* sp., *Tetragnatha* spp.), it might even be the main habitat, from which spiders regularly invade the forest during the terrestrial phase. In the samples of three different trap types in Reserva Ducke we observed a significant decrease of spiders and other arthropods on the ground through the year, sudden increases of spiders in arboreal funnel traps during periods when army ants passed through the area, and a significant decrease in medium-sized spiders on tree trunks during the year (Gasnier et al. 1995). We hypothesize repeated passing of army ants to be the main cause for the decrease in abundance and density activity on the ground (Vieira and Höfer 1994). We could not detect any correlation with rainfall or other abiotic factors. Not only were the samples of ground photoeclectors in the terra firme richer in species than in floodplain forests, but so were the samples of arboreal eclectors and canopy fogging. This shows the importance of the trunk region as microhabitat, refuge zone and pathway to the canopy in terra firme forests.

19.5 Discussion and Conclusions

Spiders are able to survive without food for a long time and profit from short-term availability of prey. It is therefore not surprising that spiders are among the first colonizers of any uninhabited area. Some species would certainly be able to colonize floodplain forests annually during the terrestrial phase. Inundation of the Central Amazon floodplain is a very regular event (Sect. 2.3). For most spiders, especially larger ones (>5 mm body length), it should be no problem to avoid being flooded by moving away short distances. Even when they are surrounded by water, most of them can escape by walking on the water surface film. Smaller spiders may avoid such flooding by climbing onto small vertical structures and then ballooning until they reach safe areas. When contact with terra firme areas exists (as in Tarumã Mirím), spiders may find refuge in these adjacent areas. Many species of ground spiders in terra firme areas make use of the lower or even upper trunk region. These spiders could use the trunk and/or canopy region in floodplain forests as refuge during inundation.

However, the comparison of both floodplain areas with terra firme areas has shown that the flood regime has an impact on the spider fauna. The impact is highest in low-lying areas, which are flooded longer, and in areas far away from terra firme (e.g., islands), where horizontal migration does not increase survival rate. A considerable portion of the spider populations dies during the aquatic phase by being flooded or by indirect consequences (see below). High water temperatures and low oxygen concentrations make it less probable that spiders survive under water, however one species of Pycnothelidae has been found in submerged litter of a floodplain forest in the Anavilhanas archipelago (Negro River, Amazonas River; see Nessimian 1985). The small-scale horizontal and vertical migrations of spiders (and other predators) to terra firme areas or higher strata, and the resulting strong concentration of animals at the beginning of the floods on the remaining ground or trunk area heavily influence the community structure. Some species or groups seem to compete better in floodplain forests, e.g., *Theotima* cf. *minutissima*, *Meioneta* sp., *Xyccarph migrans*, *Camillina taruma*, *Tricongius amazonicus*. They seem to displace the jumping spiders, which as a group of many species dominate the litter habitat in Reserva Ducke and in most tropical forests (see Jocqué 1984). Jocqué (1984) hypothesized that in the litter of tropical forests interference competition by the extremely diverse and abundant ants leads to a lower abundance and diversity of web-building spiders and favors Salticidae. In fact, ants seem to be more abundant and much more diverse in terra firme forests than in floodplain forests (Adis and Schubart 1984 and pers. observ.). In Reserva Ducke we caught seven species of army ants (including three arthropod-hunting species) in pitfall traps opened during a period of 4 weeks. One spider (arthropod) hunting species passed through our study site at least seven times within 1 year. In both floodplain forests we never observed army ant raids or caught army ants in traps.

I suppose at least some of the spider species have evolved behavior and/ or lifecycles connected with the annual inundation. They do not merely move away from the arriving water, but the whole population migrates into the trunk region (*Blechroscelis* spp., *Camillina taruma, Ischnothele guianensis, Tricongius amazonicus, Xyccarph migrans*). In the case of *Xyccarph migrans* the migration starts long before the microhabitat is actually inundated (Höfer 1990b). Secondary abiotic factors, and not the flood itself, apparently act as control mechanisms on some "terricolous migrants" (cf. Adis 1992a; Sect. 14.4).

The low number of spider species shared by terra firme forests and floodplain forests (for the capoeira see Höfer 1990a; for Reserva Ducke, Höfer et al., unpubl.) indicates that spider communities in floodplain forests are not merely impoverished terra firme communities. In many cases closely related species occur at the different sites. The low species identity observed for some spider families when comparing both floodplain forest faunas might be explained by different origins of the fauna in the two types of floodplain forest. The occurence of some species previously only known from very distant regions, *Rhoicinus gaujoni* from western South America (see Höfer and Brescovit 1994) and *Zimiromus beni* from the Beni region of Bolivia (see Brescovit and Höfer 1994a), on Ilha de Marchantaria is very interesting. Dispersal along river systems might be an explanation for such a pattern, but a better data base of the distributions of spiders in the Amazon is clearly needed before discussing biogeography and evolution of Amazon spiders. The role of floodplain forests as short-term refuges and long-term generators of species-richness has been discussed by Erwin and Adis (1982) and Adis (1984) for beetles.

For spiders the floodplain forests contribute to a high gamma diversity of spiders in central Amazonia. Whereas spider species-richness in single forest sites in central Amazonia seems to be only two to four times higher than in central European forests (Albert 1982; Höfer 1989), I would expect gamma diversity to be much higher than in temperate regions. Whereas 747 spider species are known from Bavaria ($70\,500\,km^2$, see Blick and Scheidler 1991) and this number will certainly not rise by an amount worth mentioning, 586 spider species have already been identified from the merely sporadic collections in the Manaus area ($<10\,000\,km^2$). Estimating the total number of species in these collections is rather difficult because some families (e.g., Heteropodidae, Oonopidae, Salticidae) are not yet known well enough taxonomically. In recently revised families (or genera) the portion of undescribed species varies greatly. For example 17 of 20 gnaphosid species, but only 6 of 20 ctenid species, were undescribed (Platnick and Höfer 1990; Brescovit and Höfer 1993, 1994a,b; Höfer et al. 1994a). Although the proportion cannot be given for many families, I would estimate it to be 50% or higher for all spiders of the central Amazon region.

Acknowledgments. I am grateful to the MPI for funding the study of spiders in floodplain forests and providing infrastructure in Manaus, and to the National Institute for Research in Amazonia (INPA), Manaus for research permits and the loan of material. The German Science Foundation (DFG) funded the study of spiders in terra firme forests.

I cordially thank Alexandre Bonaldo, Dr. Antonio Brescovit, Erica Buckup, Maria Aparecida Marques of the Zoological Museum of Fundação Zoobotânica de Rio Grande do Sul in Porto Alegre and Dr. Arno Lise of the PUC university in Porto Alegre, as well as Dr. Jim Carico (Lynchburg), Dr. Fred Coyle (Western Carolina University), Dr. Pablo Goloboff (San Miguel, Argentina), Dr. Pekka Lehtinen (Turku University), Dr. Herbert Levi (Harvard University) and Dr. Norman Platnick (AMNH, New York) for identification of material and taxonomic advice. Dr. Fred Coyle critically reviewed the manuscript. Dr. Joachim Adis and his collaborators povided the spider material from canopy-fogging studies on Ilha de Marchantaria.

20 The Fish

WOLFGANG J. JUNK, MARIA G.M. SOARES, and ULRICH SAINT-PAUL

20.1 Introduction

Fish represent the best-studied group of animals in Amazonian waters. Even so, many questions remain unresolved. According to Kullander (1994), 3175 species, belonging to 55 families, are known from tropical South America. This author estimated that about half of the described species may occur in Amazonia, though this could be a conservative guess. Furthermore, a considerable number of new species are to be expected. Therefore, the total number in the Amazon basin may reach about 2500 or more species. There exists a vast literature on taxonomic aspects, which will not be considered here.

Monographs by Goulding (1980) and Goulding et al. (1988) review the food and feeding habits of Amazonian fish and point to the importance of the floodplain forest in the trophic ecology of large Amazonian rivers. These qualitative studies have been recently complemented by investigations by Saint-Paul (1994) who quantified the importance of the floodplain forest for the fish communities in white and black water as a basis for estimating the value of the forest for commercial fisheries.

Lowe-McConnell (1975, 1987) discussed the Amazonian fish fauna in the context of community ecology. Welcomme (1985), in analysing tropical river fisheries, presented a comprehensive review of biological and ecological aspects of the floodplains. In recent years, several Brazilian studies have examined the impact of the construction of large reservoirs on the fish fauna (Holanda 1982; Vieira 1982). Various papers concentrate on different aspects of the Amazonian inland fisheries and their potential (Petrere Jr. 1978a,b; Smith 1981; Goulding 1981, 1983; Bayley 1982; Junk 1984; Bayley and Petrere Jr. 1989). Several studies are available on the oxygen consumption of Amazonian fishes and adaptations to low oxygen concentrations (Carter and Beadle 1931; Carter 1935; Geisler 1969; Johanson et al. 1978; Kramer et al. 1978; Kramer and McClure 1982; Junk et al. 1983; Kramer 1983; Saint-Paul 1984a, 1988; Val et al. 1986;

Ecological Studies, Vol. 126
Junk (ed) The Central Amazon Floodplain
© Springer-Verlag Berlin Heidelberg 1997

1990, 1996; Monteiro et al. 1987; Saint-Paul and Soares 1987, 1988; Soares 1993; Val and Almeida-Val 1995).

Considering the large amount of available information, this chapter concentrates on how the flood pulse affects the fish fauna in the várzea and igapó.

20.2 Food Availability and Feeding Habits

In Amazonian river–floodplain systems a wide variety of food resources are available for fish, e.g., phytoplankton, periphyton, terrestrial and aquatic herbaceous plants, plant material from the floodplain forest (leaves, flowers, and fruit), aquatic invertebrates (benthos, zooplankton, perizoon), terrestrial invertebrates and – for predators – fishes. Most of the food is produced in the floodplain, because the main river channel is a rather uniform habitat with little autochthonous primary production. Strong turbulence and heavy loads of inorganic suspended material inhibit zooplankton development. Living conditions are difficult for benthic animals in the unstable bottom deposits. Consequently, secondary production is also low.

Food availability is determined by access of the fish to the floodplain, which, in turn, depends on the water level. When the river rises, an expanded area with a large spectrum of food items becomes accessible to the fish. When the water recedes, fish are forced to leave the floodplain, thus losing their preferred feeding habitats. During low water they concentrate in the main channel and remaining floodplain pools, at which time there is a shortage of food except for predatory species. Some predatory benthic species, such as large catfishes, e.g., *Brachyplatystoma* spp. stay in the main channel most of the time because oxygen conditions near the bottom in várzea lakes are often critical. These species move to the shore area at night to prey on fish moving in and out of the floodplain.

Food items are not equally distributed in the floodplain, but are concentrated in specific habitats, e.g., the open water area, the aquatic macrophyte communities, or the floodplain forest. In várzea lakes, phytoplankton and zooplankton production is only intense when the load of suspended solids in the inflowing river water has settled down and the water has become transparent (Sects. 10.2.2, 10.3.2). Although fruit and seeds from trees of the floodplain forest ripen at different periods of the year, maximum fruiting coincides with the flood period (Ziburski 1991; Ayres 1993; Kubitzki and Ziburski 1994). Diversity of trees may reach more than 100 species ha^{-1} (Sect. 11.2). Fish must be highly mobile if they are to find the preferred fruiting trees in time. Analysis of the nutritional value of fruit showed on average soluble carbohydrates 36%, crude fiber 31%, crude protein 19%, crude fat 9.5%, and ash 3.5% dry weight, thus providing a satisfactory food source for fish (Waldhoff 1991, 1994). The highest crude protein contents were found in *Hevea brasiliensis* seeds (43%), *Pseudobombax munguba* (43%), and *Aldina latifolia* (37%), with a minimum value found in *Pisidium* sp. (11%). In feeding experiments using fruit and seeds from Amazonian floodplain forest trees, *Colossoma macropomum* showed satisfactory growth rates (Roubach and Saint-Paul 1994). Aquatic macrophytes occurring in white-water floodplains in large quantities are of poor nutritional value (Sect. 9.2) in comparison with other food items. Grasses show a crude protein content of 2.8–16.2% and a crude fat content of 0.4–2.1% (Table 9.2; Howard-Williams and Junk 1977; Ohly 1987).

Food supply and quality also depend on hydrochemical conditions and the fertility of the sediments. Water and soils of the floodplains of white-water rivers are more fertile than those of clear-water and black-water rivers. In the floodplain of the Negro River, the low primary production results in a much lower food supply for fish than in the white-water floodplain of the Amazon River. Terrestrial and aquatic herbaceous plants are much more abundant in the várzea than in the igapó (Sects. 8.3, 8.4). Flexibility in food choice is necessary in the constantly changing river and floodplain environment. Comparison of the stomach contents of specimens of the same species, but from different habitats of the Amazon floodplain, shows a great variety of food items. However, the spectrum of food resources that a species could utilize may not be apparent because the fish are usually highly selective.

Flexibility is limited by the anatomy of the mouth, including the type of dentition, the anatomy and physiology of the gastrointestinal system, and the age of the fish. Juvenile fish usually prefer food items different from those fed on by adults and select their habitats according to their requirements. Aquatic macrophytes are a preferred habitat for the juveniles of many species because they offer shelter and a large variety of food items, including detritus, periphyton, perizoon, and terrestrial invertebrates. As the juveniles become larger, they move to other habitats, such as the open water or the floodplain forest and change their diet. Some species, e.g., *Semaprochilodus* spp. live as juveniles in the várzea and later move into black-water rivers where they feed in the floodplain forest on detritus and periphyton. The periodically expanding and shrinking habitats and permanently changing food supply and food preferences force the fish to be highly mobile and guarantee optimum exploitation of the resources available in the floodplain.

Of 34 species studied in the várzea lake Camaleão during the period of rising, high and falling water levels, 16 species showed distinct feeding preferences. More than 75% of the stomach contents belonged to only one food category throughout the whole study period (Soares et al. 1986). Seven species fed predominantly on detritus (*Prochilodus nigricans, Semaprochilodus taeniurus, S. insignis, Potamorhina* cf. *altamazonica, P. latior, Psectrogaster rutiloides, P. curviventris*), four on fish (*Arapaima gigas, Hoplias malabaricus, Serrasalmus* sp., *Electrophorus electricus*), two on floodplain forest material (adult *Colossoma macropomum, C. bidens*) and one on periphyton (*Rhythiodus microlepis*). The other 18 species fed on a large variety of food items; however, comparison with the results of other authors shows that the food spectrum is even broader (Table 20.1). During low water, food uptake was greatly reduced and stomachs were mostly empty.

A semi-quantitative analysis of the different food items used by the fish community in Lago Camaleão showed that the largest part, or about 35%, was made up by detritus, followed by terrestrial invertebrates, aquatic macrophytes and periphyton (about 15% each). Perizoon, zooplankton, material of the floodplain forest, and fish contribute less than 10% each (Soares et al. 1986). This analysis is not representative of the whole várzea, because Lago Camaleão is dominated by aquatic macrophytes and wide-

Table 20.1. Main food items of some Amazonian fish species in Lago Camaleão and in other habitats

Family/species	Lago Camaleão	Other authors and other habitats
Family Anostomidae		
Schizodon fasciatus	AM, PP	AM, PP, MF
Rhythiodus microlepis	PP	PP, MF, AM
Family Serrasalmidae		
Mylossoma duriventre	AM, PP, PZ	AM, MF
Colossoma macropomum	AM, MF, TI	MF, Z, PP
Subfamily Bryconinae		
Brycon cf. *melanopterus*	MF, TI	PP, TI, MF
Triportheus angulatus	AM, PP, PZ, TI	AM, TI, MF
T. albus	AM, PP, Z, PZ, TI	AM, TI, Z
Cichlidae		
Astronotus ocellatus	AM, TI, F	MF, TI, PZ
Cichlasoma amazonarum	AM, PZ, F	PP
Heros severus	PP, PZ, TI	PP, MF
Mesonauta festivus	PP, PZ	AM, MF, PP

AM, Aquatic macrophytes; MF, material from the floodplain forest (fruits and flowers); PP, periphyton; D, detritus; Z, zooplankton; PZ, perizoon; TI, terrestrial invertebrates; F, fish.

spread hypoxia reduced the number of fish species. The low contribution of fruit from the floodplain forest is due to the fact that it covers only the margins of the lake basin. The results of a 2-year comparative study on the impact of the floodplain forest on fish in Lago do Prato, a black-water lake in the Anavilhanas archipelago of the Negro River and Lago Inácio, a white-water lake of the Solimões/Amazon River revealed that 34 species from the black water and 23 species from the white water were feeding on fruit, corresponding to 22 and 13%, respectively, of the total species captured in the area. Furthermore, an analysis of seasonal variations revealed that black-water floodplain species feed more constantly on food from the floodplain forest than do species in the white-water floodplains. This is probably due to the high diversity of food items in the várzea.

Comparing stomach contents of fish from black and white water, most Negro River fishes appear to be omnivorous on a seasonal basis, with fruit, terrestrial arthropods, crustacea, fish and detritus the most important food items. Predation was important for black-water fish during rising and high water, and for white-water fish during falling and low water periods. Zooplankton was of seasonal importance only in the white-water lake. In the black-water lake, fish fed on algae only during certain periods of the year (Table 20.2; Saint-Paul 1994).

Despite the large amounts of aquatic macrophytes in várzea lakes, their direct contribution as a food resource for fish seems to be relatively low, in

Table 20.2. Frequency of occurrence of food items in the stomach of fish from black and white water

Food item	Rising		High		Falling		Low	
	Black	White	Black	White	Black	White	Black	White
Number of species	27	41	19	40	40	41	49	35
Fruits/seeds	31	18	40	24	36	54	38	48
Allochthonous arthropods	15	24	45	22	50	53	54	44
Aquatic insect larvae	12	16	18	16	3	10	11	2
Crustacea/mollusks	18	11	5	–	31	10	17	36
Fish	37	3	31	9	19	31	45	83
Zooplankton	–	16	–	2	1	34	1	4
Algae	–	1	41	–	0	2	21	3
Detritus	41	39	55	46	67	60	41	65

The occurrence method was employed by making a list of the total number of times a particular food item occurred in the stomach in order to determine the relative importance of the various food items eaten by fish. Considering the dominance of the analyzed fish species within the community, the occurrence of each food item in this community was calculated. Stomach contents of 90 of the total number of 239 species were investigated.

Lago Camaleão being of the same order of magnitude as the contribution of periphyton. Stomach content analyses of 91 species during rising, high, falling and low water periods from várzea and igapó by Saint-Paul (1994) did not show any fish to be feeding to a major extent on aquatic macrophytes (Table 20.2). Obviously, whenever possible, herbivorous fish prefer high-protein algae with a C:N ratio of about 9.5 (Sect. 10.4) and fruit rather than low protein macrophytes with a mean C:N ratio of about 25 (Sect. 9.2). The indirect importance of macrophytes as a food resource for fish, for example, in the form of macrophyte-feeding terrestrial and aquatic invertebrates, detritus, and associated bacteria and fungi as well as in the form of dissolved organic carbon from decaying macrophytes via the microbial loop, has not yet been evaluated. Araujo-Lima et al. (1986) suggested that five adult detritivorous species fed on carbon originating in phytoplankton, based on a preliminary analysis of $\delta^{13}C$. Bayley (1989) showed that phytoplankton production is not sufficient to sustain the high fish production in the várzea, and that the floodplain forest and terrestrial and aquatic macrophytes contribute to a considerable extent to fish production.

The large number of species and individuals that specialize on detritus as food seems to contradict the hypothesis that fish prefer high quality plant material. Studies on the decomposition of herbaceous plants and the tree leaves of the floodplain forest show large losses of nutrients at the beginning of the decomposition process due to leaching; later, there is a slight rise in quality again because of an increased number of fungi and bacteria attached to the organic material (Sects. 9.2.1, 9.2.2, 9.3.1). Howard-Williams and Junk (1976) showed that detritus from different sources becomes chemically rather uniform during the decomposition process, thus facilitating processing by consumers. Due to the large amounts of detritus and its permanent availability in the water bodies, it can be an attractive food item for specialized animals. Furthermore, "detritus" in stomach content studies is a rather vague category, which includes fine organic particles that cannot easily be analyzed. Our in situ observations of detritus-feeding fish showed a rather selective feeding behavior. This suggests that fish select detritus with relatively large percentages of associated bacteria, fungi, protozoans, periphyton and detrital nonprotein amino acids, which may considerably increase the food value of this material (Bowen 1979a,b, 1980).

Terrestrial invertebrates contributed high percentages to fish diets. Most floodplain lakes shrink during low water. When refilling, large numbers of terrestrial arthropods living on or in the dry sediment, such as ants, termites and collembolans, grasshoppers, cockroaches, hemipterans, and beetles, become available for the fish. Later, the terrestrial invertebrates living on the emergent macrophytes and in the canopy of the floodplain forest become a predominant food source for some species.

Under well-oxygenated conditions, the prey to piscivore predator biomass ratio in the mouth of Lago Camaleão varied from 1.2–1.3 (Bayley 1982) which pointed to strong predation pressure and the great importance of fish as a food resource. However, in strongly hypoxic habitats, such as inside Lago Camaleão, most predatory species are absent and, thus, fish made up only a low percentage of the food resource.

Zaret and Rand (1971) found that small stream-dwelling fish became specialized in their diets during periods of low water levels when food was scarce. These authors considered that this supported the competitive-exclusion principle. Our findings suggest that in the várzea the competitive-exclusion principle for food has little impact on fish communities. During low water, when food is scarce, other factors such as increased mortality due to drought or displacement from specific habitats, play a much greater role in determining the survival rate of populations than food supply. According to Bonetto et al. (1969), the annual mortality of fish populations in the Paraná River system during the low water period reaches about 40 000 t.

Because of the reduced growth of terrestrial and aquatic herbaceous plants and low phytoplankton production, fish in the Negro River depend much more on the floodplain forest than on the Amazon floodplain. Adult fish tend to overcome food shortages by their ability to use different items. Food, however, may become a critical factor for juveniles, which prefer to feed mostly on algae and small aquatic invertebrates. Information on juveniles from the igapó is still missing.

20.3 Biomass, Growth Rates, and Production

Only a few studies on biomass, growth and production of fish communities are available from Amazonian floodplains. Bayley (1982), from his quantitative study in the bay in the mouth of Lago Camaleão, estimated an overall mean biomass for the várzea of about $160 \pm 24 \, \mathrm{g \, m^{-2}}$, corresponding to $1.6 \, \mathrm{t \, ha^{-1}}$ and a total fish production of $280 \, \mathrm{g \, m^{-2} \, year^{-1}}$. He showed that growth is strongly related to the water level. During rising water, growth was on average 60% faster than during falling water. A fast rising water level led to a higher growth rate for the species studied than a slowly rising one. Analyzing the growth rates of 11 species throughout the year, Bayley (1982) found that individual species appeared unaffected by the biomass of potentially competing species during rising and high water, and data for only 2 species out of 11 suggested density-dependent growth during falling water. This pattern is explained by the food availability. During rising water, food availability increases, whereas it is quickly reduced during falling water. These results agree with studies on the Kafue flats in Africa (Dudley 1974; Kapetsky 1974).

Similar studies do not exist from the Negro River floodplain. There are, however, comparable data on gillnet catches/unit effort in igapó and várzea from Saint-Paul (1994). In the várzea an average of $210\,\mathrm{g\,m^{-2}\,day^{-1}}$ was captured in contrast to $41\,\mathrm{g\,m^{-2}\,day^{-1}}$ in the igapó. However, there are significant seasonal differences between the fish captured within and outside of the floodplain forest in both areas (Fig. 20.1). Fish yields from the black-water floodplain forest, as compared to the open water, were be-

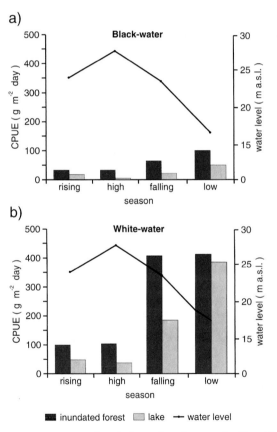

Fig. 20.1. Seasonal variations of catch per unit effort (CPUE) **a** in the black-water lake Lago do Prato (Anavilhanas) and **b** the white-water várzea lake Lago Inacio during 1990 and 1991 inside and outside a floodplain forest. (Saint-Paul 1994)

tween 183 and 550% higher, while in the white water this difference was significantly smaller (107–281%). At both sites the largest amount was caught during low water and the lowest amount during high water.

Assemblages can be characterized as relatively random associations of species with high taxonomic and abundance diversities. Very important species in the igapó are *Serrasalmus striolatus* and *Heros* sp., and in the open water *Hypophthalmus marginatus*. In the várzea *Mylossoma duriventre* and *Potamorhina altamazonica* are important in the floodplain forest and *Hypophthalmus edentatus* and *Pimelodina flavipinnis* in the open water.

20.4 Reproductive Strategies

Most information on the reproduction of Amazonian fish is derived from observations in captivity. Observations under natural conditions are very rare and are mostly related to species caught for human consumption. Commercially important Amazonian fishes are divided by the local people into a category which undergoes spawning migrations in large schools, locally called "piracema" (for instance, many large Characoids and Pimelodids), and a category which does not migrate (for instance, Cichlids and Sciaenids).

Migration patterns are complex and are not yet fully understood for many species (Goulding 1980). Some species living in the várzea migrate out of the lakes into the river to spawn, such as *Anodus* spp. The adults of *Prochilodus insignis* move downstream hundreds of kilometers from black water to white water for spawning. They than compensate for any downstream drift of eggs and larvae by migrating further upstream to the next black-water river (Fig. 20.2) (Ribeiro 1983). At the beginning of the spawning season, males develop a muscle around the swimbladder. By rhythmic contraction, the "drum muscle" causes the swimbladder to resonate, and males produce loud sounds. Generating a sound by different types of drum muscles is known for several migrating Amazonian fish species. It is also known for some species that do not migrate but spawn in schools, such as *Plagioscion squamosissimus*. The juveniles of the species migrating between black- and white-water habitats benefit from a better food supply of zooplankton and perizoon in the várzea. The adults make use of the periphyton and organic detritus in the floodplain forest of black-water rivers that are poor in suspended inorganic solids, or feed on fruit and insects.

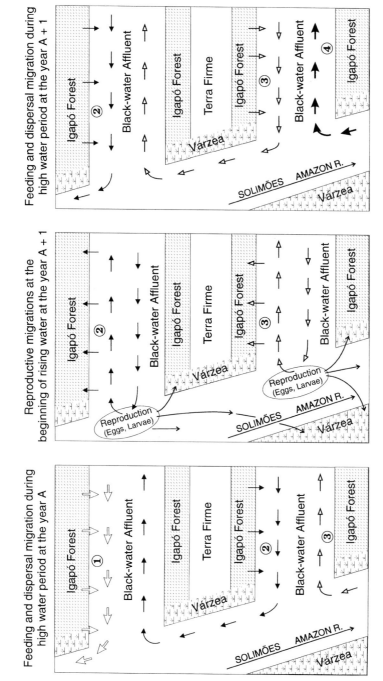

Fig. 20.2. Reproduction and feeding migrations of the jaraqui *Semaprochilodus insignis* (Ribeiro 1983). *Arrows* and the numbers *1, 2, 3,* and *4* indicate the migration of different sub-populations of jaraquis

Most of the migratory species spawn only once a year, releasing a large batch of eggs at the beginning of the rising water. The number of eggs in *Prochilodus insignis*, a fish of about 30 cm body length, may reach 300 000; in *Prochilodus nigricans* (ca. 40 cm long) and *Colossoma macropomum* (ca. 1 m long) about 500 000 and about 1 million eggs may be present, respectively. In *Prochilodus nigricans*, gonads represent 15% of the body weight, in *Prochilodus insignis* about 17%. This breeding strategy requires large amounts of energy for gonad development, migration, and spawning at low water, at a time when food is scarce and when the fish are very active. These species accumulate large amounts of fat during rising and high water when food in the floodplains is abundant (Junk 1985b; Table 20.3; Fig. 20.3). Fat accumulation may occur in the body cavity (e.g., *Colossoma macropomum*, *Piaractus brachypomum*, and many large catfish), in the muscles (e.g., *Hypophthalmus edentatus*), subcutaneously and/or in the liver (most species). Due to a greatly reduced food supply during low water periods, juvenile *Colossoma macropomum* have to metabolize glycogen as reserve material as has been demonstrated by changes of the glycogen-somatic

Table 20.3. Mean fat, water, protein and ash contents of Amazonian fish soon after spawning and during the feeding period. (Junk 1985b)

Family and species name	Fat (%)	Water (%)	Protein (%)	Ash (%)
Migrating species:				
Family Characidae				
Triportheus elongatus	4–20	62–76	15–20	1–2
Family Curimatidae				
Prochilodus nigricans	3–18	64–78	17–20	2–4
Potamorhina latior	4–17	64–76	16–20	1–2
P. cf. *altamazonica*	2–17	64–76	16–20	1–2
Family Anostomidae				
Leporinus fasciatus	6–15	66–72	17–20	2–3
Family Serrasalmidae				
Mylossoma duriventris	10–28	55–72	15–19	1–2
Family Hypophthalmidae				
Hypophthalmus edentatus[a]	4–32	56–80	13–16	1–2
Nonmigrating species:				
Family Cichlidae				
Cichla ocellaris	0.5–5	74–80	17–19	1–2
Astronotus ocellatus	1–4	74–80	16–18	2–4
Family Sciaenidae				
Plagioscion squamosissimus	0–3	76–80	16–19	1–2

All analyses were of whole fish.
[a] Fillet only.

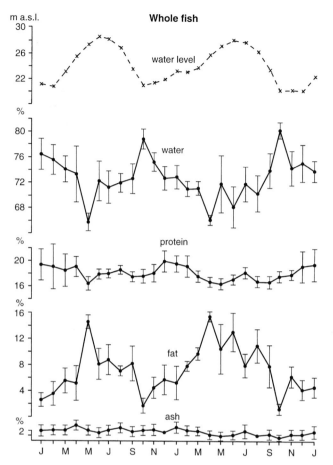

Fig. 20.3. Water, protein, fat, and ash contents of the jaraqui *Semaprochilodus insignis*) (whole fish) in comparison with the water level (44 males, 76 females, total length 22–31 cm, x = 27.5 cm, weight 120–930 g, x = 319 g). (Junk 1985b)

index (Saint-Paul 1984b). Many of the nonmigrating species, e.g., cichlids and sciaenids, reproduce several times a year, releasing only small numbers of eggs. Therefore they do not need to accumulate large amounts of fat and do not show large variations in fat content (Table 20.3).

During rising water, the spawning success of migrating species depends more on the hydrological conditions of the specific year than on food supply (Worthmann 1982). For migrating species, a fast rising river is a strong signal to start the spawning migration and to accelerate gonad development. According to the observations of the local fishermen, a slowly rising river level may lead to a reduced and delayed spawning, with negative effects on the age class of that year.

20.5 Adaptations to Hypoxia

In all tropical floodplains there is a tendency to permanent or periodical hypoxia in the water. The level of hypoxia depends on the frequency and extent of turbulence affecting the water body, on the water depth and on the presence of organic material. As shown in Section 4.4.3, all várzea lakes near Manaus suffer at least periodically from strong hypoxia because of large amounts of decomposing organic material. Input of oxygen from the air by turbulence and by phytoplankton usually only affects the upper layers of the open water (Schmidt 1973; Junk et al. 1983; Melack and Fisher 1983; MacIntyre and Melack 1984; Melack 1984). The contribution of oxygen from diffusion through the roots of emergent macrophytes (Jedicke et al. 1989) can be less than the consumption of oxygen by the easily decomposable organic matter that the macrophytes contribute to the water. Large lakes maintain better dissolved oxygen levels than small lakes because wind-induced turbulence is greater, thus increasing the input of oxygen from the air (Melack 1984). Turbulence also destroys macrophyte communities and promotes the growth of phytoplankton. This contributes less organic material and more oxygen to the water than emergent macrophytes. Large lakes in floodplains of black-water rivers suffer less hypoxia than comparable lakes of white-water rivers because of the smaller amounts of easily decomposable organic material (Fig. 20.4). Most Amazonian fish are resistant to low oxygen concentrations. Experiments in closed respiration chambers show that many species are able to tolerate oxygen concentrations lower than $0.5\,mg\,l^{-1}$, at least for some hours (Fig. 20.5). This ability allows survival during the critical hours before sunrise, when concentrations of dissolved oxygen in the várzea lakes are at their lowest levels. Many morphological, anatomical, physiological and/or ethological adaptations to low oxygen concentrations have already been described (Table 20.4), and many others probably exist (Junk et al. 1983; Soares 1993).

Different oxygen requirements and behaviors of the species lead to a distinctive distribution of fish species in the Amazon floodplain in accordance with oxygen concentrations in the respective habitats. Species poorly adapted to low oxygen concentrations (e.g., the clupeids *Pellona castelneana*, *P. flavipinnis*, *P. harrower*, *Pristigaster cayanae*, *Anchoa spinifer* (Engraulidae) and the sciaenids (*Plagioscion squamosissimus*, *P. montei*, and *P. surinamensis*) prefer well-oxygenated habitats in the open water of large lakes or in areas influenced by oxygen-rich waters of the main river. Benthic catfish, although adapted to low oxygen concentrations, stay out of floodplain lakes when the bottom is anoxic. In floodplain

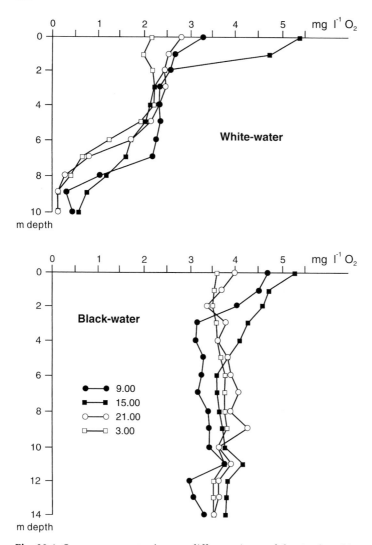

Fig. 20.4. Oxygen concentrations at different times of day in the white-water várzea lake Lago Inácio near Manacapurú and a black-water lake Lago do Prato in the Anavilhanas, Rio Negro, close to Manaus during high water level. Water temperature at noon varied in the white-water between 27.6 °C (bottom layers) and 30.6 °C (surface layers) and in the black-water between 27.7 and 28.6 °C, respectively

lakes small-scale horizontal migrations and vertical dislocations due to the diurnal cycle of oxygen concentration were observed (Saint-Paul and Soares 1987; Soares 1993).

Of the 132 species captured in Lago Camaleão and in the bay at its outlet, which is influenced by oxygen-rich river water, 40 species were frequently

Table 20.4. Fish species with adaptations to low oxygen concentrations in the hypoxic várzea lake Lago Camaleão. (Junk et al. 1983)

Family/species	Type of adaptation
Family Erythrinidae	
Hoplias malabaricus	Physiological adaptations
Hoplerithrynus unitaeniatus	Air breathing (buccal cavity, swim bladder)
Family Serrasalmidae	
Colossoma macropomum	Lip formation, physiol. adapt.
Colossoma bidens	Lip formation, physiol. adapt.
Mylossoma duriventre	Lip formation, physiol. adapt.
Family Characidae	
Brycon cf. melanopterus	Lip formation, physiol. adapt.
Triportheus angulatus	Lip formation, physiol. adapt.
Triportheus albus	Lip formation, physiol. adapt.
Family Osteoglossidae	
Arapaima gigas	Air breathing (swim bladder)
Osteoglossum bicirrhosum	Barbels, physiol. adapt.
Family Callichthyidae	
Hoplosternum thoracatum	Air breathing (intestine)
Callichthys callichthys	Air breathing (intestine)
Corydoras sp.	Air breathing (intestine)
Family Loricariidae	
Pterigoplichthys multiradiatus	Air breathing (stomach)
Family Synbranchidae	
Synbranchus marmoratus	Air breathing (gill cavity)
Family Electrophoridae	
Electrophorus electricus	Air breathing (buccal cavity)
Family Cichlidae	
Cichlasoma bimaculatum	Air breathing (stomach)
Family Lepidosirenidae	
Lepidosiren paradoxa	Air breathing (lung)

found in the strongly hypoxic inner parts of the lake. Twenty of these are known to have specific morphological, anatomical, and physiological adaptations to low oxygen concentrations (Table 20.4). Many ethological adaptations also exist, as shown by Soares (1993). During the night many fish moved from the macrophyte communities to the open water, and swam near the surface to make use of increased oxygen contents at the boundary layer until about 8 A.M. By then, oxygen conditions in the upper water layer and in the macrophytes had improved due to oxygen production by phytoplankton and periphyton. This allowed the fish to move to deeper water layers and into the macrophyte stands (Fig. 20.6).

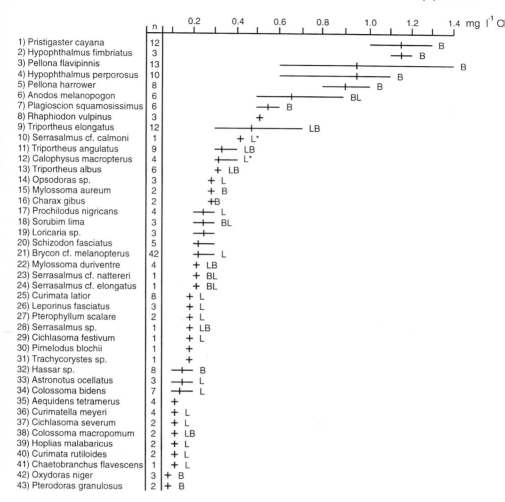

Fig. 20.5. Lethal oxygen concentrations measured in closed respiration chambers for fish from different habitats of the Ilha de Marchantaria. *B* Fish from the well-oxygenated bay at the entrance of Lago Camaleão; *L* fish from Lago Camaleão during strong hypoxia; *L** fish from Lago Camaleão during very high water levels and good oxygen conditions. **LB* Fish mostly in Lago Camaleão but sometimes also in the bay; *BL* fish mostly in the bay but sometimes also in Lago Camaleão

Under extreme hypoxic conditions some species show different behaviors, migrating from the open water into the macrophyte stands. These are able to use the small amounts of oxygen entering the water by exudation from the plant roots (Jedicke et al. 1989). In experiments, specimens in enclosures without plants died, but where plants were present they survived (Soares 1993). This behavior has also been observed in the field during the annual inflow of Antarctic air, locally called "friagem", in April/May over the South American continent. In Amazonia, for a period of 2–4

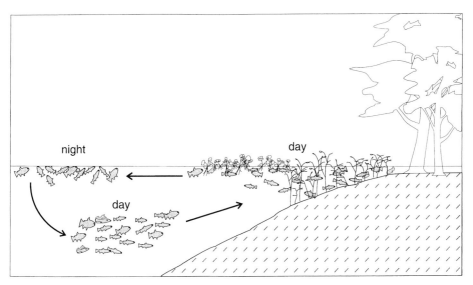

Fig. 20.6. Pattern of diurnal migrations of fish between macrophyte stands and open water in strongly hypoxic conditions. (Modified after Soares 1993)

days the temperature drops to below 20 °C and heavy rains together with strong winds lead to a complete turnover of the water column in the floodplain lakes, resulting in a sudden strong hypoxia over large areas. Severe fish kills occur (Santos 1973), but according to the observations of local fishermen fish have a chance of survival when hiding between the macrophytes.

As soon as oxygen conditions improve the species composition changes. In 1982 the Amazon River rose so high that it flooded the whole Ilha de Marchantaria, substituting the anoxic water of Lago Camaleão with oxygenated river water for a period of about 4 months. Species less tolerant to hypoxia immediately appeared, but then disappeared from the lake as soon as the water became hypoxic as the water level fell (Junk et al. 1993).

These results indicate that oxygen is an important factor in determining the species compositions of fish communities in várzea habitats. There is a gradient in hypoxia tolerance that, in addition to behavior and mobility, influences the location of the species in the different habitats and in the water column. In Lago Camaleão, most of the specimens found under extreme hypoxic conditions were juveniles of some characoid species, e.g., *Triportheus albus*, *T. elongatus*, *Prochilodus nigricans*, *Semaprochilodus taeniurus*, *S. theraponura*, *Curimata latior*, *C. ciliata*, *Colossoma macropomum*. Only a few predators were present, e.g., the pirarucú *Arapaima gigas* and *Synbranchus marmoratus*, which use atmospheric air for oxygen supply. *Raphiodon vulpinus, R, gibbus,* and *Hydrolycus scomberoides* were caught only in the bay at the mouth of the lake where oxygen conditions were better (Junk et al. 1983). Piranhas (*Serrasalmus* spp.), which in respirometer experiments have shown a high tolerance to low oxygen concentrations, also preferred to stay in the bay at the mouth of the lake. Junk et al. (1983) suggested that survival rates for juveniles may increase under strongly hypoxic conditions. Negative consequences of extreme hypoxia, such as a reduced variety of food (no benthos, reduced supply of perizoic species, less zooplankton), low water quality, and additional energy demand due to periodical need for surface respiration, are probably compensated for by reduced predation pressure. However, this situation may be misleading. Goulding (pers. comm.) points to the fact that populations of air-breathing pirarucú (*Arapaima gigas*) were much larger in former times than today, and that predation pressure in hypoxic habitats was therefore heavier.

20.6 Species Diversity

The number of fish species in Amazonian floodplains is very large. During a 2-year study on Lago Camaleão, 132 species belonging to 94 genera, 31 families and 10 orders were collected in an area of about 50 ha in size (Junk et al. 1983). The total number was certainly much greater, possibly by a factor of 2, because the sampling methods chosen were not suited for collecting the maximum number of species. For example, small species and those living in macrophyte communities were not sampled adequately. Bayley (1982), collecting fish with a fine-mesh seine net in the bay at the mouth of Lago Camaleão over a 2.5-year period, recorded more than 226 species belonging to 40 families and 132 genera.

High numbers of species have been reported from other studies too. Saint-Paul (1994) collected 153 species in Lago Inácio in the várzea and 173 species in Lago do Prato in the Anavilhanas archipelago of the Rio Negro. These species belonged to 29 families, 4 of which occurred only in black water (Cetopsidae, Chilodidae, Ctenoluciidae, and Hypopomidae) and another 4 only in white water (Callichthyidae, Electrophoridae, Gymnotidae, and Rhamphichthyidae). Here, too, the real species number is expected to be much higher.

Santos et al. (1984) collected about 300 species in the lower Tapajos. Goulding et al. (1988) carried out the most intensive collection of fish species in the lower and middle Negro River. In a stretch of 1200 km between Manaus and Barcelos, they found 450 species belonging to 202 genera and 39 families. However, considering the data available in the literature about the occurrence of species, they assumed that they had still missed many species and they estimated the total number of species in this area may reach about 700.

Due to their great mobility, many Amazonian fish species have a very wide distribution. This holds true principally for medium- and large-sized migratory species, which colonize the middle and lower courses and the respective floodplains of the Amazon River itself and its tributaries. Migrations and drift lead to an increased genetic exchange over large distances thus decreasing isolation, which is one prerequisite for speciation. Therefore, it is to be expected that the fish fauna in the Amazon River and its floodplain is rich in species but rather uniform over long distances. Goulding et al. (1988) confirm this for the main channel of the Negro River and its floodplain. Furthermore, they point to the fact that most genera found in the Negro River are also found in other large Amazonian rivers and the Orinoco. Major differences may be found between large regions, e.g., the estuary, the lower, middle, and the western Amazon (Goulding, pers. comm.).

However, the high number of fish species reported from the Amazon basin is not only the result of species diversity in the main river and its floodplain: to a large extent it also reflects the diversity of the small tributaries and headwaters. These species often have limited ranges of distribution, presumably because large rivers like the Amazon River and its main tributaries represent barriers for small species. The probability is very low that a small *Hyphessobrycon* or *Corydoras* from the headwaters of a left-bank affluent of the Amazon River could travel successfully downstream for hundreds of kilometers, cross the main channel, and then swim upstream for another few hundred kilometers to reach similar headwaters on the right bank. The large floodplains must be considered additional barriers for rheophilic headwater species, because of their lentic and often heavily hypoxic conditions.

Furthermore, physicochemical barriers exist. White-water and black-water rivers differ so much in hydrochemistry (Sects. 4.3.1–4.3.3) that specific adaptations are of evolutionary advantage for plant and animal species. After adaptation to specific hydrochemical conditions, the migration of a species from one affluent to the next one with comparable conditions may be prevented by the different hydrochemical composition of the connecting main river. *Symphysodon discus* is restricted to clear waters, whereas *S. aequifasciata* ecotype "Royal Blue" is found only in white waters (Geisler 1972). The mineral contents of the bones of these two species differ according to the hydrochemical conditions of the respective habitats. Calcium concentrations of 13.4% in the tail vertebrae of the white-water species are about twice as high as those in the clear-water species. These, in turn, had 3–4 times higher contents of magnesium and barium, probably to compensate for calcium (Geisler and Schneider 1976). The large number of variations in color of *Symphysodon* spp. coming from different regions indicates a separation of the respective populations. Goulding et al. (1988) provide evidence that at least 7 genera and 10 species living in the confluence of Amazon and Negro River do not enter the Negro River because of differences in water chemistry [*Rhytiodus argenteofuscus*, *R. microlepis*, *Schizodon fasciatus* (Anostomidae), *Pseudotylosurus microps* (Belonidae), *Mylossoma* 3 species (Characidae), *Cichlasoma* spp. (Cichlidae), *Psectrogaster* spp. (Curimatidae), and *Colomesus asellus* (Tetraodontidae)]. Similar barriers are to be expected for fishes migrating from black-water to white-water habitats.

Whether fish diversity in tropical seasonal environments is often lower than in nonseasonal water bodies as Lowe-McConnell (1987) suggested will depend on whether a connection exists with permanent water bodies from which recolonization can take place, the size of the flooded area relative to the permanent refuges, and the respective species assemblages. The large neotropical savannas, for example, are periodically inundated by rainwater to a depth of 50–150 cm for a few months during the rainy season and drained by rather small rivers. Species diversity inside the savannas may be low because the total number of species in the connected rivers is relatively small and time is too short to allow the recolonization of very large areas by the entire variety of species. Certain areas may even become connected with rivers only during extreme floods. Very few fish species can survive long periods of isolation, despite the fact that great mobility of the fish favors the faunal exchange over large distances (Barthem et al. 1991).

Large Amazonian river–floodplain systems, however, offer enough refuges for fish to overcome even severe droughts. Local extinction of species can easily be compensated for by immigration from other connected habitats, making Amazonian river–floodplains species rich habitats, despite the severe impacts of the annual drought. Interspecific competition for food or other resources is reduced by high mortality during the dry season, which does not permit the exclusive development of a few extremely competitive species eliminating other less aggressive ones. Great structural complexity in the floodplain offers very diverse habitats to the fish during high water, allowing a great number of species to share the same area. The hydrological dynamics of the river–floodplain system favor the occurrence of stochastic assemblages of species, as postulated by Lowe-McConnell (1979, 1987) and Goulding et al. (1988), if no other factors, such as severe hypoxia, force a specific community structure according to the level of tolerance against oxygen deficiency.

20.7 Discussion and Conclusions

Large river-level fluctuations greatly affect environmental conditions in the floodplain, forcing the fish to acquire flexibility in feeding habits and habitat demands (Fig. 20.7). Under such conditions, the development of

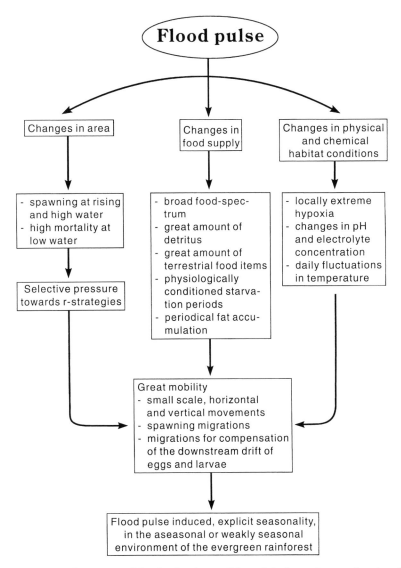

Fig. 20.7. The impact of the flood pulse on fishes of the large Amazonian river floodplains

complex behavioral patterns linked with specialization for specific nesting places, territorial behavior, parental care for juveniles or for specific food items is of advantage only in very specific conditions. Families which show these behavioral patterns (K-strategy), such as Cichlidae, Callichthyidae and Loricariidae, are represented in the floodplains by only a few species. Food availability is highly variable throughout the year and for most species includes a fasting period during low water. Many species show omnivorousness and periodic fat accumulation when they have access to the floodplain during high water period. Some perform large spawning and feeding migrations of hundereds of kilometers in extent. Most species realize small-scale migrations between the river and the floodplain, and within the floodplain, according to food availability and habitat conditions. Pronounced hypoxia is frequent and forces the animals to leave their habitats depending on diurnal or seasonal changes in oxygen concentrations, in spite of their tolerance to hypoxia.

Quick growth, early maturity, and high reproductive rates are advantageous (r-strategy) because there are annually large population losses during the low water period. Quick growth requires a good food supply. Food production depends on the fertility of floodplain soils and floodwater. In the nutrient-rich várzea, r-strategy species are very successful, as documented by large populations of characoid species. Because of their abundance some of them are of great commercial value for human consumption, e.g., *Prochilodus* spp., *Curimata nigricans*, *Brycon melanopterus*, *Triportheus* spp., *Anodus* spp. Large numbers of juveniles of this group make up a considerable part of the fish biomass in várzea lakes (Bayley 1982). The black-water river floodplain of the Negro River is of low fertility and contains only about 20% of the fish biomass found in the white-water floodplain of the Amazon River. There seems to be strong evolutionary pressure for small body size. Goulding et al. (1988) assume that in the Negro River more than 100 species are able to mature sexually before reaching 30 mm standard length. Fish of the dominant species in terms of numbers of individuals are smaller than 40 mm.

The fish fauna does not react to lower nutrient levels by a reduction in species number. The study of Goulding et al. (1988) showed great species diversity in the Negro River floodplain. Saint-Paul (1994) found that the number of species in a black-water floodplain lake is slightly greater than in a comparable white-water lake. High species numbers in várzea and igapó are maintained by great habitat diversity. The flood pulse diminishes interspecific competition because of the annual reduction of the populations during the low-water period. Extinction of species during severe droughts is avoided because the connection with permanent water bodies and the mobility of the fish allows the survival of a sufficiently large part of the populations.

The variable environmental conditions in river–floodplain systems differ fundamentally from the stable conditions in large and deep tropical lakes. Here, K-strategy, in the form of complex territorial behavior, parental care, extreme adaptation to specific habitats, nesting places, food sources, etc., is of evolutionary advantage. Cichlids show this behavior and represent the overwhelming number of species in the large African rift valley lakes. According to Daget et al. (1991), about 870 cichlid species occur in Africa. From South America 398 cichlid species are known, 285 are well described, and 171 of them occur in the Amazon basin (Kullander 1994). The attachment to one locality favors the isolation of populations and allows intralacustrine speciation. The extremely species-rich and complex cichlid community of Lake Malawi (more than 200 species) is considered an excellent example of radiation (Freyer and Iles 1972). Balon (1975) classified fish into reproductive guilds. In large African lakes guarders and bearers dominate and in large-river floodplains the nonguarders are more frequent in terms of species and individuals. The evolution of species with complex parental care and territorial behavior patterns from riverine ancestors with simpler reproductive behavior has been shown by Freyer and Iles (1972). In African rivers and their floodplains, where habitat conditions vary similarly to those in Amazonian floodplains, cichlids also occur in much lower numbers than in the lakes, and cyprinids with similar survival strategies to those of the Amazonian characoids predominate.

As shown by Welcomme (1985) and Lowe-McConnell (1979, 1987), in all tropical large-river floodplains a number of K-selected species occur, mainly in the group of "black fish", specialized to live and breed in the deoxygenated areas of floodplain lakes. Large and long-living species often occupy the position of top predators. Representatives of most of Balon's reproductive guilds are also found in várzea and igapó. However, their numbers are relatively small in comparison with the nonguarders. In Amazonia, environmental conditions for guarders, such as Loricariids and Cichlids seem to be more favorable in streams and small rivers than in the large floodplains.

However, it would not be correct to characterize the Amazonian fish fauna as an assemblage of poorly adapted r-strategists. As indicated by Balon (1975) the terms "primitive" and "complex" are rather conjectural because morphophysiological and ethological adaptations in reproductive systems in nonguarders can be as complex as in guarders and bearers. *Prochilodus insignis* shows many elements of r-strategy such as quick growth, early maturity, high reproduction rate and a lack of parental care.

At the same time, it also displays very complex migrational and schooling behavior, and has developed a specific anatomical adaptation (drum muscle) for sound production. Furthermore, fitness of the species depends not only on reproduction but also on physiological adaptations to environmental conditions, defence to predation, utilization of food resources, and tactics to excape in space and time (Southwood 1988). Adaptations of Amazonian floodplain fish include many elements of r-strategy regarding growth and reproduction, but also complex additional morphological, anatomical, physiological and ethological adaptations to cope with the conditions in a permanently changing environment.

Acknowledgments. We are very grateful to Dr. Rosemary Low-McConnell, Sussex, Dr. Michael Goulding, Gainesville, Florida, and Dr. Robin Welcomme, Fao Rome, for critical comments on the manuscript and the correction of the English.

21 Mammals, Reptiles and Amphibians

WOLFGANG J. JUNK and VERA MARIA F. DA SILVA

21.1 Introduction

When the first Europeans came to the Amazon they were deeply impressed by the rich wildlife along the river. Cristobal de Acuña, who participated in the Amazon expedition of the Portuguese General Pedro Teixeira in 1637–1638, reported the great abundance of river turtles, which the natives caught for food and stored alive behind fences. Caimans had been so frequent that they were crowded at low water level in the remaining lakes in the floodplain. The exploitation of the large stocks of manatees and turtles became economically important activities. At the end of the last century, Verissimo (1895) criticised the irresponsible fishing and hunting practises and predicted the extinction of the turtles, the manatee, and the giant osteoglossid fish Pirarucú (*Arapaima gigas*). After World War II, the hides of caimans became of interest for tanneries and the populations were greatly reduced in a few decades.

In spite of the economic importance of the species, very few efforts were made to study them. The descriptions of the first naturalists were the basis of the existing knowledge until the natural populations drastically declined. Recent research efforts, summarized by Best (1984), suffer serious problems because animal populations are very small and poaching continues in spite of the existence of protective laws. The following chapter aims to interpret the existing data in the framework of the flood-pulse concept.

21.2 Mammals

The Manatee. Manatees (*Trichechus inunguis*) occur in all major tributaries of the Amazon River, except in fast turbulent water. They belong to the few aquatic vertebrates which are strictly herbivorous and consume aquatic macrophytes in large quantities. Aquatic grasses are probably the

Ecological Studies, Vol. 126
Junk (ed) The Central Amazon Floodplain
© Springer-Verlag Berlin Heidelberg 1997

Table 21.1. Common mammals from the Amazon floodplains

Species	First sexual maturity	Number of offspring
Trichechus inunguis[a] (Amazonian manatee)	5–10 years	1 every 3 years
Hydrochoerus hydrochaeris (Capybara)	18 months	Up to 8, (mean 4) once, sometimes twice year^{-1}
Inia geoffrensis[a] (Large red river dolphin)	Females 5 years, males 10 years	1 every 2 years
Sotalia fluviatilis[a] (River dolphins)	5–10 years both sexes	1 every 2 years
Pteronura brasiliensis (Giant otter)	2–3 years	1–4 every year by the dominant female of the group
Lutra enudris[a] (Amazonian river otter)	2–3 years	1–3 once year^{-1}

[a] According to Best (1984; modified and amplified by Vera M.F. da Silva, Laboratory of Aquatic Mammals, INPA).

main food item, but they consume a wide variety of aquatic plants. Daily food uptake reaches about 8% of body weight with a digestive efficiency of 45–70% (Best 1981). In the várzea, manatees were exceptionally frequent because of the abundant aquatic plant growth (Sect. 8.5).

The animals migrate with rising water into the floodplain and there accumulate large amounts of fat. As water level falls they return to the main channel or permanent floodplain lakes. An extremely low metabolic rate facilitates their survival during a prolonged starvation period when aquatic macrophytes are not available (Gallivan and Best 1980; Best 1983). The calving and lactation period coincide with the period of highest water when availability of food for females is greatest (Table 21.1) (Best 1982a). The female produces one calf after a gestation period of 1 year.

Large-scale manatee hunting had already begun in 1580 (summarized in Best 1984). In the mid-1600s up to 20 Dutch ships filled with manatee meat were sent annually to Europe. A new era of commercial hunting began in the 1930s when tanneries developed methods to preserve the strong hide. From 1935 to 1954 between 80 000–140 000 manatees were killed in addition to the subsistence kill in the more remote areas of the Amazon (Domning 1982). Today, the manatee is a very rare and endangered species.

The Capybara. The other herbivorous mammal that occurred in large numbers in Amazonian floodplains is the capybara (*Hydrochoerus hydrochaeris*). This is a large rodent of about 50 kg of weight, which has semiaquatic habits and slightly webbed toes. It ranges from Central America to northern Argentina. According to the detailed studies of Ojasti (1973, 1983) in Venezuela, the animal reproduces after 18 months, has up to 8 offspring (average 4) and is able to reproduce throughout the year (Table 21.1). Mating takes place in the water. The birthrate peaks during rising water, after a gestation period of about 120 days. This is possible because the animals have access to food during both the aquatic and the terrestrial phases. They are very efficient grazers. During the low water period, the capybaras congregate in large groups near water bodies, and then disperse in smaller groups during the flood season (Schaller and Crawshaw 1981).

According to Ojasti (1980) and Cordeiro et al. (1981), population densities in open savanna regions of Venezuela reach about 1.8 animals ha^{-1}, and in forested habitats they reach about 2.1 animals ha^{-1}. The natural population densities in Amazonian floodplains can only be guessed at, because populations have been greatly reduced by hunting. However, food availability in the várzea is considered better than in the floodable savannas of Venezuela and in the igapó. Using various habitats, the capybara respond to environmental changes both by social structures and the use of food resources. In the várzea of the Amazon River food was probably not a limiting factor for capybara populations, which were controlled more by extreme floods or droughts and diseases than by food shortage.

The River Dolphins. The piscivorous river dolphins *Inia geoffrensis* and *Sotalia fluviatilis* still occur in large numbers in the Amazon River and its large tributaries. As they play an important role in local folklore and mythology they are not hunted and their populations can still be considered stable. *Inia reaches* a size of 2.6 m and a weight of 160 kg (Best and da Silva 1989); *Sotalia* grows to 1.5 m length and 53 kg (da Silva and Best 1994). *Sotalia* live in schools of 2 to 6 individuals, whereas *Inia* is considered a solitary dolphin (Magnusson et al. 1980), although its social structure depends on the season and habitat.

After a gestation period of about 12 months for *Inia* and 10 months for *Sotalia*, a single calf is born (Table 21.1). The breeding season starts with receding water level when fish are forced to leave the floodplain and become crowded in the main channels (Sect. 20.3). This increases food availability for the female dolphins during the last months of pregnancy and the lactation period (Table 21.1). The fact that *Inia* calves before *Sotalia* is interpreted by Best (1984) as a response to food availability. *Inia* preys more on solitary benthic fish, and *Sotalia* on schooling fish (da Silva 1983,

1986). Solitary benthic fish become more vulnerable as soon as the water recedes, whereas schooling fishes are most vulnerable at low water and as the water level starts to rise.

The Otters. The giant otter (*Pteronura brasiliensis*) and the Amazonian river otter (*Lutra longicaudis*) are not typical of the large Amazonian rivers and their floodplains. Giant otters prefer small and slow-moving rivers and streams of transparent water (Best 1984, Duplaix 1980). In seasonally flooded areas territories are defended continuously throughout the low water period. During the flood season the otters move into the flooded forest following the influx of fish (Duplaix 1980; Schweizer 1992; Carter and Rosas in prep.). Their diet is mainly of fish, especially characoids. The litter size in *Pteronura brasiliensis* varies between one and five (average two), depending on the habitat and season (Duplaix 1980). Evidence exists that breeding can occur year-round in environments with marked terrestrial and aquatic phases, with peak births occurring at the beginning of the terrestrial phase (Duplaix 1980; Brecht-Munn and Munn 1988). Cubs are kept in dens for about 5 weeks until their eyes are open and they are able to swim (Autuori and Deutsch 1977).

The otter, *Lutra longicaudis*, inhabits smaller creeks, lakes and swampy areas further from the main rivers, and apparently does not need to build a den in terra firme to raise its young (Schweizer 1992). The litter size is one to three (average two). Peak births occur mainly during the low water period, although breeding is apparently possible year-round. Their diet is much more varied than that of *P. brasiliensis*. Their main prey are small fish, crustaceans, amphibians, reptiles, birds, and small mammals (Blacher 1985).

The number of otters has been greatly reduced by hunting for the valuable fur. Official statistics indicate that over 40 000 pelts of giant otters were exported from Brazil alone between 1960 and 1967 (Best 1984), and even with the abolition of the fur trade and the significant reduction in prices giant otters are still hunted for their pelts (Carter and Rosas, in prep.). *Pteronura brasiliensis* is classified as "vulnerable" (IUCN 1990) and considered the most endangered otter species in the world by the World Conservation Union (IUCN).

21.3 Reptiles

The Turtles. The big river turtles colonize the floodplains of the large rivers. *Podocnemis expansa* is the largest species, with a mean carapace length between 64 and 80 cm (Pritchard and Trebbau 1984). Their reproductive activities depend on the hydrological cycle (Sect. 2.3). At low water each female buries between 48 and 132 eggs on a sandy beach. Thereafter, the adult animals spend, without eating, several weeks in deep pools near the nesting beaches until the beginning of the rains (Moreira and Vogt 1990). The eggs take about 45–48 days to hatch (Ojasti 1971). They hatch when the river level is already rising, which provides food and shelter on the floodplain. Hatchlings are about 5 cm long and immediately race for the water. Colony nesting of *P. expansa* and synchronous production of thousands of hatchlings during a short period is interpreted as a survival strategy, because predation pressure on the hatchlings during the first days is very heavy (Ojasti 1971; Alho and Padua 1982; Best 1984). The age of first reproduction for most species is not known. According to Alho and Padua (1982) females of *P. expansa* reproduce with a minimum carapace length of about 50 cm. Estimates of Ojasti (1971) suggest at least 5–7 years, and those of Alho et al. (1970) 15 years before this species nests (Table 21.2).

Table 21.2. Common reptiles from the Amazon floodplains

Species	First sexual maturity	Number of offspring	
Turtles:			
Podocnemis expansa	At least 5–7 years	48–134	Every 2–4 years
P. unifilis		21–40	Once year[-1]
P. erythrocephala		5–14	Once year[-1]
P. sextuberculata		6–16	Once year[-1]
Peltocephalus tracaxa		7–25	Once year[-1]
Phrynops (Batrachemys) raniceps[b]		2–4	Several times year[-1]
P. (Mesoclemys) gibbus[b]		2–4	Several times year[-1]
P. rufipes[b]		?	?
Platemys platycephala		4–7	Several times year[-1]
Chelus fimbriatus		12–28	Once year[-1]
Kinosternon scorpioides		2–4	Several times year[-1]
Rhinoclemys punctularia		2–4	Several times year[-1]
Caimans:			
Melanosuchus niger (Black caiman)[a]	?	30–60	Every 2 years
Caiman crocodilus (Spectacled caiman)[a]	5 years	14–40	Every 2 years, in captivity every year
Snakes:			
Eunectes murinus (Anaconda)	?	14–82	Every year?

[a] According to Best (1984; modified and amplified by Ronis da Silveira, pers. comm., PG/Ecology, INPA).
[b] These species seem to be restricted to small black-water rivers (Moreira, pers. comm.).

Podocnemis erythrocephala is smaller than *P. expansa*, and is restricted to black-water rivers. It nests in river banks of sand and clay with some vegetation as the water starts to recede (Moreira and Rabelo 1995). Females generally have an annual cycle. Many of the smaller turtle species prefer small rivers where the flood pulse is unpredictable. The courtship of *Platemys platycephala* takes place during the rainy season in shallow water. It nests on the ground, far away from the water to avoid flooding of the eggs (Table 21.2). *Chelus fimbriatus*, the famous mata-matá, is found in a variety of aquatic habitats and nests very close to the water. *Kinosteron scorpioides* seems to be restricted to the lower Amazon (Pritchard 1979a,b). The only record of *Rhinoclemys punctularia* in the central Amazon is in black waters, in the Uatumã River (Moreira, pers. comm.). Breeding behavior depends on the habitat, but habitat preferences have been little studied. Most species are solitary nesters with small clutch sizes. Turtles with a very small clutch size, e.g., *Phrynops* spp. may produce several clutches in a breeding season.

The commercial exploitation of eggs and adults of the large river turtles, mainly *Podocnemis expansa* and *P. unifilis*, started at the end of the sixteenth century and became a highly organized industry. According to Bates (1864) at least 6000 pots of oil from turtle eggs were exported annually from the upper Amazon River and Madeira River in the 1860s with an additional 2000 pots consumed locally. Each pot required 6000 eggs. Smith (1979, 1981) calculated the use of 208.9 million turtle eggs from historical reports. Initially, missionaries and state officials enforced conservation rules to protect parts of the beaches, but met with little success in the long term.

Snakes. A very common snake in the Amazon floodplain is the anaconda (*Eunectes murinus*). Its occurrence is closely related to rivers because its prey is mainly aquatic or semiaquatic, e.g., fish, turtles, caimans, capybaras. Mating takes place in the water, and the gestation period is about 270 days. Anacondas are viviparous. Litter size varies between 14–82 animals of about 75 cm length. The breeding season is during low water level. According to Belluomini et al. (1976/1977), after 3 years, the animals reach a length of about 3 m, then grow more slowly in length but become proportionally much heavier. Many myths about giant anacondas exist in the Amazon, including reports that they reach a length of more than 20 m. Verified records give a maximum length of about 12 m.

Other aquatic snakes abundant in the Amazon floodplain belong to the genus *Helicops* known as "Jararaca-d'agua" or "Surucucurana" and *Hydrops* (false coral snake). The species of both genera prey on small fish (Cunha and Nascimento 1993). *Helicops polyles* is viviparous.

The Caimans. There are two species of caiman in the floodplains of the middle Amazon: the black caiman (*Melanosuchus niger*) and the spectacled caiman (*Caiman crocodilus*). Two other species, *Paleosuchus palpebrosus* and *P. trigonatus*, prefer small forest streams. *Melanosuchus niger* grows to 6 m body length *Caiman crocodilus* to 2.5 m, *Paleosuchus trigonatus* to 2.3 m, and *P. palpebrosus* to 1.6 m. Juvenile caimans feed on mollusks, insects, crustaceans and small fish. Adults have a wide spectrum of prey species including mollusks, insects, crustaceans, amphibians, fish, birds and mammals (Best 1984). Snails of the family Ampullariidae seem to be a very important food item in the várzea (R. da Silveira, pers. comm.), and a preferred food item for *Caiman crocodilus* in the Pantanal of Mato Grosso.

Caimans have a relatively flexible nesting period. On Marajó Island, in the mouth area of the Amazon River, most *Caiman crocodilus* nest at high water when danger of flooding the nests is relatively small and the hatchlings find sufficient food (mostly insects and mollusks) and shelter in the floodplain (Best 1984). Near the Mamirauá Ecological Station near Tefé, between the Amazon River and the Japurá River, both species nest at low water from September to November (R. da Silveira, INPA, pers. comm.). Parental care is reported for both species. The females of *Caiman crocodilus* protect the juveniles for up to 18 months. In the lower Amazon, the black caiman is reported to estivate during the dry season (Bates 1864; Hagmann 1909).

As long as there was no commercial use for their skins the black caiman was 20 times more frequent than the spectacled caiman (Bates 1864; Hagmann 1909). The leather industry initiated the hunting of the black caiman in the 1940s and the hunting of the spectacled caiman, which was considered of poor quality, in the 1950s. Today, both species are much depleted, the spectacled caiman now being about 10 times more frequent than the black caiman (Rebelo and Magnusson 1983). In Colombia records indicate that about 10 million caiman hides were obtained between 1951 and 1976 (Medem 1982).

21.4 Amphibians

Studies and surveys of amphibians are mainly concentrated in areas of terra firme. Very little is known about the species of the large river floodplains, although amphibians are common and rich in species. Hödl (1977) lists 15 sympatric anuran species with synchronous breeding in the floating vegetation of Lago Janauari (*Sphaenorhynchus carneus, S. dorisae, S. aurantiacus, Hyla* sp. (*nana*-like), *H. haraldschultzi, H. rossalengi,*

H. triangulum, H. egleri, H. boesemani, H. punctata, H. raniceps, H. lanciformis, H. boans Fam. Hylidae, *Lysapsus lim. laevis* Fam. Pesudidae and *Leptodactylus wagneri* Fam. Leptodactylidae). The breeding season is related to the beginning of the rainy reason. Mismating is avoided by differences in mating calls. There is no further information available about specific survival strategies. Moreia (pers. comm.) observed a large breeding aggregation of *Hydrolaetane schmidti* in a stream near the mouth of Rio Jauaperí during the peak of the flood season. In *Hyla wavrini*, aggregations of males also formed at the margins of stream, rivers and lakes in igapó and várzea habitats. Most of these have been observed in habitats subjected to seasonal floods, where it is suspected that they reproduce (Martins and Moreira 1991).

21.5 Discussion and Conclusions

Historical records show that the large Amazonian rivers and their floodplains were formerly colonized by large populations of manatees, capybaras, caimans, dolphins, and river turtles. Indeed the large river floodplains, mainly the floodplains of white-water rivers, were the only ecosystems in Amazonia where wildlife was abundant in spite of the hunting pressure of a relatively dense indigenous human population of up to 28 people km^{-2} (Denevan 1976). After colonization by Europeans the indigenous population was greatly reduced, or nearly extinguished, by diseases and slave raids. At the same time the stocks of game animals were greatly reduced by hunting over a very short period of time, in spite of a smaller human population and a very large area which was hardly accessible. Today, except for the river dolphins, all other species are heavily exploited and considered endangered or vulnerable.

This catastrophic "experiment" highlights the special ecological situation of the floodplain ecosystems and the vulnerability of some of their organisms to human impact. Dolphins, manatees, turtles, otters, and caimans have to be considered as K-strategists. They are of large body size, reach sexual maturity rather late, and have a small number of offspring. Populations invest in a large biomass of adult animals which have few natural enemies and may become very old. The species are well adapted to the conditions in the floodplain. Reproductive strategies tend to optimize food availability for the females or hatchlings, availability and security of nesting places, and shelter for the hatchlings. In cases involving parental care, breeding success increases. A small number of offspring is sufficient to maintain a large population of adults in spite of the drastic changes in environmental conditions created by the flood pulse. Populations are controlled more by extreme flood events or diseases than by predators.

The long-standing legends suggest that fishing and hunting practises were strongly regulated by the Indians with taboos to minimize overexploitation of the stocks. Hunting was practised for subsistance. When the Europeans came to the Amazon, exploitation changed. Large amounts of meat, oil, turtle eggs, and hides were exported. Exploitation of all species was facilitated by the flood pulse. During low water level, animals were forced to leave their habitats in the floodplain, where they were well protected, and concentrated in the river and remaining pools where they were very vulnerable to hunters. Especially in years with a very low water level, the populations were greatly endangered by drought as well as by hunting. In 1963 for example, the Amazon River dropped to one of the lowest levels ever recorded, and many manatees, trapped in the drying lakes and channels, were killed by hunters (Thornback and Jenkins 1982). Natural recruitment did not compensate for such heavy losses and populations declined rapidly. Today, the number of individuals is so low that recovery of the populations is very slow. In the case of the manatee, the occasional poaching of a few adult animals, mainly females with calves, may eliminate the breeding success of the population in a large area.

The evaluation of the role of the large mammals and reptiles in the ecosystem is rather speculative. With the large reduction of the populations of manatees, capybaras, and turtles, man has cut important links in the food webs by which large amounts of plant material were transferred into animal protein. The animals continuously recycled considerable amounts of nutrients, stored in the plant material, into the water (Best 1982b). The large caiman populations had a similar recycling effect (Fittkau 1970b, 1973). Considering that in many floodplain lakes nitrogen is a limiting factor to phytoplankton production (Sects. 4.4, 10.2.2), we postulate that the large populations of mammals and reptiles formerly strengthened the phytoplankton pathway in the food webs.

Acknowledgments. We thank R. da Silveira and G. Moreira for sharing their data and knowledge on reptiles and amphibians. F. Rosas and B. Robertson made helpful comments on the manuscript.

22 The Birds

Peter Petermann

22.1 Introduction

Avian species richness of Amazonia is legendary. Not only is the total number of species greater than in any other rainforest area (Karr 1989), but the local diversity is also greater (Haffer 1990) despite the restricted ranges of many species. One component of this wealth of species are birds of riparian habitats. They make up some 15% of the terrestrial avifauna endemic to Amazonia (Remsen and Parker 1983).

The avifauna of the Amazon floodplain as such has rarely been addressed in ornithological literature (Remsen and Parker 1983), though parts of the floodplain are included in many regional studies. Despite the flourish of Neotropical ornithology over the last few decades (reviews in Haffer 1978, 1988, 1990; Snow 1980; Parkes 1985), it is important to be aware of some bias.

There are few long-term study sites distributed over several million km^2 of rainforest (Gentry 1990b; Haffer 1990). Bird communities of the floodplain have been the focus of research in few and widely separated sites, e.g., Lower Amazonia (Snethlage 1913; Lovejoy 1974) and Upper Amazonia (Remsen and Parker 1983; Terborgh et al. 1990; Rosenberg 1990). Even information on geographical distribution is still incomplete. Range extensions of birds continue to be reported frequently (Ridgely and Tudor 1994: e.g., p. 270, Klages' Antwren, *Myrmotherula klagesi*; p. 492, Olive-green Tyrannulet, *Phylloscartes virescens*; p. 772, Crimson Fruitcrow, *Haematoderus militaris*; Peres and Whittaker 1991). Regional differentiation might therefore be underrated in avian ecology.

Transitional forests and várzea forest adjacent to terra firme have repeatedly been investigated (Lovejoy 1974; Novaes 1980; Robinson et al. 1990), but other habitats such as lakes (Willard 1985; Bolster and Robinson 1990; Remsen 1990), sandbars (Krannitz 1989; Groom 1992), or pioneer scrub (Terborgh and Weske 1969; Rosenberg 1990) have received much less attention (e.g., floating meadows, black-water inundation forests).

Ecological Studies, Vol. 126
Junk (ed) The Central Amazon Floodplain
© Springer-Verlag Berlin Heidelberg 1997

Community structure has been at the center of most studies, as it is essential for the explanation of species diversity and for conservation measures (Terborgh 1980; Munn 1985; Bierregard and Lovejoy 1989; Remsen 1990; Johns 1991). Questions of biogeography and speciation have been discussed extensively in a historical perspective (Haffer 1985), though the investigations of the genetic basis for the biogeographic theories has only recently begun (Capparella 1991). On the other hand, autecology (e.g., Kiltie and Fitzpatrick 1984), breeding biology (e.g., Davis 1953; Oniki and Willis 1982, 1983), and ethology (e.g., Fitzpatrick 1981) of floodplain birds are much less known (Robinson et al. 1990). The influence of the flood pulse on birds has apparently not yet been the focus of research (except in the discussion of Rosenberg 1990; Haffer 1994).

Data presented here were collected during a 2-year study from September 1988 to June 1990 on Marchantaria Island (Ilha de Marchantaria) (Sect. 2.2).

22.2 The Environment of Birds: Dynamics and Structures

Environmental pulsation is the most outstanding feature of floodplains. It is related to climate or flood pulse, which are causally and temporally coupled, but not exactly synchronous. The flood pulse is of paramount ecological importance to the floodplain (Junk et al. 1989), where landscape structures are just the temporary expression of the dynamic processes involved.

The climate of the floodplains differs from that of the terra firme rainforest as shown in Section 2.3. Reasons for this are to be found in the local topographical features. Floodplains cover the valley floors, i.e., the lowest parts of a larger region. In Central Amazonia the altitudinal difference between várzea and terra firme amounts to some 50 m. Due to adiabatic warming of the air, valleys are usually climatically drier than the adjacent upland. Forests with their buffered interior microclimate cover only parts of the floodplain, frequently interrupted by river channels and lakes. Vast areas of open water, beaches, and low pioneer vegetation have rather extreme microclimates (Ribeiro and Adis 1984; Molion and Dallarosa 1990). The flood pulse of the Amazon follows the seasonality of rainfall. In Central Amazonia the Amazon reaches maximum flood level 4 to 6 weeks after the end of the rainy season (Junk et al. 1989; Sect. 2.3).

The dynamic processes of erosion and sedimentation result in a mosaic of habitats that is variable in space and time. Successional stages form distinctive bird habitats. Descriptions of bird habitats in the floodplain have been given by Remsen and Parker (1983), Terborgh (1983), and Rosenberg (1990). Their classifications are based on the situation in the Upper Amazon; it differs somewhat from that of Marchantaria Island, which is outlined in Section 2.6.

Patchy growing terrestrial stands of grasses, annual herbs, and willows (*Salix humboldtiana*) form the first-year vegetation on sandbars. Pioneer vegetation of sandbars is included in "sandbar scrub" below (Fig. 22.1a,b). Floating meadows were included in "water-edge habitats" by Remsen and Parker (1983; Fig. 22.2c) (Chap. 8).

Lakeside scrub usually borders the edge of várzea forest on the island, but also grows on narrow levees. It is completely different in floristics from sandbar scrub. In areas with high sedimentation rates it is dominated by *Alchornea castaneifolia*, with *Pseudobombax munguba*, *Cecropia* sp., mistletoes (e.g., *Psittacanthus* sp.) and species of the "mid-level tree community" (Junk 1989). It has a much more diverse vegetation, with abundant flowers and fruits, mostly of mistletoes. It is flooded annually, but is never completely submerged. This scrub might be identical to the "river-edge forest" of Remsen and Parker (1983), though it is lower in stature and different in floristic composition. *Heliconia* spp. are very rare, and *Cecropia* spp. are predominant only on disturbed sites, e.g., forest clearings (Fig. 22.1d).

The várzea forest ("high-level tree community", Junk 1989) originally covered the core and western part of the island, but has been increasingly reduced by clearings to now approximately 300 ha. The greatest part of the higher, shorter inundated parts of the forest have been transformed into plantations and man-made second-growth. The forest is at least partially flooded every year. Forests on Marchantaria Island are probably intermediate between the "river-edge forest" and "várzea forest" of Remsen and Parker (1983; Fig. 22.1e) (Chap. 11).

Man-made second-growth is extensive in the western parts of the island, which are higher and therefore more rarely inundated and were not intensively investigated in this study.

22.3 Species Numbers

A total of 204 species of birds were observed at Marchantaria Island from 1988–1990. Four to five additional species have been reported in other years or are known to occur in the close vicinity (Petermann 1996, in prep.).

a

b

Fig. 22.1. a Sandbar of Marchantaria Island, late dry season (19 October, 1988). Breeding site of Collared Plovers (*Charadrius collaris*), Terns (*Phaetusa simplex, Sterna superciliaris*), Skimmers (*Rynchops nigra*), Sand-colored Nighthawk (*Chordeiles rupestris*), Red-breasted Blackbird (*Leistes militaris*). **b** Sandbar scrub of Marchantaria Island, mid-rainy season, at the beginning of inundation (6 March 1990). Trees are 2-year-old willow (*Salix humboldtiana*). Site of Fork-tailed Flycatcher (*Tyrannus savana*), Lesser Hornero (*Furnarius minor*), White-bellied Spinetail (*Synallaxis propinqua*), River Tyrannulet (*Serpophaga hypoleuca*), Yellow-browed Sparrow (*Ammodramus aurifrons*), etc. **c** Floating meadows bordering lakeside scrub at Lago Camaleão, Marchantaria Island, late rainy season (16 May 1990), with Black-capped Donacobius (*Donacobius atricapilla*). Site of Azure Gallinule (*Porphyrio flavirostris*), Gray-breasted Crake (*Laterallus exilis*), Yellow-chinned Spinetail (*Certhiaxis cinnamomea*), Yellow-hooded Blackbird (*Agelaius*

c

d

icterocephalus), Orange-fronted Yellow-Finch (*Sicalis columbiana*), etc. **d** Open lakeside scrub of Marchantaria Island, mid-rainy season (20 April 1990). Site of Olive-spotted Hummingbird (*Leucippus chlorocercus*), Green-tailed Goldenthroat (*Polytmus theresiae*), Red-and-white Spinetail (*Certhiaxis mustelina*), Lesser Wagtail-Tyrant (*Stigmatura napensis*), Rusty-fronted Tody-Flycatcher (*Todirostrum latirostre*), Oriole Blackbird (*Gymnomystax mexicanus*), Masked Yellowthroat (*Geothlypis aequinoctialis*), Bicolored Conebill (*Conirostrum bicolor*), etc. **e** Várzea forest of Marchantaria Island, peak of an exceptionally high flood, beginning dry season (19 July, 1989). Forest edge cleared by local peasants (*caboclos*). Site of Rufescent Tiger-Heron (*Tigrisoma lineatum*), Long-billed Woodcreeper (*Nasica longirostris*), White-eyed Attila (*Attila bolivianus*), Leaden Antwren (*Mymotherula assimilis*), Chestnut-headed Oropendula (*Psarocolius angustifrons*), etc.

Fig. 22.1e

On Marchantaria Island for 57 species (28%) breeding has been confirmed by observation of nest building, feeding of young, etc. For 66 other species (32%) we suspect that they were breeding, as they were singing, defending territories, showing year-round presence in adequate habitats etc., though no nests could be found. The remaining 40% of the species thus almost certainly do not breed on Marchantaria Island.

Certainly the list of species recorded on Marchantaria Island is not yet complete. Continued observations will probably add species from several groups. Several species of migrant shorebirds have been recorded from inland Amazonia but not yet from Marchantaria Island (e.g., Buff-breasted Sandpiper, *Tryngites subruficollis*, Sick 1967; Stotz et al. 1992). Also species that are usually strictly coastal during migrations might be expected. Parrots, raptors, hummingbirds, and other species that habitually fly long distances are likely to accidentally pass through Marchantaria Island. Records of forest canopy birds are from visual observations only as mist nets could not be operated in the canopy of the inundated forest; so more species might have remained undetected there than in other habitats. The greatest part of the sandbar scrub was only 2 years old at the end of this study and additional species were still arriving. For example, the Black-and-white Antbird, *Myrmochanes hemileucus*, was found by Cohn-Haft et al. (pers. comm.) in the year following the end of this study.

Mainly the following three species groups distinguish the floodplain avifauna from terra firme avifauna: migrants, wetland species, and terrestrial species more or less completely restricted to Amazon floodplains ("várzea species") (Remsen and Parker 1983). The remaining group (excluding migrants, "várzea species", and wetland species) is composed of widespread species of secondary vegetation, forest edge, and man-made habitats (e.g., Roadside Hawk, *Buteo magnirostris*, Great Kiskadee, *Pitangus sulphuratus*, Blue-gray Tanager, *Thraupis episcopus*), with only a few rainforest species (e.g., Greater Yellow-headed Vulture, *Cathartes melambrotus*). Species living on forest floor are completely absent (e.g., Tinamidae).

22.3.1 Migrants

Twenty-one species are northern migrants (10%), about half of which stay here all winter. At least eight species are southern migrants (4%). The number of southern migrants might be greater as in some species breeding and wintering ranges overlap (Sick 1968). On Marchantaria Island, this has only been confirmed for the Fork-tailed Flycatcher (*Tyrannus savana*), which is breeding in small numbers in sandbar scrub (confirmed by the presence of nests), but appears in enormous numbers in May and June to feed on fruiting trees in the várzea (see also McNeil and Itriago 1968; Haverschmidt 1975; Hilty and Brown 1986).

There are 59 species (29%) not known to be long-distance migrants which stayed only temporarily on Marchantaria Island (including some breeding species). Seven of these species are rare visitors. This group contains species that straggle in from nearby várzea and/or anthropogenic habitats. Others fulfill different types of regular or sporadic migrations and possibly are unrecognized long-distance migrants. Among the 116 resident (year-round present) species (57%), at least nine show seasonal fluctuations in numbers, indicating migrations.

While the distribution, migration routes etc. of the Nearctic migrants are fairly well known, the austral migrations remain somewhat obscure (Ridgely and Tudor 1994). Most migrants at Marchantaria Island are shorebirds and aerial-feeding insectivores. In those groups, they outnumber residents both in species and in individuals.

Nearctic Migrants

Shorebirds. The largest group of migrants at Marchantaria Island are shorebirds, most of them on their way to wintering sites in the southern

Neotropics (Sick 1968; Myers 1980). On the migration south they find the sandbars and beaches exposed, but submerged on the return.

The Golden Plover (*Pluvialis dominica*), which rests at Marchantaria Island for a few days in large numbers in September and October (see also Sick 1968; Stotz et al. 1992), during spring migration passes via the High Andes (Fjeldså and Krabbe 1990). Other species probably use the Venezuelan Llanos as a substitute for the Amazon floodplain (Thomas 1987). However, there are more grassland species on the list for the Llanos (e.g., Buff-breasted Sandpiper, *Tryngites subruficollis*) and more sandy-shore species at Marchantaria Island (e.g., Sanderling, *Calidris alba*).

Due to the continuous and rapid decrease in the water level during the months of southward migration of the shorebirds, muddy beaches are rather narrow and ephemeral (Bolster and Robinson 1990). In Central Amazonia precipitation of the short rainy season forms pools on top of the sandbar, often overgrown by vegetation. Usually more shorebirds are found here than on the riverbank (e.g., Spotted Sandpiper, *Actitis macularia*; Solitary Sandpiper, *Tringa solitaria*; White-rumped Sandpiper, *Calidris fuscicollis*; Upland Sandpiper, *Bartramia longicauda*). This habitat is apparently exclusive to very extensive sandbars of Central Amazonia and provides the most reliable resting sites for shorebirds (together with várzea lakes). The overall importance of the Amazon in the migration systems cannot yet be evaluated.

Raptors. Two species of raptors are regular visitors. Ospreys (*Pandion haliaetus*) are most abundant in northern winter, which coincides with low water level of the Amazon, i.e., the season with high fish density in remaining pools (Junk et al. 1983). Young birds stay until the second year (Henny and van Velzen 1972). Wintering Peregrine Falcons (*Falco peregrinus*) are the only raptors of Marchantaria Island that prey exclusively on birds.

Passerines. In Central Amazonia, only swallows winter regularly in greater numbers (Stotz et al. 1992). On Marchantaria Island, they roost in floating meadows, but are rarely seen in the day, except after rainstorms, when sometimes hundreds of them rest on the wet sand of the sandbar.

Neotropical Migrants

Austral Neotropical Migrants. These include swallows, tyrants (e.g., Fork-tailed Flycatcher, *Tyrannus savana*; White-throated Kingbird, *Tyrannus albogularis*; Small-billed Elaenia, *Elaenia parvirostris*), the Lined Seedeater

(*Sporophila lineola*, Ridgely and Tudor 1989) and the Nacunda-Nighthawk (*Podager nacunda*). Data are insufficient to judge whether austral migrants winter in Amazonia or pass through on the migrations to northern Neotropics. For Fork-tailed Flycatchers (*Tyrannus savana*) wintering in northern Venezuela, McNeil and Itriago (1968) proposed that they had stored sufficient fat to cross Amazonia without needing to refuel. However, the alleged migration range of 1500 km would not be sufficient to reach the summer ranges, and it seems more likely that the northern wintering population stops over in the floodplains. Austral migrant tyrants are especially conspicuous when they gather in large flocks in fruiting trees, e.g., *Laetia corymbulosa* or *Vitex cymosa*. Some wetland species will have to be included in austral migrants as a Snowy Egret (*Egretta thula*) banded in Argentina was recovered in Amazonia (Olrog 1975).

Intra-Amazonian Migrants. Some Neotropical species migrate without leaving Amazonia (Haffer 1988). Such intra-Amazonian migrations are known for birds of sandbars (e.g., Sand-colored Nighthawk, *Chordeiles rupestris*, Collared Plover, *Charadrius collaris*; Sick 1967), dense aquatic vegetation (e.g., Azure Gallinule, *Porphyrio flavirostris*, Remsen and Parker 1990), and fruit-eating birds (e.g., White-eyed Parakeet, *Aratinga leucophthalmus*, Pale-vented Pigeon, *Columba cayennensis*, Sick 1967; Toucans, Sick 1984).

Perhaps the most remarkable example of an intra-Amazonian migrant is the Sand-colored Nighthawk (*Chordeiles rupestris*; Pinto 1947; Sick 1950, 1967). This species breeds on sandbars of most Amazonian rivers, often in colonies of terns and skimmers (Sick 1950; Groom 1992), but is missing in Lower Amazonia east of the Rio Xingu (Sick 1984). It does not range outside the Amazon basin except in the tributaries of the Upper Orinoco (Hilty and Brown 1986).

Sick (1967) hypothesized that the birds migrate regularly between those parts of the Amazon river system which have low water because their habitats are inundated every year. As the flood pulse in the various parts of the vast Amazon basin is not synchronous, there should always be some regions with sandbars.

Observations on Marchantaria Island do not support this hypothesis. The island has a small migratory breeding population. In both years of this study, the birds arrived at the end of the rainy season but before the flood reached its peak, and left when the short rainy season began while the river was still at its lowest level. The greatest danger to clutches on sandbars are the torrential rains, which frequently washed away eggs of Collared Plover (*Charadrius collaris*) and Large-billed Terns (*Phaetusa simplex*) in the short rainy season. In the Sand-colored Nighthawk no loss of clutches was observed, but the birds avoided resting on the sand when it was soaked with water soon after the recession of the flood or rain.

I presume that the migrations of Sand-colored Nighthawks are essentially between those river segments which have a flood pulse of low frequency and high predictability (i.e., the lower reaches including the Amazon/Solimões itself) and others with high flood frequency but rather low amplitude and duration. In the former they can raise their broods with little danger of flooding but encounter no adequate habitats for the rest of the year, whereas in the latter some adequate habitat is available all the time, but breeding success is threatened by unpredictable flooding. There they can molt in the relative security of small islands (Willis and Oniki 1988 found groups on islands in the Rio Uatumã in February, some 150 km northeast of Marchantaria Island). It seems likely that the migrations of Sand-colored Nighthawks are controlled by endogenous cycles, with the flood pulse as a potential synchronizer. The proximate factor triggering migrations is probably linked to precipitation.

In contrast to the Sand-colored Nighthawk, the Collared Plover (*Charadrius collaris*) depends on sandbars not only for breeding but for foraging as well. Its seasonal movements seem to be still enigmatic. Numbers of this species fluctuate throughout most of its vast range (e.g., Panama, Ridgely and Gwynne 1989; Paraguay, Hayes and Fox 1991) and it is likely that migration is not confined to Amazonia.

22.3.2 Wetland Species

Wetland species here are fish-eating birds (Podicipedidae, Ciconiiformes, Laridae, Alcedinidae), and waterfowl (Anseriformes, Heliornithidae, Rallidae). Generally, they make up a small fraction of avian species in Amazonia and none is endemic (Haffer 1985).

Nonmigratory wetland birds add 45 species (22%) to the Marchantaria list, including 33 aquatic (16%) and 12 species inhabiting swamp vegetation in- and outside Amazonia.

Swimming Birds. There are remarkably few species of swimming and diving birds on Marchantaria Island. Three are piscivorous (see below). Two duck species inhabit forest edge and sandbars rather than aquatic habitats (Muscovy Duck, *Cairina moschata*, Black-bellied Tree-Duck, *Dendrocygna autumnalis*). Horned Screamers (*Anhima cornuta*), formerly seen on Marchantaria Island (Junk, pers. comm.), are now restricted to várzea lakes south of the Solimões. The Sungrebe (*Heliornis fulica*) is a characteristic waterfowl of forest rivers in the Neotropics.

Fish-eating Birds. The composition of the guild of fish-eating birds at Marchantaria Island is somewhat depleted compared to other Neotropical (Morales et al. 1981; Dubs 1983; Kushlan et al. 1985; Thomas 1985), and even Amazonian wetlands (Willard 1985, Remsen 1990). Storks are missing, as well as species characteristic of forest rivers, or coasts (e.g., Tricolored Heron, *Egretta tricolor*).

Also missing are the species of the "small river community" (Davis 1953), characteristic of shady rivers in the rainforest (e.g., Agami Heron, *Agamia agami*, Green-and-rufous Kingfisher, *Chloroceryle inda*; Hilty and Brown 1986; Remsen 1990). Though Amazon islands of large size have forested creek-like channels (*paranás*) connecting river branches and lakes, which might have these species, similar structures do not exist on Marchantaria Island.

There are only two species that regularly fish by swimming and diving, the Neotropical Cormorant and the Anhinga (*Phalacrocorax olivaceus, Anhinga anhinga*). The Least Grebe (*Podiceps dominicus*), not known to breed in Central Amazonia, has only been recorded once as a straggler; the Sungrebe (*Heliornis fulica*) seems to dive only when in danger, and kingfishers do not swim.

Anhingas are regular visitors in small numbers (1–3), mostly in the central lake of Marchantaria Island. Along black-water rivers (e.g., Rio Negro) they are much more common than Neotropical Cormorants. Apparently, Anhingas prefer transparent water, whereas cormorants hunt in great numbers in turbid, shallow white water with greater fish biomass.

Some smaller fish-eating species breed at Marchantaria Island. Two species of terns (Large-billed Tern, *Phaetusa simplex*, Yellow-billed Tern, *Sterna superciliaris*) and the Black Skimmer (*Rynchops nigra*) form colonies on the sandbar, the characteristic species assemblage of Neotropical beaches (Preston 1962; Krannitz 1989; Groom 1992). Three species of kingfishers and the Mangrove Heron (*Butorides striatus*) breed on Marchantaria Island or in the banks of the Solimões nearby. The larger herons and egrets, which usually breed colonially, are present mainly from December until May, but then disappear to unknown breeding sites.

Most species of fishing birds show seasonal fluctuations in numbers, though few species are completely absent for some time of the year (e.g., Black Skimmer, *Rynchops nigra*).

The seasonal variability of fish-eating species is shown in Fig. 22.2. Numbers are estimates based on the maximum number counted per month 1988–1990 in the sedimentary zone downstream, and in the lakes of Marchantaria Island, but not along the banks on the west and south.

Species composition was very different in the rainy and dry season. During the first half of the year, floating meadows were crowded with flocks

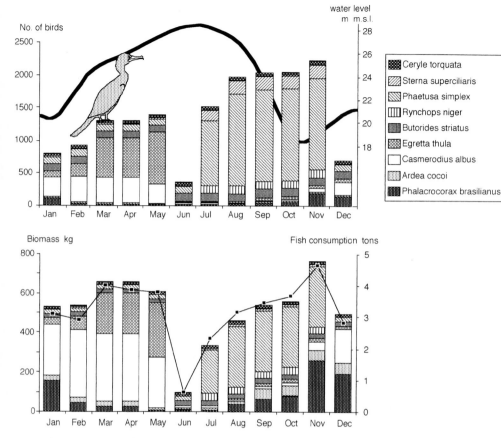

Fig. 22.2. Seasonal aspects of fish-eating birds. Species composition in individuals (*upper graph*), compared to the flood pulse (*line*). Species composition in biomass (*lower graph*) and fish consumption (*connected points*). See text for further details. Sketch: *Phalacrocorax olivaceus (= bras.) brasilianus*

of several hundred egrets (Great Egret, *Casmerodius albus*, Snowy Egret, *Egretta thula*) feeding at pools and appearing at fish-kills in lakes, which regularly occur when the water level begins to rise (not only during *friagens*; see Brinkmann and Santos 1973). Terns occur in only small numbers close to the island (and therefore in Fig. 22.2!), though they are still abundant on the middle Amazon, mainly at várzea lakes, but also entering black-water rivers. The total number of terns in the whole area does not appear to be smaller than in the breeding season. Thus neither of the terns seemed to participate in migrations to low water rivers (*contra* Sick 1967).

Ringed Kingfishers (*Ceryle torquata*) usually perched in lakeside forest, hunting in pools in floating meadows or at forest edges. Less frequently they entered the flooded forest, as Pygmy Kingfishers (*Chloroceryle aenea*) and Mangrove Herons (*Butorides striatus*) regularly do. Other fish-eating birds rarely venture inside the forest (e.g., Anhinga, *Anhinga anhinga*, Snowy Egret, *Egretta thula*).

At the end of the rainy season most egrets left. At this time the floating meadows started to disintegrate due to a stronger current at the peak of the flood and the deterioration of feeding habitats might have forced the birds to depart. Fish predation by birds was then low for some time, as the absence of the wading birds was not compensated for by terns, for as long as the sandbars were inundated.

In the second half of the year more than 1000 terns gathered on the sandbars to breed, hunting mainly over shallow water bordering the sandbars. As long as the water level was falling, shores were free of vegetation and herons and egrets were fishing from muddy shores. Cocoi Herons (*Ardea cocoi*), the largest of the piscivorous birds of the island, ventured farthest into shallow water. Species fishing from perches (e.g., Mangrove Heron, *Butorides striatus*, Black-collared Hawk, *Busarellus nigricollis*, Ringed Kingfisher, *Ceryle torquata*) convened on stranded, uprooted trees over water, though they defend feeding territories at other times. Cormorants concentrated at receding pools with very high fish biomass during low water and consequently showed a negative correlation with water level. Besides specialized fish-eating species, the abundance of fish in receding pools attracted species which are only occasionally piscivorous, e.g., Black Vultures (*Coragyps atratus*) and Yellow-headed Caracaras (*Milvago chimachima*). Black Vultures took advantage of fish, mostly Piranha, *Serrasalmus* spp., thrown on shore by fishermen, while Yellow-headed Caracaras also took (dead) fish from the surface or tried to steal prey from egrets.

The rising of the water level caused an abrupt change in species composition. As soon as the pioneer vegetation on the shores became inundated, most fishing birds changed their feeding habits. Great Egrets (*Casmerodius albus*) feeding in loose groups of some 20 birds along the shore, gathering easy prey along the shore and in shallow water rather than actually hunting, showed signs of food scarcity within a few days. At times more than ten birds were fighting over every single fish that was caught sporadically by one of them.

Obviously fish were entering the recently inundated pioneer vegetation (Soares et al. 1986), and the egrets either could not follow them as long as the floating vegetation was too weak to support them, or just took some time to adapt their behavior to the novel situation. A few weeks later, numbers of egrets increased, apparently due to birds returning from their unknown breeding colonies.

Fish consumption can be estimated as a function of biomass/bird (Fig. 22.2). Data of bird biomass were derived from literature and only the smaller species were captured in mist nets (e.g., Ringed Kingfisher, *Ceryle torquata*, Mangrove Heron, *Butorides striatus*). Daily fish consumption per bird was calculated using the formula by Nilsson and Nilsson (1976):

$$\log F = -0.293 + 0.850 \log W, \tag{1}$$

Where F = fish consumed in g, W = body mass of bird in g. A species weighing 100 g would thus consume approximately 25% of its body mass per day, a 1 kg bird approximately 16%. These values are slightly higher than estimated for African birds by Junor (1972: ≈ 16%), or Bowmaker (1963: ≈ 14%).

The monthly fish consumption (F/month) per species on Marchantaria Island was calculated as follows:

$$F/\text{month} = F \times (\text{mean numbers of birds}) \times (\text{number of days}). \tag{2}$$

For comparison, the mean biomass of birds per month (B/month) is also given in Fig. 22.2:

$$W/\text{month} = W \times (\text{number of birds}). \tag{3}$$

In the course of the year, the number of fish-eating birds is more variable than the monthly biomass or monthly fish consumption (0.6 to 4.6 t), but this is not statistically significant. Total annual fish consumption is approximately 40 t/a fresh mass (38.3 t calculated). The formulas applied by Morales et al. (1981) to piscivorous birds in the Venezuelan Llanos, based on somewhat different assumptions, result in almost identical estimation (39.0 t) for Marchantaria Island.

However, there are several potential biases in this estimation. Not all piscivorous species were included in the calculation. Omitted species are either too small (Pygmy Kingfisher, *Chloroceryle aenea*), too rare (Osprey, *Pandion haliaetus*) or too catholic in their diet (Black-crowned Night-Heron, *Nycticorax nycticorax*) to contribute predictably to fish consumption.

The numbers of individuals are conservative estimates: small species were certainly underrated. However, they contribute little to fish consumption as long as they do not occur in extremely high numbers. This is very improbable.

Not all piscivorous species feed exclusively on fish. Terns frequently forage for aerial insects (Reichholf 1982/1983). For example in December–January 1989/1990, approximately 70% of the pellets (n > 100) of Large-billed Terns (*Phaetusa simplex*) at a sandbar, where several hundred terns were molting, consisted exclusively or mainly of remains of large leafcutter ants (*Atta*), with fish bones in the rest. Large-billed Terns were often seen in flocks hawking at flying insects over floating meadows before sunset.

The impact of birds on the fish biomass and production is difficult to assess, as the fish density is extremely variable in space and time in the floodplain. Both fish production and consumption are therefore best related to a unit area of the várzea. Assuming the area of Marchantaria Island to be 3000 ha, annual fish consumption then amounts to 13 kg ha^{-1} a^{-1}. According to Welcomme (1990), the fishery yield for the Solimões floodplain is 8.1 kg ha^{-1} a^{-1}. The fish consumption of birds is of the same order of magnitude as the fishery yield.

22.3.3 Várzea Species

Twenty-five species of Marchantaria Island (12%) are restricted to "river-created habitats" in the Amazon-Orinoco lowlands (Remsen and Parker 1983: category A). When the species found in nonriparian habitats or species ranging beyond the Amazon basin are included in this group, the number rises to 67 (33%; categories A–E); these will be called "várzea species" here, as none is exclusively found in black-water inundation forests, *igapó*).

22.4 Biogeographic Relationships

For several species, the records from Marchantaria Island show a considerable extension in the known ranges. Though the rivers are traditionally the main travel routes for naturalists, bird collectors and ornithologists in Amazonia, the ranges even of abundant "várzea species" (e.g., White-bellied Spinetail, *Synallaxis propinqua*, Rosenberg 1990) are not well known. Records from the island are filling in the gaps of a seemingly disjunct distribution of some species (Least Bittern, *Ixobrychus exilis*, White-bellied Spinetail, *Synallaxis propinqua*, Yellow-bellied Elaenia, *Elaenia flavogaster*, Dull-capped Attila, *Attila bolivianus*, Bicolored Conebill, *Conirostrum bicolor*, Orange-headed Tanager, *Thlypopsis sordida*), or are extending their known ranges to the east (Olive-spotted Hummingbird, *Leucippus chlorocercus*) or west (Green-throated Mango, *Anthracothorax viridigula*, Fork-tailed Flycatcher, *Tyrannus savana*, Masked Yellowthroat, *Geothlypis aequinoctialis*) (Grantsau 1989; Ridgely and Tudor 1989, 1994).

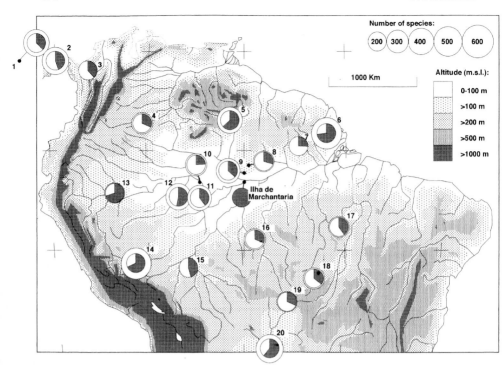

Fig. 22.3. Comparison of the avifaunas of selected sites in the Neotropics. *Larger circle* is the total number of species reported. *Segment in inner circle* is the percentage of bird species common to Marchantaria Island. Study sites and references: *1* La Selva/Costa Rica (Karr et al. 1990); *2* Barro Colorado Island/Panama (Karr et al. 1990); *3* Urabá/Colombia (Haffer 1959); *4* Vaupés/Colombia (Olivares and Hernandez 1962); *13* River islands/Peru (Rosenberg 1990); *14* Cocha Cashu/Peru (Karr et al. 1990); *15* Bení (Remsen 1986). Sites in Brazil: *5* Maracá/RR (Moskovits et al. 1985; Silva and Oren 1990); *6* Amapá (Novaes 1974, 1978); *7* Paru de Leste/PA (Novaes 1980); *8* Rio Urubu/AM (Willis and Oniki 1988); *9* M.C.S.E. Reserves near Manaus/AM (Karr et al. 1990); *10* Tefé/AM (Johns 1991); *11* Rio Urucu/AM (Peres and Whittaker 1991); *12* Juruá/AM (Gyldenstolpe 1945); *16* Aripuaná/ AM, MT (Novaes 1976); *17* southeastern Pará (Novaes 1960); *18* north-eastern Mato Grosso (Fry 1970); *19* Serra das Araras/MT (Silva and Oniki 1988); *20* southern Mato Grosso (Dubs 1983)

Most species on the island have a wide geographical distribution (Fig. 22.3) with almost half of them reaching Panama and two-thirds of them occurring in southern Mato Grosso, areas outside the Amazon rainforest, Study sites in terra firme rainforest share fewer species with Marchantaria Island than sites in savanna wetlands (e.g., the Pantanal of Mato Grosso, Dubs 1983). Naturally the overlap in species is greatest with Peruvian river islands (Rosenberg 1990) and other sites which include floodplain habitats (e.g., Cocha Cashu, Peru; Karr et al. 1990).

The relationships of the birds of Marchantaria Island to avifaunas outside Amazonia are also apparent at the generic level, as shown by the

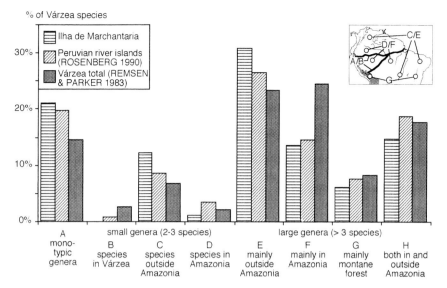

Fig. 22.4. Comparisons of the geographic distribution patterns of genera of all várzea birds (Remsen and Parker 1983), and of those "várzea species" found on Peruvian river islands (Rosenberg 1990), and on Marchantaria Island (this study). (Petermann 1996, in prep.)

distribution patterns of genera of "várzea species" (Fig. 22.4; Remsen and Parker 1983). Of all the "várzea species", about equal numbers belong to genera distributed in lowlands either inside or outside of Amazonia, with some genera that are either monotypic or contain mainly montane forest species. In contrast, the species of Marchantaria Island and Peruvian river islands (Rosenberg 1990) belong mainly to genera distributed outside Amazonia (see also Willis 1992).

Marchantaria Island shares many species with the campos of Lower Amazonia (Snethlage 1910) and savannas in the southwest of Amazonia (Remsen 1986), though several characteristic species are lacking. These species prefer short-grass vegetation (e.g., Yellowish Pipit, *Anthus lutescens*), or semi-open campos (Guira Cuckoo, *Guira guira*, Chalk-browed Mockingbird, *Mimus saturninus*, Sick 1984; Ridgely and Tudor 1989).

22.5 Habitat Preferences

General information about habitat preferences of "várzea species" exists for most species (Remsen and Parker 1983; Hilty and Brown 1986; Ridgely and Tudor 1989, 1994; Rosenberg 1990).

To evaluate the differential use of habitats, for every species an index of relative habitat preference (rhp) was calculated from census and capture data, ranging from 0 (never recorded) to 1 (exclusive habitat of the species). It should be kept in mind that the figures are rough estimates of the number of birds multiplied by the time spent by the birds in the habitat. They do not weigh the functional importance of the habitats for the bird (e.g., the Chestnut-bellied Seedeater, *Sporophila castaneiventris*, is often captured in lakeside scrub, where it apparently seeks shadow, but neither feeds nor breeds). Furthermore, habitats outside the island might be more important than an occasionally visited habitat on the island.

The means of the rhp's of all species per habitat indicates the degree of overlap in the bird community of one habitat with others. Figure 22.5 visualizes the overlap of species based on species numbers (Fig. 22.5A) and rhp (Fig. 22.5B).

Of the total of 192 Amazonian species of "river-created habitats" (Remsen and Parker 1983) only approximately 35% are found on Marchantaria Island. The portions of missing and present species differ between habitats. Figure 22.6 compares the habitat preferences (following Remsen and Parker 1983) of all species and the respective portions found on Marchantaria Island (species might be assigned to several habitats). The total number of species recorded on Marchantaria Island rises with the structural complexity of the habitats. Sandbars have the lowest species number, followed by floating meadows, sandbar and lakeside scrub, and forest. However, sandbars have almost as many species as floating meadows, and the difference between forest and lakeside scrub is also small (partially due to seemingly or really "missing" forest species).

22.5.1 Sandbar

Besides aquatic habitats, sandbars and floating meadows undergo the most profound changes in the annual cycle. In most years sandbars vanish completely for some time. Birds are confronted with the choice of emigrating or moving to other habitats, though there is no obvious potential substitute habitat for the sandbar.

Species composition is dominated by non-passerines, mostly wetland species, e.g., migrant shorebirds or breeding terns and skimmers. The number of "várzea species" is low, including occasional visitors from other habitats (e.g., Oriole Blackbird, *Gymnomystax mexicanus*). Nevertheless, most (70%) of the Amazonian species restricted to this habitat are represented (Fig. 22.6).

During high water, most long-distance migrants, some "várzea species" and other local species leave the island rather than change the habitat (e.g.,

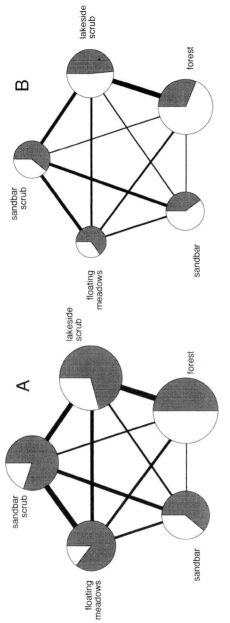

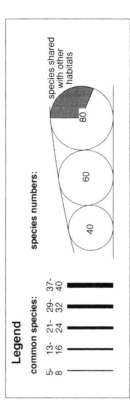

Fig. 22.5A,B. Species number per habitat (*circles*) and overlap with other habitats (*connecting lines*). The proportion of exclusive species is left *white*, species shared with other habitats are *shaded*. **A** Total number of species. **B** Numbers weighted by relative habitat preference

absolute number of species (line), and
% of spp. (bars). resp.

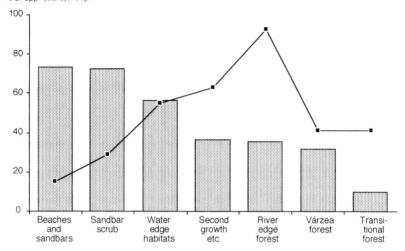

Fig. 22.6. Habitat preferences of "várzea species". The total number of species per habitat, according to Remsen and Parker (1983), *connected points* and percentages found at Marchantaria Island (*bars*) are given

Sand-colored Nighthawk, *Chordeiles rupestris*, Collared Plover, *Charadrius collaris*, Red-breasted Blackbird, *Leistes militaris*). Mean rhp therefore is high if all species are included, but reduced if restricted to "várzea species".

Ground-feeding species move to fazendas (cattle ranches) on higher ground in the floodplain (e.g., *Leistes militaris*, Yellow-browed Sparrow, *Ammodramus aurifrons*). It is not clear how they behaved before this anthropogenic habitat existed in Central Amazonia, though probably they formerly migrated to the várzea campos of Lower Amazonia, where Snethlage (1913) noticed large flocks of birds. However, it is also possible that they increased their ranges following the expansion of fazendas into Central Amazonia. The Red-breasted Blackbird is currently spreading into deforested areas in South America (Hilty and Brown 1986; Sick 1984; Ridgely and Tudor 1989).

Breeding activity of most ground breeders usually ends at the beginning of the short rainy season (e.g., Great-billed Tern, *Phaetusa simplex*, Sand-colored Nighthawk, *Chordeiles rupestris*, *Leistes militaris*, *Ammodramus aurifrons*), whereas in some it continues at a low intensity (e.g., Ladder-tailed Nightjar, *Hydropsalis climacocerca*).

22.5.2 Floating Meadows

Like sandbars, floating meadows are only available as a habitat for a part of the year (Chap. 8). Nevertheless, only few bird species characteristic for this habitat leave the island (e.g., Least Bittern, *Ixobrychus exilis*), whereas most accept pioneer vegetation on the sandbars as substitute or seasonally complementary habitat (e.g., Purple and Azure Gallinule, *Porphyrio martinicus*, *P. flavirostris*). Mean rhp of floating meadows is consequently low.

Migrants are not a characteristic element of this habitat, though some species of swallows and tyrants (Hirundinidae, Tyrannidae) roost in floating meadows in huge flocks (Stotz et al. 1992). The austral migrant Nacunda Nighthawk (*Podager nacunda*) hunts in great numbers over floating meadows during the southern winter (end of May to end of August).

Floating meadows are probably inadequate breeding sites, as the flexible stems offer little reliable support to hanging nests, and nests built close to the water surface are in danger of being drawn under water by the rhizomes of the grass. Several species that are numerically dominant in floating meadows breed in low pioneer vegetation of the sandbar (Yellow-hooded Blackbird, *Agelaius icterocephalus*, Chestnut-bellied Seedeater, *Sporophila castaneiventris*).

Floating meadows are the only of the "water-edge habitats" (Remsen and Parker 1983) on Marchantaria Island (which also include e.g., *Mauritius* palm groves, bamboo thickets, *Heliconia* thickets), hence the low representation of birds of this group (Fig. 22.6).

Jacanas (*Jacana jacana*) are among the most conspicuous birds of floating meadows. Like most species of this habitat (e.g., Yellow-hooded Blackbird, *Agelaius icterocephalus*, Purple Gallinule, *Porphyrio martinicus*, Yellow-chinned Spinetail, *Certhiaxis cinnamomea*) they are widespread throughout Neotropical wetlands. Their extraordinarily long toes enable them to walk over swimming vegetation, lawns or shores, taking up invertebrate and vegetarian food (Beltzer and Paporello de Amsler 1984). Sexual roles in breeding are reversed and females are polyandrous under certain circumstances (Jenni and Collier 1972; Osborne and Bourne 1977; Osborne 1982). This allows females to raise several clutches simultaneously or to replace them repeatedly when nest destruction is frequent (Osborne and Bourne 1977). Their clutch of usually four eggs is deposited on floating, often rather flimsy, nest platforms in aquatic vegetation (Jenni and Betts 1978).

At Marchantaria Island Jacanas breed in floating meadows. They walk with ease on horizontal mats of floating vegetation but neither cling to emergent shoots, nor enter densely growing stands of grass. Thus, they are effectively excluded from great parts of the floating meadows and habitats do not overlap with those of the *Porphyrio*-Gallinules, which prefer dense, tall grass. Jacanas started breeding (confirmed by nests) in February, soon after the floating meadows began to develop. They were usually encountered in pairs, sometimes with young birds, without indications of polyandry.

In 1988–1990 no more than five fledglings (easy to recognize, e.g., Hilty and Brown 1986) were encountered between May and August. Breeding was successful when the water level was falling and the floating meadows were reduced to mats of decaying vegetation, which gave firm support to nests. When the water level rises, growth of the floating grass just suffices to keep the shoots above water (Piedade et al. 1991). Nests are then exposed to wind, human and animal predators, currents, etc., causing nest destruction. Certainly the conditions vary interannually, so that early breeding in some years might be successful. Having a breeding system that is tolerant to nest losses, Jacanas are preadapted to this environment.

22.5.3 Sandbar Scrub and Lakeside Scrub

The two scrub communities differ remarkably in species numbers. Sandbar scrub shares most species with the floating meadows, many of which also occur in lakeside scrub at high water. Many species are exclusive to scrub habitats, both sandbar and lakeside (e.g., Bicolored Conebill, *Conirostrum bicolor*, Red-and-white Spinetail, *Certhiaxis mustelina*, White-bellied Spinetail, *Synallaxis propinqua*). The lakeside scrub is richer in species of this group (e.g., Lesser Wagtail-Tyrant, *Stigmatura napensis*, Fuscous Flycatcher, *Cnemotriccus fuscatus*). Even more important is the close proximity of the forest, as lakeside scrub is regularly visited by forest species (cuckoos, hummingbirds, tyrant flycatchers etc.). Flowers and fruits are missing in sandbar scrub, though one hummingbird species (Black-throated Mango, *Anthracothorax nigricollis*) appears regularly to hunt for small invertebrates. Flowers are abundant in lakeside scrub, where they attract most of the hummingbird species of Marchantaria Island, as well as parrots, tyrant flycatchers, etc. Species of scrub habitats contribute most to endemism of the floodplain avifauna.

22.5.4 Várzea Forest

The várzea forest of Marchantaria Island is richest in non-passerines, e.g., woodpeckers, raptors, parrots, doves etc., but is less speciose than lakeside scrub in oscine passerines. It also has a smaller proportion of "várzea species" if compared to lakeside scrub, as well as compared to the total of "várzea species" in forest habitats (Remsen and Parker 1983; Fig. 22.6). However, it is a general phenomenon that the number of "várzea species" in forest habitats is low compared to lakeside scrub as well as to the total species number of these forests (e.g., Robinson et al. 1990). These forests usually share most species with terra firme forests. At Marchantaria Island there are few terra firme birds, and the paucity of undergrowth (and probably also of canopy) birds is not compensated for by birds of scrub habitats, as can be seen from the high mean rhp. Few species of river-edge forests venture inside the forest (e.g., Leaden Antwren, *Myrmotherula assimilis*, Rusty-backed Spinetail, *Cranioleuca vulpina*).

Several characteristic species of the forest on Marchantaria Island breed within the zone regularly inundated for some months, e.g., the Long-billed Woodcreeper (*Nasica longirostris*):

The Long-billed Woodcreeper is a common bird of várzea forests and is often reported also from forest edge and second-growth on terra firme (Remsen and Parker 1983; Rosenberg 1990; Ridgely and Tudor 1994). Little is known about its breeding habits (Hilty and Brown 1986). On Marchantaria Island, a nest of the Long-billed Woodcreeper was found in the rainy season (5 April) in the várzea forest. The "nest", consisting of flakes of bark, was placed in the stump of a broken spiny palm tree (cf. *Astrocaryum* sp.) in rather open undergrowth. It contained three eggs, from which two young hatched (the third egg disappeared). At the time of egg laying the river water must have started to invade the forest; the young birds fledged successfully with a margin of about 2 weeks before the drowning of the nest site. In the previous year with exceptionally high flood (1989), the nest site had been inundated approximately 10 days earlier.

The breeding of this species was remarkable in three aspects:

1. The clutch was larger than for most woodcreeper, or Amazonian forest species (Snethlage 1935; Hilty and Brown 1986; Sick 1984).
2. The relationship of adult mass to individual egg mass did not differ from the "normal" value of this family, as calculated by the formula of Rahn et al. (1975): 9.1 and 9.5 g vs. calculated 9.1–9.7 g (depending on assumed adult mass, 82 g, Graves and Zusi 1990; or 88 and 89 g of fledging young, this study).
3. Development of young was apparently faster than expected by comparison with nestling growth rates in Ricklefs (1968), Drent and Daan (1980), O'Connor (1984).

It is tempting to view these features (time of breeding and rapid nestling development) as being adaptive to the rapidly changing floodplain environment, with the greatest amount of food available during the flooding of the forest, and simultaneously maximum shelter for the brood; however, the data can only serve to encourage further studies in that line.

Nesting close to rising water was found in species both breeding in and outside of várzea forest (e.g., Ringed Kingfisher, *Ceryle torquata*, Red-and-white and Yellow-chinned Spinetails, *Certhiaxis mustelina*, *C. cinnamomea*, White-headed Marsh-Tyrant, *Arundinicola leucocephala*, Black-backed Water-Tyrant, *Fluvicola albiventer*, Chestnut-bellied Seedeater, *Sporophila castaneiventris*, Yellow-olive Flycatcher, *Tolmomyias sulphurescens*, Social Flycatcher, *Myiozetetes similis*). Breeding at high water levels, i.e., in the rainy season, is common among várzea birds, whereas on terra firme Oniki and Willis (1982, 1983) found more nests in the climatic dry season.

22.6 Dispersal Capabilities

Floodplain birds are presented with two fundamental challenges. In the short term they must tolerate or evade the impact of the flood pulse and over longer time periods they must be able to disperse promptly to newly created patches of their habitat, as the older areas change or are eroded (Rosenberg 1990; Haffer 1994).

Of course, these constraints are habitat specific. For example, birds of sandbars have to change their breeding site repeatedly in their lifetime, while it takes many bird generations to grow a várzea forest (Irion et al. 1983; Junk 1989).

The dispersal of birds becomes manifest by the occupation of new sites. In the course of this study, more than 50 ha of barren sandbar, which in September 1988 were almost devoid of vegetation, had by December 1989 developed dense scrub dominated by approximately 5–10 m high willows, *Salix humboldtiana*.

Possible sources for colonizing birds were three small stands of *Salix humboldtiana*, several years old, but less than 20 m in diameter. They had some species typical of sandbar scrub (e.g., Rusty-backed Spinetail, *Cranioleuca vulpina*, White-bellied Spinetail, *Synallaxis propinqua*), but not the complete set. More extensive areas of sandbar scrub were found on the opposite side of the north arm of the Solimões, and along the south bank of Marchantaria Island. The distance from lakeside scrub to the island measured approximately 2 km.

After 1.5 years, in a 24-ha study plot, then mostly covered by *Salix humboldtiana* with several patches of *Paspalum* cf. *repens*, some 30 species were recorded (excluding species of bare sandbar), of which at least 7 were breeding (Table 22.1).

Repeated mapping of territories was hampered by the rapid development of the vegetation cover. Though high density was found of the Ladder-tailed Nightjar (*Hydropsalis climacocerca*; 11 clutches), most species

Table 22.1. Birds in early successional vegetation on the sandbar: species list and breeding status are differentiated for four stages of early succession on the sandbar. The sandbar scrub vegetation was less than 2 years old

	Stage 1: bare sand	Stage 2: sparsely growing herbs and scrubs	Stage 3: densely growing annual macrophytes	Stage 4: sandbar scrub	Total
Laterallus exilis			x		x
Porzana flaviventer			x		x
Porphyrio martinicus		x	x	x	x
Porphyrio flavirostris		x	x	x	x
Jacana jacana	(x)	x			x
Charadrius collaris	B	B			B
Phaetusa simplex	B				B
Sterna superciliaris	B	B			B
Rynchops niger	B				B
Columbina talpacoti		x			X
Coccyzus melacoryphus				x	X
Crotophaga ani		B	X	B	B
Crotophaga major				x	X
Tapera naevia			B?		B?
Tyto alba		x			x
Chordeiles rupestris	(x)	B			B
Hydropsalis climacocerca		(x)	B	B	B
Anthracothorax nigricollis		x	x	x	x
Furnarius minor		B?		B?	B?
Synallaxis albescens			(x)	B?	B?
Synallaxis propinqua		(x)	(x)	B?	B?
Cranioleuca vulpina		x	x	B	B
Certhiaxis cinnamomea			B	X	B
Certhiaxis mustelina				B	B
Todirostrum maculatum		x		B?	B?
Serpophaga hypoleuca		B		B	B
Knipolegus orenocensis			(x)	B?	B?
Fluvicola albiventer		B	x	B	B
Arundinicola leucocephala		(x)	x	B	B
Tyrannus melancholicus				(x)	(x)
Tyrannus savana	(x)	B	(x)	B	B
Pitangus sulphuratus		(x)			(x)
Donacobius atricapillus			(x)	(x)	(x)
Geothlypis aequinoctialis			B?	x	B?
Ammodramus aurifrons	x	B		B?	B
Conirostrum bicolor				x	x
Thlypopsis sordida				x	x
Sicalis columbiana	x	B?	B?	B?	B?
Sporophila bouvronides/lineola		x	x	x	x
Sporophila castaneiventris		B	x	B?	B

Table 22.1. *Continued*

	Stage 1: bare sand	Stage 2: sparsely growing herbs and scrubs	Stage 3: densely growing annual macrophytes	Stage 4: sandbar scrub	Total
Icterus icterus				x	x
Gymnomystax mexicanus	x	B?		x	B?
Agelaius icterocephalus		(x)	B	x	B
Leistes militaris	B	B	(x)	x	B
Molothrus bonariensis			B	x	B
Scaphidura oryzivora	x	x			x
B = Breeding confirmed	5	10	4	8	19
B? = Breeding probable	0	3	3	8	9
x = Non breeding visitor	7	15	17	17	15
(x) = Exceptional visitor	3	5	6	2	3
Σ	15	33	30	35	46

Vegetation stages are differentiated as follows:
Stage 1, bare sand: devoid of vegetation, except for stranded clumps of floating vegetation.
Stage 2, sparsely growing herbaceous plants and scrub: spotty distribution of vegetation, separated by up to 50 m of open sand.
Stage 3, densely growing annual macrophytes: herbs and grass spreading over the sandbar. Dense stands up to 2 m high, often with *Salix humboldtiana*.
Stage 4, sandbar scrub: *Salix humboldtiana* 1 to >10 m high with more or less dense undergrowth of herbaceous plants or grass.

had widely spaced territories (e.g., *Certhiaxis mustelina*, *Fluvicola albiventer*, *Tyrannus savana*), or patchy distributions (e.g., *Arundinicola leucocephala*, Riverside Black-Tyrant, *Knipolegus orenocensis*, River Tyrannulet, *Serpophaga hypoleuca*). Bicolored Conebills (*Conirostrum bicolor*) and Orange-headed Tanagers (*Thlypopsis sordida*) were seen or captured only once. For some species the area was already saturated with territories, as frequent fights between neighboring pairs indicated (e.g., *Tyrannus savana*).

The colonization of new scrub areas by birds obviously keeps pace with the development of such sites. The species involved do not have greater clutches than other Amazonian species (nests of several species had two or three eggs; e.g., *Tyrannus savana*, *Serpophaga hypoleuca*, *Leistes militaris*, see also Cruz and Andrews 1989). These results confirm the

observation of Rosenberg (1990) from Peruvian river islands that birds of
river islands are good colonizers, though this is probably not true for
species of forest habitat.

22.7 Discussion

22.7.1 Species Composition

Marchantaria Island seems poor in avian species when compared to other
Amazonian sites (Haffer 1990; Fig. 22.3). It has a high number of migrants,
though Nearctic passerine migrants, except swallows, are missing, a
somewhat lower diversity of wetland species than extra-Amazonian wetlands, and a moderate percentage of "várzea species", with only a few forest
birds.

Direct comparisons with other sites are problematic as most of these
are sections of continuous forest landscapes, including terra firme
and floodplains, or isolated rainforest reserves (e.g., Barro Colorado
Island, Karr 1990b; remanescent woodlots of Atlantic rainforest in
Brazil, Willis 1979). Only the study by Rosenberg (1990) of the avifauna
of Peruvian river islands can be compared directly. The number of
species there is slightly higher (230). Some of the difference is due to
the duration of the investigation, the participation of a greater number of
more experienced observers, and the inclusion of several islands in the
Peruvian study.

There are also ecological reasons for differences in species number and
composition. Some important bird habitats found on Peruvian river islands are nonexistent on Marchantaria Island, e.g., *Tessaria* scrub and
Heliconia thickets. They are apparently replaced ecologically by lakeside scrub, e.g., *Alchornea castaneifolia* with a heavy load of mistletoes
(*Psittacanthus* sp.), a type of habitat that has not been described in ornithological studies from the upper Amazon. On the other hand, várzea forest is
missing on Peruvian river islands (Rosenberg 1990).

However, in species composition the várzea forest of Marchantaria Island resembles Peruvian river-edge forest (Remsen and Parker
1983), and is otherwise rather poor in species (even though some species
probably have been overlooked; Sect. 22.3). Several families of forest
birds are completely absent, e.g., specialized frugivores (Ramphastidae,
Pipridae, Cotingidae), or non-passerine insectivores (e.g., Bucconidae,
Galbulidae).

Migrants

The percentage of nonbreeding birds at Marchantaria Island is rather high, chiefly due to migrant or vagrant wetland species. In other Amazonian sites, non-breeding birds play a minor role, but precise numbers of breeding species are rarely encountered (e.g., Robinson et al. 1990). The percentage of migrants is similar to West Amazonian sites (Pearson 1980), but higher than in rainforest (Bierregard 1990; Robinson and Terborgh 1990).

Most migrant species apparently depend on seasonal resources, which are available during certain phases of the flood pulse. An exception are the swallows, as migrant species either from the north (Barn Swallow, *Hirundo rustica*) or from the south (Brown-chested Martin, *Phaeoprogne tapera*) fill the niche of aerial feeders for most of the year almost exclusively.

Most North American migrant passerines winter in Central America and the Caribbean (Keast and Morton 1980; Rappole et al. 1983). Few species reach the equatorial Amazon rainforest. This is an obvious parallel to palaearctic migrants, which also shun equatorial rainforests (Brosset 1990; Robinson et al. 1990). Assuming that migrants tend to winter in a climate that is least different from the climate of their breeding grounds, we would least expect them to choose equatorial rainforests as winter sites because of the great differences in temperature, humidity, etc. Open or disturbed areas in the Amazon, e.g., riverside scrub and second-growth, i.e., habitats with less buffered temperatures and humidity, are more readily accepted wintering sites (Pearson 1980).

Wetland Species

Species numbers of this group are lower on Marchantaria Island (and in Central Amazonia) than in other Neotropical wetlands. This is most conspicuous in waterfowl and storks.

It has been argued that the global diversity of waterbirds follows an antitropical trend, caused by the competitive exclusion of waterfowl by a highly diverse tropical fish fauna (Reichholf 1983). The alleged trend includes piscivorous species (e.g., Podicipedidae), benthivores (e.g., diving ducks, Aythyini), herbivores (e.g., geese) and omnivores (e.g., coots, *Fulica* spp.). In the Central Amazon floodplain there is no lack of fish or of aquatic vegetation with high nutritional value (Sects. 9.2.1 and 20.3; Howard-Williams and Junk 1977; Junk et al. 1983; Junk 1989; Piedade et al. 1991), even taking into consideration that the area of floating meadows has been somewhat increased artificially (Soares et al. 1986). Competitive exclusion by fish thus cannot account exclusively, if at all, for the trend.

Amazon wetlands are almost exclusively riparian. Neither in Amazonia, nor in temperate zones are there many species of waterfowl well-adapted to riparian habitats; in Amazonia probably only the Finfoot (*Heliornis fulica*) is so adapted.

While raising their precocial young and during the molt of the flight feathers, when they are temporarily flightless, waterfowl depend on habitats with predictable qualities. Due to the almost continuously changing water level, these requirements are hardly met in Amazonian floodplains.

Waterbirds of Marchantaria Island have some preadaptive features that help to minimize the impact of the flood pulse. Black-bellied Tree-Ducks (*Dendrocygna autumnalis*) breed in holes in trees and lead their recently hatched offspring to sandbars, even crossing river branches several 100 m wide on their way. There they complete the molt of body plumage and leave with their fledged young to unknown further molting sites to the east, where they change their flight feathers (see Haverschmidt 1947). They return at the peak of the flooding (i.e., at the end of the rainy season). They never seem to take up food from water, which they normally only enter to bathe or to escape from danger, but rather take fruit from trees or graze on sandbars. The Sungrebe (*Heliornis fulica*) has a remarkably short incubation period and transports its young not only when swimming, but also in flight (del Toro 1971). It thus achieves the mobility to evade sudden changes of the environment even with young birds.

The three species of Neotropical storks are found in most savanna wetlands (e.g., Snethlage 1935; Dubs 1983; Kushlan et al. 1985; Thomas 1985; Remsen 1986), but not in Central Amazonia (Hilty and Brown 1986; Willard 1985). The Wood Stork (*Mycteria americana*) is an exception as it is the most gregarious of these species (Sick 1967, 1984; Hilty and Brown 1986). Its occurrence at Marchantaria Island does not prove the existence of a resident population in Central Amazonia, though it is more widespread than the other species (Sick 1984).

Reasons for the absence of storks are to be found in their morphology. Storks walk on (inundated) ground, they usually do not climb through vegetation when feeding. Their toes are rather short and thus adapted to stand and walk, but not to grasp. They could potentially profit from receding pools during low water, but certainly not from floating meadows, as they are too heavy to be supported. Floating meadows can barely support the Cocoi Heron *Ardea cocoi*, which is the size of *Mycteria americana* (and is better adapted to grasp). The other two species are much larger (Thomas 1985).

The composition of the wetland species group at Marchantaria Island is thus ultimately determined by the flood pulse and the dynamics of the environment, precluding the presence of some species completely (e.g., Storks) and others temporarily (see Fig. 22.2). Studies of one site cannot answer the question as to whether the spatial-temporal behavior of the species is adaptive to annually recurring situations or whether they opportunistically gather at favorite sites found by chance.

Forest Species

There are probably two main reasons as to why rather few forest species are found on Marchantaria Island.

Limited Forest Area. The forest area of Marchantaria Island is naturally and artificially limited and interrupted by lakes and canals. This certainly has diminished the diversity of forest microhabitats, increased the area influenced by forest edge climate, and fragmented the microclimatically buffered forest interior.

The low number of species in the várzea forest of Marchantaria Island is not surprising, if taken into consideration with the structure of bird communities in continuous rainforests: Spatial organization is characterized by patchy distribution (Terborgh et al. 1990), temporal movements to upland to evade flooding (Robinson et al. 1990), or spatial shifting following distinctive microclimates (Karr and Freemark 1983). Most forest species have low densities, e.g., two-thirds of the forest species of Cocha Cashu have less than three pairs per 100 ha (Munn 1985; Bierregard and Lovejoy 1989; Robinson et al. 1990; Terborgh et al. 1990). Therefore an isolated patch of forest will have only a fraction of the species encountered in a similar-sized section of continuous forest (Terborgh et al. 1990). The forest area of Marchantaria Island would be insufficient to support viable populations of the bulk of "rare" species.

Isolated Forest Area. The forest on Marchantaria Island is naturally isolated from similar habitats. Though there is extensive floodplain on both banks of the river, undisturbed várzea forests today are found only along backwater areas, while the banks themselves are inhabited by rural people. North of Marchantaria Island, where a narrow peninsula separates the Solimões and Negro Rivers, only pastures and *Cecropia*-dominated second-growth persists.

The limited dispersal activity of forest birds is highlighted by the fact that no species of the forest interior were encountered on the sandbar, except for the Rusty-backed Spinetail (*Cranioleuca vulpina*). However, birds of this species caught in the forest were significantly smaller and lighter than individuals caught on the sandbar, indicating that separate populations might be involved.

Almost no species could be classified as stragglers from terra firme forest. This is hardly surprising as the island is even more isolated from upland rainforest than other parts of the várzea. It is situated at the junction of the valleys of the Rio Negro and Solimões with broad rivers and open floodplains stretching out in three directions from Marchantaria Island. To the northeast, the urban area of Manaus has long since replaced rainforest. Even if terra firme birds were ever to cross the river, this would be the most unlikely point for them to do so.

Both factors, small size and isolation of island forest, have been reinforced by deforestation in the várzea. However, it is not clear whether the low number of forest species is attributable to the natural geographic situation or to anthropogenic changes in the landscape.

Várzea Species

One-third of the species of Marchantaria Island belong to species listed by Remsen and Parker (1983) as having probably originated in river-created habitats in Amazonia. The reasons for the high endemism in the floodplain have not yet been satisfactorily explained. Of course, it is a prerequisite to the survival of this endemic avifauna that floodplain habitats persisted over long geological epochs (Remsen and Parker 1983). However, there must also be ecological reasons for the separation of the avifaunas of terra firme and várzea.

The floodplain differs from terra firme forest and savannas chiefly in the flood pulse, climatic peculiarities and habitat structures (Sects. 1.5 and 22.2).

Other factors, e.g., regional floristic composition of the vegetation or nutrient levels in ecosystems, cannot be decisive for the observed patterns of distribution of várzea birds, as bird species of the floodplain forests are generally distributed in forests along both black- and white-water rivers. Those forests differ more between one another in floristic composition and nutrients, than they differ from terra firme (Kubitzki 1989).

Climate. Relationships between climatic differentiation and species diversity in Amazonia have repeatedly been emphasized (Gentry 1982; Haffer 1987, 1990), though the mechanisms are still poorly understood. No data are available to determine if certain regional climatic variations influence avian biogeography and ecology, e.g., if precipitation is 1800 or 3500 mm/

a, if the dry season lasts for 3 weeks or 6 months, if the minimum temperatures falls as low as 7 or 17 °C, etc. (but see Willis 1977) just to mention some climageographical differences encountered inside Amazonia (Salati and Marques 1984).

Some patterns of correlation between climate and distribution emerge. In the tropics, the seasonal range of climatic conditions at any site is usually smaller than in temperate latitudes. Local geographical peculiarities result in distinctive local climates with little overlap with neighboring sites (Janzen 1967). Species with narrow climatic tolerance are consequently more common in the tropics (Stevens 1989).

Direct evidence for responses of birds to different climates is rare. Birds of forest undergrowth react to temporal and spatial differences in the microclimate (Pearson 1977; Greenberg 1981; Karr and Freemark 1983). Forest interiors have higher diversity than forest edges (Terborgh et al. 1990).

The floodplains of Central Amazonia are presumed to have relatively high climatic variabilities (Sect. 2.3). Climatic conditions should thus favor species with greater climatic tolerance, e.g., birds of extra-Amazonian savannas (examples in Willis 1992), or of forest canopy or second-growth (Snethlage 1913). Some species of floodplains are known to expand their ranges into terra firme when competing species are missing (Terborgh and Weske 1969; Remsen and Parker 1983).

Species of terra firme undergrowth will probably encounter familiar microclimates in the floodplain only in várzea and transitional forests. Numbers of endemic "várzea species" consequently are lower in these habitats (Fig. 22.6) as terra firme species occupy most of the niches (Robinson et al. 1990).

It thus can be argued that the common feature of all "várzea species" is the fact that they are restricted to or have evolved in a part of Amazonia that can be delimited climatologically from the rest of Amazonia. However, the same is apparently true for the flood pulse.

Flood Pulse. It is obvious that the flood pulse all but excludes several groups of species from the várzea or at least river islands (e.g., tinamous, Tinamidae; birds of undergrowth; Sick 1967; Remsen and Parker 1983; Haffer 1990). Species of sandbars are forced to emigrate temporarily when their habitat is inundated. A temporary shift to other habitats can also be observed in the birds of floating meadows.

However, these habitats have only small number of "várzea species", while most endemic species are found in scrub and forest, i.e., habitats less affected by the flood pulse. There the impact of the flood pulse results mainly in seasonal quantitative variations of resources (Sect. 22.2). Seasonality of resources as such is not an exclusive feature of the floodplain (e.g., Karr and Freemark 1983; Karr 1990a) and therefore not a sufficient explanation for the degree of endemism of the várzea avifauna.

Birds are highly mobile, and can thus evade the flooding (Sick 1967; Remsen and Parker 1983; Rosenberg 1990). However, reproduction fixes the birds to the nest site for several weeks. To ensure reproduction, breeding should either take place at times and sites with low probability of inundation, or should be attempted more frequently to compensate for failures (e.g., Jacana, Sect. 22.5.2). Many species choose breeding sites within the zone of inundation (Snethlage 1913; Sect. 22.5.4). Reproductive success there depends on the rate of change in the water level at the time of breeding. For breeding birds inundation offers the possibility to breed in relative safety from predators above the water. However, breeding in the rainy high-water season, confirmed for several species of Marchantaria Island, has probably more than one advantage. The availability of food for both insectivorous and frugivorous species is highest in that season as well. There is no way to disentangle the relative importances of both factors.

Habitats. Most "várzea species" are restricted to one or a few distinctive habitats (Sect. 22.5). The habitat definition will often have to emphasize microclimatological features, including light, rather than floristics or structure. For example species of lakeside scrub are found in sites with predominant *Cecropia* sp., *Alchornea castaneifolia*, *Salix humboldtiana*, as well as in coastal mangroves (Lefebvre et al. 1992). These plants have quite different architectures, but form habitats of similar height and exposure to light. A species from lakeside scrub could undoubtedly find similarly structured undergrowth scrub in terra firme forests, but would have to adapt to a great reduction in light and a different microclimate.

22.7.2 Conclusions

1. The flood pulse is the most important factor determining the spatio-temporal behavior of birds on Marchantaria Island, acting both as a proximate and an ultimate factor. However, the existence of an avifauna endemic to the Amazon floodplain cannot be attributed exclusively to adaptations to the flood pulse, as climatic factors might be equally important.
2. The avifauna of Marchantaria Island has fewer species than would be expected. This is partially due to the flood pulse (responsible for the absence of tinamous, storks, waterfowl) but is more a result of the island situation of the study site (responsible for the low number of forest

species). This effect has been aggravated by anthropogenic habitat changes.

3. Species of várzea forest seem to be most vulnerable to human interference, while some birds of sandbars or scrub readily colonize anthropogenic habitats, even urban centers (Manaus: Yellow-browed Sparrow, *Ammodramus aurifrons*, Red-capped Cardinal, *Paroaria gularis*, Chestnut-bellied Seedeater, *Sporophila castaneiventris*).

4. The Amazon floodplain has a great, but as yet nonquantifiable importance for migrating birds of both hemispheres, especially wading birds from the Nearctic (Charadriiformes), swallows from both hemispheres (Hirundinidae) and tyrant flycatchers from the austral Neotropics (Tyrannidae).

Part V
Conclusions

23 Structure and Function
of the Large Central Amazonian River Floodplains: Synthesis and Discussion

Wolfgang J. Junk

According to the flood pulse concept (Junk et al. 1989; Sect. 1.3) the flood pulse is the main driving variable in large river–floodplain systems. It leads to varying environmental conditions, periodic changes in plant and animal communities, and intensive multiple interactions between the terrestrial and the aquatic phases in the Aquatic Terrestrial Transition Zone (ATTZ). Major biotic processes which determine the function of the river–floodplain system in respect to production and decomposition of organic material and related processes are concentrated in the floodplain. The main river and its affluents influence the floodplain by the hydrological regime, erosion and sediment deposition, the input and export of dissolved substances, and the exchange of organisms. In the previous chapters, the available information on central Amazonian river floodplains has been presented and discussed under specific aspects. In this chapter an attempt is made to synthesize the results and draw some generalities.

23.1 Habitat Diversity and Dynamics

Floodplain habitats are situated along a gradient of permanently aquatic to permanently terrestrial conditions. The permanently aquatic habitats of the Amazon floodplains are river channels and perennial floodplain lakes. Permanently terrestrial habitats are Pleistocene sediment depositions above the highest flood level, the uplands (terra firme), bordering the floodplain and the canopy of the floodplain forest. Edaphic conditions, i.e., grain size of soil particles, water retention capacity, and content of organic matter often vary over short distances (Chap. 2). Different plant communities and communities of different successional stages develop and result in a complex and highly diverse horizontal and vertical spatial structure of the floodplain habitats during the aquatic and the terrestrial phases (Chap. 8). Changes in water level, erosion, and sediment deposition modify the habitats and disturb plant and animal communities.

Ecological Studies, Vol. 126
Junk (ed) The Central Amazon Floodplain
© Springer-Verlag Berlin Heidelberg 1997

However, the habitat type is always present during certain periods of time and is available for colonizers, especially if they have strategies to reach the habitat in time or to maintain their presence during unfavorable conditions. Sandbars or mudflats surface only at low water. They change size, shape and position in the river channel and become submerged again when the river rises. But, at low water, sandbars and mudflats are always present in the river and play an important role as habitats and nesting places for river turtles and birds (Chaps. 21, 22).

Strategies to cope with this situation depend upon the life-history traits of organisms. The formation of large numbers of flood or drought resistant seeds and resting eggs is one of the most common and simple adaptations which allows the recolonization of a periodically available habitat. This strategy is frequently found in terrestrial and aquatic invertebrates and herbaceous plants. The other strategy is great mobility, which enables the organisms to reach the habitat when it becomes available and to leave it when it disappears, e.g., by active migration or passive transport by wind and water.

The predictable flood pulse enables plants and animals to select suitable habitats for living and breeding. Predictable aquatic and terrestrial phases of several months duration allow for better exploitation of the resources by producers and consumers than very short and unpredictable phases. Communities of terrestrial arthropods, which play an essential role in litter decomposition can develop only when they have sufficient time to become established (Sect. 9.3.3; Chaps. 14–19). Periphyton and perizoon need time to colonize submerged surfaces (Chaps. 10, 13), aquatic and terrestrial herbaceous plants can develop only when the respective periods are long enough (Chap. 8).

In the humid tropics, a regular monomodal flood pulse creates a distinct annual seasonality in a nonseasonal or weakly seasonal evnironment. Most plant and animal species living in the floodplain show a pronounced seasonal behavior. In many cases, this behavior is synchronized with the flood pulse but is apparently not genetically fixed. Without the flood pulse, the life cycle becomes nonseasonal or slightly seasonal, as shown for many annual herbaceous plants. However, there are also many species strictly dependent on the flood pulse, for example many migrating fish species, which do not reproduce in habitats with a stable water level. The extent to which seasonal behavior is triggered by the flood pulse itself or by other secondary factors such as temperature and rainfall, as shown for some terrestrial invertebrates (Chap. 14), is still an open question. According to Erwin and Adis (1982) the adaptation of carabid beetles to periodic flooding facilitated their colonization of temperate regions with a pronounced climatic seasonality. On the other hand, adaptation to seasonality such as periodic drought seems to facilitate the colonization of floodplains as shown by tree species that grow well in both habitats (Chap. 11).

23.2 The Impact of Global Paleoclimatic Changes on Amazonian Floodplains

The great number of plant and animal species and the large number of adaptations to the specific conditions in the Amazonian floodplains give an evolutionary dimension to the stability aspect of habitats. Sedimentological evidence indicates that the central Amazonian várzea in its present form is only about 5000 years old. The river level and the surface area of the várzea were established following the fall and rise of sea levels during glacial and interglacial periods. In periods of low sea level pronounced erosion, and during high sea level sedimentation, took place in the Amazon valley and in the lower sections of the valleys of its affluents. These affected not only the dynamics in the floodplains but also their extent. During the last glacial period, the sea level was about 120 m lower than today. Erosional processes prevailed and the floodplain was smaller and some tens of meters below the present level (Sect. 2.1).

According to the refuge theory summarized in Prance (1982), there were climatic changes in the Amazon basin during the glacial periods, which resulted in a cooler, drier climate with pronounced dry and rainy seasons, and a reduction and partitioning of the rainforest area. These climatic changes probably also led to a change in the discharge of the Amazon River system and to a change in flood amplitude, although not to a very large extent as shown by Müller et al. (1995). The size and shape of the paleochannel found in old river sediments of Rio Preto da Eva, an affluent of the Amazon River downstream of Manaus, point to a discharge pattern similar to the existing one (Müller et al. 1995; Sect. 2.7). Climatic changes also did not change the predictability of the flood pattern of the main river and its large tributaries, because this was related to the precipitation in large catchment areas. In spite of worldwide climatic changes, large Amazonian river floodplains have existed as predictable, slowly pulsating tropical systems, for at least 2.4 million years or since Mid-Tertiary times, when the Amazon River opened to the Atlantic Ocean. Sediments of fluviatile-lacustrine origin of the Barreiras Formation and the Solimões Formation also point to periodical flooding of large areas with freshwater during the Cretaceous, Tertiary, and Pleistocene (Grabert 1983; Putzer 1984). Kubitzki

(1989) relates the larger number of flood-resistant trees in the igapó forest in comparison to the várzea forest to the existence of flood-resistant forests adapted to nutrient-poor conditions since the Upper Cretaceous.

The organisms colonizing the floodplains had little difficulty accompanying the slow, long-term changes in discharge, erosion and sedimentation behavior of the rivers because they were already adapted to annual variations in the flood pulse. The large size and heterogeneity of the Amazonian river–floodplain system increased the buffer capacity against loss of species. Local elimination of organisms could be compensated for by immigration from other less affected areas. Periodic increases and reductions in the extent of the floodplains accompanying global changes in climate and sea level, favored the development of adaptations and increased habitat and species diversity (Salo et al. 1986; Salo 1990; Kalliola et al. 1993). Plants and animals both reacted differently to the flood pulse. Therefore the number of endemic species and the variability of adaptations vary considerably between different groups as documented for terrestrial invertebrates (Chaps. 14–19). We think that relatively high species diversity (e.g., in tree species, Chap. 11), high number of endemic species (e.g., in Pseudoscorpiones, Adis and Mahnert 1990), and adaptations to the flood pulse (e.g., in some terrestrial invertebrates (Chaps. 14, 15)) are the result and the proof of continuous evolution in a very dynamic but relatively old pulsing tropical ecosytem.

In the arid southern and northern belts adjacent to central Amazonia, the impact of global climatic changes was much more severe. Dry periods in the area of the Pantanal of Mato Grosso resulted in a different discharge and sedimentation pattern of the affluents (Ab'Saber 1988). Wetlands in the Pantanal and the surrounding uplands became strongly reduced in size, and extinction rates surpassed rates of speciation. When climatic conditions during interglacial periods became more humid and more favorable, relatively small numbers of wetland plants and animals were left to colonize the increasing wetland areas of the Pantanal. Today, the fauna and flora of the Pantanal is a mixture of species from Amazonia, the Cerrado and the Chaco with a very low number of endemics (Adamoli 1981; Brown Junior 1986). Species numbers of many plant and animal groups are much smaller than in the floodplains of the large Amazonian rivers (Junk and da Silva 1995). The total number of fish species reported from the Pantanal is about 250, corresponding to the number of species collected by Bayley (1982) in Lago Camaleão, a small várzea lake. The number of flood-resistant tree species growing in 1 ha of an Amazonian floodplain forest is higher than the total number of flood-resistant tree species in the Pantanal. Exceptions are groups of organisms which are less affected by extreme dry periods or are very mobile. Terrestrial herbaceous plants are more resistant to a dry climate than trees, therefore the Pantanal and the surrounding area are richer in herbaceous plant species than central Amazonia and its river floodplains, which are dominated by trees. Schessl (pers. comm.) reports about 600 species in the Pantanal near Pocone. Aquatic birds are very mobile and occur in greater species numbers in the Pantanal.

23.3 Species Diversity

At the very beginning of the study of limnology, Thienemann formulated two general principles to explain species diversity. Species number in a biocenosis increases with habitat diversity (Thienemann 1918). Species number of a biocenose decreases but abundance increases when environmental conditions become extreme for the majority of organisms (Thienmann 1920). Franz linked the recent status of ecosystems with their history, stressing the importance of time and time-dependent environmental changes for species diversity. He formulated a third general principle: Species number and stability of a biocenosis increases with the level and the period of continuity of the environmental conditions (Franz 1952–1953; Thienemann 1954).

Extreme environmental conditions put rigid limits on speciation and species diversity. Therefore species diversity is low near the poles and increases as we move toward the equator. In deserts, species diversity is also small because abiotic environmental stress factors are dominant. In most ecosystems of the humid tropics environmental parameters are not limiting for plant and animal life. Coevolution is the driving force for speciation and theoretically puts no limits on species diversity. The great age of tropical ecosystems, increased mutation rates, and a higher number of generations per unit time favor diversification (Mayr 1969). High species diversity in tropical ecosystems may also be the response by organisms to competition for limited resources. The highly diversified Amazonian rainforest optimizes the use of the small amounts of nutrients in the soil because of slightly different nutrient requirements and uptake efficiencies of the different species (Klinge and Fittkau 1972; Fittkau 1973a, 1982; Klinge 1973).

High species diversity, considered characteristic for the tropics, is also found in some plant and animal groups in Amazonian floodplains. Clark and Benforado (1981) report a total of about 100 tree species from all bottomland hardwood forests in southeastern USA, including those with a flood tolerance of only a few days. The same number of tree species can be found on 1 ha of Amazonian floodplain forest (Chap. 11). The number of fish species in Amazonian rivers is higher than in comparable rivers of temperate regions (Lowe-McConnell 1987).

These examples are not necessarily representative of all plant and animal groups, nor of all habitats. Considering the size of the floodplain area and the habitat diversity, the number of mammal species is small (Chap. 21). Patrick (1966), studying Peruvian streams near Tingo Maria in the foothills of the Andes stated that the total number of species of aquatic organisms of small body size, such as algae, protozoa and insects, is similar or sometimes even lower than those in comparable streams of temperate regions. They related this surprising result to the size of the system and the number of available ecological niches.

According to Fittkau (1973a, 1982), species diversity for Plecoptera and Ephemeroptera may be smaller in tropical water bodies. Species diversity for Odonata and decapod Crustacea are greater and for Trichoptera and Chironomidae similar to species diversity in temperate regions. In Diatomeae, species diversity is greater in temperate regions, whereas Desmidiaceae reach their greatest diversity in nutrient-poor acid tropical waters. Fittkau considers the cold stenothermic lightly shaded streams and lakes of temperate regions better suited for the evolution of a species-rich aquatic flora and fauna. For instance, species diversity of benthic diatoms in small cold streams can be extremely high. Furthermore, autochthonous production by algae and aquatic macrophytes is greater due to better light conditions. The greater and qualitatively better autochthonous nutrient supply forms a better basis for a larger variety of aquatic invertebrates, than does the nutrient-poor terrestrial detritus of the Amazonian forest (Chap. 9).

Due to cyclical and predictable annual changes in light and temperature, highly variable food supply, and a large variety of substrates and habitats, the number of ecological niches available for successive colonization by algae and aquatic invertebrates is higher in temperate zones than in equivalent tropical ones. Both a central Amazonian and a central European stream 3-km-long harbor about 150 species of chironomids. Most of the species in the central European stream produce only one or two generations per year, their development being strictly timed by environmental factors. Small chironomids in central Amazonia need only 2–3 weeks for development (Chap. 13). All species can occur together at different stages throughout the year, competing for food and habitats. In the small Central European coastal region comprising the area from Belgium to southern Scandinavia up to 500 m above sea level, about 5000 species of aquatic invertebrates have been recorded, corresponding to 40% of Europe's aquatic invertebrates (Illies 1968). Fittkau (1973a) stressed that such a concentration of aquatic invertebrates is unknown in the tropics and postulates that it would not be expected to occur.

According to Thieneman's second principle, species number should be reduced in floodplains when environmental conditions become extreme. Grime's model of stress, competition and disturbance (Grime 1979) indicates that species diversity is low at sites where there are low or high degrees of stress and disturbance. The Intermediate Disturbance hypothesis of Connell (1978) postulates that species number increases when an intermediate disturbance factor increases habitat diversity and diminishes interspecific competition.

Pickett and White (1985) defined a disturbance as "any relatively discrete event in time that disrupts ecosystem, community, or population structure, and that changes resources, availability of substratum, or the physical environment". Resh et al. (1988) considered unpredictability to be an important characteristic of a disturbance. Sparks et al. (1990) modified the definition of a disturbance for river–floodplain systems as an "unpredictable, discrete or gradual, event (natural or man-made) that disrupts structure or function at the ecosystem, community, or population level". In river–floodplain systems annual flooding is the normal behavior of the system and predictability of the floods allows the development of strategies to deal with environmental changes. Therefore, floods are not considered disturbances by Sparks et al. (1990), unless they are so amplified, reduced, or mistimed, that they produce significant changes as defined above. Poff (1992) suggested that disturbances (including predictable ones) always have ecological effects; however, evolutionary adjustments to a predictable disturbance factor may constrain the magnitude of ecological responses of the biota.

Our data show that these general statements have to be adjusted according to the species and the position of their populations on the flood gradient, because adaptations and survival strategies of different species to flooding and drought vary greatly. The predictable flood pulse acts every year as a disturbance factor for many free-floating aquatic macrophytes, disrupting structure and function at the population level at low water because the response is quick vegetative propagation to reestablish the populations. Losses are large enough to diminish interspecific competition, allowing the coexistence of a large variety of free-floating aquatic macrophytes. The same pulse affects trees only during the period of seedling establishment but does not affect the established floodplain forest, because adult trees do not die in normal flood periods. For organisms colonizing the levees or the edges of the floodplain, the predictable annual flood pulse becomes an unpredictable disturbance factor that occurs at 1- to 5-year intervals because of the annual varibility of the flood amplitude.

The stress related to the oscillation between a pronounced aquatic and a pronounced terrestrial phase is heavy and results in a considerable reduction in species number of most terrestrial plant and animal groups in comparison with those on the surrounding terra firme. About 100 tree species ha^{-1} occur in an undisturbed várzea forest subject to about 3 months of inundation near Manaus, in comparison with up to 176 species ha^{-1} in a terra firme forest (Chap. 11). Inside the floodplain the number of terrestrial plants and terrestrial invertebrate species decreases as flood stress increases (Chaps. 8, 16–19). Tree species in the várzea decrease from about 100 species ha^{-1} on the levees to about 5 species ha^{-1} in areas flooded for 9 months $year^{-1}$ (Chap. 11).

With 388 recorded species belonging to 182 genera and 64 Families, the herbaceous vegetation is rich in species. There are 330 terrestrial species, 43 aquatic species, and 24 species which have an intermediate status. The total number is nearly as large as the number reported from some Brazilian savanna areas. Species number is highest in disturbed high-lying areas, where the flood stress is low. The large number of ruderal species indicates that a majority of the herbaceous species are not specific to the floodplain and can be found in disturbed areas all over Amazonia, or even throughout the tropics (Chap. 8).

In Lago Camaleão, 14 planktonic cladoceran species and seven copepod species were found. These numbers are only slightly higher than the numbers reported for temperate lakes. Krambeck et al. (1994) studied the Plusssee and found eight cladoceran and six copepod species that were abundant and typical for eutrophic north German lakes. The number of rotifer species in plankton samples of Lago Camaleão reached 175; however, most of them were epizoic (Chap. 13). Koste (1974) found 152 epizoic species in the roots of floating aquatic macrophytes in the mouth area of the Tapajos River. In Alfsee, an artificial reservoir for flood control of the Hase River in Germany, Koste and Poltz (1987) found 136 species, including 34 planktonic species, 12 semiplanktonic species and 90 epiphytic species. Chengalath et al. (1990) reported 138 species from 228 localities sampled in Sri Lanka including lakes, ponds, rivers, rice fields, and others. A similar situation is encountered for phytoplankton. Krambeck et al. (1993) indicate at least 180 species for the small Plusssee. Uherkovich and Schmidt (1974) found 209 species in the much larger Lago do Castanho and Rodrigues (1994) reported 262 species in Lago Camaleão.

Diversity decreases when additional stress factors such as large sediment deposition, pronounced hypoxia, extremely low nutrient status, or extreme drought during the terrestrial phase are added to the flood stress. The definition of "extreme conditions" is relative in respect to the requirements of different groups of organisms. Species number of herbaceous aquatic plants in the Negro River floodplain is small because of low nutrient availability and low pH values in the water (Chap. 8), but the number of tree species is high because nutrient levels for trees are obviously sufficient (Chap. 11). Fish react to the low food supply of the habitat by a decrease in biomass but not by a decrease in species number. Data show even slightly higher species numbers in the igapó than in the várzea (Chap. 20).

High species diversity is maintained by permanent aquatic and terrestrial habitats which are important refuges for many species that colonize the ATTZ (Amoros and Roux 1988; Sedell et al. 1990). Mobile species escape flooding by horizontal and lateral migration. Migrants may be endemic to floodplains, e.g. many terrestrial invertebrate species colonize the litter of the floodplain forest and migrate on to the tree trunks or into the canopy to survive the aquatic phase. Others immigrate obligatorily or occasionally from the surrounding uplands. In both cases the availability of permanently terrestrial habitats, i.e., trees, and the distance from, and connection with, the surrounding upland will influence species diversity. This indicates the importance of the flood plain forest in the maintenance of species diversity of the soil fauna. Furthermore, the canopy of the floodplain forest represents a habitat for a specific canopy fauna including invertebrates, birds and others animals (Chaps. 14, 18, 19, 22).

Lowe-McConnell (1987) showed that periodically drying habitats often show reduced numbers of fish species. Species numbers in Amazonian river floodplains are very high because species find refuge during the low-water period in permanent water bodies such as remaining lakes, the river channels, and terra firme affluents. Floodplains not connected with permanent water bodies are poor in aquatic species and mostly harbor species that produce resting stages or are able to migrate by air or land.

There is little information available about the differences in species composition between the different parts of the Amazon floodplain and between the floodplains of different affluents. Genetic exchange occurs mostly longitudinally and is hindered laterally by the large river channels. Differences in sediment load, sediment fertility and water chemistry of the respective rivers increase habitat diversity and favor species diversity. They may represent barriers for isolated populations. The exclusion of species due to hydrochemical conditions has been exemplified in aquatic macrophytes, algae, trees, freshwater shrimps, mollusks, and fish (Chaps. 8, 10, 11, 13, 20). The limited data on tree species composition (Kubitzki 1989) and the fish fauna (Goulding et al. 1988, pers. comm.) of the floodplains in different parts of the major Amazonian rivers indicate regional differences in species composition (β-diversity).

23.4 Adaptations to the Flood Pulse

Organisms colonizing the central Amazon floodplain can be divided into species that occasionally invade or visit the floodplain, species that colonize floodplains but are also frequently found in other habitats, and species that are restricted to floodplains. Species of the first group have no adaptations to floodplain conditions. Species of the second group show general adaptations that allow the colonization of a wide range of habitats including floodplains. For instance, ruderal plants occurring in large numbers in the várzea are adapted to any type of disturbance but not specifically to disturbance by floods. They are r-strategists, which are selected for high population growth in an uncrowded population and are characterized by quick growth, early maturity, and high numbers of offspring to balance the costs of short life cycles and low competitive ability. K-strategists are species that are selected for competitive ability in crowded populations. They opt for the long-term occupation of the habitat by long life spans, low numbers of offspring, parental care, and great competitive strength (MacArthur and Wilson 1967; Pianka 1970; Stearns 1976). A rigid differentiation between r-strategists and K-strategists is often difficult because there are many ways to optimize strategies by combining different levels of r- and K-strategies with each other.

The flood pulse concept postulates that periodic flooding favours the development of r-strategists because of the large population losses related to the change from the terrestrial to the aquatic phase and vice versa. Most terrestrial and aquatic invertebrates (Chaps. 13, 14, 17–19), many fish species (Chap. 20), algae (Chap. 10), most herbaceous plants (Chap. 8), and some tree species (Chap. 11) use this strategy. An exception may be oribatid mites, which seem to be little affected by the flood. Many oribatids show rather long life spans and juvenile specimens make up only a low percentage of the populations (Chap. 16). Some highly adapted K-strategists are found among the fish (Chap. 20), the large mammals and reptiles (Chap. 21), and the birds (Chap. 22).

Quick growth is related to good nutrient supply in water and soils. In the igapó of the Negro River plant growth is reduced and many herbaceous species are absent because of the lack of nutrients. Many fast-growing common pioneer tree species of the várzea, e.g., *Cecropia* spp., *Pseudobombax munguba*, and *Salix humboldtiana* rarely occur or are totally absent from the igapó. Recovery of the várzea forest from natural or man-induced disturbance is quick, recovery of the igapó forest is very slow. Total animal biomass is smaller because of low food supply. Many r-strategy species occur frequently in the várzea but are rare or missing in the igapó and are sometimes substituted by K-strategy species (Chap. 13).

There are various other aspects, in addition to reproduction, that determine the survival of species and populations in their habitats, e.g., adaptations to inclement physical conditions, avoidance of predation (including herbivory), utilization of available food, and tactics to escape in space or in time (Southwood 1988). The survival of plant and animal species in the Amazon floodplain depends on the ability to overcome periodic flooding and drying and to survive other unsuitable environmental conditions, such as low dissolved oxygen concentrations in water and soils. The combination of strategies determines the fitness of the species or population to cope with their environment. The fish species *Semaprochilodus insignis* combines quick growth, early maturity, and high reproduction rates with a pronounced resistance to hypoxia and a very complex schooling, migrational, and spawning behavior coupled with the flood pulse.

Environmental conditions in floodplains differ fundamentally from conditions in large lakes. In the shore area of the large East African rift valley lakes there is great habitat diversity and a large number of highly adapted fish species. The habitats are stable over very long periods of time and populations reach and maintain maximum densities. Adaptations of organisms are related to the permanent occupation and the defense of the habitats against relatives or other species (K-strategy). In such systems escape strategies in space and time are of little value and are restricted even in highly mobile organisms, i.e., fish. Large-scale intralacustrine speciation is possible because fish do not often move from one bay of the lake to the next one.

23.5 Biomass and Primary Production

The river continuum concept (Vannote et al. 1980) postulates that river systems depend to a large extent on allochthonous carbon from the catchment area which is processed by aquatic consumers during downstream transport to the sea. Downstream consumers capitalize on the inefficiencies in nutrient processing of upstream consumers. According to the flood pulse concept (Junk et al. 1989), most organic carbon processed in river floodplains is produced inside the floodplains. Main primary producer communities in central Amazonian floodplain ecosystems are trees, terrestrial and aquatic herbaceous plants, phytoplankton, and periphyton (Chaps. 8, 10, 11). Most of the material cycles inside the floodplain between the terrestrial and the aquatic phase (Junk 1985a). Primary production depends on the availability of dissolved nutrients, the amount and fertility of suspended matter, and, for algae, on light availability. Estimates of primary production (P) and biomass (B) are given in Fig. 23.1. Highest production rates are reported for terrestrial and aquatic herbaceous plant communities and the floodplain forest. Total productivity and nutrient flux in floodplain lakes are directly correlated with the ATTZ/LBA ratio, i.e., the size of the ATTZ in comparison with the lake basin area (LBA) (Furch and Junk 1992; Junk and Weber 1996; Chaps. 4, 5, 6).

The differences in the P/B ratio of the different plant communities indicate different functions in the floodplain. The low biomass and large P/B ratios of algae show that they quickly recycle a small nutrient pool, but their importance for nutrient accumulation and retention in the floodplain is small (Bayley 1989). Annual herbaceous plants have a higher biomass and a lower P/B ratio. Their function in nutrient cycling is seen in the succession of several populations during the hydrological cycle. Annual terrestrial plants take up nutrients from the soil and deliver them into the water when decomposing during the aquatic phase. Aquatic macrophytes grow during the aquatic phase, take up these nutrients, maintain them in the system and deliver them to the soil when they die at low water. Highly productive aquatic macrophyte communities can capture additional nutrients from the river water and increase the nutrient level of the system (Fig. 23.2a). The amount of nutrients captured by the perennial semiaquatic grass *Echinochloa polystachya* can have a strong influence on the nutrient budget of floodplain lakes because of the large biomass (up to $80\,t\,ha^{-1}$ dry weight) and an annual production of up to $100\,t\,ha^{-1}$ (Chap. 5).

In black-water habitats, nutrients released into the water from decomposing terrestrial plants can be washed out of the system because of the low numbers of aquatic and semiaquatic macrophytes, which would take up these nutrients and maintain them in the system. However, export is small because the biomass of terrestrial herbaceous plants is small and nutrient losses from this source are low. The different strategies of the igapó trees also point to the need for nutrient retention, e.g., low productivity, an increase in litter fall during the end of the aquatic phase to reduce loss by lixiviation, dense superficial root mats to maximize nutrient uptake from the litter, and high nutrient-use efficiency (Meyer 1991; Chap. 3; Fig. 23.2b). In the várzea nutrient shortage seems to be less critical. In spite of the flood stress, productivity is high, nutrient levels in the leaves are high and special adaptations of the roots to maximize nutrient uptake are not found (Chap. 9).

The flood pulse modifies oxygen availability and causes important changes in the biogeochemical cycles in the aquatic and the terrestrial phases. The large amounts of decomposing organic material cause anoxic conditions in the flooded soil and water and the formation of H_2S and

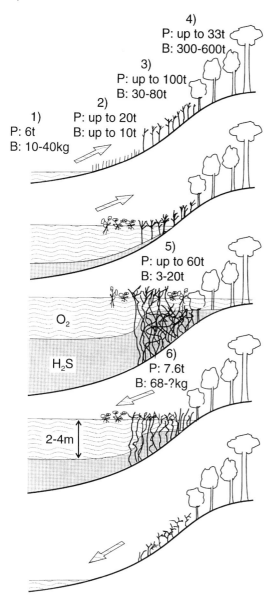

Fig. 23.1. Biomass (B) and net primary production (P) per annum of different plant communities in the floodplain of the Amazon River according to Junk et al. (1989). Estimates are given as dry weight per hectare. *1* Phytoplankton, *2* annual terrestrial plants; *3* perennial grasses; *4* várzea forest; *5* emergent aquatic macrophytes; *6* periphyton. (Putz and Junk, Chap. 10)

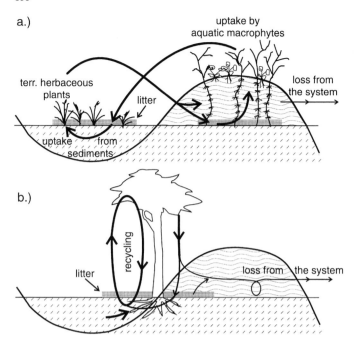

Fig. 23.2a,b. The role of different plant communities in the nutrient exchange between the terrestrial and the aquatic phase. **a** The exchange between terrestrial and aquatic herbaceous plants in the várzea leads to a retention of nutrients in the system. **b** The floodplain forest of the igapó reduces nutrient loss by increased litter fall at the end of the aquatic phase and intensive nutrient recycling during the terrestrial phase by a dense superficial root system and mycorrhiza

methane (Sect. 4.4.3). During floods, the Amazon várzea is an important natural source of methane for the atomosphere (Chap. 7). Denitrification in water is strongly reduced by lack of oxygen. It becomes an important process when the sediments are exposed to air. Losses of nitrogen are not balanced by nitrogen fixation by cyanobacteria but are compensated for by the nitrogen input from the river (Chap. 6). The contribution of higher plants to the nitrogen budget is not yet quantified, but considering the large number of Leguminosae in the floodplain forest nitrogen fixation is expected to be considerable (Chap. 6). Nitrogen is often a limiting factor for plant growth in the water (Chap. 10) and is probably a limiting factor in the soils of the várzea (Chap. 3).

23.6 Food Webs

According to the trophic-dynamic concept of Lindeman (1942) the structures and functions of ecosystems depend to a large extent on the transfer of nutrients and energy through the food webs. Data on the food utilization of consumers can be converted into energy equivalents, which allows the calculation of the efficiencies of energy flow (Kozlovsky 1968; Welch 1968). However, the relationship between organisms and their food is not adequately described by the energy flow. Food utilization depends on the species, number and growth of consumer and decomposer organisms, and the amount, availability and quality of the food (Boyd and Goodyear 1971).

In Amazonian floodplains, as elsewhere, the amount of food and food quality depend on nutrient availability in soils and water (Chaps. 3, 9). Food supply and food quality are much better in nutrient-rich white-water floodplains (várzea) than in nutrient-poor black-water floodplains (igapó). In the várzea the low use of herbaceous plant material by large primary consumers is not the result of a low food quality (Sect. 9.2) but the result of a low number of flood-adapted animal species. The number of herbivorous mammals in Amazonian floodplains is small in comparison to African and Asian wetlands. There are only two species, the manatee *Trichechus inunguis* and the capybara *Hydrochoerus hydrochaeris*, which occurred in large numbers in the floodplains. The periodic use of the plants by immigrating mammals from the terra firme is also small, because the number of herbivorous species in Amazonia is low and the size of their populations is small. Probably, the high amplitude of the flood pulse hinders mammals from the terra firme from using the food resources of the várzea on a large scale.

Food supply is related to habitat availability and varies strongly with the flood pulse. Omnivory, mobility and periodic fat accumulation allow consumers to optimize the exploitation of the different resources. Data on fish production show that in nutrient-rich várzeas, populations are controlled more by the changing environmental conditions than by food supply (Chap. 20). During rising and high water levels, fish growth is not food limited. Drought, loss of habitats and concomitant piscivory at low water has a greater influence on fish mortality than lack of food. Because plenty of food is available, the decline of one consumer population does not lead necessarily to an increase in another with similar food requirements. The strong reduction in the populations of large herbivorous mammals and turtles by man did not result in the increase of other herbivorous species (Chap. 21). The data suggest that a very large part of the primary production of the floodplain forest and terrestrial and aquatic macrophytes is not utilized by primary consumers but passes directly into the microbial food webs. Therefore, cattle or water buffaloes in the várzea do not compete with native herbivorous animals for food. Negative side effects of ranching will, however, arise by large-scale clearing of the floodplain forest for pasture plantation on the levees to provide food for poorly flood-adapted domestic animals during high water.

In spite of the ability of plants to accumulate nutrients, organic material produced by herbaceous plant communities and the forest in the igapó contains lower amounts of mineral elements and proteins than material produced in the várzea. The low food value, often in combination with high concentrations of toxic secondary metabolic products, is considered to be a very efficient plant strategy to reduce losses to herbivorous animals. Exceptions are fruits and seeds which are of high nutritional value. Because of the large amounts of fruit, the floodplain forest has a great importance for the food webs. Clear-cutting of the forest will eliminate or reduce the populations of many frugivorous species, some of them of commercial value, e.g. the tambaqui (*Colossoma macropomum*), a large characoid fish.

Terrestrial and aquatic organisms that decompose the large amounts of organic material have a key function in floodplain ecosystems. Termites and wood-feeding beetles play a major role in the fragmentation of wood, and terrestrial soil invertebrates are important for the decomposition of leaf litter during the terrestrial phase. In várzea habitats the aquatic larvae of *Asthenopus curtus* (Ephemeroptera, Polymitarcyidae) destroy wood during the aquatic phase. Data suggest that terrestrial invertebrates are more important in litter decomposition in the igapó than in the várzea, because mechanical fragmentation of the nutrient-poor and resistant material facilitates further processing by microorganisms. The exclusion of terrestrial soil invertebrates during the aquatic phase may slow down decomposition of organic matter in the igapó (Chaps. 9, 12). Soil invertebrates are fewer in species number and less abundant in the várzea habitat on Ilha da Marchantaria than in the igapó of Tarumã Mirim, because the deposition of fine inorganic sediments degrades habitat conditions (Chaps. 14–19). It is assumed that their abundance and diversity, as well as their role in litter decomposition, may increase near the terra firme and in habitats with low sediment deposition rates.

The large biomass produced by herbaceous plants and the floodplain forest and the time lag between primary production and the development of populations of invertebrate primary consumers places special importance on the detritus pathway. The data suggest that detritus and associated fungi, and bacteria, and algae are major food items for fish and terrestrial and aquatic invertebrates. Leaching plays a major role in decomposition processes of herbaceous plant material in the várzea (Chap. 9). We postulate that part of the dissolved organic carbon enters higher trophic levels through the microbial loop. The floodplain forests and terrestrial and aquatic macrophytes provide large surface areas during the aquatic phase. These are colonized by periphytic algae, fungi, bacteria, and aquatic invertebrates, which use dissolved and particulate material from the water and considerably increase primary and secondary production (Chaps. 10, 13).

Ecology of Floodplains

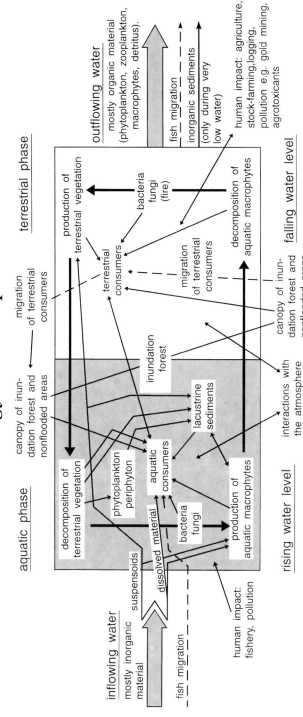

Fig. 23.3. Schematic description of the flux of organic and inorganic nutrients between the Amazon River and the floodplain and between the terrestrial and the aquatic phases

The filtering effect of aquatic macrophytes has also been reported from the Orinocco floodplain by Lewis et al. (1990).

The flood pulse allows a large-scale transfer of food items between the terrestrial and the aquatic phase in the ATTZ (Fig. 23.3). A major portion of the food items taken up by the fish is material that was produced during the terrestrial phase, such as flowers and fruit from the floodplain forest, terrestrial insects and seeds from terrestrial herbs. Welcomme (1985) showed that fish production in river–floodplain systems is positively correlated with the size of the floodplain and the extent of flooding. Our findings suggest that aquatic food webs are strongly influenced by the input of terrestrial material and that the terrestrial food webs are strongly influenced by aquatic material.

Acknowledgments. I am very grateful to Prof. Dr. Harald Sioli, Max-Planck-Insitute for Limnology, Plön, Dr. Rosemary Lowe-McConnell, Sussex, Dr. Robin Welcomme, FAO Rome, Dr. Keith Brown, Universidade Estadual de Campinas, São Paulo, and Dr. Peter Bayley, Oregon State University, Nash Hall, for critical comments on the manuscript and the correction of the English.

References

Ab'Saber AN (1988) O Pantanal Mato-Grossense e a teoria dos refúgios. Rev Bras Geogr 50:9–57

Abe T, Matsumoto T (1979) Studies on the distribution and ecological role of termites in a lowland rain forest of West Malaysia (3) Distribution and abundance of termites in Pasoh Forest Reserve. Jpn J Ecol 29:337–351

Adamoli J (1981) O Pantanal e suas relações fitogeográficas com os cerrados. Discussão sobre o conceito "Complexo do Pantanal". Congr Nacional de Botânica, Teresina, 1981. Soc Bras de Bot 32:109–119

Adams ES, Levings SC (1987) Territory size and population limits in mangrove termites. J J Anim Ecol 56(3):1069–1081

Adis J (1977) Programa mínimo para análises de ecossistemas: Artrópodos terrestres em florestas inundáveis da Amazônia Central. Acta Amazonica 7(2):223–229

Adis J (1981) Comparative ecological studies of the terrestrial arthropod fauna in Central Amazonian inundation forests. Amazoniana 7(2):87–173

Adis J (1982a) Eco-entomological observations from the Amazon: III. How do leafcutting ants of inundation forests survive flooding? Acta Amazonica 12(4):839–840

Adis J (1982b) Zur Besiedlung zentralamazonischer Überschwemmungswälder (Várzea-Gebiet) durch Carabiden (Coleoptera). Arch Hydrobiol 95(1/4):3–15

Adis J (1984) "Seasonal Igapó"-forests of Central Amazonian blackwater rivers and their terrestrial arthropod fauna. In: Sioli H (ed) The Amazon. Limnology and landscape ecology of a mighty tropical river and its basin. Junk, Dordrecht, pp 245–268

Adis J (1986) An "aquatic" millipede from a Central Amazonian floodplain forest. Oecologia 68(3):347–349

Adis J (1987) Extraction of arthropods from Neotropical soils with a modified Kempson apparatus. J Trop Ecol 3(2):131–138

Adis J (1988) On the abundance and density of terrestrial arthropods in Central Amazonian dryland forests. J Trop Ecol 4(1):19–24

Adis J (1990) Thirty million arthropod species – too many or too few? J Trop Ecol 6: 115–118

Adis J (1992a) Überlebensstrategien terrestrischer Invertebraten in Überschwemmungswäldern Zentralamazoniens. Verh Naturwiss Ver Hamb 33(NF):21–114

Adis J (1992b) How to survive six months in a flooded soil: strategies in Chilopoda and Symphyla from Central Amazonian floodplains. In: Adis J, Tanaka S (eds) Symposium on life-history traits in tropical invertebrates. (INTECOL, Yokohama, Japan 1990). Stud Neotrop Fauna Environ 27(2–3):117–129

Adis J (1992c) On the survival strategy of *Mestosoma hylaeicum* Jeekel (Paradoxomatidae, Polydesmida, Diplopoda), a millipede from Central Amazonian floodplains. (Proc 8th Int Congr Myriapodology, Innsbruck). Ber Nat-Med Ver Innsbruck Suppl 10: 183–187

Ecological Studies, Vol. 126
Junk (ed) The Central Amazon Floodplain
© Springer-Verlag Berlin Heidelberg 1997

Adis J, Arnett RH Jr (1987) Eco-entomological observations from the Amazon: VI. Notes on the natural history and flood resistance of *Sisenopiras gounellei* Pic (Coleoptera: Oedemeridae). Coleopt Bull 41(2):171-172

Adis J, Latif M (1996) Amazonian arthropods respond to El Niño. Biotropica 28(3):403-408

Adis J, Mahnert V (1985) On the natural history and ecology of Pseudoscorpiones (Arachnida) from an Amazonian black-water inundation forest. Amazoniana 9(3): 297-314

Adis J, Messner B (1990) On the species composition of Pseudoscorpiones from Amazonian dryland forests and inundation forests in Brazil. Rev Suisse Zool 97(1):49-53

Adis J, Messner B (1991) Langzeit-Überflutungsresistenz als Überlebensstrategie bei terrestrischen Arthropoden. Beispiele aus zentralamazonischen Überschwemmungs-gebieten. Dtsch Entomol Z NF 38(1-3):211-223

Adis J, Ribeiro MO de A (1989) Impact of deforestation on soil invertebrates from Central Amazonian inundation forests and their survival strategies to long-term flooding. Water Qual Bull 14(2):88-98 + 104

Adis J, Righi G (1989) Mass migration and life cycle adaptation a survival strategy of terrestrial earthworms in Central Amazonian inundation forests. Amazoniana 11(1): 23-30

Adis J, Scheller U (1984) On the natural history and ecology of *Hanseniella arborea* Scheller (Myriapoda, Symphyla, Scutigerellidae), a migrating symphylan from an Amazonian black-water inundation forest. Pedobiologia 27(1):35-41

Adis J, Schubart HOR (1984) Ecological research on arthropods in Central Amazonian forest ecosystems with recommendations for study procedures. In: Cooley JH, Golley FB (eds) Trends in ecological research for the 1980s. Plenum Press, New York, pp 111-144

Adis J, Sturm H (1987a) On the natural history and ecology of Meinertellidae (Archaeognatha, Insecta) from dryland and inundation forests of Central Amazonia. Amazoniana 10(2):197-218

Adis J, Sturm H (1987b) Flood-resistance of eggs and life-cycle adaptation, a survival strategy of *Neomachilellus scandens* (Meinertellidae, Archaeognatha) in Central Amazo-nian inundation forests. Insect Sci Appl 8(4-6):523-528

Adis J, Furch K, Irmler U (1979) Litter production of a Central Amazonian inundation forest. Trop Ecol 20:236-245

Adis J, Lubin YD, Montgomery GG (1984) Arthropods from the canopy of inundated and terra firme forests near Manaus, Brazil, with critical considerations on the pyrethrum-fogging technique. Stud Neotrop Fauna Environ 19(4):223-236

Adis J, Paarmann W, Erwin TL (1986) On the natural history and ecology of small terrestrial ground-beetles (Col.: Bembidiini: Tachyina: *Polyderis*) from an Amazonian black-water inundation forest. In: den Boer PJ, Luff ML, Mossakowski D, Weber F (eds) Carabid beetles. Their adaptations and dynamics. G. Fischer, Stuttgart, pp 413-427

Adis J, Morais JW de, Mesquita HG de (1987a) Vertical distribution and abundance of arthropods in the soil of a neotropical secondary forest during the rainy season. Stud Neotrop Fauna Environ 22(4):189-197

Adis J, Morais JW de, Ribeiro EF (1987b) Vertical distribution and abundance of arthropods in the soil of a neotropical secondary forest during the rainy season. Trop Ecol 28(1):174-181

Adis J, Mahnert V, Morais JW de, Rodrigues JMG (1988) Adaptation of an Amazonian pseudoscorpion (Arachnida) from dryland forests to inundation forests. Ecology 69(1):287-291

Adis J, Messner G, Groth I (1989a) Zur Überflutungsresistenz und zum Spinnvermögen von Japygiden (Diplura). Zool Jahrb Anat 119(3/4):371-382

Adis J, Morais JW de, Ribeiro EF, Ribeiro JC (1989b) Vertical distribution and abundance of arthropods in the soil of a neotropical campinarana forest during the rainy season. Stud Neotrop Fauna Environ 24(4):193–200

Adis J, Platnick NI, Morais JW de, Rodrigues JMC (1989c) On the abundance and ecology of Ricinulei (Arachnida) from Central Amazonia, Brazil. J NY Entomol Soc 97(2): 133–140

Adis J, Ribeiro EF, Morais JW de, Cavalcante ETS (1989d) Vertical distribution and abundance of arthropods in the soil of a neotropical campinarana forest during the dry season. Stud Neotrop Fauna Environ 24(4):201–211

Adis J, Paarmann W, Höfer H (1990) On phenology and life-cycle of *Scarites* sp. (Scaritini, Carabidae) from Central Amazonian floodplains. In: Stork NE (ed) Ground beetles: their role in ecological and environmental studies. Intercept Ltda., Andover, pp 269–275

Adis J, Messner B, Hirschel K, Ribeiro MO de, Paarmann W (1993) Zum Tauchvermögen eines Sandlaufkäfers (Coleoptera: Carabidae: Cicindelinae) im Überschwemmungsgebiet des Amazonas bei Manaus, Brasilien. Verh Westd Entom Tag 1992 (Löbbecke-Museum, Düsseldorf), pp 51–62

Adis J, Golovatch S, Hamann S (1996a) Survival-strategy of the terricolous millipede *Cutervodesmus adisi* (Fuhrmannodesmidae, Polydesmida) in a blackwater inundation forest of Central Amazonia (Brazil) in response to the flood pulse. Acta Myriapodologica. Mém Mus Nat Hist Nat 169:523–532

Adis J, Morais JW de, Scheller U (1996b) On abundance, phenology and natural history of Symphyla from a mixedwater inundation forest in Central Amazonia, Brazil. Acta Myriapodologica. Mém Mus Nat Hist Nat 169:607–616

Albert R (1982) Untersuchungen zur Struktur und Dynamik von Spinnengesellschaften verschiedener Vegetationstypen im Hoch-Solling. Hochschulsammlung Naturwissenschaft Biologie 16:1–146

Alho CJR (1982) Sincronia entre o regime de vazante do rio e o comportamento de nidificação da tartaruga da Amazônia *Podocnemis expansa* (Testudinata: Pelomedusidae). Acta Amazonica 12:323–326

Alho CJR, Pádua LFM (1982) Reproductive parameters and nesting behavior of the Amazon turtle *Podocnemis expansa* (Testudinata: Pelomedusidae) in Brazil. Can J Zool 60: 97–103

Alho CJR, Carvalho AG, Pádua LFM (1970) Ecologia da tartaruga da Amazônia e avaliação de seu manejo na Reserva Biológica do Trombetas. Bras Flor 38:29–47

Alves LF (1993) The fate of streamwater nitrate entering littoral areas of an Amazonian floodplain lake: the role of plankton, periphyton, inundated soils, and sediments. PhD Thesis, University of Maryland-CEES, College Park

Alvim PT, Alvim R (1978) Relation of climate to growth periodicity in tropical trees. In: Tomlinson PB, Zimmermann MH (eds) Tropical trees as living systems. Cambridge University Press, Cambridge, pp 449–464

Amoros C, Roux AL (1988) Interaction between water bodies within the floodplains of large rivers: function and development of connectivity. In: Schreiber KF (ed) Connectivity in landscape ecology. Proc 2nd Int Sem Int Assoc Landscape Ecol, Münster, pp 125–130

Amoros C, Roux AL, Reygrobellet JL, Bravard JP, Pautou G (1986) A method for applied ecological studies of fluvial hydrosystems. Regulated Rivers 1:17–36

Anderson JM (1973) The breakdown and decomposition of sweet chestnut (*Castanea sativa* MILL.) and beech (*Fagus sylvatica* L.) leaf litter in two deciduous woodland soils. II. Changes in the carbon, hydrogen, nitrogen and polyphenol content. Oecologia 12: 275–288

Anderson JM, Swift MJ (1983) Decomposition in tropical forests. In: Sutton SL, Whitmore TC, Chadwick AC (eds) Tropical rain forest: ecology and management, Spec Publ No 2 of the British Ecological Society. Blackwell, Oxford, pp 287–309

André HM (1979) Notes on the ecology of corticolous epiphyte dwellers. Recent Adv Acarol 1:551–557

André HM (1984) Notes on the ecology of corticolous epiphyte dwellers. 3. Oribatida. Acarologia 25(4):385–396

André HM (1985) Associations between corticolous microarthropod communities and epiphytic cover of bark. Holarct Ecol 8:113–119

André HM, Lebrun P (1979) Quantitative comparison of the funnel and the brushing methods for extracting corticolous microarthropods. Entomol Exp Appl 26:252–258

Andreae MO, Andreae TW (1988a) The cycle of biogenic sulfur compounds over the Amazon basin. 1. Dry season. J Geophys Res 93:1489–1497

Andreae MO, Talbot RW, Andreae TW, Harriss RC (1988b) Formic and acetic acid over the Central Amazon region. J Geophys Res 93:1616–1624

Andreae MO, Talbot RW, Berresheim H, Beecher KM (1990) Precipitation chemistry in Central Amazonia. J Geophys Res 95:16987–16999

Anonymous (1972a) Regenwasseranalysen aus Zentralamazonien, ausgeführt in Manaus, Amazonas, Brasilien, von Dr. Harald Ungemach. Amazoniana 3:186–198

Anonymous (1972b) Die Ionenfracht des Rio Negro, Staat Amazonas, Brasilien, nach Untersuchungen von Dr. Harald Ungemach. Amazoniana 3:175–185

Aranha C, Bacchi O, Leitão, Filho H de F (1982) Plantas invasoras de culturas 2. Instituto Campineiro de Ensino Agricola, Campinas, São Paulo, pp 297–597

Araújo RL (1970) Termites of the Neotropical region. In: Krishna K, Weesner FM (eds) Biology of termites. Academic Press, London, pp 527–576

Araujo-Lima CARM, Forsberg BR, Vitoria R, Martinelli L (1986) Energy sources for detritivorous fishes in the Amazon. Science 234:1256–1258

Arber A (1920) Water plants: a study of aquatic angiosperms. University Press, Cambridge

Ashton PS (1989) Species richness in tropical forests. In: Holm-Nielson LB, Nielsen I, Balslev H (eds) Tropical forests. Academic Press, London, pp 239–252

Autuori MP, Deutsch LA (1977) Contribution to the knowledge of the giant Brazilian otter, *Pteronura brasiliensis* (Gmelin 1788), Carnivora, Mustelidae. Zool Gart 47(1):1–8

Ayres JMC (1986) Uakaris and Amazonian flooded forest. PhD Thesis, Cambridge

Ayres JM (1993) As matas de várzea do Mamirauá. In: Sociedade civil Mamirauá (ed) Estudos de Mamirauá, Mamirauá, vol 1, pp 1–123

Bakwin PS, Wofsy SC, Fan S-M (1990a) Measurements of reactive nitrogen oxides (NOy) within and above a tropical forest canopy in the wet season. J Geophys Res 95: 16765–16772

Bakwin PS, Wofsy SC, Fan S-M, Keller M, Trumbore SE, da Costa JM (1990b) Emission of nitric oxide (NO) from tropical forest soils and exchange of NO between the forest canopy and atmospheric boundary layers. J Geophys Res 95:16755–16764

Ball L (1973) Micromorphological analysis of soils. Lower levels in the organization of organic soil materials. Soil Sur Pap no 6, Netherlands Soil Surv Inst, Wageningen

Balon EK (1975) Reproductive guilds of fishes: a proposal and definition. J Fish Res Board Can 32(6):821–864

Balslev H, Lutteyn J, Yllgaard B, Holm-Nielsen LB (1987) Composition and structure of adjacent unflooded and floodplain forest in Amazonian Ecuador. Opera Bot 92: 37–57

Bandeira AG (1978) Ecologia de térmitas da Amazônia Central: Efeitos do desmatamento sobre as populações e fixação de nitrogênio. MSc Thesis, INPA/FUA, Manaus

Bandeira AG (1983) Estrutura ecológica de comunidades de cupins (Insecta, Isoptera) na Zona Bragantina, Estado do Pará. PhD Thesis, CNPq-INPA-FUA, Manaus

Bandeira AG (1989) Análise da termitofauna (Insecta: Isoptera) de uma floresta primária e de uma pastagem na Amazônia Oriental, Brasil. Bol Mus Para Emílio Goeldi Sér Zool 5(2):225–241

Bandeira AG, Harada AY (1991) Cupins e formigas na Amazônia. In: Val AL, Figliuolo R, Feldberg E (eds) Bases científicas para estratégias de preservaçõ e desenvolvimento da Amazônia: Fatos e perspectivas, vol 1. INPA, Manaus, pp 387–395

Bandeira AG, Macambira MLJ (1988) Térmitas de Carajás, Estado do Pará, Brasil: Composição faunística, distribuição e hábito alimentar. Bol Mus Para Emílio Goeldi Sér Zool 4(2):175–190

Bandeira AG, Torres MFP (1985) Abundância e distribuição de invertebrados do solo em ecossistemas da Amazônia Oriental – o papel ecológico dos cupins. Bol Mus Para Emílio Goeldi Sér Zool 2(1):13–38

Baroni-Urbani C, Josens G, Peakin GJ (1978) Empirical data and demographic parameters. In: Brian MV (ed) Production ecology of ants and termites. International Biological Programme 13. Cambridge University Press, Cambridge, pp 5–44

Barthem RB, de Brito Ribeiro MCL, Petrere M Jr (1991) Life strategies of some long-distance migratory catfish in relation to hydroelectric dams in the Amazon Basin. Biol Conserv 55:339–345

Bartlett KB, Harriss RC (1993) Review and assessment of methane emissions from wetlands. Chemosphere 26(1–4):261–320

Bartlett KB, Crill P, Sebacher DI, Harriss RC, Wilson JO, Melack JM (1988) Methane flux from Central Amazonian floodplain. J Geophys Res 93:1573–1582

Bartlett KB, Crill PM, Bonassi JE, Richey RC, Harriss RC (1990) Methane flux from Central Amazonian floodplains: emission during rising water. J Geophys Res 95(10): 16733–16788

Bates HW (1864) The naturalist on the river Amazonas, 2nd edn. Murray, London

Baumeister W, Ernst W (eds) (1978) Mineralstoffe und Pflanzenwachstum. G Fischer Verlag, Stuttgart

Bayley PB (1980) The limits of limnological theory and approaches as applied to river – floodplain systems and their fish production. In: Furtado JI (ed) Tropical ecology and development. Proc 5th Intern Symp Trop Ecol, Intern Soc Trop Ecol, Kuala Lumpur, pp 739–746

Bayley PB (1983) Central Amazon fish populations: biomass, and some dynamic characteristics. – PhD Thesis, Dalhousie University, Halifax, Canada

Bayley PB (1989) Aquatic environments in the Amazon basin, with an analysis of carbon sources, fish production, and yield. In: Dodge DP (ed) Proc Int Large River Symp (LARS). Can Spec Publ Fish Aquat Sci 106:399–408

Bayley PB, Petrere M Jr (1989) Amazon fisheries: assessment methods, current status, and management options. In: Dodge DP (ed) Proc Int Large River Symp (LARS). Can Spec Publ Fish Aquat Sci 106:385–398

Beadle NCW (1966) Soil phosphate and its role in molding segments of the Australian flora and vegetation, with special reference to xeromorphy and sclerophylly. Ecology 47: 992–1007

Beck L (1969) Zum jahreszeitlichen Massenwechsel zweier Oribatidenarten (Acari) im neotropischen Überschwemmungswald. Verh Dtsch Zool Ges (Innsbruck 1968):535–540

Beck L (1971) Bodenzoologische Gliederung und Charakterisierung des amazonischen Regenwaldes. Amazoniana 3(1):69–132

Beck L (1972) Der Einfluß der jahresperiodischen Überflutungen auf den Massenwechsel der Bodenarthropoden im zentralamazonischen Regenwaldgebiet. Pedobiologia 12: 133–148

Beck L (1976) Zum Massenwechsel der Makro-Arthropodenfauna des Bodens in Überschwemmungswäldern des zentralen Amazonasgebietes. Amazoniana 6(1):1–20

Beck L (1983) Bodenzoologie der amazonischen Überschwemmungswälder. Amazoniana 8(1):91–99

Beck S (1983) Vegetationsökologische Grundlagen der Viehwirtschaft in den Überschwemmungs-Savannen des Rio Yacuma (Departamento Beni, Bolivien). Diss Botanicae 80, Cramer, Vaduz

Begon M, Harper JL, Townsend CR (eds) (1990) Ecology – individuals, populations, communities. Blackwell, Oxford

Behan-Pelletier VM, Paoletti MG, Bissett B, Stinner BR (1993) Oribatid mites of forest habitats in northern Venezuela. Trop Zool 1:39–54

Belluomini HE, Veinert T, Dissmann F, Hoge AR, Penha AM (1976/77) Notas biológicas a respeito do gênero *Eunectes* Wagler, 1830 "sucuris" (Serpentes: Bionae). Mem Inst Butantan 40/41:79115

Beltzer AH, Paporello de Amsler G (1984) Food and feeding habits of the Wattled Jacana *Jacana jacana* (Charadriiformes: Jacanidae) in Middle Parana River floodplain. Stud Neotrop Fauna Environ 19:195–200

Berg A (1961) Role écologique des eaux de la cuvette congolaise sur la croissance de la jacinthe d'eau [*Eichhornia crassipes* (Mart.) Solms]. Academie Royale des Sciences d'Outre-Mer, Brüssel

Best RC (1981) Foods and feeding habits of captive and wild Sirenia. Mammal Rev 11:3–29

Best RC (1982a) Seasonal breeding in the Amazonian manatee, *Trichechus inunguis* (Mammalia: Sirenia). Biotropica 14:76–78

Best RC (1982b) A salvação de uma espécie: novas perspectivas para o peixe-boi da Amazônia. Spec Publ of IBM – Brasil, INPA, Manaus

Best RC (1983) Apparent dry-season fasting in Amazonian manatees (Mammalia. Sirenia). Biotropica 15:61–64

Best RC (1984) The aquatic mammals and reptiles of the Amazon. In: Sioli H (ed) The Amazon: limnology and landscape ecology of a mighty tropical river and its basin. Junk, Dordrecht, pp 371–412

Best RC, da Silva VMF (1984) Preliminary analysis of reproductive parameters in Amazonian freshwater dolphins, *Inia* and *Sotalia*. In: Perrin WF, DeMaster D, Brownell RL Jr (eds) Reproduction of Cetacea: with special reference to stock assessment. Int Whaling Comm Spec Issue 6:361–369

Best RC, da Silva VMF (1989) The Amazon river dolphin, boto, *Inia geoffrensis* (de Blainville, 1917). In: Ridgway SH, Harrison R (eds) Handbook of marine mammals, vol 4. River dolphins and the large toothed whales. Academic Press, London, pp 1–23

Bhowmik NG, Stall JB (1979) Hydraulic geometry and carrying capacity of floodplains. Univ of Illinois Water Resources Center, Res Rep 145 UI LU-WRC-79-0145, Champaign

Bierregard RO Jr (1990) Species composition and trophic organization of the understory bird community in a Central Amazonian Terra firme forest. In: Gentry AH (ed) Four Neotropical rainforests. Yale University Press, New Haven, pp 217–236

Bierregard RO Jr, Lovejoy TE (1989) Effects of forest fragmentations on Amazonian understory bird communities. Acta Amazonica 19:215–241

Blacher C (1987) Ocorrência e preservação de *Lutra longicaudis* (Mammalia: Mustelidae) no litoral de Santa Catarina. Bol Fund Bras Conserv Nat 22:105–117

Black GA, Dobzhansky T, Pavan C (1950) Some attempts to estimate species diversity and population density of trees in Amazonian forests. Bot Gaz 111:413–425

Blick T, Scheidler M (1991) Kommentierte Artenliste der Spinnen Bayerns (Araneae). Arachnol Mitt 1:27–80

Bluntschli H (1921) Die Amazonasniederung als harmonischer Organismus. Geogr Z 27: 49–67

Bolster DC, Robinson SK (1990) Habitat use and relative abundance of migrant shorebirds in a western Amazonian site. Condor 92:239–242

Bolt GH, Bruggenwert MGM (eds) (1976) Soil chemistry. A. Basic elements. Elsevier, Amsterdam

Bonetto A, Cordiviola de Yuan E, Pignalberi C, Oliveros O (1969) Ciclos hidrologicos del Rio Paraná y las poblaciones de peces contenidas en las cuencas temporarias de su valle de inundacion. Physis 29(78):213–223

Bouwman AF (1989) Soils and the greenhouse effect. Wiley, Chichester

Bowen SH (1979a) A nutritional constraint in detritivory by fishes: the stunted population of *Sarotherodon mossambicus* in Lake Sibaya, South Africa. Ecol Monogr 49(1):17–31

Bowen SH (1979b) Determinants of the chemical composition of periphytic detrital aggregate in a tropical lake (Lake Valencia, Venezuela). Arch Hydrobiol 87(2):166–177

Bowen SH (1980) Detrital nonprotein amino acids are the key to rapid growth of *Tilapia* in Lake Valencia, Venezuela. Science 207:1216–1218

Bowmaker AP (1963) Cormorant predation on two Central African lakes. Ostrich 34:2–26

Boyd CE (1968) Freshwater plants: a potential source of protein. Econ Bot 22:359–368

Boyd CE (1970a) Chemical analysis of some vascular aquatic plants. Arch Hydrobiol 67:78–85

Boyd CE (1970b) Amino acid, protein and caloric contents of vascular aquatic marophytes. Ecology 51:902–906

Boyd CE, Goodyear CP (1971) Nutritive quality of food in ecological systems. Arch Hydrobiol 69:256–270

Braga RAP (1979) Contribuição à biologia e ecologia de *Asthenopus curtus* (Insecta Ephemeroptera) dos arredores de Manaus (Amazônia Central). MSc Thesis, INPA-FUA, Manaus

Brandorff, GO (1977) Untersuchungen zur Populations-Dynamik des Crustaceenplanktons im tropischen Lago Castanho (Amazonas, Brasilien). PhD Thesis, Universität Kiel

Brandorff GO (1978) Preliminary comparison of the Crustacean plankton of a white water and a black water lake in Central Amazon. Verh Int Ver Limnol 20:1198–1202

Brandorff GO, Andrade ER (1978) The relationship between the water level of the Amazon River and the fate of the zooplankton population in Lago Jacaretinga, a várzea lake in the Central Amazon. Stud Neotrop Fauna Environ 13:63–70

Braum E, Bock R (1985) Funktions-morphologische Untersuchungen über die Barteln von *Osteoglossum bicirrhosum* (Pisces, Osteoglossidae) während der Notatmung. Amazoniana 9(3):353–370

Braum E, Junk WJ (1982) Morphological adaptation of two Amazonian characoids (Pisces) for surviving in oxygen deficient waters. Int Rev Gesamten Hydrobiol 67:869–886

Braun R (1952) Limnologische Untersuchungen an einigen Seen im Amazonasgebiet. Schweiz Z Hydrol 14(1):128

Bravard JP, Amoros C, Pautou G (1986) Impact of civil engineering works on the succession of communities in a fluvial system. Oikos 47:92–111

Brecht-Munn M, Munn CA (1988) The Amazon's gregarious giant otters. Anim Kingdom 91(5):34–41

Brescovit AD, Höfer H (1993) Aranhas dos gêneros Lygromma e Eilica, da Amazônia Central, Brasil (Araneae, Gnaphosoidae). Iheringia Sér Zool 74:103–107

Brescovit AD, Höfer H (1994a) On the spider genus *Zimiromus* (Araneae: Gnaphosidae) in Central Amazonia. Bull Br Arachnol Soc 9(8):262–266

Brescovit AD, Höfer H (1994b) *Amazoromus,* a new genus of the spider family Gnaphosidae (Araneae) from central Amazonia, Brazil. Andrias 13:65–70

Bretfeld G, Gauer U (1994) Diagnostic description of the males of new *Sphaeridia* species (Insecta, Collembola) from South America. Andrias 13:113–136

Breznak JA, Brill WJ, Mertins JW, Coppel JC (1973) Nitrogen fixation in termites. Nature 244:577–580

Brian MV (1978) Production ecology of ants and termites. International Biological Programme 13. Cambridge University Press, Cambridge

Brinkmann WLF (1983) Hydrogeochemistry of groundwater resources in Central Amazonia. In: Groundwater in water resources planning. Proc Int Symp convened by UNESCO in cooperation with the National Committee FRG for the International Hydrological Programme, Koblenz, 28 Aug-3 Sept 1983, pp 67-83

Brinkmann WLF (1989) System propulsion of an Amazonian lowland forest: an outline. GeoJournal 19:369-380

Brinkmann WLF, Santos A dos (1971) Natural waters in Amazonia. V. Soluble magnesium properties. Turrialba 21:459-465

Brinkmann WLF, Santos A dos (1973) Natural waters in Amazonia. VI. Soluble calcium properties. Acta Amazonica 3:33-40

Brinkmann WLF, Santos UM (1973) Heavy fish-kill in unpolluted floodplain-lakes of Central Amazonia, Brasil. Biol Conserv 5:146-147

Broecker WS, Peng TH (1974) Gas exchange rates between air and sea. Tellus 26(12):21-35

Brosset A (1990) A long term study of the rain forest birds in M'Passa (Gabon). In: Keast A (ed) Biogeography and ecology of forest bird communities. SPB Acad Publ, The Hague, pp 259-274

Brown KS Jr (1986) Zoogeografia da região do Pantanal Mato-Grossense. Anais do Primeiro Simpósio sobre os Recursos Naturais e Sócio-econômicos do Pantanal. Corumbá 1984 Departamento de Difusão e Tecnologia, Documentos No 5 EMBRAPA Brasilia, pp 137-178

Brown RH (1978) A difference in N use efficiency in C3 and C4 plants and its implications in adaptation and evolution. Crop Sci 18:93-98

Budowski G (1961) Studies on forest succession in Costa Rica and Panama. PhD Thesis, Yale University, New Haven

Bustamante NC (1993) Preferências alimentares de 5 espécies de cupins do gênero Nasutitermes Dudley, 1890 (Termitidae: soptera) por 7 espécies de madeiras da várzea na Amazônia Central. MSc Thesis. INPA/FUA, Manaus

Butler JHA, Buckerfield JC (1979) Digestion of lignin by termites. Soil Biol Biochem 11: 507-513

Camargo JMF (1970) Ninhos e biologia de algumas espécies de Meliponideos (Hymenoptera: Apidae) da região de Pôrto Velho, Território de Rondônia, Brasil. Rev Biol Trop 16(2):207-239

Camargo JMF (1984) Notas sobre hábitos de nidificação de Scaura (Scaura) latitarsis (Friese) (Hymenoptera, Apidae, Meliponinae). Bol Mus Para Emílio Goeldi Sér Zool 1(1):89-95

Campbell DG, Daly DC, Prance GT, Maciel UN (1986) Quantitative ecological inventory of terra firme and várzea tropical forest on the Rio Xingú, Brazilian Amazon. Brittonia 38(4):369-393

Campos ZES de (1994) Parâmetros físico-químicos em igarapés de agua clara e preta ao longo da rodovia BR-174 entre Manaus e Presidente Figueiredo-AM. MSc Thesis, INPA/FUA, Manaus

Capparella AP (1991) Neotropical avian diversity and riverine barriers. Acta XX Congr Int Ornithol, Christchurch, (1990):307-316

Carbiener R, Schnitzler A, Walter JM (1988) Problèmes de dynamique forestière et de définition des stations en millieu alluvial. Colloq Phytosociol XIV 14:655-686

Carico J (1993) Revision of the genus Trechalea Thorell (Araneae, Trechaleidae) with a review of the taxonomy of the Trechaleidae and Pisauridae of the western hemisphere. J Arachnol 21:226-257

Carico J, Adis J, Penny ND (1985) A new species of Trechalea (Pisauridae: Araneae) from Central Amazonian inundation forests and notes on its natural history and ecology. Bull Br Arachnol Soc 6(7):289-294

Carignan R, Neiff JJ (1994) Limitation of water hyacinth by nitrogen in subtropical lakes of the Paraná floodplain (Argentina). Limnol Oceanogr 39(2):439–443

Carter GS (1935) Reports on the Cambridge Expedition to British Guiana, 1933. Respiratory adaptations of the fishes of the forest waters, with descriptions of the accessory respiratory organs of *Electrophorus electricus* and *Plecostomus plecostomus*. J Linn Soc Lond Zool 39:219–233

Carter GS, Beadle LC (1931) The fauna of the swamps of the Paraguayan chaco in relation to its environment. II. Respiratory adaptations in the fishes. J Linn Soc Lond Zool 37:327–366

Carvalho ML (1981) Alimentação do tambaquí jovem (*Colossoma macropomum* Cuvier, 1818) e sua relação com a communidade zooplanktônica do Lago Grande-Manaquirí, Solimões, AM. MSc Thesis, INPA/FUA, Manaus

Caswell H (1976) Community structure: a neutral model analysis. Ecol Monogr 46:327–354

Causey NB (1943) Studies on the life history and the ecology of the hothouse millipede, *Orthomorpha gracilis* (C.L. Koch 1874). Am Midl Nat 29:670–682

Chanton JP, Smith LK (1993) Seasonal variations in the isotopic composition of methane associated with aquatic macrophytes. In: Oremland RS (ed) Biogeochemistry of global change. Chapman and Hall, New York, pp 619–632

Chanton JP, Crill P, Bartlett K, Martens C (1989) Amazon capins (floating grassmats): a source of ^{13}C enriched methane to the troposphere. Geophys Res Lett 16:799–802

Chapleau F, Johansen PH, Williamson M (1988) The distinction between pattern and process in evolutionary biology: the use and abuse of the term "strategy". Oikos 53(1):136–138

Chapman PJ (1976) Speleobiology. In: Brook D (ed) The British New Guinea speleological expedition, 1975. Trans Br Cave Res Assoc 3:113–243

Chase A (1944) Grasses of Brazil and Venezuela. Agric Am IV(7):123–126

Chengalath R, Fernando CH, Koste W (1990) Rotifera from Sri Lanca (Ceylon) 3. New species and records with a list of species recorded and their distribution in different habitats from Sri Lanka. Bull Fish Res St Ceylon 12:327–340

Christensen O (1977) Estimation of standing crop and turnover of dead wood in a Danish oak forest. Oikos 28:177–196

Church TM, Tramontano JM, Whelpdale DM, Andreae MO, Galloway J, Keene WC, Knap AH, Tokos J Jr (1984) Atmospheric and precipitation chemistry over the North Atlantic Ocean: shipboard results, April–May 1984. J Geophys Res 96:18705–18725

Cicerone RJ, Oremland RS (1988) Biogeochemical aspects of atmospheric methane. Global Biochem Cycles 2:299–327

Clark JR, Benforado J (1981) Wetlands of bottomland hardwood forests. Developments in Agriculture and Managed Forest Ecology 11. Elsevier, Amsterdam

Clements FE (1905) Research methods in ecology. University Publ Co, Lincoln, Nebraska

Cochrane TT, Sanchez PA (1982) Land resources, soil and their management in the Amazon region: a state of knowledge report. In: Hecht SB (ed) Amazonia. Agriculture and land use research. Centro-Internacional de Agricultura Tropical Series 03, E-03, Cali, Columbia, pp 137–209

Collins NM (1988) Termites. In: Earl of Cranbrook (ed) Key environments: Malaysia. Pergamon, London, pp 196–211

Conforti V (1993) Study of the Euglenophyta from Camaleão Lake (Manaus-Brazil). Rev Hydrobiol Trop 26(1):3–18

Connell JH (1978) Diversity in tropical rain forests and coral reefs. Science 199:1302–1310

Conrad R (1989) Control of methane production in terrestrial ecosystems. In: Andrae MO, Schimmel DS (eds) Exchange of trace gases between terrestrial ecosystems and the atmosphere. Wiley, Chichester, pp 39–58

Constantino R (1992) Abundance and diversity of termites (Insecta: Isoptera) in two sites of primary rain forest in Brasilian Amazonia. Biotropica 24(3):420–430

Constantino R, Cancello E (1992) Cupins da Amazônia brasileira. Rev Bras Biol 52(3): 401–413

Cookson LJ (1987) ^{14}C-Lignin degradation by three Australian termite species (Isoptera: Mastotermitidae, Rhinotermitidae, Termitidae). Wood Sci and Technol 21(1):11–25

Cookson LJ (1988) The site and mechanism of ^{14}C-lignin degradation by *Nasutitermes exitiosus*. J Insect Physiol 34(5):409–414

Cordeiro R, Ojasti GA, Ojasti J (1981) Comparison of capybara populations of open and forested habitats. J Wildl Manage 45:267271

Costanza R, Maxwell T (1991) Spatial ecosystem modelling using parallel processors. Ecol Model 58:159–183

Coudreau HA (1897a) Voyage au Tapajós. A Lahure, Paris

Coudreau HA (1897b) Voyage au Tocantins – Araguaya. A Lahure, Paris

Coudreau HA (1897c) Voyage au Xingú. A Lahure, Paris

Coudreau HA (1899) Voyage au Yamundá. A Lahure, Paris

Coudreau HA (1900) Voyage au Trombetas. A Lahure, Paris

Coudreau O (1901) Voyage au Guminá. A Lahure, Paris

Coudreau O (1903a) Voyage à la Mapuera. A Lahure, Paris

Coudreau O (1903b) Voyage au Maycurú. A Lahure, Paris

Coudreau O (1903c) Voyage au Rio Curuá. A Lahure, Paris

Coudreau O (1906) Voyage au Canumá. A Lahure, Paris

Coventry RJ, Holt JA, Sinclair DF (1988) Nutrient cyling by mound-building termites in low-fertility soils of semi-arid tropical Australia. Aust J Soil Res 26: 375–390

Cowgill UM (1974) The hydrogeochemistry of Linsley Pond, North Branford, Connecticut. II. The chemical composition of aquatic macrophytes. Arch Hydrobiol (Suppl) 45: 1–119

Cowling EB, Merrill W (1966) Nitrogen in wood and its role in wood deterioration. Can J Bot 44:1539–1554

Cox CF (1988) Estudo de migrações laterais de peixes no sistema Lago do Rei (Ilha do Careiro) Am, BR. MSc Thesis, INPA/FUA, Manaus

Coyle F (1995) A revision of the funnelweb mygaomorph spider subfamily Ischnothelinae (Araneae, Dipluridae). Bull Am Mus Nat Hist 226:1–133

Crawford RMM (1977) Tolerance of anoxia and ethanol metabolism in germinating seeds. New Phyt 79:551–517

Crawford RMM (1982) Physiological responses to flooding. In: Pirson A, Zimmermann MH (eds) Encyclopedia of plant physiology, new Series. Springer, Berlin Heidelberg New York, pp 453–477

Crawford RMM (1983) Root survival in flooded soils. In: Gore AJP (ed) Ecosystems of the world 4A: mires: swamp, bog, fen and moor. Elsevier, Amsterdam, pp 257–283

Crawford RMM (1989) Studies in Plant Survival. Blackwell Scientific Publ, Oxford

Crill PM, Bartlett KB, Wilson JO, Sebacher Harriss RC (1988) Tropospheric methane from an Amazonian floodplain lake. J Geophys Res 93:1564–1570

Crow TR (1980) A rainforest chronicle: a 30-year record of change in structure and composition at El Verde, Puerto Rico. Biotropica 1:42–55

Crutzen PJ (1991) Methane's sinks and sources. Nature 350:380–381

Cruz A, Andrews RW (1989) Observations on the breeding biology of passerines in a seasonally flooded savanna in Venezuela. Wilson Bull 101:62–76

Cummins MP (1992) Amphibious behaviour of a tropical, adult tiger beetle, *Oxycheila polita* Bates (Coleoptera: Cicindelidae). Coleopt Bull 46(2):145–151

Cunha OR da, Nascimento FP do (1993) Ofídos da Amazônia. As cobras da região leste do Pará. Bol Mus Para Emilio Goeldi Zool 9(1):1–191

Curtis DJ (1980) Pitfalls in spider community studies (Arachnida, Araneae). J Arachnol 8:271–280

Curtis WF, Meade RH, Nordin CF, Price NB, Sholkovitz ER (1979) Non-uniform vertical distribution of fine sediment in the Amazon River. Nature 280:381–383

Daget J, Gosse J-P, Teugels GG, Thys van den Audenaerde DFE (eds) (1991) Check-list of the freshwater fishes of Africa 4. Orstom, Paris

Damuth JE, Fairbridge RW (1970) Equatorial atlantic deep – sea arkosic sands and ice – age aridity in tropical South America. Geol Soc Am Bull 81:189–206

Davies BR, Walker KF (1986) River systems as ecological units: An introduction to the ecology of river systems. In: Davies BR, Walker FF (eds) The ecology of river systems. Junk, Dordrecht, pp 1–8

Davis TAW (1953) An outline of the ecology and breeding seasons of birds of the lowland forest region of British Guiana. Ibis 95:450–467

Day FP (1983) Effects of flooding on leaf litter decomposition in microcosms. Oecologia (Ber) 56:180–184

Deacon EL (1977) Gas transfer to and across an air-water interface. Tellus 29(4):363–374

Deharveng L, Oliveira EP de (1990) Isotomiella (Collembola: Isotomidae) d'Amazonie: les espéces du groupe delamarei. Ann Soc Entomol Fr 26:185–201

del Toro MA (1971) On the biology of the American Finfoot in Southern Mexico. Living Bird 10:79–88

Demaree D (1932) Submerging experiments with *Taxodium*. Ecology 13:258–262

Denevan WM (1966) The aboriginal cultural geography of the Llanos de Mojos of Bolivia. Ibero-Americana 48, University of California Press, Berkeley

Denevan WM (1976) The aboriginal population of Amazonia. In: Denevan WM (ed) The native population of the Americas. University of Wisconsin Press, Madison, pp 205–234

Denich M (1989) Untersuchungen zur Bedeutung junger Sekundärvegetation für die Nutzungssystemproduktivität im östlichen Amazonasgebiet, Brasilien. Gött Beitr Land- und Forstwirtschaft Tropen Subtropen 46:1–265

Devol AH, Richey JF, Clark WA, King SL, Martinelli LA (1988) Methane emissions to the troposphere from the Amazon river floodplain. J Geophys Res 93:1583–1592

Devol AH, Richey JF, Forsberg BR, Martinelli LA (1990) Seasonal dynamics in methane emissions from the Amazon river floodplain to the troposphere. J Geophys Res 95:16417–16426

Devol AH, Forsberg BR, Richey JE, Pimentel TP (1995) Seasonal variation in chemical distributions in the Amazon (Solimoes) river: a multiyear time series. Global Biogeochem Cycle 9:307–328

di Castri F, Hansen AJ, Holland MM (1988): A new look at ecotones: emerging international projects on landscape boundaries. Biol Int, Special Issue 17:1–163

Dister E (1980) Geobotanische Untersuchungen in der hessischen Rheinaue als Grundlage für die Naturschutzarbeit. PhD Thesis, Universität Göttingen

Dister E (1985) Lebensräume und Retentionsfunktion. In: Akademie für Naturschutz und Landschaftspflege (ed) Die Zukunft der ostbayerischen Donaulandschaft. Laufener Seminarbeiträge 3/85, Laufen, pp 74–89

Dister E (1988) Ökologie der mitteleuropäischen Auenwälder. In: Wilhelm-Münker-Stiftung (ed) Die Auenwälder gestern und heute und morgen? No 4, Siegen, pp 6–30

Domning DP (1982) Commercial exploitation of manatees *Trichechus* in Brasil c 1785–1973. Biol Conserv 22:101–126

Doyle RD (1991) Primary production and nitrogen cycling within the periphyton community associated with emergent aquatic macrophytes in an Amazon floodplain lake. PhD Thesis, University of Maryland-CEES, College Park

Doyle RD, Fisher TR (1994) Nitrogen fixation by periphyton and plankton on the Amazon floodplain at Lake Calado. Biogeochemistry 26:41–66

Drent RH, Daan S (1980) The prudent parent: energetic adjustments in avian breeding. Ardea 68:225–252

Dubs B (1983) Die Vögel des südlichen Mato Grosso. Verbands-Verlag Betadruck, Bern

Ducke A, Black GA (1953) Phytogeographical notes on the Brazilian Amazon. An Acad Bras Ciênc 25:1–46

Dudel G, Schlangstedt M, Heder A, Kohl JG (1990) Measurement and calculation of dinitrogen fixation in water bodies. Arch Hydrobiol (Suppl) 33:723–731

Dudley RG (1974) Growth of Tilapia of the Kafue floodplain, Zambia: predicted effects of the Kafue Gorge Dam. Trans Am Fish Soc 103(2):281–291

Duplaix N (1980) Observations on the ecology and behavior of the giant river otter *Pteronura brasiliensis* in Suriname. Rev Ecol (Terre Vie) 34:495–620

Dwarakanath SK, Job SV (1974) The effect of aqueous submersion on respiration in the millipede *Spirostreptus asthenes*. Monit Zool Ital NS 8:11–18

Easley JF, Shirley RL (1974) Nutrient elements for livestock in aquatic plants. Hyacinth Control Journal 12:82–85

Eisenbeis G, Wichard W (1985) Atlas zur Biologie der Bodenarthropoden. Fischer, Stuttgart

Ellenberg H (1988) Vegetation ecology of Central Europe. University Press, Cambridge

Ellenberg H, Mayer R, Schauermann J (eds) (1986) Ökosystemforschung – Ergebnisse des Sollingprojektes 1966–1986. Ulmer, Stuttgart

Ellenbroek GA (1987) Ecology and productivity of an African wetland system. The Kafue flats, Zambia. Geobotany 9, Junk, Dordrecht

Emerson S (1975) Gas exchange rates in small Canadian shield lakes. Limnol Oceanogr 20(5):754–761

Engle DL (1993) Ecological consequences of riverine flooding in an Amazon floodplain lake: effects of sediment and nutrient inputs on seston dynamics and the epiphytic algae and aquatic invertebrates of floating meadows. PhD Thesis, University of California, Santa Barbara

Engle DL, Melack JM (1990) Floating meadow epiphyton: biological and chemical features of epiphytic material in an Amazon floodplain lake. Freshwater Biol 23:479–494

Engle DL, Melack JM (1993) Consequences of riverine flooding for seston and the periphyton of floating meadows in an Amazon floodplain lake. Limnol Oceanogr 38:1500–1520

Engle DL, Sarnelle O (1990) Algal use of sediment phosphorus from an Amazon floodplain lake – implications for total phosphorus analyses in turbid waters. Limnol Oceanogr 35:483–490

Ertel JR, Hedges JI, Devol AH, Richey JE (1986) Dissolved humic substances of the Amazon River system. Limnol Oceanogr 31:739–754

Erwin TL (1979) Thoughts on the evolutionary history of ground beetles: hypothesis generated from comparative fauna analysis of lowland forest sites in temperate and tropical regions. In: Erwin TL, Ball GE, Whitehead DR, Halpern AL (eds) Carabid beetles: Their evolution, natural history, and classification. Junk, The Hague, pp 539–592

Erwin TL (1981) Taxon pulses, vicariance, and dispersal: an evolutionary synthesis illustrated by carabid beetles. In: Nelson G, Rosen G (eds) Vicariance biogeography. Columbia University Press, New York, pp 159–196

Erwin TL (1983) Beetles and other insects of tropical forest canopies at Manaus, Brazil, sampled by insecticidal fogging. In: Sutton SL, Whitmore TC, Chadwick AC (eds) Tropical rain forest: ecology and management. Blackwell, Oxford, pp 59–75

Erwin TL (1988) The tropical forest canopy: the heart of biotic diversity. In: Wilson EO (ed) Biodiversity. National Academic Press, Washington, pp 123–129

Erwin TL, Adis J (1982) Amazonian inundation forests. Their role as short-term refuges and generators of species richness and taxon pulses. In: Prance GT (ed) Biological diversification in the tropics. Proc V Int Symp Assoc Trop Biol (Caracas 1979). Columbia University Press, New York, pp 358–371

Eutick ML, Veivers P, O'Brien RW, Slaytor M (1978) Dependence of the higher termite, *Nasutitermes exitiosus* and the lower termite, *Coptotermes lacteus* on their gut flora. J Insect Physiol 24:363–368

Fairbanks RG (1989) A 17000 – year glacio – eustatic sea level record: influence of glacial melting rates on the Younger Dryas event and deep – ocean circulation. Nature 342: 637–642

Falesi IC, Rodrigues TE, Morikawa IK, Reis RS (1971) Solos do distritio agropecuário da SUFRAMA. IPEAAOc, Serie Solos 1(1):1–99

Favareto L, Machado Z, Santos ES dos (1976) Consumo de oxigênio em *Macrobrachium amazonicum* (Heller, 1862). Efeito da saturação de oxigênio dissolvido. Acta Amazonica 6(4):449–453

Fearnside PM (1985) Agriculture in Amazonia. In: Prance GT, Lovejoy TE (eds) Amazonia. Pergamon Press, Oxford, pp 393–418

Ferreira LV (1991) O efeito do periodo de inundação, fenelogia e regeneração de clareiras em uma floresta de igapó na Amazônia Central. MSc Thesis, INPA, Manaus

Figueiredo A (1986) O papel dos sedimentos em dois lagos da Amazônia (Lago Calado e Lago Cristalino). MSc Thesis, INPA/University of Amazonas, Manaus

Fisher TR (1979) Plankton and primary production in aquatic systems of the central Amazon basin. Comp Biochem Physiol 62A:31–38

Fisher TR, Parsley PE (1979) Amazon lakes: water storage and nutrient stripping. Limnol Oceanogr 24:547–553

Fisher TR, Melack JM, Robertson B, Hardy ER, Alves LF (1983) Vertical distribution of zooplankton and physico-chemical conditions during a 24-hour period in an Amazon floodplain lake – Lago Calado, Brazil. Acta Amazonica 13(3–4):475–487

Fisher TR, Lesack LFW, Smith LK (1991) Input, recycling, and export of N and P on the Amazon floodplain at Lake Calado. In: Tiessen H, López-Hernández D, Salcedo IH (eds) Phosphorus cycles in terrestrial and aquatic ecosystems. Regional workshop 3: South and Central America, University Saskatchewan, pp 34–53

Fittkau E-J (1964) Remarks on limnology of Central-Amazonian rain-forest streams. Verh Int Ver Limnol 15:1092–1096

Fittkau E-J (1967) On the ecology of Amazonian rain-forest streams. In: Lent H (ed) Atas do Simpósio sôbre a Biota Amazônica, vol 3 (Limnologia). Conselho Nacional des Pesquisas, Rio de Janeiro, pp 97–108

Fittkau E-J (1968) *Chironomus strenzkei* n. sp. (Chironomidae, Dipt.), ein neues Laboratoriumstier. Z Morphol Tiere 63:239–250

Fittkau E-J (1970) Role of caimans in the nutrient regime of mouth-lakes of Amazon affluents. Biotropica 2:138–142

Fittkau E-J (1971) Ökologische Gliederung des Amazonas-Gebietes auf geochemischer Grundlage. Münster Forsch Geol Paläontol 20/21: 35–50

Fittkau E-J (1973) Artenmannigfaltigkeit amazonischer Lebensräume aus ökologischer Sicht. Amazoniana 4(3):321–340

Fittkau E-J (1973) Crocodiles and nutrient metabolism of Amazonian waters. Amazoniana 4:103–133

Fittkau E-J (1982) Struktur, Funktion und Diversität zentralamazonischer Ökosysteme. Arch Hydrobiol 95:29–45

Fittkau E-J, Irmler U, Junk WJ, Reiss F, Schmidt GW (1975) Productivity, biomass and population dynamics in Amazonian water bodies. In: Golley FB, Medina E (eds) Tropical ecological systems. Springer, Berlin Heidelberg New York, pp 289–311

Fitzpatrick JW (1981) Search strategies of tyrant flycatchers. Anim Behavior 29: 810–821

Fjeldsa J, Krabbe N (1990) Birds of the High Andes. Apollo Books, Svendborg

Forsberg BR (1984) Nutrient processing in Amazon floodplain lakes. Verh Int Ver Limnol 22:1294–1298

Forsberg BR, Devol AH, Richey JE, Martinelli LA, Santos H dos (1988) Factors controlling nutrient concentrations in Amazon floodplain lakes. Limnol Oceanogr 33(1):41–56

Forsberg BR, Pimentel TP, Nobre AD (1991) Photosynthetic parameters for phytoplankton in Amazon floodplain lakes, April – May 1987. Verh Int Ver Limnol 24:1188–1191

Förster K (1969) Amazonische Desmidieen. 1. Areal Santarém. Amazoniana 2:5–232

Förster K (1974) Amazonische Desmidieen. 2. Areal Maués Abacaxis. Amazoniana 5: 135–242

Forti MC, Neal C (1992) Hydrochemical cycles in tropical rainforests: an overview with emphasis on Central Amazonia. J Hydrol 134:103–115

Foster RB, Brokaw NVL (1983) Structure and history of the vegetation of Barro Colorado Island. In: Leigh EG, Rand AS, Windsor DM (eds) The ecology of a tropical forest. Oxford University Press, Oxford, pp 67–81

Franco W (1979) Die Wasserdynamik einiger Waldstandorte der West-Llanos Venezuelas und ihre Beziehungen zur Saisonalität des Laubfalles. Thesis, Universität Göttingen

Franken M, Irmler U, Klinge H (1979) Litterfall in inundation, riverine and terra firme forests of Central Amazonia. Trop Ecol 20:225–235

Franken W, Leopoldo PR (1984) Hydrology of catchment areas of Central-Amazonian forest streams. In: Sioli H (ed) The Amazon. Limnology and landscape ecology of a mighty tropical river and its basin. Monographiae Biologicae. Junk, Dordrecht, pp 501–519

Franken W, Leopoldo PR, Bergamin H (1985) Nutrient flow through natural waters in "terra firme" forest in Central Amazonia. Turrialba 35:383–393

Franklin E (1994) Ecologia de oribatídeos (Acari: Oribatida) em florestas inundáveis da Amazônia Central. PhD Thesis, INPA/University Amazonas, Manaus

Franklin E, Woas S (1992) Some basic opiid-like taxa (Acari, Oribatei) from Amazonia. Andrias 9:57–74

Franz H (1952–1953) Dauer und Wandel der Lebensgemeinschaften. Schriften des Vereins zur Verbreitung naturwiss. Kenntnisse in Wien, Selbstverlag Wien Ber 93:27–45

French JR (1975) The role of termite hindgut bacteria in wood decomposition. Mater Org 10:1–13

French JRJ, Bland DE (1975) Lignin degradation in the termites Coptotermes lacteus and Nasutitermes exitiosus. Mater Org 10:281–286

Freyer G, Iles TD (eds) (1972) The cichlid fishes of the great lakes of Africa. Oliver & Boyd, Edinburgh

Friebe B, Adis J (1983) Entwicklungszyklen von Opiliones (Arachnida) im Schwarzwasser-Überschwemmungswald (Igapó) des Rio Tarumã Mirím (Zentralamazonien, Brasilien). Amazoniana 8(1):101–110

Fry CH (1970) Ecological distribution of birds in north-eastern Mato Grosso State, Brazil. An Acad Brasil Ci 42(2):275–313

Funke W (1971) Food and energy turnover of leaf-eating insects and their influence on primary production. In: Ellenberg H (ed) Integrated experimental ecology. Ecological studies, vol 2. Springer, Berlin Heidelberg New York, pp 81–93

Furch B, Otto K (1987) Characterization of light regime changes (PAR) by irradiance reflectance in two Amazonian water bodies with different physico-chemical properties. Arch Hydrobiol 110(4):579–587

Furch B, Correa AFF, Mello JASN, Otto K (1985) Lichtklimadaten in drei aquatischen Ökosystemen verschiedener physikalisch-chemischer Beschaffenheit. I. Abschwächung, Rückstreuung, und Vergleich zwischen Einstrahlung, Rückstrahlung und sphärisch gemessener Quantenstromdichte (PAR). Amazoniana 9(3):411–430

Furch K (1976) Haupt- und Spurenelementgehalte zentralamazonischer Gewässertypen (erste Ergebnisse). Biogeographica 7:27–43

Furch K (1982) Seasonal variation of water chemistry of the middle Amazon várzea-lake Lago Calado. Arch Hydrobiol 95:47–67

Furch K (1984a) Interannuelle Variation hydrochemischer Parameter auf der Ilha de Marchantaria. Biogeographica 19:85–100

Furch K (1984b) Seasonal variation of the major cation content of the várzea-lake Lago Camaleão, middle Amazon, Brazil, in 1981 and 1982. Verh Int Ver Limnol 22:1288–1293

Furch K (1984c) Water chemistry of the Amazon basin: The distribution of chemical elements among freshwaters. In: Sioli H (ed) The Amazon. Limnology and landscape ecology of a mighty tropical river and its basin. Junk, Dordrecht, pp 167–200

Furch K (1985) Hydrogeochemie von Fließgewässern im Bereich der Transamazônica (Nordbrasilien). Amazoniana 9(3):371–410

Furch K (1986) Hydrogeochemistry of Amazonian freshwater along the Transamazônica in Brazil. Zbl Geol Paläont 1(9/10):1485–1493

Furch K (1987) Amazonian rivers: their chemistry and transport of dissolved solids through their basins. Mitt Geol Paläont Inst Univ Hamburg, SCOPE-UNEP Sonderbd 64:311–323

Furch K, Junk WJ (1980) Water chemistry and macrophytes of creeks and rivers in Southern Amazonia and the Central Brasilian shield. In: Furtado JI (ed) Tropical ecology and development, part 2. The International Society of Tropical Ecology, Kuala Lumpur, pp 771–796

Furch K, Junk WJ (1985) Dissolved carbon in a floodplain lake of the Amazon and in the river channel. In: Degens ET, Kempe S, Herrera R (eds) Transport of carbon and minerals in major world rivers, part 3. Mitt Geol Paläont Inst Univ Hamburg, SCOPE/UNEP Sonderbd 58:285–298

Furch K, Junk WJ (1992) Nutrient dynamics of submersed decomposing Amazonian herbaceous plant species Paspalum fasciculatum and Echinochloa polystachya. Rev Hydrobiol Trop 25(2):75–85

Furch K, Junk WJ (1993) Seasonal nutrient dynamics in an Amazonian floodplain lake. Arch Hydrobiol 128(3):277–285

Furch K, Klinge H (1978) Toward a regional characterization of the biogeochemistry of alkali and alkali-earth metals in northern South America. Acta Cient Venez 29:434–444

Furch K, Klinge H (1989) Chemical relationships between vegetation, soil and water in contrasting inundation areas of Amazonia. In: Proctor J (ed) Mineral nutrients in tropical forest and savanna ecosystems. Blackwell, Oxford, pp 189–204

Furch K, Junk WJ, Klinge H (1982) Unusual chemistry of natural waters from the Amazon region. Acta Cient Venez 33:269–273

Furch K, Junk WJ, Dieterich J, Kochert N (1983) Seasonal variation in the major cation (Na, K, Mg and Ca) content of the water of Lago Camaleão, an Amazon floodplain lake near Manaus, Brazil. Amazoniana 8(1):75–90

Furch K, Junk, WJ, Campos ZES (1988) Release of major ions and nutrients by decomposing leaves of Pseudobombax munguba, a common tree in the Amazonian floodplain. Verh Int Ver Limnol 23:642–646

Furch K, Junk, WJ, Campos ZES (1989) Nutrient dynamics of decomposing leaves from Amazonian floodplain forest species in water. Amazoniana 11(1):91–116

Gallivan GJ, Best RC (1980) Metabolism and respiration of the Amazonian manatee (Trichechus inunguis). Physiol Zool 53:245–253

Galloway JN, Likens GE (1976) Calibration of collection procedures for the determination of precipitation chemistry. Water Air Soil Pollut 6:241–258

Galloway JN, Likens GE, Keene WC, Miller JM (1982) The composition of precipitation in remote areas of the world. J Geophys Res 87:8771–8786

Garcia-Méndez G, Maass JM, Matson PA, Vitousek PM (1991) Nitrogen transformations and nitrous oxide flux in a tropical deciduous forest in México. Oecologia 88:362–366

Gasnier T, Höfer H, Brescovit AD (1995) Factors affecting the "activity density" of spiders on tree trunks in an Amazonian rainforest. Ecotropica 1(2):69–77

Geisler R (1969) Untersuchungen über den Sauerstoffgehalt, den biochemischen Sauerstoffbedarf und den Sauerstoffverbrauch von Fischen in einem tropischen Schwarzwasser (Rio Negro, Amazonien, Brasilien). Arch Hydrobiol 66(3):307–325

Geisler R, Schneider J (1976) The element matrix of Amazon waters and its relationship with the mineral content of fishes (determination using neutron activation analysis). Amazoniana 6(1):47–65

Gentry AH (1982) Patterns of Neotropical plant species diversity. Evol Biol 15:1–84

Gentry AH (1990a) Floristic similarities and differences between southern central America and upper and central Amazonia. In: Gentry AH (ed) Four neotropical rainforests. Yale University, New Haven, pp 141–157

Gentry AH (1990b) Four neotropical rainforests. Yale University Press, New Haven

Gentry AH, Lopez-Parodi J (1980) Deforestation and increased flooding of the upper Amazon. Science 210:1354

Gessner F (ed) (1959) Das Durchlüftungssystem der Wasser- und Sumpfpflanzen. In: Hydrobotanik II. Deutscher Verlag des Wissens, Berlin

Gibbs RJ (1967a) The geochemistry of the Amazon river system. I. The factors that control the salinity and the composition and concentration of the suspended solids. Geol Soc Am Bull 78:1203–1232

Gibbs RJ (1967b) Amazon river: environmental factors that control its dissolved and suspended load. Science 156:1734–1737

Gibbs RJ (1972) Water chemistry of the Amazon river. Geochim Cosmochim Acta 36: 1061–1066

Gill RH (1989) Ritmicidad en el crecimento de Vallea stipularis L. Pitteria (Merida, Venezuela) 18:44–56

Goeldi EA (1904) Against the destruction of White Herons and Red Ibises on the lower Amazon, especially on the Island of Marajó. Pará, Brazil

Golley FB, Richardson T (1977) Chemical relationships in tropical forests. Geo Eco Trop 1: 35–44

Golovatch SI (1992) Review of the Neotropical fauna of the millipede family Fuhrmannodesmidae, with the description of four new species from near Manaus, Central Amazonia, Brazil (Diplopoda, Polydesmida). Amazoniana 12(2):207–226

Golovatch SI (1994) Further new Fuhrmannodesmidae from the environs of Manaus, Central Amazonia, Brazil, with a revision of Cryptogonodesmus Silvestri, 1898 (Diplopoda, Polydesmida). Amazoniana 13(1/2):131–161

Goreau TJ, De Mello WZ (1988) Tropical deforestation: some effects on atmospheric chemistry. Ambio 17:275–281

Gosselink JG, Bayley SE, Conner WH, Turner RE (1981) Ecological factors in the determination of riparian wetland boundaries. In: Clark JR, Benforado J (ed) Wetlands of bottomland hardwood forests. Development in agricultural and managed forest ecology 11. Elsevier, Amsterdam, pp 197–219

Gottsberger G (1978) Seed dispersal by fish in the inundated regions of Humaitá, Amazonia. Biotropica 10:170–183

Goulding M (1980) The fishes and the forest: explorations in Amazonia natural history. California University Press, Berkeley

Goulding M (1981) Man and fisheries on an Amazon frontier. Junk, The Hague

Goulding M (1983a) Amazonian fisheries. In: Moran EF (ed) The dilemma of Amazonian Development. Westview Press, Boulder, pp 189–210

Goulding M (1983b) The role of fishes in seed dispersal and plant distribution in Amazonian floodplain ecosystems. In: Kubitzki K (ed) Dispersal and distribution. Sonderb Naturwiss Ver Hamb 7:271–283

Goulding M (1985) Forest fishes of the Amazon. In: Prance GT, Lovejoy TE (eds) Amazonia. Pergamon Press, Oxford, pp 267–276

Goulding M, Carvalho ML, Ferreira EG (1988) Rio Negro: rich life in poor water. SPB Academic Publishing, The Hague

Gourou P (1950) Observações geográficas na Amazônia. Rev Bras Geogr 11(3):355–408

Grabert H (1983) Das Amazonas-Entwässerungssystem in Raum und Zeit. Geol Rundsch 72:671–683

Grantsau R (1989) Die Kolibris Brasiliens, 2nd edn. Expressão e Cultura, Rio de Janeiro

Granville JJ (1974) Aperçu sur la structure des pneumatophores de deux espèces des sols hydromorphes en Guyane *Mauritia flexuosa* L. et *Euterpe oleracea* Mart. (Palmae). Généralisation au système respiratoire racinaire d'autres palmiers. Cah ORSTOM, Ser Biol 23:3–23

Grassé PP (1984) Termitologia tome 2. Fondation des Sociétés – Construction. Masson, Paris

Grassé PP (1986) Termitologia Tome 3. Comportement – Socialité – Écologie – Évolution – Systématique. Masson, Paris

Grasshoff K (1976) Methods of seawater analysis. Verlag Chemie, Weinheim

Graves GR, Zusi RL (1990) Avian body weights from the lower Rio Xingu, Brazil. Bull Br Ornithol Club 110:20–25

Greenberg R (1981) The abundance and seasonality of forest canopy birds on Barro Colorado Island, Panama. Biotropica 13:241–251

Grime JP (1979) Plant strategies and vegetation processes. Wiley, Chichester

Grönblad R (1945) De algis Brasiliensibus praecipue Desmidiacei in regione inferiore fluminis Amazonas a Prof. August Ginzberger anno 1927 collectis. Acta Soc Sci Fenn Nov Ser B 26:1–43

Groom MJ (1992) Sand-colored Nighthawks parasitize the antipredator behavior of three nesting bird species. Ecology 73:785–793

Grosse W, Büchel HB, Tiebel H (1991) Pressurized ventilation in wetland plants. Aquat Bot 39:89–98

Guest WC, Durocher PP (1979) Palaemonid shrimp, *Macrobrachium amazonicum*: effects of salinity and temperature on survival. Prog Fish Cult 41(1):14–18

Guimarães RL, Franklin E, Adis J (1993) Estratégias de sobrevivência de ácaros oribatídeos (Acari: Oribatida) à inundação periódica em florestas inundáveis da Amazônia Central. Resumos 2° Congr Iniciação Científica do Amazonas (INPA, Manaus), pp 164–165

Gyldenstolpe N (1945) The bird fauna of Rio Juruá in Western Brazil. K Sven Vetenskapsakad Handl 22(3):1–388

Haase R, Beck SG (1989) Structure and compositions of savanna vegetation in northern Bolivia: a preliminary report. Brittonia 41:80–100

Haffer J (1959) Notas sobre las aves de la región de Urabá (1). Lozania 12:1–49

Haffer J (1978) Distribution of Amazon forest birds. Bonn Zool Beitr 29:38–78

Haffer J (1982) General aspects of the refuge theory. In: Prance GT (ed) Biological diversification in the tropics. Columbia University Press, New York, pp 6–24

Haffer J (1985) Avian zoogeography of the Neotropical lowlands. Ornithol Monogr 36:113–146

Haffer J (1987) Biogeography of neotropical birds. In: Whitmore TC, Prance GT (eds) Biogeography and quaternary history in tropical America. Oxford Monogr Biogeography No 3. Clavendon Press, Oxford, pp 105–150

Haffer J (1988) Vögel Amazoniens: Ökologie, Brutbiologie und Artenreichtum. J Ornithol 129:1–53

Haffer J (1990) Avian species richness in tropical South America. Stud Neotrop Fauna Environ 25:157–183

Haffer J (1994) "Very small" bird populations in Amazonia. In: Remmert H (ed) Minimal animal populations. Ecological Studies, vol 106. Springer, Berlin Heidelberg New York, pp 105–117

Hagmann G (1909) Die Reptilien der Insel Mexiana, Amazonenstrom. Zool Jahrb Syst 28:473–504

Haines B, Jordan C, Clark H, Clark K (1983) Acid rain in an Amazon rainforest. Tellus 35B:77–80

Hallé F, Martin R (1968) Etude de la croissance rythmique chez l'Hévéa (Hevea brasiliensis Müll. Arg.). Adansonia ns 8:474–503

Hallé F, Oldemann RAA, Tomlinson PB (1978) Tropical trees and forests. An architectural analysis. Springer, Berlin Heidelberg New York

Hamilton SK, Lewis WM (1987) Causes of seasonality in the chemistry of a lake on the Orinoco river floodplain, Venezuela. Limnol Oceanogr 32:1277–1290

Hammerton D (1972) The Nile river – a case study. In: Oglesby RT, Carlson CA, McCann MJ (eds) River ecology and man. Academic Press, New York, pp 171–214

Hansen AJ, di Castri F (eds) (1992) Landscape boundaries: Consequences for biotic diversity and ecological flows. Ecological studies 92, Springer, Berlin Heidelberg New York

Hardy ER, Robertson B, Koste W (1984) About the relationship between the zooplankton and fluctuating water levels of Lago Camaleão, a Central Amazonian várzea lake. Amazoniana 9(1):43–52

Hardy RWF, Burns RC, Holsten RD (1973) Applications of the acetylene-ethylene assay for measurements of nitrogen fixation. Soil Biol Biochem 5:47–81

Harmon ME, Franklin JF, Swanson FJ, Sollins P, Gregory SV, Lattin JD, Anderson NH, Cline SP, Aumen NG, Sedell JR, Lienkaemper GW, Cromack K Jr, Cummins KW (1986) Ecology of coarse woody debris in temperate ecosystems. Adv Ecol Res 15:133–302

Harriss RC, Garstang M, Wofsy SC, Beck SM, Bendura RJ et al. (1988) The Amazon boundary layer experiment: dry season 1985. J Geophys Res 93:1351–1360

Harriss RC, Garstang M, Wofsy SC, Beck SM, Bendura RJ et al. (1990) The Amazon boundary layer experiment: wet season 1987. J Geophys Res 95:16721–16736

Haverschmidt F (1947) Field notes on the Black-bellied Tree Duck in Dutch Guiana. Wilson Bull 59:209

Haverschmidt F (1975) More bird records from Surinam. Bull Br Ornithol Club 95:74–77

Hayes FE, Fox JA (1991) Seasonality, habitat use, and flock size of shorebirds at the Bahía de Asunción, Paraguay. Wilson Bull 103:637–649

Heckman CW (1994) The seasonal succession of biotic communities in wetlands of the tropical wet-and-dry climatic zone: I. Physical and chemical causes and biological effects in the Pantanal of Mato Grosso, Brazil. Int Rev Gesamten Hydrobiol 79(3):397–421

Henderson PA, Walker I (1986) On the leaf litter community of the Amazonian blackwater stream Tarumazinho. J Trop Ecol 2:1–17

Henny CJ, Velzen WT van (1972) Migration patterns and wintering localities of American Ospreys. J Wildlife Manage 36:1133–1141

Hettler J, Irion G (1994) Umweltprobleme von Großbergbau: Die Kupfer und Goldmine Ok Tedi in Papua Neuguinea. In: Niedermeyer RO, Hüneke H, Scholle T (eds): Kurzfassung – Sediment 94 – Reihe A, Band 2, Universität Greifswald, pp 75–76

Heyer J (1990) Der Kreislauf des Methans. Mikrobiologie, Ökologie, Nutzung. Akademie-Verlag, Berlin

Hilty SL, Brown W (eds) (1986) A guide to the birds of Colombia. Princeton University Press, Princeton

Hödl WW (1977) Call differences and calling site segregation in anuran species from Central Amazonian floating meadows. Oecologia 28:351–363

Höfer H (1989) Beiträge zur Wirbellosenfauna der Ulmer Region: I. Spinnen (Arachnida:Araneae). Mitt Ver Naturw Math Ulm 35:157–176

Höfer H (1990a) Zur Ökologie der Spinnengemeinschaft eines Schwarzwasser-Überschwemmungswaldes (Igapó) in Zentralamazonien. PhD Thesis, Universität Ulm

Höfer H (1990b) The spider community (Arachnida, Araneae) of a Central Amazonian blackwater inundation forest (Igapó). Acta Zool Fenn 190:173–179

Höfer H, Brescovit AD (1994) On the spider genus *Rhoicinus* (Araneae, Trechaleidae) in a Central Amazonian inundation forest. J Arachnol 22(1):54–59

Höfer H, Brescovit AD, Gasnier T (1994a) The wandering spiders of the genus *Ctenus* (Ctenidae, Araneae) of Reserva Ducke, a rainforest reserve in Central Amazonia. Andrias 13:81–98

Höfer H, Brescovit AD, Adis J, Paarmann W (1994b) The spider fauna of neotropical tree canopies in Central Amazonia: first results. Stud Neotrop Fauna Envrion 29(1): 23–32

Hoffman RL (1977/78) Diplopoda from Papuan caves (Zoological results of the British Speleological Expedition to Papua-New Guinea, 1975,4). Int J Speleol 9:281–307

Hoffman RL (1995) Redefinition of the millipede genus *Pycnotropis*, and description of a new species from Manaus, Brazil (Polydesmida: Platyrhacidae: Euryurinae). Amazoniana 13(3/4): 283–292

Holanda OM (1982) Captura, distribuição, alimentação e aspectos reprodutivos de *Hemiodus unimaculatus* (Bloch, 1794) e *Hemiodopsis* sp. (Osteichthyes, Characoidei, Hemiodidae) na represa hidrelétrica de Curuá-Una, Pará. MSc Thesis, CNPq, INPA, FUA, Manaus

Holdrige LR (1966) The life zone systems. Adansonia 6:199–203

Holland MM, Risser PG, Naiman RJ (eds) (1991) Ecotones. The role of landscape boundaries in the management and restoration of changing environments. Chapman and Hall, New York

Holm LG, Plucknett DL, Pancho JV, Herberger IP (eds) (1977) The world's worst weeds. Distribution and biology. Honolulu University Press, Hawaii

Holm L, Pancho JV, Herberger JP, Plucknett DL (eds) (1979) A geographical atlas of world weeds. Wiley, New York

Holt DM (1983) Bacterial degradation of lignified wood cell wall in aerobic aquatic habitats: decay patterns and mechanisms proposed to account for their formation. Jour Inst of Wood Sci 9(5):212–223

Holt JA (1987) Carbon mineralization in semi-arid northeastern Australia: the role of termites. J Trop Ecol 3:255–263

Hopkin SP, Read H (eds) (1992) The biology of millipedes. Oxford University Press, Oxford

Howard-William C (1974) Nutritional quality and calorific value of Amazonian forest litter. Amazoniana 5:67–75

Howard-Williams C, Junk WJ (1976) The decomposition of aquatic macrophytes in the floating meadows of a Central Amazonian várzea lake. Biogeographica 7:115–123

Howard-Williams C, Junk WJ (1977) The chemical composition of Central Amazonian aquatic macrophytes with special reference to their role in the ecosystem. Arch Hydrobiol 79(4):446–464

Howard-Williams C, de Esteves F, Santos JE, Downes MT (1989) Short term nitrogen dynamics in a small Brazilian wetland (Lago Infernão, São Paulo). J Trop Ecol 5:323–335

Huber J (1910) Matas e madeiras amazônica. Bol Mus Goeldi Belém 6:91–203

Hueck K (1966) Die Wälder Südamerikas. Fischer, Stuttgart

Hustedt F (1965) Neue und wenig bekanne Diatomeen. IX. Süßwasserdiatomeen aus Brasilien, insbesondere des Amazonasgebietes. Int Rev Gesamten Hydrobiol 50(3): 391–410

Illies J (1968) Limnofauna Europaea. Fischer, Stuttgart

Irion G (1976a) Die Entwicklung des zentral- und oberamazonischen Tieflands im Spät-Pleistozän und im Holozän. Amazoniana 6(1):67–79

Irion G (1976b) Mineralogisch-geochemische Untersuchungen an der pelitischen Fraktion amazonischer Oberböden und Sedimente. Biogeographica 7:7–25

Irion G (1978) Soil infertility in the Amazon rain forest. Naturwissenschaften 65:515–519

Irion G (1982) Mineralogical and geochemical contribution to climate history in Central Amazonia during Quarternary time. Trop Ecol 23:76–85

Irion G (1984a) Sedimentation and sediments of Amazonian rivers and evolution of the Amazonian landscape since Pliocene times. In: Sioli H (ed) The Amazon – Limnology and landscape ecology of a mighty tropical river and its basin. Monographiae Biologicae. Junk, Dordrecht, pp 201–214

Irion G (1984b) Clay minerals of Amazonian soils. In: Sioli H (ed) The Amazon – Limnology and landscape ecology of a mighty tropical river and its basin. Monographiae Biologicae. Junk, Dordrecht, pp 537–579

Irion G (1989) Quaternary geological history of the Amazon lowlands. Tropical Forests: 23–34

Irion G, Adis J (1979) Evolução de florestas amazônicas inundadas, de igapó – um exemplo do rio Tarumã Mirím. Acta Amazonica 9:299–303

Irion G, Zöllmer V (1990) Pathways of fine-grained clastic sediments. Examples from the Amazon, the Weser estuary, and the North Sea. In: Heling D, Rothe P, Förstner U, Stoffers P (eds) Sediments and environmental geochemistry. Springer, Berlin Heidelberg New York, pp 351–366

Irion G, Adis J, Junk WJ, Wunderlich F (1983) Sedimentological studies of the Ilha de Marchantaria in the Solimões/Amazon river near Manaus. Amazoniana 8(1):1–18

Irion G, Adis, J, Junk WJ, Wunderlich F (1984) Sedimentaufbau einer Amazonas-Insel. Natur Museum 114:1–13

Irmler U (1975) Ecological studies of the aquatic soil invertebrates in three inundation forests of Central Amazonia. Amazoniana 5:337–409

Irmler U (1976) Zusammensetzung, Besiedlungsdichte und Biomasse der Makrofauna des Bodens in der emersen und submersen Phase dreier zentralamazonischer Überschwemmungswälder. Biogeographica 7:79–99

Irmler U (1977) Inundation–forest types in the vicinity of Manaus. Biogeographica 8:17–29

Irmler U (1979a) Abundance fluctuations and habitat changes of soil beetles in Central Amazonian inundation forests (Coleoptera: Carabidae, Staphylinidae). Stud Neotrop Fauna Environ 14:1–16

Irmler U (1979b) Considerations on structure and function of the "Central Amazonian inundation forest ecosystem" with particular emphasis on selected soil animals. Oecologia 43:1–18

Irmler U (1979c) Die Bedeutung der Schaben bei der Streuzersetzung in zentralamazonischen Überschwemmungswäldern. Gesellsch Ökologie, Verhandlungen 7:483–486

Irmler U (1981) Überlebensstrategien von Tieren im saisonal überfluteten amazonischen Überschwemmungswald. Zool Anz Jena 206(1/2):26–38

Irmler U (1982) Litterfall and nitrogen turnover in an Amazonian blackwater inundation forest. Plant Soil 67:355–358

Irmler U, Furch K (1979) Production, energy and nutrient turnover of the cockroach Epilampra irmleri Rocha e Silva and Aguiar in a Central-Amazonian inundation forest. Amazoniana 6:497–520

Irmler U, Furch K (1980) Weight, energy and nutrient changes during the decomposition of leaves in the emersion phase of Central-Amazonian inundation forest. Pedobiologia 20:118–130

Irmler U, Junk WJ (1982) The inhabitation of artifically exposed leaf samples by aquatic marco-invertebrates at the margin of Amazonian inundation forest. Trop Ecol 23(1): 64–75

IUCN (1990) Otters. IUCN, Species Survival Commission, Gland, Switzerland

Jacob DJ, Wofsy SC (1990) Budgets of reactive nitrogen, hydrocarbons, and ozone over the Amazon forest during the wet season. J Geophys Res 95:16737–16754

Janssen A (1986) Flora und Vegetation der Savannen von Humaitá und ihre Standortbedingungen. Diss Botanicae 93. Cramer, Berlin

Janzen DH (1967) Why mountain passes are higher in the tropics. Am Nat 101:233–249

Janzen DH (1985) Plant defences against animals in the Amazonian rainforest. In: Prance GT, Lovejoy TE (eds) Key environments: Amazon. Pergamon Press, Oxford, pp 166–191

Jedicke A, Furch B, Saint-Paul U, Schlüter UB (1989) Increase in the oxygen concentration in Amazon waters resulting from the root exudation of two notorious water plants, *Eichhornia crassipes* (Pontederiaceae) and *Pistia stratiotes* (Araceae). Amazoniana 11(1):53–69

Jenni DA, Betts BJ (1978) Sex differences in nest construction, incubation and parental behaviour in the polyandrous American Jaçana (*Jacana spinosa*). Anim Behav 26:207–218

Jenni DA, Collier C (1972) Polyandry in the American Jaçana (*Jacana spinosa*). Auk 89:743–765

Jocqué R (1984) Considerations concernant l'abondance relative des araignées errantes et des araignées à toile vivant au niveau du sol. Rev Arachnol 5(4):193–204

Johansen K, Mangum CM, Lykkeboe G (1978) Respiratory properties of the blood of Amazon fishes. Can J Zool 56:898–906

Johns AD (1991) Responses of Amazonian rain forest birds to habitat modification. J Trop Ecol 7:417–437

Johnson DS (1967) On the chemistry of freshwaters in southern Malaya and Singapore. Arch Hydrobiol 63:477–496

Johnson MJ, Meade RH (1990) Chemical weathering of fluvial sediments during alluvial storage: the Macuapanim Island point bar, Solimões River, Brazil. J Sediment Petrol 60:827–842

Johnsson DL (1983) The Californian continental borderland: landbridges, watergaps and biotop dispersal. In: Quaternary coastlines, Academic Press, London, pp 481–527

Jordan CF (1982) The nutrient balance of an Amazonian rain forest. Ecology 63:647–654

Jordan CF (1983) Productivity of tropical rain forest ecosystems and the implications for their use as future wood and energy sources. In: Golley FB (ed) Tropical rain forest ecosystems. Ecosystems of the world 14 A. Elsevier, Amsterdam, pp 117–136

Jordan CF, Escalante G (1980) Root productivity in an Amazonian rain forest. Ecology 61:14–18

Jordan CF, Uhl C (1978) Biomass of a "terra firme" forest of the Amazon basin calculated by a refined allometric relationship. Oecol Plant 13:387–400

Jordan CF, Golley F, Hall J, Hall J (1980) Nutrient scavenging of rainfall by the canopy of an Amazonian rain forest. Biotropica 12:61–66

Jordan CF, Caskey W, Escalante G, Herrera R, Montagnini F, Todd R, Uhl C (1982) The nitrogen cycle in a "Terra Firme" rain forest on oxisol in the Amazon territory of Venezuela. Plant Soil 67:325–332

Jordan CF, Caskey W, Escalante G, Herrera R, Montagnini F, Todd R, Uhl C (1983) Nitrogen dynamics during conversion of primary Amazonian rain forest to slash and burn agriculture. Oikos 40:131–139

Junk WJ (1970) Investigations on the ecology and production biology of the "floating meadows" Paspalo-Echinochloetum on the Middle Amazon. I. The floating vegetation and its ecology. Amazoniana 2(4):449–495

Junk WJ (1973) Investigation of the ecology and production-biology of the "Floating meadows" Paspalu-Echinochloetum on the Middle Amazon. II. The aquatic fauna in the root zone of floating vegetation. Amazoniana 4(1):9–112

Junk WJ (1980) Áreas inundáveis – um desafio para limnologia. Acta Amazonica 10(4): 775–795

Junk WJ (1982) Amazonian floodplains: their ecology, present and potential use. Rev Hydro Trop 15(4): 285–301

Junk WJ (1983a) Ecology of swamps on the middle Amazon. In: Gore AJP (ed) Ecosystems of the world. Mires, swamp, bog, fen and moor. 3. Regional studies. Elsevier, Amsterdam, pp 269–294

Junk WJ (1983b)As águas da região amazônica. In: Salati E, Schubart H, Junk WJ, Oliveira AR (eds): Amazonia: desenvolvimento, integração e ecologia. CNPq, editora brasiliense, Brasilia, pp 45–100

Junk WJ (1984a) Ecology, fisheries and fish culture in Amazonia. In: Sioli H (ed) The Amazon – Limnology and landscape ecology of a mighty tropical river and its basin. Monographiae Biologicae. Junk, Dordrecht, pp 443–476

Junk WJ (1984b) Ecology of the Várzea, floodplain of Amazon white water rivers. In: Sioli H (ed) The Amazon – Limnology and landscape ecology of a mighty tropical river and its basin. Monographiae Biologicae. Junk, Dordrecht, pp 215–243

Junk WJ (1985a) The Amazon floodplain – a sink or source of organic carbon? In: Degens ET, Kempe S, Herrera R (eds) Transport of carbon and minerals in major world rivers, part 3. Mitt Geol Paläontol Inst Univ Hamb 58:267–283

Junk WJ (1985b) Temporary fat storage, an adaptation of some fish species to the waterlevel fluctuations and related environmental changes of the Amazonian rivers. Amazoniana 9(3):315–352

Junk WJ (1986) Aquatic plants of the Amazon system. In: Davies BR, Walker KF (eds) The ecology of river systems. Junk, Dordrecht, pp 319–337

Junk WJ (1989) Flood tolerance and tree distribution in central Amazonian floodplains. In: Holm-Nielsen LB, Nielsen IC, Balslev H (eds) Tropical forests: botanical dynamics, speciation and diversity. Academic Press, New York, pp 47–64

Junk WJ (1990) Die Krautvegetation der Überschwemmungsgebiete des Amazonas (Várzea) bei Manaus und ihre Bedeutung für das Ökosystem. Habilitationsschrift, Universität Hamburg

Junk WJ (1993) Wetlands of tropical South America. In: Whigham DF, Hejny S, Dykyjova D (eds) Wetlands of the world IC. Kluwer, Dordrecht, pp 679–739

Junk WJ (1995) Human impact on neotropical wetlands: historical evidence, actual status and perspectives. In: Heinen HD, San Jose JJ, Arais HC (eds) Scientia Guianae 5: 299–311

Junk WJ, da Silva CJ (1995) Neotropical floodplains: a comparison between the Pantanal of Mato Grosso and the large Amazonian river floodplains. In: Tundisi JG, Bicudo CEM, Tundisi TM (eds) Limnology in Brazil. Brazilian Academy of Sciences, Brazilian Limnological Society, Rio de Janeiro, pp 195–217

Junk WJ, Furch K (1985) The physical and chemical properties of Amazonian waters and their relationships with the biota. In: Prance GT, Lovejoy TE (eds) Amazonia. Pergamon Press, Oxford, pp 3–17

Junk WJ, Furch K (1991) Nutrient dynamics in Amazonian floodplains: Decomposition of herbaceous plants in aquatic and terrestrial environments. Verh Int Ver Limnol 24:2080–2084

Junk WJ, Howard-Williams C (1984) Ecology of aquatic macrophytes in Amazonia. In: Sioli (ed) The Amazon – Limnology and landscape ecology of a mighty tropical river and its basin. Monographiae Biologicae. Junk, Dordrecht, pp 269–293

Junk WJ, Mello JASN (1987) Impactos ecológicos das represas hidrelétricas na bacia amazônica brasileira. In: Kohlhepp G, Schrader A (eds) Homem e natureza na Amazônia. Tübinger Geogr Studien 95:367–385

Junk WJ, Piedade MTF (1993a) Herbaceous plants of the Amazon floodplain near Manaus: species diversity and adaptations to the flood pulse. Amazoniana 12(3/4):467–484

Junk WJ, Piedade MTF (1993b) Species diversity and distribution of herbaceous plants in the floodplain of the middle Amazon. Verh Int Ver Limnol 25:1862–1865

Junk WJ, Piedade MTF (1993c) Biomass and primary-production of herbaceous plant communities in the Amazon floodplain. Hydrobiologia 263:155–162

Junk WJ, Weber GE (1996) Amazonian floodplains: a limnological perspective. Verh Int Ver Limnol 26:149–157

Junk WJ, Welcomme RL (1990) Floodplains. In: Patten BC, Jørgensen SE, Dumont H (eds) Wetlands and shallow continental water bodies 1. SPB Acad Publ, The Hague, pp 491–524

Junk WJ, Robertson BA, Darwich AJ, Vieira L (1981) Investigações limnológicas e ictiológicas em Curuá-Una, a primeira represa hidrelétrica na Amazônica Central. Acta Amazonica 11(4):689–716

Junk WJ, Soares MGM, Carvalho FM (1983) Distribution of fish species in a lake in the Amazon river floodplain near Manaus Lago Camaleão with special reference to extreme oxygen conditions. Amazoniana 7(4):397–431

Junk WJ, Bayley PB, Sparks RE (1989) The flood pulse concept in river – floodplain systems. In: Dodge DP (ed) Proc Int Large River Symp (LARS). Can Spec Publ Fish Aquat Sci 106:110–127

Junor F (1972) Estimation of the daily food intake of piscivorous birds. Ostrich 43: 193–205

Kalliola R, Puhakka M, Danjoy W (eds) (1993) Amazonia peruana: vegetación húmeda tropical en el Llano subandino. Projeto Amazonia Universidad de Turku e Oficina Nacional de Evaluación de Recoursos Naturales, Lima

Karr JR (1989) Birds. In: Lieth H, Werger MJA (eds) Tropical rain forest ecosystems. Biogeographical and ecological studies. Ecosystems of the World 14B. Elsevier, Amsterdam, pp 401–416

Karr JR (1990a) Birds of tropical rainforest: comparative biogeography and ecology. In: Keast A (ed) Biogeography and ecology of forest bird communities. SPB Acad Publ, The Hague, pp 215–228

Karr JR (1990b) The avifauna of Barro Colorado Island and the pipeline road, Panama. In: Gentry AH (ed) Four neotropical rainforests. Yale University Press, New Haven, pp 183–198

Karr JR, Freemark KE (1983) Habitat selection and environmental gradients: dynamics in the "stable" tropics. Ecology 64:1481–1494

Karr JR, Robinson SK, Blake JG, Bierregard RO (1990) Birds of four Neotropical forests. In: Gentry AH (ed) Four neotropical rainforests. Yale University Press, New Haven, pp 237–269

Käärik AA (1974) Decomposition of wood. In: Dickinson CH, Pugh GJF (eds) Biology of plant litter decomposition. Academic Press, London, pp 129–174

Kapetsky JM (1974) Growth, mortality, and production of five fish species of the Kafue river floodplain, Zambia. PhD Thesis, University of Michigan, Ann Arbor

Keast A, Morton ES (eds) (1980) Migrant birds in the neotropics. Ecology, behavior, distribution and conservation. Smithsonian Institution Press, Washington DC

Keel SH, Prance GH (1979) Studies of the vegetation of a white-sand black-water igapó (Rio Negro, Brazil). Acta Amazonica 9:645–655

Kempe S (1982) Long-term records of CO_2 pressure fluctuations in fresh waters. In: Degens ET (ed) Transport of carbon and minerals in major world rivers, Pt 1. Mitt Geol Paläon Inst Univ Hamburg, SCOPE/UNEP Sonderbd 52:91–332

Keene WC, Galloway JN (1984) A note on acid rain in an Amazon rainforest. Tellus 36B: 137–138

Kern J (1995) Die Bedeutung der N_2-Fixierung und der Denitrifikation für den Stickstoffhaushalt des amazonischen Überschwemmungssees Lago Camaleão. PhD Thesis, Universität Hamburg

Kern J, Darwich A, Furch K, Junk WJ (1996) Seasonal denitrification in flooded and exposed sediments from the Amazon floodplain at Lago Camaleão. Microb Ecol 32(1):47–57

Kerr WE, Sakagami SF, Zucchi R, Araujo VP, Camargo JMF (1967) Observações sobre a arquitetura dos ninhos e comportamento de algumas espécies de abelhas sem ferrão das vizinhanças de Manaus, Amazonas (Hymenoptera, Apoidea). Atas Simpósio sobre a Biota Amazônica 5 (Zoologia):255–309

Keyes MR, Grier CC (1981) Below- and aboveground biomass and net production in two contrasting Douglas-fir stands. Can J For Res 11:599–605

Kiltie R, Fitzpatrick JW (1984) Reproduction and social organization of the Black-capped Donacobius (Donacobius atricapillus). Auk 101:804–811

Kira T (1978) Community architecture and organic matter dynamics in tropical lowland rain forests of Southeast Asia with special reference to Pasoh forest, West Malaysia. In: Tomlinson PB, Zimmermann MH (eds) Tropical trees as living systems. Cambridge University Press, Cambridge, pp 561–590

Kira T, Yoda K (1989) Vertical stratification in microclimate. In: Lieth H, Werger MJA (eds) Ecosystems of the world 14B. Tropical rain forest ecosystems. Elsevier, Amsterdam, pp 55–71

Kira T, Ogawa H, Yoda K, Ogino K (1964) Primary production by a tropical rain forest of southern Thailand. Bot Mag Tokyo 77:428–429

Kira T, Ogawa H, Yoda K, Ogino K (1967) Comparative ecological studies on three main types of forest vegetation in Thailand. Nature Life SE Asia 5:149–174

Kirk JTO (1983) Light and photosynthesis in aquatic ecosystems. Cambridge University Press, Cambridge

Klammer G (1984) The relief of the extra-Andean Amazon basin. In: Sioli H (ed) The Amazon – Limnology and landscape ecology of a mighty tropical river and its basin. Monographiae Biologicae. Junk, Dordrecht, pp 47–83

Klenke M, Ohly JJ (1993) Wood from the floodplains. In: Junk WJ, Bianchi HK (eds) 1st SHIFT Worksh, Belém, 1993. GKSS-Research Center, Geesthacht

Klinge H (1967) Podzol soils: a source of blackwater rivers in Amazonia. Atas do Simposio sobre a Biota Amazônica 3:117–125

Klinge H (1973a) Root mass estimation in lowland tropical rain forests of Central Amazonia, Brazil, I. Trop Ecol 14:29–33

Klinge H (1973b) Struktur und Artenreichtum des zentralamazonischen Regenwaldes. Amazoniana 4(3):283–292

Klinge H (1975) Bilanzierung von Hauptnährstoffen im Ökosystem tropischer Regenwald (Manaus) – vorläufige Daten. Biogeographica 7:59–76

Klinge H (1977) Fine litter production and nutrient return to the soil in three natural forest stands of Eastern Amazonia. Geo-Eco-Trop 1:159–167

Klinge H (1985) Foliar nutrient levels of native tree species from Central Amazonia. 2. Campina. Amazoniana 9:281–295

Klinge H, Fittkau EJ (1972) Filterfunktion im Ökosystem des zentral-amazonischen Regenwaldes. Mitt Dtsch Bodenkundl Ges 16:130–135

Klinge H, Furch K (1991) Towards the classification of Amazonian floodplains and their forests by means of biogeochemical criteria of river water and forest biomass. Interciencia 16:196–201

Klinge H, Herrera R (1978) Biomass studies in Amazon Caatinga forest in southern Venezuela. 1. Standing crop of composite root mass in selected stands. Trop Ecol 19:93–110

Klinge H, Ohle W (1964) Chemical properties of rivers in the Amazonian area in relation to soil conditions. Verh Int Ver Limnol 15:1067–1076

Klinge H, Rodrigues W (1968) Litter production in an area of Amazonian terra firme forest. Part I. Litterfall, organic and total nitrogen contents of litter. Amazoniana 1:287–302

Klinge H, Rodriguez WA, Brünig E, Fittkau EJ (1975) Biomass and structure in a Central Amazonian rain forest. In: Golley FB, Medina E (eds) Tropical ecological systems. Springer, Berlin Heidelberg New York, pp 115–122

Klinge H, Furch K, Harms E, Revilla J (1983) Foliar nutrient levels of native tree species from Central Amazonia. 1. Inundation forests. Amazoniana 8:19–45

Klinge H, Furch K, Harms E (1984) Selected bioelements in bark and wood of native tree species from Central-Amazonian inundation forests. Amazoniana 9(1):105–117

Klinge H, Junk WJ, Revilla JC (1990) Status and distribution of forested wetlands in tropical South America. For Ecol Manage 33/34:81–101

Klinge H, Adis J, Worbes M (1996) The vegetation of a seasonal várzea forest in the lower Solimões River, Amazon region of Brazil. Acta Amazonica 25(3–4):201–220

Koch W (1969) Einfluß von Umweltfaktoren auf die Samenphase annueller Unkräuter, insbesondere unter dem Gesichtspunkt der Unkrautbekämpfung. Arbeiten aus der Universität Hohenheim, 50, Ulmer Verlag, Stuttgart

Koepcke M (1972) Über die Resistenzformen der Vogelnester in einem begrenzten Gebiet des tropischen Regenwaldes in Peru. J Ornithol 113(2):145–160

Koslowski TT (1984) Flooding and plant growth. Academic Press, New York

Koste W (1974) Zur Kenntnis der Rotatorienfauna der "schwimmenden Wiese" einer Uferlagune in der Várzea Amazoniens, Brasilien. Amazoniana 5(1):25–59

Koste W (1996) Additions to the checklist of rotifers Monogononta recorded from Neotropics (in press)

Koste W, Poltz J (1987) Über die Rädertiere (Rotatoria, Phylum Aschelminthes) des Alfsees, eines Hochwasser-Rückhaltebeckens der Hase, NW-Deutschland, FRG. Osnabrücker Naturwiss Mitt 13:185–220

Koste W, Robertson B (1983) Taxonomic studies of the Rotifera (Phylum Aschelminthes) from a Central Amazonian Várzea lake, Lago Camaleão (Ilha de Marchantaria, Rio Solimões, Amazonas, Brazil). Amazoniana 8(2):225–254

Koste W, Hardy E, Robertson B (1984) Further taxonomical studies of the rotifera (Phylum Aschelminthes) from a Central Amazonian várzea lake, Lago Camaleão (Ilha de Marchantaria, Rio Solimões, Amazonas, Brazil). Amazoniana 8(4):555–576

Kozlovsky DG (1968) A critical evaluation of the trophic level concept. I. Ecological efficiencies. Ecology 49(1):48–60

Krambeck HJ, Albrecht D, Hickel B, Hofmann W, Arzbach H-H (1994) Limnology of the Plußsee. In: Overbeck J, Chróst RJ (eds) Microbial ecology of Lake Plußsee. Ecological Studies, vol 105. Springer, Berlin Heidelberg New York, pp 1–23

Kramer DL (1983) Aquatic surface respiration in the fishes of Panama: distribution in relation to risk of hypoxia. Environ Biol Fishes 8:49–54

Kramer DL, McClure M (1982) Aquatic surface respiration, a widespread adaptation to hypoxia in tropical freshwater fishes. Environ Biol Fishes 7:47–55

Kramer DL, Lindsey CC, Moodie GEE, Stevens ED (1978) The fishes and the aquatic environment of the Central Amazon basin, with particular reference to respiratory patterns. Can J Zool 56:717–729

Krannitz PG (1989) Nesting biology of Black Skimmers, Large-billed Terns, and Yellow-billed Terns in Amazonian Brazil. J Field Ornithol 60:216–223

Kubitzki K (1989) The ecogeographical differentiation of Amazonian inundation forests. Plant Syst Evol 162:285–304

Kubitzki K, Ziburski A (1994) Seed dispersal in flood plain forest of Amazonia. Biotropica 26:30–43

Kunz E (1975) Von der Tulla'schen Rheinkorrektion bis zum Oberrheinausbau. 150 Jahre Wasserbau am Oberrhein. Jahrb Natursch Landschaftspfl 24:59–78

Kushlan JA, Morales G, Frohring PC (1985) Foraging niche relations of wading birds in tropical wet Savannas. Ornithol Monogr 36:663–682

Kullander SO (1994) Amazonische Cichliden – jenseits der Flußbiegung. DATZ Sonderh:53–59

La Fage JP, Nutting WL (1978) Nutrient dynamics of termites. In: Brian MV (ed) Production ecology of ants and termites, International Biological Programme 13. Cambridge University Press, Cambridge, pp 165–232

LaBaugh JW (1986) Wetland ecosystem studies from a hydrologic perspective. Water Resour Bull 22:1–10

Lacerda LD, Pfeiffer WC, Ott AT, Silveira EG (1989) Mercury contamination in the Madeira river, Amazon: Hg inputs to the environment. Biotropica 21:91–93

Lal R (1987) Tropical ecology and physical edaphology. Wiley-Interscience, Chichester

Lang GE, Knight DH (1979) Decay rates for boles of tropical trees in Panama. Biotropica 2(4):316–317

Lang GE, Knight DH (1983) Tree growth, mortality, recruitment, and canopy gap formation during a 10-year period in a tropical moist forest. Ecology 64:1075–1080

Larson JS, Bedinger MS, Bryan CF, Brown S, Huffmann RT, Miller EL, Rhodes DG, Tuchet BA (1981) Transition from wetlands to uplands in south-eastern bottomland hardwood forests. In: Clark JR, Benforado J (eds) Wetlands of bottomland hardwood forests. Development in agricultural and managed forest ecology 11. Elsevier, Amsterdam, pp 225–273

Larson PR (1964) Some indirect effects of environment on wood formation. In: Zimmermann MH (ed) The formation of wood in forest trees. Academic Press, New York, pp 345–365

Law CS, Rees AP, Owens NJP (1993) Nitrous oxide production by estuarine epiphyton. Limnol Oceanogr 38:435–441

Lee KE, Wood TG (eds) (1971) Termites and soils. Academic Press, London

Lee RB (1977) Effect of organic acids in the loss of ions from barley roots. J Exp Bot 28:578–587

Lee RB (1978) Inorganic nitrogen metabolism in barley roots under poorly aerated conditions. J Exp Bot 29:693–708

Leenheer JA, Santos UM (1980) Considerações sobre os processos de sedimentação na água preta ácida do Rio Negro (Amazônia Central). Acta Amazonica 10:343–355

Lefebvre, G, Poulin B, McNeil R (1992) Settlement period and function of long-term territory in tropical mangrove passerines. Condor 94:83–92

Leitão Filho H d F, Aranha C, Bacchi O (1982) Plantas invasoras de culturas 1. Instituto Campineiro de Ensino Agrícola, Campinas, São Paulo

Lenz PH, Melack JM, Robertson B, Hardy EA (1986) Ammonium and phosphate regeneration by zooplankton of an Amazonian floodplain lake. Freshwater Biol 16:821–830

Lesack LFW (1988) Mass balance of nutrients, major solutes, and water in an Amazon floodplain lake and biogeochemical implications for the Amazon basin. PhD Thesis, University of California, Santa Barbara

Lesack LFW (1993) Export of nutrients and major ionic solutes from a rain forest catchment in the Central Amazon basin. Water Resour Res 29:743–758

Lesack LFW (1995) Seepage exchange in an Amazon floodplain lake. Limnol Oceanogr 40:598–609

Lesack LFW, Melack JM (1991) The deposition, composition, and potential sources of major ionic solutes in rain of the Central Amazon basin. Water Resour Res 27:2953–2977

Lesack LFW, Melack JM (1995) Flooding hydrology and mixture dynamics of lake water derived from multiple sources in an Amazon floodplain lake. Water Resour Res 31:329–345

Levings SC, Adams ES (1984) Intra- and interspecific territoriality in *Nasutitermes* (Isoptera: Termitidae) in a Panamanian mangrove forest. J Anim Ecol 53:705–714

Lewis WM Jr, Weibezahn FH, Saunders JF, Hamilton SK (1990) The Orinoco River as an ecological system. Interciencia 15(6):346–357

Lieberman M, Hartshorn G, Peralta R (1985) Growth rates and age – size relationships of tropical wet forest trees in Costa Rica. J Trop Ecol 1:97–109

Lieberman D, Lieberman M, Peralta R, Hartshorn GS (1985) Mortality patterns and stand turnover rates in a wet tropical forest in Costa Rica. J Ecol 73(3):915–924

Lieth H, Whittaker RH (1975) Primary productivity of the biosphere. Springer, Berlin Heidelberg New York

Likens GE, Keene WC, Miller JM, Galloway JN (1987) Chemistry of precipitation from a remote, terrestrial site in Australia. J Geophys Res 92:13299–13314

Lima IMB (1996) Uma nova espécie de *Prosekia* Vandel, 1968 da Amazônia Brasileira (Crustaceae, Isopoda, Philosciidae). Amazoniana 14:101–108

Lima RR (1986) Várzeas da Amazônia Brasileira e sua potencialidade agropecuária. 1° Simpósio dó Trópico Úmido Proc, Belém 1984, EMBRAPA 1986, pp 141–164

Lindeman RL (1942) The trophic-dynamic aspect of ecology. Ecology 23:399–418

Livingstone DA (1963) Chemical composition of rivers and lakes. In: Fleischer M (ed) Data of geochemistry. US Geol Surv Prof Pap 440 G:1–64

Lorenzi H (1982) Plantas daninhas do Brasil. Edição do autor, Nova Odessa, São Paulo

Lovejoy TE (1974) Bird diversity and abundance in Amazon forest communities. Living Bird 13:127–191

Loveless AR (1961) A nutritional interpretation of sclerophyllous and mesophytic leaves. Ann Bot 25:164–168

Loveless AR (1962) Further evidence to support a nutritional interpretation of sclerophylls. Ann Bot 26:549–561

Lowe-McConnell RH (1975) Fish communities in tropical freshwaters. Longman, London

Lowe-McConnell RH (1979) Ecological aspects of seasonality in fish of tropical waters. Symp Zool Soc Lond 44:219–241

Lowe-McConnell RH (1987) Ecological studies in tropical fish communities. Tropical Biology Ser, University Press, Cambridge

Lubin YD, Young OP (1977) Food resources of ant-eaters (Edentata: Myrmecophagidae) 1. A year's census of arboreal nests of ants and termites on Barro Colorado Island, Panamá Canal Zone. Biotropica 9:26–36

MacArthur R, Wilson EO (eds) (1967) The theory of island biogeography. Princeton University Press, Princeton

MacIntyre S, Melack JM (1984) Vertical mixing in Amazon floodplain lakes. Verh Int Ver Limnol 22:1283–1287

MacIntyre S, Melack JM (1988) Frequency and depth of vertical mixing in an Amazon floodplain lake (L. Calado, Brazil). Verh Int Ver Limnol 23:80–85

Magalhães C (1984) Desenvolvimento larval em cativeiro e influência do pH e tipo de agua na sobrevivência dos adultos de *Macrobrachium amazonicum* (Heller, 1862) Crustacea, Decapoda, Palaemonidae). MSc Thesis, INPA, Manaus

Magalhães C (1985) Desenvolvimento larval obtido em laboratorio de palaemonideos da Regiao Amazônica. I. *Macrobrachium amazonicum* (Heller, 1862) (Crustacea, Decapoda). Amazoniana 9(2):247–274

Magalhães C, Walker I (1986) Larval development and ecological distribution of central Amazonian palaemonid shrimps (Decapoda, Caridea). Crustaceana 55(3):279–292

Magalhães FMM (1986) O estado atual do conhecimento sobre fixação biológica de nitrogênio na Amazônia. In: Embrapa-CPATU, Anais Vol 1. Clima e solo, Belém 1986. Simpósio do Trópico Umido, Belém 1984, pp 499–512

Magalhães FMM, Döbereiner J (1984) Ocorrência de *Azospirillum amazonense* em alguns ecossistemas da Amazônia. Rev Brasil Microbiol 15:246–252

Magnusson WE, Best RC, Silva VMF da (1980) Numbers and behaviour of Amazonian dolphins, *Inia geoffrensis* and *Sotalia fluviatilis*, in the Rio Solimões, Brasil. Aquat Mammals 8:27–32

Mahnert V, Adis J (1985) On the occurrence and habitat of Pseudoscorpiones (Arachnida) from Amazonian forests of Brazil. Stud Neotrop Fauna Environ 20(4):211–215

Maldague ME (1964) Importance des populations des termites dans les sols équatoriaux. Trans 8th Int Congr Soil Sci Bukarest 1964(3), pp 743–751

Mandl K (1981) Eine Cicindeliden-Ausbeute aus dem nördlichen Brasilien und Beschreibung neuer Arten aus dieser Familie (Col., Cicindelidae). Entomol Bras 6:154–160

Manokaran N, Kochummen KM (1987) Recruitment, growth and mortality of tree species in a lowland dipterocarp forest in Peninsular Malaysia. J Trop Ecol 3:315–330

Mariaux A (1967) Les cernes dans les bois tropicaux africains, nature et périodicité. Rev Bois For Trop 113:3–14

Marlier G (1965) Etude sur les lacs de l'amazonie central. Cadern da Amazônia, vol 5. INPA, Manaus

Marlier G (1967) Ecological studies on some lakes of the Amazon valley. Amazoniana 1(2):91–115

Marmillod D (1982) Methodik und Ergebnisse von Untersuchungen über Zusammensetzung und Aufbau eines Terrassenwaldes im peruanischen Amazonien. PhD Thesis, Universität Göttingen

Martinelli LA, Victoria RL, Devol AH, Forsberg BR (1989) Suspended sediment load in the Amazon basin. GeoJournal 19:381–389

Martinelli LA, Victoria RL, Dematte JLI, Richey JE, Devol AH (1993) Chemical and mineralogical composition of Amazon river floodplain sediments, Brazil. Appl Geoch 8:391–402

Martins DV (1980) Desmidioflórula dos lagos Cristalino e São Sebastião, Estado do Amazonas. PhD Thesis, INPA/FUA, Manaus

Martins M, Moreira G (1991) The nest and the tadpole of *Hyla wavrini*, Parker (Amphibia, Anura). Mem Inst Butantan 53(2):197–204

Martius C (1987) The adaptation of termites (*Nasutitermes* sp. – Termitidae, Nasutitermitinae) to Amazonian inundation forests. In: Eder J, Rembold H (eds) Chemistry and biology of social insects. Peperny, München, pp 609–610

Martius C (1989) Untersuchungen zur Ökologie des Holzabbaus durch Termiten (Isoptera) in zentralamazonischen Überschwemmungswäldern (Várzea). AFRA, Frankfurt

Martius C (1990) The influence of geophagous termites on soils of inundation forests in Amazonia – first results. In: Veeresh GK, Mallik B, Viraktamath CA (eds) Social insects and the environment. Oxford & IBH Publishing, New Dehli, pp 209–210

Martius C (1992a) Food provision storing by xylophagous termites in Amazonia (Isoptera: Nasutitermitidae). Entomol Gen 17(4):296–276

Martius C (1992b) Density, humidity, and nitrogen content of dominant wood species of floodplain forests (várzea) in Amazonia. Holz Roh- Werkstoff 50:300–303

Martius C (1994a) Termite nests as structural elements of the Amazon floodplain forest. Andrias 13:137–150

Martius C (1994b) Diversity and ecology of termites in Amazonian forests. Pedobiologia 38:407–428

Martius C, Wassmann R, Thein U, Bandeira A, Rennenberg H, Junk WJ, Seiler W (1993) Methane emission from wood-feeding termites in Amazonia. Chemosphere 26(1–4):623–632

Martius C, Höfer H, Verhaagh M, Adis J, Mahnert V (1994) Terrestrial arthropods colonizing an abandoned termite nest in a floodplain forest of the Amazon River during the flood. Andrias 13:17–22

Martius CFP von, Eichler AG, Urban I (1840–1869) Tabulae physiognomicae Flora Brasiliensis. Reprint Cramer (1967), München

Matson PA, Vitousek PM, Livingston GP, Swanberg NA (1990) Sources of variation in nitrous oxide flux from Amazonian Ecosystems. J Geophys Res 95 D10:16789–16798

Matsumoto T (1976) The role of termites in an equatorial rain forest ecosystem of West Malaysia. Oecologia (Ber) 22(2):153–178

Mayr E (1969) Bird speciation in the tropics. Biol J Linn Soc 1:1–17

McIntire CD (1973) Periphyton dynamics in laboratory streams: a simulation model and its implications. Ecol Monogr 43:399–420

McNeil R, Itriago MC de (1968) Fat deposition in the Scissors-tailed Flycatcher (*Muscivora t. tyrannus*) and the Small-billed Elaenia (*Elaenia parvirostris*) during the Austral migratory period in northern Venezuela. Can J Zool 46:123–128

Meade RH (1985) Suspended sediment in the Amazon river and its tributaries in Brazil during 1982–84. US Geol Surv Open-File Rep, pp 85–492

Meade RH (1994) Suspended sediments of the modern Amazon and Orinco rivers. Quat Int 21:29–39

Meade RH, Nordin CF, Curtis WF, Mahoney HA, Delaney BM (1979a) Suspended-sediment and velocity data, Amazon river and its tributaries, June-July 1976 and May-June 1977. US Geolog Surv Open-File Rep 79-515, pp 42

Meade RH, Nordin CF Jr, Curtis WF, Rodriguez FMC, Vale RM, Edmond JM (1979b) Sediment loads in the Amazon river. Nature 278:161–163

Meade RH, Dunne T, Richey JE, Santos U de M, Salati E (1985) Storage and remobilization of suspended sediment in the lower Amazon river of Brazil. Science 228:488–490

Medem F (1982) Sanctioned exportation of caiman hides from Colombia. In: Crocodiles. Proc 5th Work Meet Crocodile Spec Group. International Union for the Conservation of Nature and Natural Resources (IUCN) New Ser, IUCN Pub, Gland, Switzerland, pp 121–126

Medina E (1981) Significacíon ecológica del contenido foliar de nutrientes y e área foliar específica en ecosistemas tropicales. Proc 2nd Congr Latinamer Bot, Brasilia 1978

Medina E (1984) Nutrient balance and physiological processes at the leaf level. In: Medina E, Mooney HA, Vásquez-Yanes C (eds) Physiological ecology of plants of the wet tropics. Junk, The Hague, pp 139–154

Medina E, Cuevas E (1989) Patterns of nutrient accumulation and release in Amazonian forests of the upper Rio Negro basin. In: Proctor J (ed) Mineral nutrients in tropical forest and savanna ecosystems. Blackwell, Oxford, pp 217–240

Medina E, Klinge H (1983) Productivity of tropical forests and tropical woodlands. Encyclopedia of plant physiology, new series 12D. Springer, Berlin Heidelberg New York, pp 281–303

Medina E, Bifano de T, Delgado M (1976) Differenciacion fotossintetica en plantas superiores. Interciencia 1:96–104

Medina E, Herrera R, Jordan CF, Klinge H (1977) The Amazon project of the Venezuelan Institute for Scientific Research. Nature Resou UNESCO 13(3):4–6

Meggers BJ (1984) The indigenous peoples of Amazonia, their cultures, land use patterns and effects on the landscape and biota. In: Sioli H (ed) The Amazon – Limnology and landscape ecology of a mighty tropical river and its basin. Monographiae Biologicae. Junk, Dordrecht, pp 627–648

Meggers BJ (1985) Aboriginal adaptation to Amazonia. In: Prance GT, Lovejoy TE (eds) Amazonia. Pergamon Press, Oxford, pp 307–327

Meggers BJ (1987) The early history of man in Amazonia. In: Whitmore TC, Prance GT (eds) Biogeography and quaternary history in tropical America. Clarendon Press, Oxford, pp 151–174

Meggers BJ (1992) Amazonia: real or counterfeit paradise? Rev of Archaeol 13(2):25–40

Melack JM (1984) Amazon floodplain lakes: shape, fetch, and stratification. Verh Int Ver Limnol 22:1278–1282

Melack JM, Fisher TR (1983) Diel oxygen variations and their ecological implications in Amazon floodplain lakes. Arch Hydrobiol 98:422–442

Melack JM, Fisher TR (1988) Denitrification and nitrogen fixation in an Amazon floodplain lake. Verh Int Ver Limnol 23:2232–2236

Melack JM, Fisher TR (1990) Comparative limnology of tropical floodplain lakes with an emphasis on the Central Amazon. Acta Limnol Brasil 3:1–48

Melendez ME (1978) Aspectos agronomicos del pasto aleman (*Echinochloa polystachya*). In: Aniversario del Fondo Nacional de Investigaciones Agropecuarias, Estacion Experimental Calabozo, Venezuela:1–11

Mengel K, Kirkby EA (eds) (1978) Principles of plant nutrition. International Potash Institute, Bern

Merck AMT (1994) Aspectos taxonômicos, morfológicos, e autoecológicos de *Pomacea lineata* (Philippi, 1851) e *Pomacea papyracea* (Spix, 1827) Mollusca, Prosobranchia, nas áreas alagáveis nos rios Amazonas e Negro, AM. PhD Thesis, University São Carlos

Merrill W, Cowling EB (1966) Role of nitrogen in wood deterioration: amounts and distribution of nitrogen in tree stems. Can J Bot 44(10–12):1555–1580

Mertes LAK (1985) Floodplain development and sediment transport in the Solimões-Amazon River, Brazil. MSc Thesis, University of Washington, Seattle

Mertes LAK (1994) Rates of floodplain sedimentation on the Central Amazon river. Geology 22:171–174

Mertes LAK, Meade RH (1985) Particle size of sands collected from the bed of the Amazon River and its tributaries in Brazil during 1982–84. US Geol Surv Open-File Rep, pp 85–333

Mertes LAK, Smith MO, Adams JB (1993) Estimating suspended sediment concentrations in surface waters of the Amazon river wetlands from Landsat images. Remote Sens Environ 43:281–301

Messner B (1988) Vorschlag für die Neufassung des Begriffes "Plastron" bei den Arthropoden. Dtsch Entomol Z NF 35(45):379–381

Messner B, Adis J (1988) Die Plastronstrukturen der bisher einzigen submers lebenden Diplopodenart *Gonographis adisi* Hoffman 1985 (Pyrgodesmidae, Diplopoda). Zool Jahrb Anat 117:277–290

Messner B, Adis J (1992a) Kutikuläre Wachsausscheidungen als plastronhaltende Strukturen bei Larven von Schaum- und Singzikaden (Auchenorrhyncha: Cercopidae und Cicadidae). Rev Suisse Zool 99(3):713–720

Messner B, Adis J (1992b) Die Plastronatmung bei aquatischen und flutresistenten terrestrischen Arthropoden (Acari, Diplopoda und Insecta). Mitt Dtsch Ges Allg Angew Entomol 8:325–327

Messner B, Adis J (1994) Funktionsmorphologische Untersuchungen an den Plastronstrukturen der Arthropoden. Verh West Entom Tag 1993 (Löbbecke-Museum Düsseldorf), pp 51–56

Messner B, Lunk A, Groth I, Subklew HJ, Taschenberger D (1981) Neue Befunde zum Atmungssystem der Grundwanze *Aphelocheirus aestivalis* (Heteroptera, Hydrocorisae). I. Imagines. Zool Jahrb Anat 105:474–496

Messner B, Groth I, Messner U, Geisel T (1987) Die Plastronstrukturen der Larven, der Puppe und des submers lebenden Weibchens von *Acentria nivea* (Olivier 1791) (Lepidoptera, Pyralidae). Zool Jahrb Anat 115:163–180

Messner B, Adis J, Ribeiro EF (1992) Eine vergleichende Untersuchung über die Plastronstrukturen bei Milben (Acari). Dtsch Entomol Z NF 39(1–3):159–176

Messner B, Adis J, Zulka KP (1996) Stigmale Plastronstrukturen, die einigen Diplopoden-Arten eine submerse Lebensweise in kaltem und in fließendem Wasser ermöglichen. Rev Suisse Zool 103(3):613–622

Meyer U (1991) Feinwurzelsysteme und Mykorrhizatypen als Anpassungsmechanismen in zentralamazonischen Überschwemmungswäldern – Igapó und Várzea. PhD Thesis, Universität Hohenheim

Mill AE (1982) Populações de térmitas (Insecta: Isoptera) em quatro habitats no baixo Rio Negro. Acta Amazonica 12(1):53–60

Mill AE (1991) Termites as structural pests in Amazônia, Brazil. Sociobiology 19(2):339–348

Milliman JD (1990) River discharge of water and sediment to the oceans: variation in space and time. In: Ittekkot V, Kempe S, Michaelis W, Spitzy A (eds) Facets of modern biogeochemistry. Springer, Berlin Heidelberg New York, pp 83–90

Milliman JD, Meade (1983) Worldwide delivery of river sediment to the oceans. J Geol 91:1–21

Mitchell DS (1974) The development of excessive populations of aquatic plants. In: Mitchell DS (ed) Aquatic vegetation and its use and control. UNESCO, Paris, pp 38–49

Mitsch WJ, Fennessy MS (1991) Modelling nutrient cycling in wetlands. In: Jorgensen SE (ed) Modelling in environmental chemistry. Elsevier, Amsterdam, pp 249–275

Molion LCB, Dallarosa R (1992) Pluviometria da Amazônia. São os dados confiaveis? Climanálise São José dos Campos, SP 5(3):40–43

Monteiro PJC, Val AL, Almeida-Val VMF (1987) Biological aspects of Amazonian fishes. Hemoglobin, hematology, intraerythocytic phosphates and whole blood Bohr effects of *Mylossoma duriventre*. Can J Zool 65:1805–1811

Monteith JL (1978) Reassessment of maximum growth rates for C3 and C4 crops. Exp Agric 14:1–5

Moorhead KK, Reddy KR (1988) Oxygen transport through selected aquatic macrophytes. J Environ Qual 17:138–142

Morales G, Pinowski J, Pacheco J, Madriz M, Gomez F (1981) Densidades poblacionales, flujo de energía y hábitos alimentarios de las aves ictiofagas de los módulos de Apure, Venezuela. Acta Biol Venez 11:1–45

Moreira G, Rabelo J (1995) Estudos das comunidades de quêlonios e crocodilianos do Parque Nacional do Jaú. Julho 1993 – novembro 1994. Final Rep of Fundação Vitória Amazônica, pp 1–26

Moreira G, Vogt R (1990) Movements of *Podocnemis expansa* before and after nesting in the Trombetas river, Brazil. Annu Joint Meet of the Herpetologists' League and the Society for the Study of Amphibians and Reptiles, Aug 5–9, Tulane University, New Orleans

Morrissey KM, Fisher TR (1988) Regeneration and uptake of ammonium by plankton in an Amazon floodplain lake. J Plankton Res 10:31–48

Moskovits D, Fitzpatrick JW, Willard DE (1985) Lista preliminar das aves da Estação Ecológica de Maracá, Território de Roraima, Brasil, e áreas adjacentes. Pap Avulsos Zool (São Paulo) 36(6):51–68

Müller J, Irion G, de Mello JN, Junk WJ (1995) Hydrological changes of the Amazon during the last glacial-interglacial cycle in Central Amazonia (Brazil). Naturwissenschaften 82:232–235

Müller W (1982) Arbeitsgruppe Bodenkunde der geol. Landesämter und der Bundesanstalt für Geowissenschaften und Rohstoffe in der BRD. Bodenkundliche Kartieranleitung, Hannover

Munn CA (1985) Permanent canopy and understory flocks in Amazonia. Species composition and population density. Ornithol Monogr 36:683-712

Murphy PG (1975) Net primary productivity in tropical terrestrial ecosystems. In: Lieth H, Whittaker RH (eds) Primary productivity of the biosphere. Springer, Berlin Heidelberg New York, pp 217-231

Myers JP (1980) The Pampas shorebird community: interactions between breeding and nonbreeding members. In: Keast A, Morton ES (eds) Migrant birds in the neotropics. Smithsonian Institution Press, Washington DC, pp 37-49

Nadelhoffer KJ, Aber JM, Melillo JM (1985) Fine roots, net primary production, and soil nitrogen availability: a new hypothesis. Ecol 66(4):1377-1390

Naiman RJ, Décamps H (eds) (1990) The ecology and management of aquatic-terrestrial ecotones. Parthenon, Carnforth Hall, UK

Nascimento C, Homma A, (1984) Amazonia: meio ambiente e tecnologia agricola. EMBRAPA-CPATU documentos 27:1-182

Nessimian JL (1985) Estudo sobre a ecologia da fauna invertebrada aquática na liteira submersa das margens de dois lagos do arquipélago de Anavilhanas (Rio Negro, Amazonas, Brasil). MSc Thesis, INPA/FUA, Manaus

Nicolai V (1986) The bark of trees: thermal properties, microclimates and fauna. Oecologia 69:148-160

Nicolai V (1989) Thermal properties and fauna on the bark of trees in two different African ecosystems. Oecologia 80:421-430

Nielsen MG, Josens G (1978) Production by ants and termites. In: Brian MV (ed) Production ecology of ants and termites. International Biological Programme 13. Cambridge University Press, Cambridge, pp 45-54

Nilsson SG, Nilssen IN (1976) Numbers, food consumption, and fish predation by birds in Lake Möckeln, southern Sweden. Ornis Scand 7:61-70

Novaes FC (1960) Sobre uma coleção de aves do sudeste do estado do Pará. Arq Zool (São Paulo) 11:133-146

Novaes FC (1974) Ornitologia do Território do Amapá, I. Publ Avulsas Mus Para Goeldi 25:1-121

Novaes FC (1976) As aves do Rio Aripuana, Estados de Mato Grosso e Amazonas. Acta Amazônica 6(4):61-85

Novaes FC (1978) Ornitologia do Território do Amapá, II. Publ Avulsas Mus Para Goeldi 29:1-75

Novaes FC (1980) Observações sobre a avifauna do alto curso do rio Paru de leste, Estado do Pará. Bol Mus Para Goeldi NS Zool 100:58

Nolte U (1986) Erstbesiedlung künstlicher Kleingewässer und primäre Sukzession des Makrozoobenthos unter besonderer Berücksichtigung der Chironomidae (Diptera). Feldexperimente in Zentralamazonien. PhD Thesis, Universität Göttingen

Nordin CF, Meade RH (1982) Deforestation and increased flooding of the upper Amazon. Science 215:426-427

Nortcliff S, Thornes JB (1978) Water cation movement in a tropical rainforest environment. 1 Objectives, experimental design and preliminary results. Acta Amazonica 8:245-258

NRC (1971) Atlas of nutritional data on United States and Canadian feeds. National Research Council United States and Department of Agriculture Canada, National Academy of Science, Washington DC

NRC (1984) Nutrient requirements of beef cattle. National Research Council United States, Washington DC

O'Connor R (1984) The growth and development of birds. Wiley, Chichester

Odelson DA, Breznak JA (1985) Cellulase and other polymer-hydrolyzing activities of *Trichomitopsis termopsidis*, a symbiotic protozoan from termites. Appl Environ Microbiol 49:622-626

Odum EP (1959) Fundamentals of ecology. Saunders, London

Odum EP (1981) Foreword. In: Clark JR, Benforado J (eds) Wetlands of bottomland hardwood forests. Development in agricultural and managed forest ecology 11. Elsevier, Amsterdam

Odum HT (1970a) Holes in leaves and the grazing control mechanism. In: Odum HT, Pigeon RF (eds) A tropical rain forest: a study of irradiation and ecology at El Verde, Puerto Rico. Division of Technical Information, US Atomic Energy Commision, Oak Ridge, Tennessee, pp I-6:69–80

Odum HT (1970b) Summary: an emerging view of the ecological system at El Verde. In: Odum HT, Pigeon RF (eds) A tropical rain forest. A study of irradiation and ecology at El Verde, Puerto Rico. Division of Technical Information, US Atomic Energy Commission, Oak Ridge, Tennessee, pp I-10:191–281

Ohly J (1987) Untersuchungen über die Eignung der natürlichen Pflanzenbestände auf den Überschwemmungsgebieten (várzea) am mittleren Amazonas, Brasilien, als Weide für den Wasserbüffel (*Bubalus bubalis*) während der terrestrischen Phase des Ökosystems. PhD Thesis, Universität Göttingen

Ohly J, Junk WJ (1996) Multiple use of central Amazon floodplains: combining ecological conditions, requirements for environmental protection and socioeconomic needs. Adv Econ Bot (in press)

Ojasti J (1971) La tortuga arrau del Orinoco. Def Nat 1:3–10

Ojasti J (1973) Estudo Biológico del Chigüire o Capibara. Fondo Nacional de Investigaciones Agropecuarias, Venezuela

Ojasti J (1980) Ecology of capybara raising on inundated savannas of Venezuela. Trop Ecol Dev 1980:287–293

Ojasti J (1983) Ungulates and large rodents of South America. In: Bourliere F (ed) Ecosystems of the world. The tropical savannas. Elsevier, Amsterdam, pp 427–439

Olivares A, Hernandez J (1962) Aves de la comisaría del Vaupés, Colombia. Rev Biol Trop 10(1):61–90

Oliveira EP de, Daherveng L (1990) Isotomiella (Collembola: Isotomidae) d'Amazonie: les espéces du groupe minor. Bull Mus Natl Hist Nat Paris, 4e Sér, 12, Section A, 1:75–93

Olrog CC (1975) Vagrancy of Neotropical Cormorant, egrets, and White-faced Ibis. Bird Banding 46:207–212

Olson JS (1963) Energy storage and the balance of producers and decomposers in ecological systems. Ecology 44:322–331

Oniki Y, Willis EO (1982) Breeding records of birds from Manaus, Brazil: I–III. Rev Bras Biol 42:563–569, 733–740, 745–752

Oniki Y, Willis EO (1983) Breeding records of birds from Manaus, Brazil: IV–V. Rev Bras Biol 43:45–54, 55–64

Ordinez-Collart O (1988) Ecologie de la crevette d'Amazonie, *Macrobrachium amazonicum*. In: Condicions ecologiques et economiques de la production dune ile de Várzea: l'ile du Careiro. Rapport terminal, Orstom, INPA, pour la Commission des Communautés Européennes (CEE), pp 52–71, 313–319

Osborne DR (1982) Replacement nesting and polyandry in the Wattled Jacana. Wilson Bull 94:206–208

Osborne DR, Bourne GR (1977) Breeding behavior and food habits of the Wattled Jacana. Condor 79:98–105

Ovington JD, Madgwick HAI (1959) Distribution of organic matter and plant nutrients in a plantation of Scots Pine. For Sci 5:344–355

Paarmann W, Irmler U, Adis J (1982) *Pentacomia egregia* Chaud. (Carabidae, Cicindelinae), an univoltine species in the Amazonian inundation forest. Coleopt Bull 36(2):183–188

Parkes KC (1985) Neotropical ornithology – an overview. Ornithol Monogr 36:1025–1036

Pasti MB, Pometto AL, Nuti MP, Crawford DL (1990) Lignin-solubilizing ability of actino-mycetes isolated from termite (Termitidae) gut. Appl Environ Microbiol 56(7):2213–2218

Patrick WH Jr, Dissmeyer G, Hook DD, Lambou VW, Leitman HM, Wharton CH (1981) Characteristics of wetland ecosystems of southeastern bottomland hardwood forests. In: Clark JR, Benforado J (ed) Wetlands of bottomland hardwood forests. Development in agricultural and managed forest ecology 11. Elsevier, Amsterdam, pp 276–300

Patrick R (1966) The Catherwood Foundation Peruvian-Amazon Expedition I. Limnological observations and discussion of results. Monogr Acad Nat Sci Phila 14:5–28

Pearcy RW, Ehleringer J (1984) Comparative ecophysiology of C_3 and C_4 plant. Plant Cell Environ 7:1–13

Pearson DL (1977) Ecological relationships of small antbirds in Amazonian bird communities. Auk 94:283–292

Pearson DL (1980) Bird migration in Amazonian Ecuador, Peru, and Bolivia. In: Keast A, Morton ES (eds) Migrant birds in the Neotropics. Symp Nat Zool Park, Smithsonian Institution Press, Washington DC, pp 273–283

Pearson DL (1984) The tiger beetles (Coleoptera: Cicindelidae) of the Tambopata Reserved Zone, Madre de Dios, Peru. Rev Peru Entomol 27:15–24

Pearson DL (1988) Biology of tiger beetles. Annu Rev Entomol 33:123–147

Peres CA, Whittaker A (1991) Annotated checklist of the bird species of the upper Rio Urucú, Amazonas, Brazil. Bull Br Ornithol Club 111:156–171

Perez-Iñigo C (1987) Contribucion al conocimiento de los oribatidos (Acari, Oribatei) que viven sobre citricos (I). Graellsia 43:127–137

Petrere M Jr (1978a) Pesca e esforço de pesca no Estado do Amazonas. I. Esforço e captura por unidade de esforço. Acta Amazonica 8:439–454

Petrere M Jr (1978b) Pesca e esforço de pesca no Estado do Amazonas. II. Locais, aparelhos de captura e estatística de desembarque. Acta Amazonica 8(Suppl 2):1–54

Petrick C (1978) The complementary function of floodlands for agricultural utilization. Appl Sci Dev 12:26–46

Petrusewicz K, Macfadyen A (eds) (1970) Productivity of terrestrial animals – principles and methods. IBP Handbook 13. Blackwell, Oxford

Pianka ER (1970) On r- and K-selection. Am Nat 104:592–597

Picket STA, White PS (eds) (1985) The ecology of natural disturbance and patch dynamics. Academic Press, Orlando, Florida

Piedade MTF (1985) Ecologia e biologia reproductiva de *Astrocaryum jauari* Mart. (Palmae) como exemplo de população adaptada às áreas inundáveis do Rio Negro (igapó). PhD Thesis INPA, Manaus

Piedade MTF, Junk WJ, Long SP (1991) The productivity of the C_4 grass *Echinochloa polystachya* on the Amazon floodplain. Ecology 72:1456–1463

Piedade MTF, Junk WJ, Mello JAN de (1992) A floodplain grassland of the Central Amazon. In: Long SP, Jones MB, Roberts MJ (eds) Primary production of grass ecosystems of the tropics and sub-tropics. Chapman and Hall, London, pp 127–158

Piedade MTF, Long SP, Junk WJ (1994) Leaf and canopy photosynthetic CO_2 uptake of a stand of *Echinochloa polystachya* on the Central Amazon floodplain. Oecologia 97:193–201

Pinheiro P (1985) Estudo sazonal dos efeitos das adições de nutrientes sobre o crescimento do fitoplâncton em um lago de várzea (Lago Calado – Amazonia central). MSc Thesis, INPA/FUA Manaus

Pires JM, Koury HM (1959) Estudo de um trecho de mata de várzea próximo a Belém. Bol Tecn IAN 36:3–44

Pinto OM de O (1947) Contribuição à ornitologia do baixo Amazonas. Arq Zool (São Paulo) 5(6):311–482

Platnick NI, Höfer H (1990) Systematics and ecology of ground spiders (Araneae, Gnaphosidae) from Central Amazonian inundation forests. Am Mus Novi 2971:1–16

Plonczak M (1989) Struktur und Entwicklungsdynamik eines Naturwaldes unter Konzessionsbewirtschaftung in den westlichen Llanos Venezuelas. Gött Beitr Land- und Forstwirtschaft Tropen Subtropen 43

Poff NL (1992) Why disturbances can be predictable: a perspective on the definition of disturbance in streams. J N Am Benthol Soc 11(1):86–92

Polinisi JM, Boyd CE (1972) Relationships between cell-wall fractions, nitrogen and standing crop in aquatic marcophytes. Ecology 53:484–488

Ponnamperuma FN (1972) The chemistry of submerged soils. In: Brady NC (ed) Advances in agronomy, vol 24. Academic Press, New York, pp 29–88

Ponnamperuma FN (1984) Effects of flooding on soils. In: Kozlowski TT (ed) Flooding and plant growth. Academic Press, New York, pp 10–46

Potts RC, Hewitt PH (1973) The distribution of intestinal bacteria and cellulase activity in the harvester termite *Trinervitermes trinervoides* (Nasutitermitinae). Insectes Soc 20(3):215–220

Prado AL do, Heckman CW, Martins FR (1994) The seasonal succession of biotic communities in wetlands of the tropical wet-and-dry climatic zone: II. The aquatic macrophyte vegetation in the Pantanal of Mato Grosso, Brazil. Int Rev Gesamten Hydrobiol 79(4):569–589

Prance GT (1979) Notes on the vegetation of Amazonia III. The terminology of Amazonian forest types subject to inundation. Brittonia 31:26–38

Prance GT (1982) Biological diversification in the tropics. Columbia University Press, New York

Prance GT (1989) American tropical forests. In: Lieth H, Werger MJA (eds) Tropical rain forest ecosystems. Ecosystems of the world 14B. Elsevier, Amsterdam, pp 99–132

Prance GT, Schaller GB (1982) Preliminary study of some vegetation types of the Pantanal, Mato Grosso, Brazil. Brittonia 2:228–251

Prance GT, Rodrigues WA, da Silva MF (1976) Inventário florestal de um hectare de mata de terra firme km 30 da Estrada Manaus – Itacoatiara. Acta Amazonica 6(1):9–35

Preston FW (1962) A nesting of Amazonian terns and skimmers. Wilson Bull 74:286–287

Primack RB, Ashton PS, Chai P, Lee HS (1985) Growth rates and population structure of Moraceae trees in Sarawak, East Malaysia. Ecology 66:577–588

Pritchard PCH (1979a) Encyclopedia of turtles. TFH, New York

Pritchard PCH (1979b) Taxonomy, evolution and zoogeography. In: Harless M, Morlock H (eds) Turtles, perspectives and research. Wiley, New York, pp 1–42

Pritchard PCH, Trebbau P (1984) The turtles of Venezuela. Society for the study of amphibians and reptiles. Contrib Herpetol 2:1–403

Projeto Radambrasil (1972) Mosaicos semi – controlados de Radar 1:250000. Departamento Nacional da Produção Mineral DNPM, Brasília and Rio de Janeiro

Putz R (1992) Aufwuchsdiatomeen im Rhein und in der Rheinaue bei Rastatt: Sukzession, Biomasse und Produktivität. PhD Thesis, Universität Freiburg

Putz R (1997) Periphyton communities in Amazonian black- and whitewater habitats: Community structure, biomass and productivity. Aquatic Sciences 59(1) (in press)

Putzer H (1984) The geological evolution of the Amazon basin and its mineral resources. In: Sioli H (ed) The Amazon – Limnology and landscape ecology of a mighty tropical river and its basin. Monographiae Biologicae. Junk Publishers, Dordrecht, pp 15–46

Py-Daniel L (1984) Sistematica dos Loricariidae (Ostariophysi, Siluroidei) do complexo de lagos do Janauacá, Rio Solimões, AM, e aspectos da sua biologia e ecologia. MSc Thesis, INPA/FUA, Manaus

Rahn H, Paganelli CV, Ar A (1975) Relation of avian egg weight to body weight. Auk 92:750–65

Rai H (1978) Distribution of carbon, chlorophyll-a and pheo-pigments in the blackwater lake ecosystem of Central Amazon Region. Arch Hydrobiol 682:74–87

Rai H (1981) Physical and chemical studies of Lago Tupé; a Central Amazonian black water "Ria Lake". Int Rev Gesamten Hydrobiol 66:37–82

Rai H, Hill G (1984) Primary production in the Amazonian aquatic ecosystem. In: Sioli H (ed) The Amazon – Limnology and landscape ecology of a mighty tropical river and its basin. Monographiae Biologicae. Junk, Dordrecht, pp 311–335

Rankin de Merona JM (1988) Conditions écologiques et économiques de la production d'une ile de várzea: L'ile du Careiro, Part 4: Les relations poisson-forêt. Rapport Terminal. ORSTOM, Paris, INPA, Manaus, pp 202–228

Rappole JR, Morton ES, Lovejoy TE, Ruos JL (eds) (1983) Nearctic avian migrants in the neotropics. US Dept Int, Fish and Wildlife Service, Washington DC

Rauh W (1939) Über die Gesetzmäßigkeit der Verzweigung und deren Bedeutung für die Wuchsform der Pflanzen. Mitt Dtsch Dendrol Ges 52:86–111

Rebello AMC, Martius C (1994) Dispersal flight of termites in Amazonian forests. Sociobiology 24:127–146

Rebelo GH, Magnusson WE (1983) An analysis of the effect of hunting on *Caiman crocodilus* and *Melanosuchus niger* based on the sizes of confiscated skins. Biol Conserv 26:95–104

Reckow KH (1994) Water quality simulation modeling and uncertainty analysis for risk assessment and decision making. Ecol Model 72:1–20

Reichholf J (1982/83) Extreme Wasservogelarmut am Rio Negro, Amazonien. Verh Ornithol Ges Bayern 23:525–528

Reichholf J (1983) Analyse von Verbreitungsmustern der Wasservögel und Säugetiere in Südamerika. Spixiana Suppl 9:167–178

Reichle DE, Dinger BE, Edwards NT, Harris WF, Solling P (1973) Carbon flow and storage in a forest ecosystem. In: Woodwell GM, Pecan EV (eds) Carbon and the biosphere. Proc 24th Brookhaven Symp Biology. US Atomic Energy Commision, Washington, pp 345–365

Reid J (1989) The distribution of the genus *Thermocyclops* (Copepoda, Cyclopoida) in the western hemisphere, with description of *T. parvus*, new species. Hydrobiologia 175:149–174

Reiss F (1974) Vier neue *Chironomus*-Arten (Chironomidae, Diptera) und ihre ökologische Bedeutung für die Benthosfauna zentralamazonischer Seen und Überschwemmungswälder. Amazoniana 5(1):3–23

Reiss F (1976a) Charakterisierung zentralamazonischer Seen aufgrund ihrer Makrobenthosfauna. Amazoniana 6(1):123–134

Reiss F (1976b) Die Benthoszoozönosen zentralamazonischer Várzeaseen und ihre Anpassungen an die jahresperiodischen Wasserstandsschwankungen. Biogeographica 7:125–135

Reiss F (1977) Qualitative and quantitative investigations on the macrobenthic fauna of Central Amazon lakes.I. Lago Tupé, a black water lake on the lower Rio Negro. Amazoniana 6(2):203–235

Remsen JV Jr (1986) Aves de una localidad en la sabana húmeda del norte de Bolivia. Ecol Bolivia 8:21–35

Remsen JV Jr (1990) Community ecology of neotropical kingfishers. Univ Calif Publ Zool 124:1–116

Remsen JV Jr, Parker III TA (1983) Contribution of river-created habitats to bird species richness in Amazonia. Biotropica 15:223–231

Remsen JV Jr, Parker III TA (1990) Seasonal distribution of the Azure Gallinule (*Porphyrula flavirostris*), with comments on vagrancy in rails and gallinules. Wilson Bull 102:380–399

Resh VH, Brown AV, Covich AP, Gurtz ME, Li HW, Minshall GW, Reice SR, Sheldon AL,

Wallace JB, Wissmar R (1988) The role of disturbance in stream ecology. J N Am Benthol Soc 7:433–455

Resh VH, Barnes JR, Craig DA (1990) Distribution and ecology of benthic macroinvertebrates in the Opunohu river catchment, Moorea, French Polynesia. Ann Limnol 26(2–3):195–214

Revilla JD (1981) Aspectos floristicos e fitossociologicos da floresta inundavel (igapó) Praia Grande, Rio Negro, Amazonas, Brasil. MSc Thesis, Manaus

Revilla JD (1991) Aspectos floristicos e estruturais da floresta inundável (várzea) do Baixo Solimões, Amazonas Brasil. PhD Thesis, Manaus

Ribeiro EF, Schubart HOR (1989) Oribatídeos (Acari: Oribatida) colonizadores de folhas em decomposição sobre o solo de três sítios florestais da Amazônia Central. Bol Mus Para Emilio Goeldi Ser Zool 5(2):243–276

Ribeiro JA, Martius C (1996) On population size of *Anoplotermes banksi* Emerson (Isoptera: Termitidae) in Central Amazonian rainforests. Stud Neotrop Fauna Environ (in press)

Ribeiro M de NG, Adis J (1984) Local rainfall variability – a potential bias for bioecological studies in the Central Amazon. Acta Amazonica 14(1–2):159–174

Ribeiro MCL de B (1983) As migrações dos jaraquís (Pisces, Prochilodontidae) no Rio Negro, Amazonas – Brasil. MSc Thesis, INPA/FUA, Manaus

Ribeiro MO de A (1994) Abundância, distribuição vertical e biomassa de artrópodos do solo em uma capoeira na Amazônia Central. MSc Thesis, INPA/FUA, Manaus

Richey JE (1982) The Amazon river system: a biogeochemical model. Mitt Geol Paläontol Inst Univ Hamb SCOPE/UNEP Sonderb 52:365–378

Richey JE (1983) Interactions of C, N, P, and S in river systems: a biogeochemical model. In: Bolin B, Cook RB (eds) The major biogeochemical cycles and their interactions. Wiley, New York, pp 365–383

Richey JE, Brock JT, Naiman RJ, Wissmar RC, Stallard RF (1980) Organic carbon: oxidation and transport in the Amazon River. Science 207:1348–1351

Richey JE, Meade RH, Salati E, Devol AH, Nordin CF, Santos UM (1986) Water discharge and suspended sediment concentrations in the Amazon river. Water Resour Res 22:756–764

Richey JE, Nobre C, Deser C (1989) Amazon river discharge and climatic variability: 1903–1985. Science 246:101–103

Ricklefs RE (1968) Patterns of growth in birds. Ibis 110:419–451

Ridgely RS, Gwynne JA Jr (eds) (1989) A guide to the birds of Panama, with Costa Rica, Nicaragua and Honduras, 2nd edn. Princeton University Press, Princeton

Ridgely RS, Tudor G (eds) (1989) The birds of South America, volume I. Oxford University Press, Oxford

Ridgely RS, Tudor G (eds) (1994) The birds of South America, volume II. Oxford University Press, Oxford

Robertson BA, Hardy ER (1984) Zooplankton of Amazonian lakes and rivers. In: Sioli H (ed) The Amazon – Limnology and landscape ecology of a mighty tropical river and its basin. Monographiae Biologicae. Junk, Dordrecht, pp 337–352

Robinson SK, Terborgh J (1990) Bird communities of the Cocha Cashu biological station in Amazonian Peru. In: Gentry AH (ed) Four neotropical rainforests. Yale University Press, New Haven, pp 199–216

Robinson SK, Terborgh J, Munn CA (1990) Lowland tropical forest bird communities of a site in Western Amazonia. In: Keast A (ed) Biogeography and ecology of forest bird communities. SPB Acad Publ, The Hague, pp 229–258

Rodrigues MS (1994) Biomassa e produção fitoplanctônica do Lago Camaleão (Ilha de Marchantaria, Amazonas). PhD Thesis, INPA/FUA, Manaus

Rodriguez JP, Joly LJ, Pearson DL (1994) Los escarabajos tigre de Venezuela: su identificación, distribución e historia natural (Coleoptera: Cicindelidae). Bol Entomol Venez NS 9(1):55–120

Roosevelt AC (1991) Moundbuilders of the Amazon: geophysical archaeology on Marajó Island, Brazil. Academic Press, San Diego

Ropelewski CF, Halpert MS (1987) Global and regional scale precipitation patterns associated with the El Niño/Southern Oscillation. Mon Wea Rev 115:1606–1626

Rosenberg GH (1990) Habitat specialization and foraging behavior by birds of Amazonian river islands in Northeastern Peru. Condor 92:427–443

Roubach R, Saint-Paul U (1994) Use of fruits and seed from Amazonian inundated forest in feeding trials with *Colossoma macropomum* (Cuvier 1818) (Pisces, Characidae). J Appl Ichthyol 10:134–140

Rouland C, Hararas C, Renoux J (1989a) Les osidases digestives présentes dans l'intestin moyen, l'intestin postérieur et les glandes salivaires du termite humivore *Crenetermes albotarsalis*. C R Acad Sci Paris 308 Sér III:281

Rouland C, Lenoir-Rousseaux JJ, Mora P, Renoux J (1989b) Origin of the exocellulase and the B-glucosidase purified from the digestive tract of the fungus-growing termite *Macrotermes muelleri*. Sociobiology 15(2):237

Saint-Paul U (1984a) Physiological adaptations to hypoxia of a neotropical characoid fish *Colossoma macropomum*, Serrasalmidae. Environ Biol Fishes 11:53–62

Saint-Paul U (1984b) Investigations on the seasonal changes in the chemical composition of liver and condition from a neotropical fish *Colossoma macropomum* (Serrasalmidae). Amazoniana 9:147–158

Saint-Paul U (1988) Diurnal routine O_2 consumption at different O_2 concentration by *Colossoma macropomum* and *Colossoma brachypomum* (Teleostei, Serrasalmidae). Comp Biochem Physiol 89A:675–682

Saint-Paul U (1994) Der neotropische Überschwemmungswald: Beziehung zwischen Fisch und Umwelt. Final Rep BMFT No 0339366A, BMFT, Bonn

Saint-Paul U, Soares MGM (1987) Diurnal distribution and behavioral responses of fishes to extreme hypoxia in an Amazon floodplain lake. Environ Biol Fishes 20:91–104

Saint-Paul U, Soares MGM (1988) Ecomorphological adaptation to oxygen deficiency in Amazon floodplains by Serrasalmidae fish of the genus *Mylossoma*. J Fish Biol 32:231–236

Salati E, Marques J (1984) Climatology of the Amazon region. In: Sioli H (ed) The Amazon – Limnology and landscape ecology of a mighty tropical river and its basin. Monographiae Biologicae. Junk, Dordrecht, pp 85–126

Salati E, Dall'Olio A, Matsui E, Bat JA (1979) Recycling of water in the Amazon basin. An isotopic study. Water Resour Res 15(5):1250–1258

Salati E, Sylvester-Bradley R, Victoria L (1982) Regional gains and losses of nitrogen in the Amazon basin. Plant Soil 67:367–376

Salick J, Herrera R, Jordan CF (1983) Termitaria: nutrient patchiness in nutrient-deficient rain forest. Biotropica 15(1):1–7

Salo J (1990) External processes influencing origin and maintenance of inland water-land ecotones. In: Naiman RJ, Decamps H (eds) The ecology and management of aquatic-terrestrial ecotones, MAB series vol 4. UNESCO Paris and The Parthenon Publishing Group, pp 37–64

Salo J, Kalliola R, Häkkinen I, Mäkinen Y, Niemelä P, Puhakka M, Coley PD (1986) River dynamics and the diversity of Amazon lowland forest. Nature 322:254–258

Santos GM, Jegu M, Merona B de (1984) Catálogo de peixes comerciais do baixo Rio Tocantins. ELETRONORTE/INPA Manaus-AM

Santos U de M (1973) Beobachtungen über Wasserbewegungen, chemische Schichtung und Fischwanderungen in Várzea-Seen am mittleren Solimões (Amazonas). Oecologia 13:239–246

Santos U de M, Bringel SRB, Bergamin Filho H, Ribeiro M de NG, Bananeira M (1984) Rios da bacia Amazonica. I. Afluentes do Rio Negro. Acta Amazonica 14(1/2):222–237

Sattler W (1967) Über die Lebensweise, insbesondere das Bauverhalten, neotropischer Eintagsfliegen-Larven (Ephemeroptera, Polymitarcidae). Beitr Neotrop Fauna 5(2):88–110

Sayles FL, Mangelsdorf PC (1979) Cation-exchange characteristics of Amazon River suspended sediment and its reaction with seawater. Geochim Cosmochim Acta 43:767–779

Schachtschabel P, Blume HP, Hartge KH, Schwertmann U (eds) (1982) Lehrbuch der Bodenkunde. Ferdinand Enke Verlag, Stuttgart

Schaden R (1978) Zur Diversität und Identität Amazonischer Rotatortienzoome. Amazoniana 6:347–371

Schaefer M (1992) Wörterbuch der Biologie. UTB, Fischer, Jena

Schaefer M, Tischler W (1983) Wörterbuch der Biologie: Ökologie, 2. Aufl. UTB, Fischer, Stuttgart

Schaller F (1969) Zur Frage des Formensehens bei Collembolen. Verh Dtsch Zool Ges 1968, Innsbruck, pp 368–375

Schaller F (1971) Über den Lautapparat von Amazonasfischen I. Naturwissenschaften 58(11):573–574

Schaller F (1972) Über den Lautapparat von Amazonasfischen II, III. Naturwissenschaften 59(4):169–170

Schaller GB, Crawshaw PG Jr (1981) Social organization in a capybara population. Säugetierk Mitt Sonderdruck 29(1):3–16

Scherfose V (1990) Feinwurzelverteilung und Mykorrhizatypen von *Pinus sylvestris* in verschiedenen Bodentypen. Ber Forschz Waldöko Göttingen, Reihe A, Bd 62

Schlegel HG (1976) Allgemeine Mikrobiologie, 4th edn. Thieme, Stuttgart

Schlüter UB (1985) Der Einfluß verschiedener Umweltbedingungen auf die Bildung nicht-photosynthetisch aktiver Farbstoffe in den Wurzeln der amazonischen Wasserhyazinthe *Eichhornia crassipes*. MSc Thesis, Universität Kiel

Schlüter UB (1989) Morphologische, anatomische und physiologische Untersuchungen zur Überflutungstoleranz zweier charakteristischer Baumarten (*Astrocaryum jauari* und *Macrolobium acaciifolium*) des Weiß- und Schwarzwasserüberschwemmungswaldes bei Manaus. PhD Thesis, Universität Kiel

Schlüter UB, Furch B (1987) Ecological and physiological investigations on *Eichhornia crassipes* (MART) SOLMS. 1. The effect of different environmental conditions on the development of root colour. Amazoniana 10(2):163–171

Schlüter UB, Furch B, Joly CA (1993) Physiological and anatomical adaptations by young *Astrocaryum jauari* Mart. (Arecaceae) in periodically inundated biotopes of Central Amazonia. Biotropica 25(4):384–396

Schmidt GW (1972a) Chemical properties of some waters in the tropical rain-forest region of Central Amazonia along the new road Manaus-Caracarai. Amazoniana 3:199–207

Schmidt GW (1972b) Amounts of suspended solids and dissolved substances in the middle reaches of the Amazon over the course of one year (August 1969–July 1970). Amazoniana 3:208–223

Schmidt GW (1972c) Seasonal changes in water chemistry of a tropical lake (Lago do Castanho, Amazonia, South America). Verh Int Ver Limnol 18:613–621

Schmidt GW (1973a) Primary production of phytoplankton in the three types of Amazonian waters. II. The limnology of a tropical floodplain lake in Central Amazonia (Lago do Castanho). Amazoniana 4:139–203

Schmidt GW (1973b) Primary production of phytoplankton in the three types of Amazonian waters. III. Primary production of phytoplankton in a tropical floodplain lake of Central Amazonia, Lago do Castanho, Amazon, Brazil. Amazoniana 4:379–404

Schmidt GW (1976) Primary production of phytoplankton in three types of Amazonian waters. IV. On the primary productivity of phytoplankton in a bay of the lower Rio Negro (Amazonas, Brazil). Amazoniana 5:517–528

Schulz MW, Slaytor M, Hogan M, O'Brien RW (1986) Components of cellulase from the higher termite, *Nasutitermes walkeri*. Insect Biochem 16(6):929–932

Schuster G (1978) Soil mites in the marine environment. Recent Adv Acarol 1:593–602

Schütz H, Holzapfel-Pschorn A, Conrad R, Rennenberg H, Seiler W (1989) A three years continuous record on the influence of daytime, season, and fertilizer treatment on methane emission rates from an Italian rice paddy field. J Geophys Res 94:16405–16416

Schweizer J (1992) Ariranhas no Pantanal: ecologia e comportamento da *Pteronura brasiliensis*. Edibran-Editora Brasil Natureza, Curitiba

Scott AM, Grönblad H, Croasdale H (1965) Desmids from the Amazon Basin, Brazil. Acta Bot Benn 69:1–94

Sculthorpe CD (1985) The biology of aquatic vascular plants. Koeltz, Königstein

Sedell JR, Reeves GH, Hauer FR, Stanford JA, Hawkins CP (1990) Role of refugia in recovery from disturbances: modern fragmented and disconnected river systems. Environ Manage 14(5):711–724

Seidenschwarz F (1986) Pioniervegetation im Amazonasgebiet Perus. Ein Pflanzensoziologischer Vergleich von vorandinem Flußufer und Kulturland. Monographs on agriculture and ecology of warmer climates, vol 3. J Margraf, Triops Verlag, Langen

Setaro FV (1983) Responses of phytoplankton to experimental fertilization with nitrogen and phosphorus in an Amazon floodplain lake. PhD Thesis, University of California, Santa Barbara

Setaro FV, Melack JM (1984) Responses of phytoplankton to experimental nutrient enrichment in an Amazon floodplain lake. Limnol Oceanogr 29:972–984

Shackleton JN, Opdyke ND (1973) Oxygen isotope and paleomagnetic stratigraphy of equatorial pacific core V28–V238: oxygen isotope temperatures and ice volumes on a 10^5 and 10^6 year scale. Quat Res 3:39–55

Shevenell BJ, Shortle WC (1986) An ion profile of wounded Red Maple. Phytopathology 76(2):132–135

Shigo AL, Marx HG (1977) Compartmentalization of decay in trees. US Dep Agric Tech Bull 405:1–73

Shigo AL, Shortle WC (1985) Shigometry: a reference guide. Agric Handb 646:3–48

Sick H (1950) Contribuição ao conhecimento da ecologia de "*Chordeiles rupestris*" (Spix) (Caprimulgidae, Aves). Rev Bras Biol 10(3):295–306

Sick H (1967) Rios e enchentes na Amazônia como obstaculo para a avifauna. In: Lent H (ed) Atas do simpósio sôbre a Biota Amazônica, vol 5 Zool, CNPq, Rio de Janeiro, pp 495–520

Sick H (1984) Ornitologia brasileira, uma introdução. Editora Univ Brasilia, 2 vol, Brasilia

Siepe A (1989) Untersuchungen zur Besiedlung einer Auen-Catena am südlichen Oberrhein durch Laufkäfer (Coleoptera: Carabidae) unter besonderer Berücksichtigung der Einflüsse des Flutgeschehens. PhD Thesis, Universität Freiburg

Silva JMC da, Oren DC (1990) Resultados de uma excursão ornitológica à Ilha de Maracá, Roraima, Brasil. Goeldiana Zoologia 5:1–8

Silva JMC da, Oniki Y (1988) Lista preliminar da avifauna da Estação ecológica Serra das Araras, Mato Grosso, Brasil. Bol Mus Para Goeldi, NS Zool 4:123–143

Silva VMF da (1983) Ecologia alimentar dos golfinhos da Amazônia. MSc Thesis, University of Amazônia, Manaus

Silva VMF da (1986) Separação ecológia dos golfinhos de água doce da Amazônia. Prim Reun de Exp Mam Acuát Am del Sur 25–29 Jun 1984, Buenos Aires, pp 215–227

Silva VMF da, Best RC (1994) Tucuxi, *Sotalia fluviatilis* (Gervais, 1853). In: Ridgway SH, Harrison R (eds) Handbook of marine mammals. The first book of dolphins, vol 5. Academic Press, London, pp 43–69

Singer R, Aguiar IA (1986) Litter decomposing and ecomycorrhizal Basidiomycetes in an igapó forest. Plant Syst Evol 153:107–117

Sioli H (1950) Das Wasser im Amazonasgebiet. Forsch Fortschr 26:274–280

Sioli H (1951) Zum Alterungsprozeß von Flüssen und Flußtypen im Amazonasgebiet. Arch Hydrobiol 45:267–283

Sioli H (1954a) Betrachtungen über den Begriff "Fruchtbarkeit" eines Gebietes anhand der Verhältnisse in Böden und Gewässern Amazoniens. Forsch Fortschr 28:65–72

Sioli H (1954b) Beiträge zur regionalen Limnologie des Amazonasgebietes. Arch Hydrobiol 49:441–518

Sioli H (1955) Die Bedeutung der Limnologie für die Erforschung wenig bekannter Großräume zu praktischen Zwecken, anhand der Erfahrungen im Amazonas-Gebiet. Forsch Fortschr 29(3):73–84

Sioli H (1956) Über Natur und Mensch im brasilianischen Amazonasgebiet. Erdkunde 10(2):89–109

Sioli H (1965a) Bemerkungen zur Typologie amazonischer Flüsse. Amazoniana 1(1):74–83

Sioli H (1965b) Zur Morphologie des Flußbettes des unteren Amazonas. Naturwissenschaften 52:104

Sioli H (1968) Hydrochemistry and geology in the Brazilian Amazon region. Amazoniana 1(3):267–277

Sioli H (1969) Entwicklung und Aussichten der Landwirtschaft im brasilianischen Amazonasgebiet. Erde 100:307–326

Sioli H (1975) Tropical rivers as expressions of their terrestrial environments. In: Golley FB, Medina E (eds) Tropical ecological systems. Trends in terrestrial and aquatic research. Springer Heidelberg Berlin New York, pp 275–288

Sioli H (1984) The Amazon and its main affluents: hydrography, morphology of the river courses, and river types. In: Sioli H (ed) The Amazon – Limnology and landscape ecology of a mighty tropical river and its basin. Monographiae Biologicae. Junk, Dordrecht, pp 127–165

Smith DR, Adis J (1984) Notes on the systematics and natural history of *Dielocerus fasciatus* (Enderlein) and key to species of the genus (Hymenoptera: Argidae). Proc Entomol Soc Wash 86(3):720–721

Smith LK, Fisher TR (1985) Nutrient fluxes and sediment oxygen demand associated with the sediment water interface of two aquatic environments. In: Hatcher KJ (ed) Sediment oxygen demand: processes, modelling, and measurement. Institute of Natural Resources, University of Georgia, Athens, pp 343–366

Smith-Morrill L (1987) The exchange of carbon, nitrogen and phosphorus between the sediments and water-column of an Amazon floodplain lake. PhD Thesis, University of Maryland

Smith NJH (1979) Quelônios aquáticos da Amazônia: um recurso ameaçado. Acta Amazonica 9:87–97

Smith NJH (1981a) Caimans, capybaras, otters, manatees, and man in Amazonia. Biol Conserv 19:177–187

Smith NJH (1981b) Man, fishes and the Amazon. Columbia University Press, New York

Snethlage E (1910) Sobre a distribuição da avifauna campestre na Amazônia. Bol Mus Para Goeldi 6(1909):226–235

Snethlage E (1913) Über die Verbreitung der Vogelarten in Unteramazonien. J Ornithol 61:469–539

Snethlage E (1935) Beiträge zur Brutbiologie brasilianischer Vögel. J Ornithol 83:1–24, 532–562

Snow DW (1980) Ornithological research in tropical America – the last 35 years. Bull Br Ornithol Club 100:123–131

Soares MGM (1993) Estratégias respiratórias em peixes do Lago Camaleão (Ilha da Marchantaria) – AM, Brasil. PhD Thesis, INPA/FUA, Manaus

Soares MGM, Almeida RG, Junk WJ (1986) The trophic status of the fish fauna in Lago Camaleão, a macrophyte dominated floodplain lake in the Middle Amazon. Amazoniana 9(4):511–526

Sobrado MA, Medina E (1980) General morphology, anatomical structure, and nutrient content of sclerophyllous leaves of the "Bana" vegetation of Amazonas. Oecologia 45:341–345

Sombroek WG (1984) Soils of the Amazon region. In: Sioli H (ed) The Amazon – Limnology and landscape ecology of a mighty tropical river and its basin. Monographiae Biologicae. Junk, Dordrecht, pp 521–535

Sømme L, Conradi-Larsen EM (1977) Anaerobiosis in overwintering collembolans and oribatid mites from windswept mountain ridges. Oikos 29:127–132

Southwood TRE (1977) Habitat, the templet for ecological strategies? J Anim Ecol 46:337–365

Southwood TRE (1988) Tactics, strategies and templets. Oikos 52(1):3–18

Sparks RE, Bayley PB, Kohler SL, Osborne LL (1990) Disturbance and recovery of large floodplain rivers. Environ Manage 14(5):699–709

Spix JB, Martius CFP von (eds) (1823–1831) Reise in Brasilien in den Jahren 1817–1820. Lindauer, München

Spruce R (1908) Notes of a botanist on the Amazon and Andes. MacMillan, London

Stallard RF, Edmond JM (1981) Geochemistry of the Amazon. 1. Precipitation chemistry and the marine contribution to the dissolved load at the time of peak discharge. J Geophys Res 86:9844–9858

Stallard RF, Edmond JM (1983) Geochemistry of the Amazon. 2. The influence of geology and weathering environment on the dissolved load. J Geophys Res 88:9671–9688

Stark N, Holley C (1975) Final report on studies of nutrient cycling on white and black water areas in Amazonia. Acta Amazonica 5:51–76

Stauder A (1990) Untersuchungen des Makrozoobenthos in einem Bach auf Madeira mit zoogeographischen Aspekten. MSc Thesis, Universität Freiburg

Stearns SC (1976) Life-history tactics: a review of the ideas. Q Rev Biol 51:3–47

Steinmetzger K (1982) Die Diplopoden des Waldgebiets Hakel im nordöstlichen Harzvorland der DDR. Hercynia NF 19(2):197–205

Sternberg HOR (1960a) Die Viehzucht im Careiro-Cambixegebiet: ein Beitrag zur Kulturgeographie der Amazonasniederung. Heidelb Geogr Arb 15:171–197

Sternberg HOR (1960b) Radiocarbon dating as applied to a problem of Amazonian morphology. C R XVIII Congr Int Geogr, Rio de Janeiro 2:399–423

Sternberg HOR (1987) Aggravation of floods in the Amazon river as consequence of deforestation? Geogr Annlr 69A:201–219

Stevens GC (1989) The latitudinal gradient in geographical range: how so many species coexist in the tropics. Am Nat 133(2):240–256

Stotz DF, Bierregard RO, Cohn-Haft M, Petermann P, Smith J, Whittaker A, Wilson SV (1992) The status of North American migrants in Central Amazonian Brazil. Condor 94:608–621

Stowe LG, Teeri JA (1978) The geographic distribution of C4 species of the Dicotyledonae in relation to climate. Am Nat 112:609–623

Strickland AH (1944) The arthropod fauna of some tropical soils. Trop Agric 21:107–114

Su NY, Scheffrahn RH, Ban PM (1991) Uniform size particle barrier: a physical exclusion device against subterranean termites (Isoptera, Rhinotermitidae). J Econ Entomol 84(3):912–916

Suguio K, Martin S, Flexor J-M (1988) Quarternary sea levels of the Brazilian coast: Recent Progress. Episodes 11(3):203–208

Swaine MD, Whitmore TC (1988) On the definition of ecological species groups in tropical rain forests. Vegetatio 75:81–86

Swift MJ (1977) The ecology of wood decomposition. Sci Prog 64:175–199

Sylvester-Bradley R, Bandeira AG, Oliveira LA (1978) Fixação de nitrogenio (redução de acetileno) em cupins (Insecta: Isoptera) da Amazonia Central. Acta Amazonica 8(4):621–627

Sylvester-Bradley R, Oliveira LA de, Podestá Filho JA de, John TVS (1980) Nodulation of legumes, nitrogenase activity of roots and occurrence of nitrogen-fixing *Azospirillum* spp. in representative soils of Central Amazonia. Agro-Ecosystems 6:249–266

Sylvester-Bradley R, Oliveira LA, Bandeira AG (1983) Nitrogen fixation in *Nasutitermes* in Central Amazonia. In: Jaisson P (ed) Social insects in the tropics, vol 2. Université Paris-Nord, Paris, pp 236–244

Tadler A, Thaler K (1993) Genitalmorphologie, Taxonomie und geographische Verbreitung ostalpiner Polydesmida (Diplopoda: Helminthomorpha). Zool Jahrb Syst 120:71–128

Takeuchi M (1962) The structure of the Amazonian vegetation VI. Igapó. J Fac Sci, Univ Tokyo

Tarasevich YL (1992) Diplopoda in the association of mixed forests in Byelorussia. (Proc 8th Int Congr Myriapodology, Innsbruck). Ber Nat-Med Ver Innsbruck Suppl 10:213–218

Tathy JP, Delmas RA, Marenco A, Cros B, Labat M, Servant J (1992) Methane emission from flooded forest in Central Africa. J Geophys Res 97:6159–6168

Taylor HM (1987) Minirhizotron observation tubes: methods and application for measuring rhizosphere dynamics. Am Soc Agr Spec Publ 50, Madison

Teeri JA, Stowe LG (1976) Climatic patterns and the distribution of C_4 grasses in N. America. Oecologia 23:1–12

Terborgh J (1980) Causes of tropical species diversity. Acta 17 Congr Int Ornithol Berlin, 1978, pp 955–961

Terborgh J (1983) Five new world primates. A study on comparative ecology. Princeton University Press, Princeton

Terborgh J, Robinson SK, Parker III TA, Munn CA, Pierpont N (1990) Structure and organization of an Amazonian forest bird community. Ecol Monogr 60:213–238

Terborgh J, Weske JS (1969) Colonization of secondary habitats by Peruvian birds. Ecology 50:765–782

Thatcher V, Robertson B (1984) The parasitic crustaceans of fishes from the Brazilian Amazon. 11. Vaigamidae fam. nov. (Copepoda, Poecilostomatoida) with males and females of *Vaigamus retrobarbatus* gen. et sp. nov. and *V. spinicephalus* sp. nov. from plankton. Can J Zool 62:716–729

Thienemann A (1918) Lebensgemeinschaft und Lebensraum. Naturwiss Wochenschrift NF 17:282–290, 297–303

Thienemann A (1920) Die Grundlagen der Biozönotik und Monards faunistische Prinzipien. Festschrift für Zschokke 4, Kober, Basel, pp 1–14

Thienemann A (1954) Ein drittes biozönotisches Grundprinzip. Arch Hydrobiol 49:421–422

Thomas BT (1985) Coexistance and behavior differences among the three western hemisphere storks. Ornithol Monogr 36:921–931

Thomas BT (1987) Spring shorebird migration through Central Venezuela. Wilson Bull 99:571–578

Thomasson K (1971) Amazon algae. Mem Inst R Sci Nat Belg Ser 2 86:1–57

Thomasson K (1976) Algen aus den Flüssen Rio Negro und Rio Tapajós. Amazoniana 5:465–515

Thompson K (1985) Emergent plants of permanent and seasonally flooded wetlands. In: Denny P (ed) The ecology and management of African wetland vegetation: a botanical account of African swamps and shallow waterbodies. Junk, Dordrecht, pp 43–107

Thompson K, Shewry PR, Woolhouse HW (1979) Papyrus swamp development in the Upemba Basin, Zaire: studies of population structure in *Cyperus papyrus* stands. Bot J Linn Soc 78:299–316

Thornback J, Jenkins M (eds) (1982) The IUCN Red Data Book Part 1. Threatened mammalian taxa of the Americas and Australasian zoogeographic region (excluding Cetacea). Int Union Conserv Nat Nat Res, Gland, Switzerland

Tilman D (1982) Resource competition and community structure. Princeton University Press, Princeton

Tischler W (1984) Einführung in die Ökologie, 3rd edn. Fischer, Stuttgart

Tomiuk J, Vohland K, Bachmann L, Adis J (1996) Ecological and genetic studies of the millipede *Pycnotropis epiclysmus*: Influence of flood water seasons on the evolution of ecotypes in tropical forests. Hereditas 124:298-299

Tuomisto H (1993) Classificación de vegetatión en la selva baja de la Amazonia. In: Kalliola R, Puhakka M, Danjoy W (eds) Amazonia Peruana. Vegetatión humeda tropical en el Llano Subandino. Gummerus Printing, Jyväskyla, Finland, pp 103-112

Uehara G, Gillman G (eds) (1981) The mineralogy, chemistry, and physics of tropical soils with variable charge clays. Westview Press, Boulder

Uetz G, Unzicker JD (1975) Pitfall trapping in ecological studies of wandering spiders. J Arachnol 3:101-111

Uherkovich G (1976) Algen aus den Flüssen Rio Negro und Rio Tapajós. Amazoniana 5(4):465-515

Uherkovich G (1981) Algen aus einigen Gewässern Amazoniens. Amazoniana 7:191-219

Uherkovich G (1984) Phytoplankton. In: Sioli H (ed) The Amazon – Limnology and landscape ecology of a mighty tropical river and its basin. Monographiae Biologicae. Junk, Dordrecht, pp 295-310

Uherkovich G, Franken M (1980) Aufwuchsalgen aus zentralamazonischen Regenwaldbächen. Amazoniana 7(1):49-79

Uherkovich G, Rai H (1979) Algen aus dem Rio Negro und seinen Nebenflüssen. Amazoniana 6:611-638

Uherkovich G, Schmidt GW (1974) Phytoplanktontaxa in dem zentralamazonischen Schwemmlandsee Lago do Castanho. Amazoniana 5:243-283

Uhl C (1980) Studies of forest, agricultural and successional environments in the upper Rio Negro region of the Amazon basin. PhD Thesis, Michigan State University

Ungemach H (1971) Chemical rain water studies in the Amazôn region. Asociación pro Biología Tropical, II. Simposio y Foro de Biología Tropical Amazônica, Florencia (Caquetá) y Leticia (Amazonas), Columbia, enero 1969. Bogotá, pp 354-358

US Soil Taxonomy (1990) Keys to soil taxonomy by soil survey staff. Agency for international development United States department of agriculture soil management support services. Cornell University

Usher MB (1970) Seasonal and vertical distribution of a population of soil arthropods: Collembola. Pedobiologia 10:224-236

Val AL, Almeida-Val VM (1995) Fishes of the Amazon and their environment. Physiological and biochemical aspects. Springer Berlin, Heidelberg New York

Val AL, Schwantes AR, Almeida-Val VMF (1986) Biological aspects of Amazonian fishes. VI. Hemoglobins and whole blood properties of *Semaprochilodus* species (Prochilodontidae) at two phases of migration. Comp Biochem Physiol 83B:659-667

Val AL, Almeida-Val VMF, Affonso EG (1990) Adaptative features of Amazon fishes; Hemoglobins, hematology, intraerythrocytic phosphates and whole blood Bohr effects of *Pterygoplichthys multiradiatus* (Siluriformes). Comp Biochem Physiol 97B(3):435-440

Val AL, Almeida-Val VM, Randall DJ (eds) (1996) Physiology and biochemistry of the fishes of the Amazon. INPA, Manaus

Vannote RL, Minshall GM, Cummins KW, Sedell JR, Cushing CE (1980) The river continuum concept. Can J Fish Aquat Sci 37:130-137

Veillon JP (1985) El crecimiento de algunos bosques naturales de Venezuela en relación con los parámetros del medio ambiente. Revista For Venez, Merida 29:1-120

Veríssimo J (1895) A pesca na Amazônia. Livr Clássica Alves, Rio de Janeiro (Reprinted 1970, Universidade Federal do Pará)

Vieira I (1982) Aspectos sinecológicos da ictiofauna de Curuá- Una, represa hidrelétrica da região Amazônica. MSc Thesis, Univ de Juiz de Fora

Vieira M de F, Adis J (1992) Abundância e biomassa de *Paulinia acuminata* (DE GEER, 1773) (Orthoptera: Pauliniidae) em um lago de várzea da Amazônia Central. Amazoniana 12(2):337–352

Vieira RS, Höfer H (1994) Prey spectrum of two army ant species in central Amazonia, with special attention on their effect on spider populations. Andrias 13:189–198

Viets K (1954) Wassermilben aus dem Amazonasgebiet (Hydrachnellae, Acari). (Systematische und ökologische Untersuchungen.) Bearbeitung der Sammlungen Dr. R. Braun, Aarau, und Dr. H. Sioli, Belém. Schweiz Zschr Hydrol, Basel 16(1):78–151; 16(2):161–247

Viner AB (1982) Nitrogen fixation and denitrification in sediments of two Kenyan lakes. Biotropica 14:91–98

Vitousek P (1984) Litterfall, nutrient cycling, and nutrient limitation in tropical forests. Ecology 65(1):285–298

Voigtländer K, Dunger W (1992) Long-term observations of the effects of increasing dry polution on the myriapod fauna of the Neisse valley (East Germany). (Proc 8th Int Congr Myriapodology, Innsbruck). Ber Nat-Med Ver Innsbruck Suppl 10:251–256

Wahlen M, Tanaka N, Henry R, Deck B, Zeglen J, Vogel S, Southon J, Shemesh A, Fairbanks R, Broecker W (1989) Carbon-14 in methane sources and in atmospheric methane: the contribution from fossil carbon. Science 245:286–290

Waldhoff D (1991) Morphologie, Nährwert und Bioelementgehalt hydrochor und zoochor verbreiteter Früchte und Samen aus amazonischen Überschwemmungswäldern bei Manaus. MSc Thesis, Universität Kiel

Waldhoff D (1994) Analyse von Früchten aus den Überschwemmungswäldern Zentralamazoniens und dem Pantanal Matogrossense in Hinsicht auf ihren Wert als Nahrung für kommerziell zu nutzende Fischarten. PhD Thesis, Universität Kiel

Walker I (1985) On the structure and ecology of the micro-fauna in the Central Amazonian forest stream Igarapé da Cachoeira. Hydrobiologia 122:137–152

Walker I (1986) Experiments on colonization of small waterbodies by Culicidae and Chironomidae as a function of decomposing plant substrates and their implications for natural Amazonian ecosystems. Amazoniana 10(1):113–125

Walker I (1988) Study of benthic micro-faunal colonization of submerged litter leaves in the Central Amazonian blackwater stream Tarumã-Mirim (Tarumãzinho). Acta Limnol Brazil 2:623–648

Walker I (1992a) The benthic litter habitat with its sediments load in the inundation forest of the Central Amazonian blackwater river Tarumã Mirim. Amazoniana 12:143–153

Walker I (1992b) Life history traits of shrimps (Decapoda: Palaemonidae) of Amazonian inland waters and their phylogenetic interpretation. In: Adis J, Tanaka S (eds) Symposium on life-history traits in tropical invertebrates. (INTECOL, Yokohama, Japan 1990). Stud Neotrop Fauna Environ 27(2–3):131–143

Walker I, Henderson PA, Sterry P (1991) On the patterns of biomass transfer in the benthic fauna of an amazonian black-water river, as evidenced by ^{32}P label experiment. Hydrobiologia 215:153–162

Wallace AR (1889) A narrative of travels on the Amazon and Rio Negro, with an account of the native tribes. Ward, Lock and Co, London

Walter H, Breckle SW (eds) (1991) Ökologie der Erde, Band 2: Spezielle Ökologie der Tropen und Subtropen. Fischer, Stuttgart

Ward JV (1989) The four-dimensional nature of lotic ecosystems. J N Am Benthol Soc 8(1):2–8

Ward JV, Stanford JA (1995) Ecological connectivity in alluvial river ecosystems and its disruption by flow regulation. Regulated Rivers Res Manage 10:159–168

Ward JV, Wiens JA (1995) Ecotones of riverine ecosystems: rôle and typology, spatio-temporal dynamics, and river regulation. In: Zalewski M, Thorge JE (eds) Fish and land – inland water ecotones. Man and the biosphere series. Parthenon, Casterton Hall, UK

Wassmann R, Thein UG (1996) Spatial and seasonal variation of methane emission in an Amazonian floodplain lake. Mitt Int Ver Limnol 25:179–185

Wassmann R, Thein UG, Whiticar MJ, Rennenberg H, Seiler W, Junk WJ (1992) Methane emissions from the Amazon floodplain: characterization of production and transport. Global Biogeochem Cycles 6:3–13

Weber GE (1996) Causes of hydrochemical seasonality of major cations in Lago Camaleão, a Central Amazonian floodplain lake. Verh Inter Ver Limnol (in press)

Weber GE, Furch K, Junk WJ (1996) A simple modelling approach towards hydrochemical seasonality of major cations in a Central Amazonian floodplain lake. Ecol Model 91:39–56

Wehsarg O (1954) Ackerunkräuter: Biologie, allgemeine Bekämpfung und Einzelbekämpfung. Akademie-Verlag, Berlin

Weigmann G (1973) Zur Ökologie der Collembolen und Oribatiden im Grenzbereich Land-Meer (Collembola, Insecta – Oribatei, Acari). Z Wiss Zool 186(3/4):295–391

Welch HE (1968) Relationships between assimilation efficiencies and growth efficiencies for aquatic consumers. Ecology 49(4):755–759

Welcomme RL (1979) Fisheries ecology of floodplain rivers. Longmann, London

Welcomme RL (1985) River fisheries. FAO Fish Tech Pap 262, Food and Agriculture Organisation of the United Nations, Rome

Welcomme RL (1990) Status of fisheries in South American rivers. Interciencia 15(6):337–345

Werner D (1977) The biology of diatoms. Botanical monographs, vol 13. Blackwell, Oxford

West-Eberhard MJ (1989) Phenotypic plasticity and the origins of diversity. Annu Rev Ecol Syst 20:249–278

Westlake DF (1963) Comparisons of plant productivity. Biol Rev 38:385–425

Wharton CH, Lambou VW, Newson J, Winger PV, Gaddy LL, Mancke R (1981) The fauna of bottomland hardwoods in Southeastern United States. In: Clark JR, Benforado J (eds) Wetlands of bottomland hardwood forests. Development in agricultural and managed forest ecology 11. Elsevier, Amsterdam, pp 87–160

Whitaker V, Matvienko B (1992) A method for the study of N_2O evolution in tropical wetlands. Hydrobiologia 230:213–218

White J (1981) The allometric interpretation of the self thinning rule. J Theoretical Biol 89:475–500

Whitmore TC (1984) Tropical rain forest of the Far East. Clarendon Press, Oxford

Whitmore TC (1990) An introduction to tropical rain forests. Oxford University, Oxford

Whitmore TC (1993) Tropische Regenwälder: Eine Einführung. Spektrum Akad Verlag, Heidelberg, Berlin, New York

Wiegert RG (1970) Energetics of the nest-building termite, Nasutitermes costalis (Holmgren), in a Puerto Rican forest. In: Odum HT, Pidgeon RF (eds) A tropical rain forest. A study of irradiation and ecology at El Verde, Puerto Rico. Division of Technical Information. US Atomic Energy Commission, Oak Ridge, pp I.57–I.64

Wiens JA, Crawford CS, Gosz JR (1985) Boundary dynamics: a conceptual framework for studying landscape ecosystems. Oikos 45:421–427

Willard DE (1985) Comparative feeding ecology of twenty-two tropical piscivores. Ornithol Monogr 36:788–797

Williams MAJ (1968) Termites and soil development near Brock's Creek, Northern Territory. Aust J Sci 31:63–124

Williams MR (1993) Nutrient inputs to an Amazon floodplain stream following deforestation. MSc Thesis, University of Maryland-CEES

Williams MR, Fisher TR, Melack JM (1996) Chemical composition and deposition of rain in central Amazonas, Brazil. Atmos Environ (in press)

Willis EO (1977) Effects of a cold wave on an Amazonian avifauna in the upper Paraguay drainage, western Mato Grosso, and suggestions on oscine-suboscine relationships. Acta Amazonica 6(3):379–394

Willis EO (1979) The composition of avian communities in remanescent woodlots in southern Brasil. Pap Avulsos Zool, (São Paulo) 33(1):1–25

Willis EO (1992) Zoogeographical origin of eastern Brazilian birds. Ornithol Neotrop 3:1–15

Willis EO, Oniki Y (1988) Aves observados em Balbina, Amazonas, e os possiveis efeitos da barragem. Ciênc Cult (São Paulo) 40(3):280–285

Wilson DA (1974) Survival of cicindelid larvae after flooding. Cicindela 6:79–82

Winter Ch (1962) Zur Ökologie und Taxonomie der neotropischen Bodentiere II. Zur Collembolen-Fauna Perus. Zool Jahrb Syst 90:393–520

Wissmar RC, Richey JE, Stallard RF, Edmond JM (1981) Plankton metabolism and carbon processes in the Amazon River, its tributaries, and floodplain waters, Peru-Brazil, May-June 1977. Ecology 62:1622–1633

Woas S (1981) Die Arten der Gattung *Hermannia* Nicolet, 1955 (Acari: Oribatei) II. Andrias 1:89–100

Woas S (1986) Beitrag zur Revision der Oppioidea sensu Balogh, 1972 (Acari, Oribatei). Andrias 5:21–224

Woas S (1990) Die phylogenetischen Entwicklungslinien der Höheren Oribatiden (Acari) I. Zur Monophylie der Poronota Grandjean, 1953. Andrias 7:91–168

Woas S (1997) Die Organisationsstufen des Bauplangefüges der Oribatiden. Andrias (in press)

Wolf HG, Adis J (1992) Genetic differentiation between populations of *Neomachilellus scandens* (Meinertellidae, Archaeognatha, Insecta) inhabiting neighbouring forests in Central Amazonia. Verh Naturwiss Ver Hamb 33(NF):5–13

Woltemade H (1982) Zur Ökologie baumrindenbewohnender Hornmilben (Acari, Oribatei). Sber Ges Naturf Freunde Berl 22:118–139

Wood TG (1976) The role of termites (Isoptera) in decomposition processes. In: Anderson JM, Macfadyen A (eds) The role of terrestrial and aquatic organisms in decomposition process. Blackwell, Oxford, pp 145–168

Wood TG (1978) Food and feeding habits of termites. In: Brian MV (ed) Production ecology of ants and termites. International Biological Programme 13. Cambridge University Press, Cambridge, pp 55–80

Wood TG, Sands WA (1978) The role of termites in ecosystems. In: Brian MV (ed) Production ecology of ants and termites. International Biological Programme 13. Cambridge University Press, Cambridge, pp 245–292

Worbes M (1983) Vegetationskundliche Untersuchungen zweier Überschwemmungswälder in Zentralamazonien – vorläufige Ergebnisse. Amazoniana 8(1):47–65

Worbes M (1984) Periodische Zuwachszonen an Bäumen zentralamazonischer Überschwemmungswälder. Naturwissenschaften 71:157

Worbes M (1985) Structural and other adaptations to long-term flooding by trees in Central Amazonia. Amazoniana 9:459–484

Worbes M (1986) Lebensbedingungen und Holzwachstum in zentralamazonischen Überschwemmungswäldern. Scr Geobot 17:1–112

Worbes M (1989) Growth rings, increment and age of trees in inundation forests, savannas and a mountain forest in the Neotropics. IAWA Bull 10(2):109–122

Worbes M (1994a) Bestimmung der Holzproduktion in neotropischen Waldbeständen mit Hilfe von Jahresringuntersuchungen. Appl Bot Rep Hamburg 5:31–35

Worbes M (1994b) Grundlagen und Anwendungen der Jahresringforschung in den Tropen. Habilitationschrift, Univ Hamburg

Worbes M, Junk WJ (1989) Dating tropical trees by means of ^{14}C from bomb tests. Ecology 70:503–507

Worbes M, Klinge H, Revilla JD, Martius C (1992) On the dynamics, floristic subdivision and geographical distribution of várzea forests in Central Amazonia. J Veg Sci 3:553–564

Worbes M, Klosa D, Lewark S (1995) Rohdichtestruktur von Jahresringen tropischer Hölzer aus zentralamazonischen Überschwemmungswäldern. Holz Roh- Werkstoff 53:63–67

Worthmann HOW (1982) Aspekte der Biologie zweier Sciaenidenarten, der Pescadas *Plagioscion squamosissimus* (Heckel) and *Plagioscion monti* (Soares) in verschiedenen Gewässertypen Zentralamazoniens. PhD Thesis, Universität Kiel

Wunderle I (1992a) Arboricolous and edaphic Oribatei (Acari) in the lowland rain forest of Panguana, Peru. Amazoniana 12(1):119–142

Wunderle I (1992b) Die Oribatiden-Gemeinschaften (Acari) der verschiedenen Habitate eines Buchenwaldes. Carolinea 50:79–144

Zaleskaja NT (1994) The centipede genus *Lamyctes* MEINERT, 1868, in the environs of Manaus, Central Amazonia, Brazil (Chilopoda, Lithobiomorpha, Henicopidae). Amazoniana 13(1/2):59–64

Zaret TM, Rand S (1971) Competition in tropical stream fishes. Support for the competitive exclusion principle. Ecology 52(2):336–342

Zaret TM, Dewoll AH, Santos A dos (1981) Nutrient addition experiments in Lago Jacaretinga, Central Amazon Basin, Brazil. Verh Int Ver Limnol 21:721–724

Zelitch I (1971) Photosynthesis, photorespiration, and plant productivity. Academic Press, New York

Ziburski A (1990) Ausbreitungs- und Reproduktionsbiologie einiger Baumarten der amazonischen Überschwemmungswälder. PhD Thesis, Universität Hamburg

Ziburski A (1991) Dissemination, Keimung und Etablierung einiger Baumarten der Überschwemmungswälder Amazoniens. Trop Subtrop Pflanzenwelt 77:1–96

Zimmermann PR, Greenberg SO, Wandiga SO, Crutzen PJ (1982) Termites: a potentially large source of atmospheric methane, carbon dioxide, and molecular hydrogen. Science 218:563–565

Zulka KP (1989) Einfluss der Hochwässer auf die epigäische Arthropodenfauna im Überschwemmungsbereich der March (Niederösterreich). Mitt Dtsch Ges Allg Angew Entomol 7:74–75

Zulka KP (1991) Überflutung als ökologischer Faktor: Verteilung, Phänologie und Anpassungen der Diplopoda, Lithobiomorpha und Isopoda in den Flußauen der March. PhD Thesis, Universität Wien

Zulka KP (1996) Submersion tolerance of some diplopod species. Acta Myriapodologica. Mém Mus Nat Hist Nat 169:477–481

Subject Index

Printing: Saladruck, Berlin
Binding: Buchbinderei Lüderitz & Bauer, Berlin

Ecological Studies

Volumes published since 1992